A First Course in Continuum Mechanics

Presenting a concise account of various classic theories of fluids and solids, this book is designed for courses in continuum mechanics for graduate students and advanced undergraduates. Thoroughly class-tested in courses at Stanford University and the University of Warwick, it is suitable for both applied mathematicians and engineers. The only prerequisites are an introductory undergraduate knowledge of basic linear algebra and differential equations.

Unlike most existing works at this level, this book covers both isothermal and thermal theories. The theories are derived in a unified manner from the fundamental balance laws of continuum mechanics.

Intended both for classroom use and for self-study, each chapter contains a wealth of exercises, with fully worked solutions to odd-numbered questions. A complete solutions manual is available to instructors upon request. Short bibliographies appear at the end of each chapter, pointing to material which underpins or expands upon the material discussed.

OSCAR GONZALEZ is an Associate Professor of Mathematics at the University of Texas. His research interests cover computational and applied mathematical problems related to the large-scale deformations of thin rods and ribbons, and more general three-dimensional bodies. He has contributed articles to numerous journals across mathematics, engineering and chemistry. His current research efforts are directed toward understanding the mechanical properties of DNA at various length scales.

ANDREW M. STUART is Professor of Mathematics at the University of Warwick. His general research interests cover computational stochastic processes and dynamical systems and his current research efforts are directed mainly towards problems at the interface of applied mathematics and statistics. He has contributed articles to numerous journals across mathematics, engineering and physics and is the recipient of six prizes for his work in applied mathematics.

Cambridge Texts in Applied Mathematics

All titles listed below can be obtained from good booksellers or from Cambridge University Press.
For a complete series listing, visit http:publishing.cambridge.org/uk/series/sSeries.asp?code=CTAM

Thinking about Ordinary Differential Equations
ROBERT E. O'MALLEY JR

A Modern Introduction to the Mathematical Theory of Water Waves
ROBIN STANLEY JOHNSON

Rarefied Gas Dynamics: From Basic Concepts to Actual Calculations
CARLO CERCIGNANI

Symmetry Methods for Differential Equations: A Beginner's Guide
PETER E. HYDON

High Speed Flow
C. J. CHAPMAN

Wave Motion
J. BILLINGHAM, AND A. C. KING

An Introduction to Magnetohydrodynamics
P. A. DAVIDSON

Linear Elastic Waves
JOHN G. HARRIS

Vorticity and Incompressible Flow
ANDREW J. MAJDA, AND ANDREA L. BERTOZZI

Infinite-Dimensional Dynamical Systems
JAMES C. ROBINSON

Introduction to Symmetry Analysis
BRAIN J. CANTWELL

Bäcklund and Darboux Transformations
C. ROGERS, AND W. K. SCHIEF

Finite Volume Methods for Hyperbolic Problems
RANDALL J. LEVEQUE

Introduction to Hydrodynamic Stability
P. G. DRAZIN

Theory of Vortex Sound
M. S. HOWE

Scaling
GRIGORY ISAAKOVICH BARENBLATT

Complex Variables: Introduction and Applications (2nd Edition)
MARK J. ABLOWITZ, AND ATHANASSIOS S. FOKAS

A First Course in Combinatorial Optimization
JONE LEE

Practical Applied Mathematics: Modelling, Analysis, Approximation
SAM HOWISON

An Introduction to Parallel and Vector Scientific Computation
RONALD W. SHONKWILER, AND LEW LEFTON

A First Course in Continuum Mechanics

OSCAR GONZALEZ AND ANDREW M. STUART

CAMBRIDGE
UNIVERSITY PRESS

CAMBRIDGE
UNIVERSITY PRESS

University Printing House, Cambridge CB2 8BS, United Kingdom

Cambridge University Press is part of the University of Cambridge.

It furthers the University's mission by disseminating knowledge in the pursuit of education, learning and research at the highest international levels of excellence.

www.cambridge.org
Information on this title: www.cambridge.org/9780521714242

First published 2008
Reprinted 2010

A catalogue record for this publication is available from the British Library

ISBN 978-0-521-88680-2 Hardback
ISBN 978-0-521-71424-2 Paperback

Dedicated to our parents,
our mentors, Juan Simo and John Norbury,
and our partners, Martha and Anjum.

Contents

Preface

This book is designed for a one- or two-quarter course in continuum mechanics for first-year graduate students and advanced undergraduates in the mathematical and engineering sciences. It was developed, and continually improved, by over four years of teaching of a graduate engineering course (ME 238) at Stanford University, USA, followed by over four years of teaching of an advanced undergraduate mathematics course (MA3G2) at the University of Warwick, UK. The resulting text, we believe, is suitable for use by both applied mathematicians and engineers. Prerequisites include an introductory undergraduate knowledge of linear algebra, multivariable calculus, differential equations and physics.

This book is intended both for use in a classroom and for self-study. Each chapter contains a wealth of exercises, with fully worked solutions to odd-numbered questions. A complete solutions manual is available to instructors upon request. A short bibliography appears at the end of each chapter, pointing to material which underpins, or expands upon, the material discussed here. Throughout the book we have aimed to strike a balance between two classic notational presentations of the subject: coordinate-free notation and index notation. We believe both types of notation are helpful in developing a clear understanding of the subject, and have attempted to use both in the statement, derivation and interpretation of major results. Moreover, we have made a conscious effort to include both types of notation in the exercises.

Chapters 1 and 2 provide necessary background material on tensor algebra and calculus in three-dimensional Euclidean space. Chapters 3–5 cover the basic axioms of continuum mechanics concerning mass, force and motion, the balance laws of mass, momentum, energy and entropy, and the concepts of frame-indifference and material constraints. Chapters 6–9 cover various classic theories of inviscid and viscous fluids, and linear and nonlinear elastic solids. Chapters 6 and 7 cover isothermal theories, whereas Chapters 8 and 9 cover corresponding thermal theories. We emphasize the formulation of typical initial-boundary value problems for the various material models, study important qualitative properties, and, in several cases, illustrate how the technique of linearization can be used to simplify problems under appropriate assumptions.

For reasons of space and timeliness the scope of material covered in this book is necessarily limited. For further general reading we recommend the books by Gurtin (1981), Chadwick (1976), Mase (1970) and Malvern (1969). For further reading in the area of fluid mechanics we suggest the texts by White (2006), Anderson (2003), Chorin and Marsden (1993) and Temam (1984). In the area of solid mechanics we suggest the texts by Antman (1995), Ciarlet (1988), Ogden (1984) and Marsden and Hughes (1983). For a wealth of information on both the subject and history of various classic field theories of continuum mechanics we recommend the encyclopedia articles by Gurtin (1972), Truesdell and Noll (1965), Truesdell and Toupin (1960) and Serrin (1959).

We are indebted to many of our colleagues at Stanford and Warwick, especially to Huajian Gao who gave a version of the course which we sat through in the 1993-94 and 1994-95 academic years, to Tom Hughes who gave us considerable encouragement to develop the notes into a book, as well as guidance on the choice of material, and to Robert Mackay who read and commented upon an early draft of the book. It is also a pleasure to thank the many students at Stanford and Warwick who helped produce and type solutions. Special thanks go to Nuno Catarino, Doug Enright, Gonzalo Feijoo, Liam Jones, Teresa Langlands, Matthew Lilley and Paul Lim.

Note to instructors Chapters 1–5, together with selected topics from Chapters 6–9, can form a standard one-semester or one-quarter course. This format has been used as the basis for an advanced undergraduate course in applied mathematics. Chapters 1–9 in their entirety, together with supplemental material from outside sources, can form a standard two-semester or two-quarter course. This format has been used as the basis for a graduate engineering course sequence. Example supplemental material might include the finite element method, fluid-solid interactions, the balance laws of electro-magnetism and their coupling to the thermo-mechanical balance laws, and topics such as piezo-electricity. Fully worked solutions to all exercises are available to instructors who are using the book as part of their teaching. A copy of the solutions manual can be obtained via the web site www.cambridge.org/9780521714242 or by emailing solutions@cambridge.org.

Notational conventions Mathematical statements such as definitions, axioms, theorems and results are numbered consecutively within each chapter. Thus Definition 2.3 precedes Result 2.4 in Chapter 2. The symbol □ is used to denote the end of each such statement, including proofs, remarks and examples. Equation numbers, when assigned, are also in consecutive order within each chapter. Thus equations (3.6) and (3.7) are the sixth and seventh numbered equations in Chapter 3. Often, a single number is assigned to a displayed group of equations, and in this case, subscripts are used to refer to independent equations from the group. Thus $(3.7)_1$ and $(3.7)_2$ denote the first and second independent equations from the equation group (3.7). The ordering of independent equations within a group will always follow the standard reading order: left to right, top to bottom.

1
Tensor Algebra

Underlying everything we study is the field of real numbers $\mathbb{R}$ and three-dimensional Euclidean space $\mathbb{E}^3$. We refer to elements of $\mathbb{R}$ as **scalars** and denote them by light-face symbols such as a, b, x and y. We refer to elements of $\mathbb{E}^3$ as **points** and denote them by bold-face symbols such as $\boldsymbol{a}$, $\boldsymbol{b}$, $\boldsymbol{x}$ and $\boldsymbol{y}$.

The aim of this chapter is to build up a mathematical structure for $\mathbb{E}^3$ and various other sets associated with it, such as the set of vectors $\mathcal{V}$. The important ideas that we introduce are: (i) the distinction between points and vectors; (ii) the distinction between a vector and its representation in a coordinate frame; (iii) index notation; (iv) the set of second-order tensors $\mathcal{V}^2$ as linear transformations in $\mathcal{V}$; (v) the set of fourth-order tensors $\mathcal{V}^4$ as linear transformations in $\mathcal{V}^2$.

1.1 Vectors

By a **vector** we mean a quantity with a specified magnitude and direction in three-dimensional space. For example, a line segment in space can be interpreted as a vector as can other things such as forces, velocities and accelerations. We denote a vector by a bold-face symbol such as $\boldsymbol{v}$ and denote its magnitude by $|\boldsymbol{v}|$. Notice that for any vector $\boldsymbol{v}$ we have $|\boldsymbol{v}| \geq 0$. It is useful to define a **zero vector**, $\mathbf{0}$, as a vector with zero magnitude and no specific direction. Any vector having unit magnitude is called a **unit vector**.

A vector may be represented graphically by an arrow as depicted in Figure 1.1. The orientation of the arrow represents the direction of the vector and the length of the arrow represents the magnitude. We will always identify vectors with arrows in this way. Moreover, two vectors

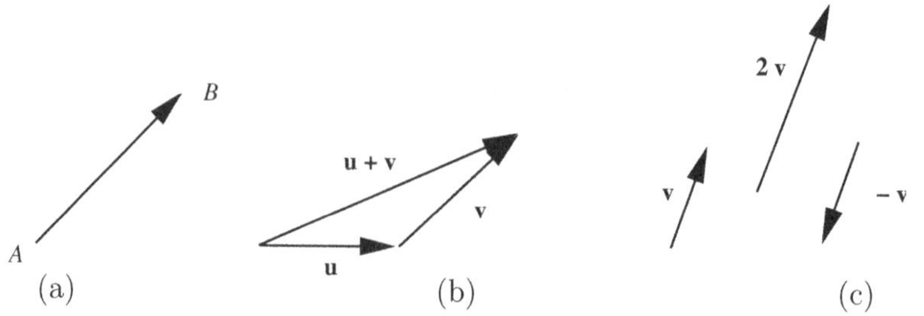

Fig. 1.1 Graphical illustration of a vector and basic vector operations. (a) the tail A and tip B of a vector. (b) the sum of two vectors. (c) the multiplication of a vector by a scalar.

will be considered equal if they have the same direction and magnitude, regardless of their location in space.

Scalars and vectors are examples of more general objects called **tensors**. For example, a scalar is a zeroth-order tensor, while a vector is a first-order tensor. Later we will consider second-order and fourth-order tensors, but for now we concentrate on first-order ones, that is, vectors.

1.1.1 Vector Algebra

If we let $\mathcal{V}$ be the set of all vectors, then $\mathcal{V}$ has the structure of a real vector space since

$$\boldsymbol{u}+\boldsymbol{v} \in \mathcal{V} \quad \forall \boldsymbol{u}, \boldsymbol{v} \in \mathcal{V} \quad \text{and} \quad \alpha \boldsymbol{v} \in \mathcal{V} \quad \forall \alpha \in \mathbb{R},\ \boldsymbol{v} \in \mathcal{V}.$$

In particular, we define the sum $\boldsymbol{u}+\boldsymbol{v}$ as that element of $\mathcal{V}$ which completes the triangle when $\boldsymbol{u}$ and $\boldsymbol{v}$ are placed tip-to-tail as shown in Figure 1.1(b). Given any scalar $\alpha \in \mathbb{R}$ we define the product $\alpha\boldsymbol{v}$ as that element in $\mathcal{V}$ with magnitude $|\alpha||\boldsymbol{v}|$, where $|\alpha|$ denotes the absolute value of α, and direction the same as or opposite to $\boldsymbol{v}$ depending on whether α is positive or negative, respectively, as depicted in Figure 1.1(c). If $\alpha = 0$, then $\alpha\boldsymbol{v}$ is the zero vector.

With the vector space $\mathcal{V}$ in hand we can define some useful mathematical operations between elements of $\mathbb{E}^3$ and $\mathcal{V}$. Given any two points $\boldsymbol{x}$ and $\boldsymbol{y}$ we define their difference $\boldsymbol{y}-\boldsymbol{x}$ to be the unique vector which points from $\boldsymbol{x}$ to $\boldsymbol{y}$, as shown in Figure 1.2(a). Hence given $\boldsymbol{x}, \boldsymbol{y} \in \mathbb{E}^3$ we have $\boldsymbol{y}-\boldsymbol{x} \in \mathcal{V}$. Given any point $\boldsymbol{z}$ and any vector $\boldsymbol{v}$ we define the sum $\boldsymbol{z}+\boldsymbol{v}$ to be the unique point such that $(\boldsymbol{z}+\boldsymbol{v})-\boldsymbol{z}=\boldsymbol{v}$, as shown in Figure 1.2(b). Hence given $\boldsymbol{z} \in \mathbb{E}^3$ and $\boldsymbol{v} \in \mathcal{V}$ we have $\boldsymbol{z}+\boldsymbol{v} \in \mathbb{E}^3$.

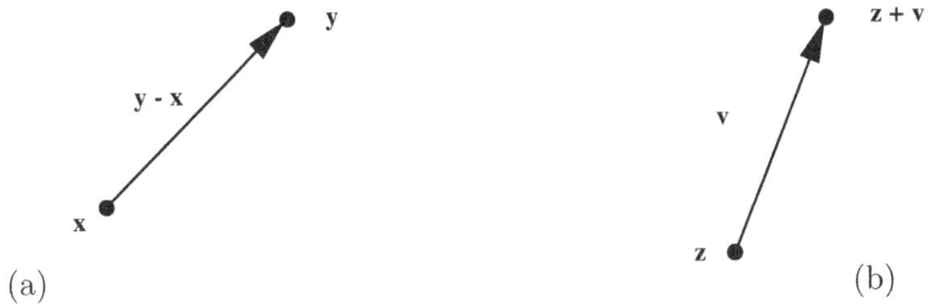

Fig. 1.2 Graphical interpretation of operations between points and vectors. (a) the difference between two points is a vector. (b) the sum of a point and a vector is a point.

We next define some geometrical operations between vectors which will allow us to construct mathematical representations for $\mathcal{V}$ and $\mathbb{E}^3$.

1.1.2 Scalar and Vector Products

The **scalar** or **dot product** of two vectors $\boldsymbol{a}$ and $\boldsymbol{b}$ is defined geometrically as

$$\boldsymbol{a} \cdot \boldsymbol{b} = |\boldsymbol{a}||\boldsymbol{b}| \cos\theta,$$

where $\theta \in [0, \pi]$ is the angle between the tips of $\boldsymbol{a}$ and $\boldsymbol{b}$. Two vectors $\boldsymbol{a}$ and $\boldsymbol{b}$ are said to be **orthogonal** or **perpendicular** if $\boldsymbol{a} \cdot \boldsymbol{b} = 0$. For any vector $\boldsymbol{a}$ notice that $\boldsymbol{a} \cdot \boldsymbol{a} = |\boldsymbol{a}|^2$.

The **vector** or **cross product** of two vectors $\boldsymbol{a}$ and $\boldsymbol{b}$ is defined geometrically as

$$\boldsymbol{a} \times \boldsymbol{b} = (|\boldsymbol{a}||\boldsymbol{b}| \sin\theta)\boldsymbol{e},$$

where, as before, $\theta \in [0, \pi]$ is the angle between the tips of $\boldsymbol{a}$ and $\boldsymbol{b}$, and $\boldsymbol{e}$ is a unit vector perpendicular to the plane containing $\boldsymbol{a}$ and $\boldsymbol{b}$. The orientation of $\boldsymbol{e}$ is defined such that a right-handed rotation about $\boldsymbol{e}$, through an angle θ, carries $\boldsymbol{a}$ to $\boldsymbol{b}$. If $\boldsymbol{a} \times \boldsymbol{b} = \boldsymbol{0}$, then $\boldsymbol{a}$ and $\boldsymbol{b}$ are said to be **parallel**. Notice that the magnitude of $\boldsymbol{a} \times \boldsymbol{b}$ is the same as the area of a parallelogram with sides defined by $\boldsymbol{a}$ and $\boldsymbol{b}$.

Example: Let $V = |\boldsymbol{a} \times \boldsymbol{b} \cdot \boldsymbol{c}|$. Then V is the volume of the parallelepiped defined by $\boldsymbol{a}$, $\boldsymbol{b}$ and $\boldsymbol{c}$. In particular, we have

$$V = |\boldsymbol{a}||\boldsymbol{b}||\boldsymbol{c}||\sin\theta||\cos\phi|,$$

where θ is the angle between $\boldsymbol{a}$ and $\boldsymbol{b}$, and ϕ is the angle between $\boldsymbol{c}$ and

$\boldsymbol{a} \times \boldsymbol{b}$. The factor $|\boldsymbol{a}||\boldsymbol{b}||\sin\theta|$ is the area of the parallelogram defined by $\boldsymbol{a}$ and $\boldsymbol{b}$, and the factor $|\boldsymbol{c}||\cos\phi|$ gives the height of the parallelepiped with respect to the plane of $\boldsymbol{a}$ and $\boldsymbol{b}$. □

1.1.3 Projections, Bases and Coordinate Frames

Let $\boldsymbol{e}$ be a unit vector. Then any vector $\boldsymbol{v}$ can be written in the form

$$\boldsymbol{v} = \boldsymbol{v}_e + \boldsymbol{v}_e^{\perp}, \tag{1.1}$$

where $\boldsymbol{v}_e$ is parallel to $\boldsymbol{e}$ and $\boldsymbol{v}_e^{\perp}$ is perpendicular to $\boldsymbol{e}$. The vector $\boldsymbol{v}_e$ is called the **projection** of $\boldsymbol{v}$ parallel to $\boldsymbol{e}$ and is defined as

$$\boldsymbol{v}_e = (\boldsymbol{v} \cdot \boldsymbol{e})\boldsymbol{e}.$$

The vector $\boldsymbol{v}_e^{\perp}$ is called the projection of $\boldsymbol{v}$ perpendicular to $\boldsymbol{e}$. Since $\boldsymbol{v}_e^{\perp} = \boldsymbol{v} - \boldsymbol{v}_e$ we find that

$$\boldsymbol{v}_e^{\perp} = \boldsymbol{v} - (\boldsymbol{v} \cdot \boldsymbol{e})\boldsymbol{e}.$$

Thus any vector $\boldsymbol{v}$ can be uniquely decomposed into parts parallel and perpendicular to a given unit vector $\boldsymbol{e}$.

By a right-handed orthonormal **basis** for $\mathcal{V}$ we mean three mutually perpendicular unit vectors $\{\boldsymbol{i}, \boldsymbol{j}, \boldsymbol{k}\}$ which are oriented in the sense that

$$\boldsymbol{i} \times \boldsymbol{j} = \boldsymbol{k}, \qquad \boldsymbol{j} \times \boldsymbol{k} = \boldsymbol{i}, \qquad \boldsymbol{k} \times \boldsymbol{i} = \boldsymbol{j}.$$

While any three mutually perpendicular unit vectors necessarily satisfy $|(\boldsymbol{i} \times \boldsymbol{j}) \cdot \boldsymbol{k}| = 1$, a right-handed basis has the property that $(\boldsymbol{i} \times \boldsymbol{j}) \cdot \boldsymbol{k} = 1$. In contrast, a left-handed basis satisfies $(\boldsymbol{i} \times \boldsymbol{j}) \cdot \boldsymbol{k} = -1$.

Using the notation $\{\boldsymbol{e}_1, \boldsymbol{e}_2, \boldsymbol{e}_3\}$ in place of $\{\boldsymbol{i}, \boldsymbol{j}, \boldsymbol{k}\}$ we find, by repeated application of (1.1), that any vector $\boldsymbol{v}$ can be uniquely decomposed as

$$\boldsymbol{v} = v_1\boldsymbol{e}_1 + v_2\boldsymbol{e}_2 + v_3\boldsymbol{e}_3 \quad \text{where} \quad v_i = \boldsymbol{v} \cdot \boldsymbol{e}_i \in \mathbb{R}.$$

The numbers v_1, v_2, v_3 are called the **components** of $\boldsymbol{v}$ in the given basis $\{\boldsymbol{e}_1, \boldsymbol{e}_2, \boldsymbol{e}_3\}$. For brevity we will often denote the set of components by v_i and a given basis by $\{\boldsymbol{e}_i\}$, where it is understood that the subscript i ranges from 1 to 3.

We will frequently find it useful to arrange the components of a vector $\boldsymbol{v}$ into a 3×1 column matrix $[\boldsymbol{v}]$, in particular

$$[\boldsymbol{v}] = \begin{Bmatrix} v_1 \\ v_2 \\ v_3 \end{Bmatrix} \in \mathbb{R}^3,$$

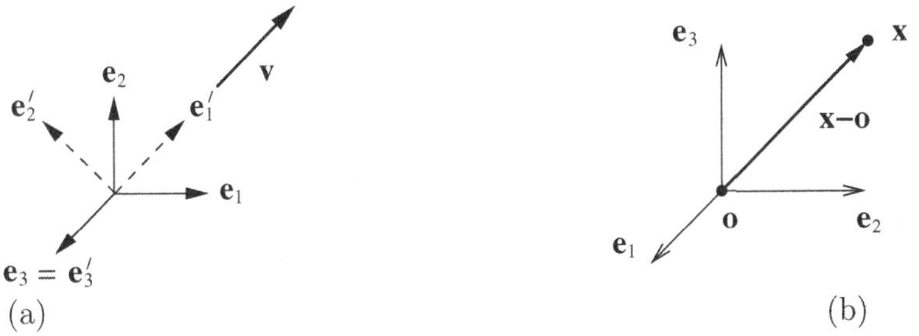

Fig. 1.3 Graphical illustration of orthonormal bases and a coordinate frame. (a) two orthonormal bases for $\mathcal{V}$. (b) Cartesian coordinate frame for $\mathbb{E}^3$.

where $\mathbb{R}^3$ denotes the set of all ordered triplets of real numbers. We will also have need to consider the **transpose** of $[\boldsymbol{v}]$, which is a 1×3 row matrix defined as $[\boldsymbol{v}]^T = (v_1, v_2, v_3)$. Notice that by properties of the transpose we may write $[\boldsymbol{v}] = (v_1, v_2, v_3)^T$. We call $[\boldsymbol{v}]$ the matrix **representation** of $\boldsymbol{v}$ in the given basis.

It is important to distinguish between vectors $\boldsymbol{v} \in \mathcal{V}$ (arrows in space) and their representations $[\boldsymbol{v}] \in \mathbb{R}^3$ (triplets of numbers). In particular, a given vector can have different representations depending on the choice of basis for $\mathcal{V}$.

Example: Let $\{\boldsymbol{e}_i\}$ be a basis for $\mathcal{V}$ and consider a vector $\boldsymbol{v}$ whose representation in this basis is $[\boldsymbol{v}] = (1, 1, 0)^T$ as shown in Figure 1.3(a). Thus

$$\boldsymbol{v} \cdot \boldsymbol{e}_1 = 1, \ \boldsymbol{v} \cdot \boldsymbol{e}_2 = 1, \ \boldsymbol{v} \cdot \boldsymbol{e}_3 = 0 \quad \text{or} \quad \boldsymbol{v} = 1\,\boldsymbol{e}_1 + 1\,\boldsymbol{e}_2 + 0\,\boldsymbol{e}_3.$$

Next, consider the same vector $\boldsymbol{v}$, but rotate the basis $\{\boldsymbol{e}_i\}$ through an angle of $\pi/4$ about the axis defined by $\boldsymbol{e}_3$. The result of this rotation is a new basis $\{\boldsymbol{e}_i'\}$, and in this basis the components of $\boldsymbol{v}$ are

$$\boldsymbol{v} \cdot \boldsymbol{e}_1' = \sqrt{2}, \ \boldsymbol{v} \cdot \boldsymbol{e}_2' = 0, \ \boldsymbol{v} \cdot \boldsymbol{e}_3' = 0 \quad \text{or} \quad \boldsymbol{v} = \sqrt{2}\,\boldsymbol{e}_1' + 0\,\boldsymbol{e}_2' + 0\,\boldsymbol{e}_3'.$$

The representation of $\boldsymbol{v}$ in the new basis is $[\boldsymbol{v}]' = (\sqrt{2}, 0, 0)^T$ and we see that $[\boldsymbol{v}]' \neq [\boldsymbol{v}]$. □

By a Cartesian **coordinate frame** for $\mathbb{E}^3$ we mean a reference point $\boldsymbol{o} \in \mathbb{E}^3$ called an **origin** together with a right-handed orthonormal basis $\{\boldsymbol{e}_i\}$ for the associated vector space $\mathcal{V}$ as depicted in Figure 1.3(b). To any point $\boldsymbol{x} \in \mathbb{E}^3$ we then ascribe **coordinates** x_i according to the expression

$$x_i = (\boldsymbol{x} - \boldsymbol{o}) \cdot \boldsymbol{e}_i.$$

Thus the coordinates of $\boldsymbol{x}$ are the unique numbers x_i such that

$$\boldsymbol{x} - \boldsymbol{o} = x_1\boldsymbol{e}_1 + x_2\boldsymbol{e}_2 + x_3\boldsymbol{e}_3.$$

Throughout our developments we will use the term coordinate frame for $\mathbb{E}^3$ without making explicit reference to the point $\boldsymbol{o} \in \mathbb{E}^3$ chosen as the origin. Moreover, we will frequently identify a point $\boldsymbol{x} \in \mathbb{E}^3$ with its position vector $\boldsymbol{x} - \boldsymbol{o} \in \mathcal{V}$ and use the same symbol $\boldsymbol{x}$ for both.

1.2 Index Notation

In this section we express various operations between vectors in terms of their components in a given frame. We begin with the summation convention for index notation and then introduce the Kronecker delta and permutation symbols which will be used throughout the remainder of our developments.

1.2.1 Summation Convention

The representation of vectors and vector operations in terms of components naturally involves sums. For example, let $\boldsymbol{a}$ and $\boldsymbol{b}$ be vectors with components a_i and b_i in a frame $\{\boldsymbol{e}_i\}$. Then

$$\boldsymbol{a} = a_1\boldsymbol{e}_1 + a_2\boldsymbol{e}_2 + a_3\boldsymbol{e}_3 = \sum_{i=1}^{3} a_i\boldsymbol{e}_i,$$

$$\boldsymbol{b} = b_1\boldsymbol{e}_1 + b_2\boldsymbol{e}_2 + b_3\boldsymbol{e}_3 = \sum_{j=1}^{3} b_j\boldsymbol{e}_j,$$

$$\boldsymbol{a}\cdot\boldsymbol{b} = \left(\sum_{i=1}^{3} a_i\boldsymbol{e}_i\right)\cdot\left(\sum_{j=1}^{3} b_j\boldsymbol{e}_j\right) = \sum_{i=1}^{3}\sum_{j=1}^{3} a_i b_j \boldsymbol{e}_i\cdot\boldsymbol{e}_j = \sum_{i=1}^{3} a_i b_i,$$

where the last line follows from the fact that $\boldsymbol{e}_i\cdot\boldsymbol{e}_j = 1$ or 0 depending on whether $i = j$ or $i \neq j$.

The expressions for $\boldsymbol{a}$, $\boldsymbol{b}$ and $\boldsymbol{a}\cdot\boldsymbol{b}$ each involves a sum over a pair of indices. Because sums of this type occur so often we adopt a convention and abbreviate the above three equations as

$$\boldsymbol{a} = a_i\boldsymbol{e}_i, \quad \boldsymbol{b} = b_j\boldsymbol{e}_j \quad \text{and} \quad \boldsymbol{a}\cdot\boldsymbol{b} = a_i b_i.$$

The convention is: *whenever an index occurs twice in a term a sum is implied over that index.* In particular, $a_i\boldsymbol{e}_i$ is shorthand for the sum $\sum_{i=1}^{3} a_i\boldsymbol{e}_i$. Unless mentioned otherwise we will always assume that the summation convention is in force. In this case any pair of repeated indices in a term represents a sum, for example

$$a_2 b_j b_j b_3 = a_2(b_1 b_1 + b_2 b_2 + b_3 b_3) b_3,$$
$$a_i b_i + b_i b_i = (a_1 b_1 + a_2 b_2 + a_3 b_3) + (b_1 b_1 + b_2 b_2 + b_3 b_3),$$
$$a_i a_j b_j a_i = a_i a_i a_j b_j = (a_1 a_1 + a_2 a_2 + a_3 a_3)(a_1 b_1 + a_2 b_2 + a_3 b_3).$$

Any repeated index, such as i or j above, which is used to represent a sum is called a **dummy index**. This terminology reflects the fact that the actual symbol used for a repeated index is immaterial since

$$a_i b_i = a_1 b_1 + a_2 b_2 + a_3 b_3 = a_j b_j.$$

Notice that the summation convention applies only to *pairs* of repeated indices. In particular, no sums will be implied in expressions of the form a_i, $a_i b_i b_i$, and so on.

Any index which is not a dummy index is called a **free index**. For example, in the equation

$$a_i = c_j b_j b_i,$$

the index i is a free index, while j is a dummy index. Free indices can take any of the values 1, 2 or 3, and are used to abbreviate groups of similar equations. For example, the above equation is shorthand for the three equations

$$a_1 = c_j b_j b_1, \qquad a_2 = c_j b_j b_2, \qquad a_3 = c_j b_j b_3.$$

Similarly, the equation

$$a_i b_j = c_i c_k c_k d_j$$

is shorthand for nine equations since both i and j are free indices. The nine equations are

$$a_1 b_1 = c_1 c_k c_k d_1, \qquad a_1 b_2 = c_1 c_k c_k d_2, \qquad \dots \qquad a_3 b_3 = c_3 c_k c_k d_3.$$

Notice that each term in an equation should have the same free indices, and the same symbol cannot be used for both a dummy and free index. For example, equations of the form

$$a_i = b_j, \qquad a_i b_j = c_i d_j d_j \qquad \text{and} \qquad a_i b_j = c_i c_k d_k d_j + d_p c_l c_l d_q$$

are all ambiguous and are not permissible in our use of the summation convention.

1.2.2 Kronecker Delta and Permutation Symbols

To any frame $\{\boldsymbol{e}_i\}$ we associate a **Kronecker delta** symbol δ_{ij} defined by

$$\delta_{ij} = \boldsymbol{e}_i \cdot \boldsymbol{e}_j = \begin{cases} 1, & \text{if } i = j, \\ 0, & \text{if } i \neq j, \end{cases}$$

and a **permutation symbol** ϵ_{ijk} defined by

$$\epsilon_{ijk} = (\boldsymbol{e}_i \times \boldsymbol{e}_j) \cdot \boldsymbol{e}_k = \begin{cases} 1, & \text{if } ijk = 123,\ 231 \text{ or } 312, \\ -1, & \text{if } ijk = 321,\ 213 \text{ or } 132, \\ 0, & \text{otherwise (repeated index).} \end{cases}$$

Notice that the numerical values of δ_{ij} and ϵ_{ijk} are the same for any right-handed orthonormal frame.

The Kronecker delta and permutation symbols enjoy certain symmetry properties under index transposition and permutation. In particular, we find that δ_{ij} is invariant under transposition of indices in the sense that $\delta_{ij} = \delta_{ji}$, whereas ϵ_{ijk} changes sign under (pairwise) transposition, namely, $\epsilon_{ijk} = -\epsilon_{jik}$, $\epsilon_{ijk} = -\epsilon_{ikj}$ and $\epsilon_{ijk} = -\epsilon_{kji}$. However, ϵ_{ijk} is invariant under circular or cyclic permutation of indices in the sense that $\epsilon_{ijk} = \epsilon_{jki} = \epsilon_{kij}$.

The permutation symbol can alternatively be defined as

$$\epsilon_{ijk} = \det\left([\boldsymbol{e}_i], [\boldsymbol{e}_j], [\boldsymbol{e}_k]\right),$$

where $([\boldsymbol{e}_i], [\boldsymbol{e}_j], [\boldsymbol{e}_k])$ is the 3×3 matrix with columns $[\boldsymbol{e}_i]$, $[\boldsymbol{e}_j]$ and $[\boldsymbol{e}_k]$. In particular, all the properties outlined above for ϵ_{ijk} under index transposition and permutation can be deduced from properties of the determinant.

1.2.3 Frame Identities

Various useful identities can be established between the vectors of a frame $\{\boldsymbol{e}_i\}$ and the Kronecker delta δ_{ij} and the permutation symbol ϵ_{ijk}. In particular, from the definition of δ_{ij} we deduce the identity

$$\boldsymbol{e}_i = \delta_{ij} \boldsymbol{e}_j.$$

That is, $\delta_{1j}\boldsymbol{e}_j = \boldsymbol{e}_1$, $\delta_{2j}\boldsymbol{e}_j = \boldsymbol{e}_2$ and $\delta_{3j}\boldsymbol{e}_j = \boldsymbol{e}_3$. This identity shows that, when δ_{ij} is summed with another quantity, the net effect is a "transfer of index". We call this the **transfer property** of δ_{ij}.

From the definition of ϵ_{ijk} we deduce the identity

$$\boldsymbol{e}_i \times \boldsymbol{e}_j = \epsilon_{ijk} \boldsymbol{e}_k,$$

which provides a concise way to express the nine possible vector products between the vectors of a frame, that is, $\boldsymbol{e}_1 \times \boldsymbol{e}_1 = \boldsymbol{0}$, $\boldsymbol{e}_1 \times \boldsymbol{e}_2 = \boldsymbol{e}_3$ and so on. A similar identity which can easily be verified is

$$\boldsymbol{e}_i = \tfrac{1}{2}\epsilon_{ijk}\boldsymbol{e}_j \times \boldsymbol{e}_k.$$

1.2.4 Vector Operations in Components

The Kronecker delta and permutation symbols naturally arise when various operations between vectors are expressed in terms of their components in a given frame.

Scalar product. Let $\boldsymbol{a} = a_i\boldsymbol{e}_i$ and $\boldsymbol{b} = b_j\boldsymbol{e}_j$. Then

$$\begin{aligned} \boldsymbol{a} \cdot \boldsymbol{b} &= (a_i\boldsymbol{e}_i) \cdot (b_j\boldsymbol{e}_j) \\ &= a_i b_j\, \boldsymbol{e}_i \cdot \boldsymbol{e}_j \\ &= a_i b_j \delta_{ij} \\ &= a_i b_i, \end{aligned}$$

where the last line follows from the transfer property of δ_{ij}, that is, $b_j\delta_{ij} = b_i$. Thus, from the geometrical definition of the scalar product we find $|\boldsymbol{a}|^2 = \boldsymbol{a} \cdot \boldsymbol{a} = a_i a_i$.

Vector product. Let $\boldsymbol{a} = a_i\boldsymbol{e}_i$, $\boldsymbol{b} = b_j\boldsymbol{e}_j$ and $\boldsymbol{d} = d_k\boldsymbol{e}_k$. Then

$$\boldsymbol{a} \times \boldsymbol{b} = a_i b_j\, \boldsymbol{e}_i \times \boldsymbol{e}_j = a_i b_j \epsilon_{ijk}\boldsymbol{e}_k,$$

where the last equality follows from the frame identity $\boldsymbol{e}_i \times \boldsymbol{e}_j = \epsilon_{ijk}\boldsymbol{e}_k$. Thus, if we let $\boldsymbol{d} = \boldsymbol{a} \times \boldsymbol{b}$, then $d_k = a_i b_j \epsilon_{ijk}$.

Triple scalar product. Let $\boldsymbol{a} = a_i\boldsymbol{e}_i$, $\boldsymbol{b} = b_j\boldsymbol{e}_j$ and $\boldsymbol{c} = c_m\boldsymbol{e}_m$. Then

$$\begin{aligned} (\boldsymbol{a} \times \boldsymbol{b}) \cdot \boldsymbol{c} &= (a_i b_j \epsilon_{ijk}\boldsymbol{e}_k) \cdot (c_m\boldsymbol{e}_m) \\ &= \epsilon_{ijk} a_i b_j c_m \delta_{km} \qquad (1.2) \\ &= \epsilon_{ijk} a_i b_j c_k. \end{aligned}$$

Moreover, by properties of the permutation symbol under cyclic permutation of its indices we find

$$(\boldsymbol{a} \times \boldsymbol{b}) \cdot \boldsymbol{c} = (\boldsymbol{b} \times \boldsymbol{c}) \cdot \boldsymbol{a} = (\boldsymbol{c} \times \boldsymbol{a}) \cdot \boldsymbol{b}.$$

From elementary vector analysis we recall that

$$(\boldsymbol{a} \times \boldsymbol{b}) \cdot \boldsymbol{c} = \det([\boldsymbol{a}], [\boldsymbol{b}], [\boldsymbol{c}]), \qquad (1.3)$$

where $([\boldsymbol{a}], [\boldsymbol{b}], [\boldsymbol{c}])$ is the 3×3 matrix with columns $[\boldsymbol{a}]$, $[\boldsymbol{b}]$ and $[\boldsymbol{c}]$. When (1.3) is combined with (1.2) we obtain

$$\det([\boldsymbol{a}], [\boldsymbol{b}], [\boldsymbol{c}]) = \epsilon_{ijk} a_i b_j c_k, \tag{1.4}$$

which provides an explicit expression for the determinant in terms of a triple sum.

Triple vector product. Let $\boldsymbol{a} = a_q \boldsymbol{e}_q$, $\boldsymbol{b} = b_i \boldsymbol{e}_i$, $\boldsymbol{c} = c_j \boldsymbol{e}_j$ and $\boldsymbol{d} = d_p \boldsymbol{e}_p$. Then

$$\begin{aligned} \boldsymbol{a} \times (\boldsymbol{b} \times \boldsymbol{c}) &= (a_q \boldsymbol{e}_q) \times (b_i c_j \epsilon_{ijk} \boldsymbol{e}_k) \\ &= \epsilon_{ijk} a_q b_i c_j \, \boldsymbol{e}_q \times \boldsymbol{e}_k \\ &= \epsilon_{qkp} \epsilon_{ijk} a_q b_i c_j \boldsymbol{e}_p \\ &= \epsilon_{pqk} \epsilon_{ijk} a_q b_i c_j \boldsymbol{e}_p, \end{aligned} \tag{1.5}$$

where the last line follows from the fact that $\epsilon_{qkp} = \epsilon_{pqk}$. Thus, if we let $\boldsymbol{d} = \boldsymbol{a} \times (\boldsymbol{b} \times \boldsymbol{c})$, then $d_p = \epsilon_{pqk} \epsilon_{ijk} a_q b_i c_j$. Below we show that this expression can be used to derive a fundamental identity for the triple vector product.

Remark: The geometrical definitions of the scalar and triple scalar products imply that these quantities are frame-independent. That is, the scalar and triple scalar products can be computed using components in any coordinate frame, with the same value obtained in each frame. Identifying scalar quantities with this property will be an important theme in the sequel. For other examples see the discussion of the trace, determinant and principal invariants of second-order tensors. □

1.2.5 Epsilon-Delta Identities

By virtue of their definitions in terms of a right-handed orthonormal frame, the permutation symbol and the Kronecker delta satisfy the following identities.

Result 1.1 ***Epsilon-Delta Identities.*** *Let ϵ_{ijk} be the permutation symbol and δ_{ij} the Kronecker delta. Then*

$$\epsilon_{pqs}\epsilon_{nrs} = \delta_{pn}\delta_{qr} - \delta_{pr}\delta_{qn} \quad \textit{and} \quad \epsilon_{pqs}\epsilon_{rqs} = 2\delta_{pr}.$$

□

Proof See Exercise 18. □

As an application of the above result we establish an identity involving the triple vector product

$$\boldsymbol{a} \times (\boldsymbol{b} \times \boldsymbol{c}) = (\boldsymbol{a} \cdot \boldsymbol{c})\boldsymbol{b} - (\boldsymbol{a} \cdot \boldsymbol{b})\boldsymbol{c}, \quad \forall \boldsymbol{a}, \boldsymbol{b}, \boldsymbol{c} \in \mathcal{V}.$$

To see this notice that, by (1.5) and Result 1.1, we have

$$\begin{aligned}\boldsymbol{a} \times (\boldsymbol{b} \times \boldsymbol{c}) &= \epsilon_{pqk}\epsilon_{ijk}a_q b_i c_j \boldsymbol{e}_p \\ &= (\delta_{pi}\delta_{qj} - \delta_{pj}\delta_{qi})a_q b_i c_j \boldsymbol{e}_p \\ &= (a_q c_q)b_p \boldsymbol{e}_p - (a_q b_q)c_p \boldsymbol{e}_p \\ &= (\boldsymbol{a} \cdot \boldsymbol{c})\boldsymbol{b} - (\boldsymbol{a} \cdot \boldsymbol{b})\boldsymbol{c},\end{aligned}$$

which establishes the result.

1.3 Second-Order Tensors

Many physical quantities in mechanics, such as forces, velocities and accelerations, are well-represented by the notion of a vector (first-order tensor). In any Cartesian coordinate frame for $\mathbb{E}^3$ such quantities are described by three components. In our studies we will encounter other physical quantities, such as stress and strain, which are not well-represented by vectors, but rather by *linear transformations between vectors*, leading to the notion of a second-order tensor as defined below. As we will see, such quantities will be described by nine components in any Cartesian coordinate frame for $\mathbb{E}^3$.

1.3.1 Definition

By a **second-order tensor** $\boldsymbol{T}$ on the vector space $\mathcal{V}$ we mean a mapping $\boldsymbol{T} : \mathcal{V} \to \mathcal{V}$ which is *linear* in the sense that:

(1) $\boldsymbol{T}(\boldsymbol{u} + \boldsymbol{v}) = \boldsymbol{T}\boldsymbol{u} + \boldsymbol{T}\boldsymbol{v}$ for all $\boldsymbol{u}, \boldsymbol{v} \in \mathcal{V}$,

(2) $\boldsymbol{T}(\alpha \boldsymbol{u}) = \alpha \boldsymbol{T}\boldsymbol{u}$ for all $\alpha \in \mathbb{R}$ and $\boldsymbol{u} \in \mathcal{V}$.

We denote the set of all second-order tensors on $\mathcal{V}$ by the symbol $\mathcal{V}^2$. Analogous to the zero vector, we define a **zero tensor** $\boldsymbol{O}$ with the property $\boldsymbol{O}\boldsymbol{v} = \boldsymbol{0}$ for all $\boldsymbol{v} \in \mathcal{V}$, and we define an **identity tensor** $\boldsymbol{I}$ with the property $\boldsymbol{I}\boldsymbol{v} = \boldsymbol{v}$ for all $\boldsymbol{v} \in \mathcal{V}$. Two second-order tensors $\boldsymbol{S}$ and $\boldsymbol{T}$ are said to be equal if and only if $\boldsymbol{S}\boldsymbol{v} = \boldsymbol{T}\boldsymbol{v}$ for all $\boldsymbol{v} \in \mathcal{V}$.

Examples:

(1) **Projections**. The action of projecting a vector $\boldsymbol{v}$ onto a given unit vector $\boldsymbol{e}$ can be expressed in terms of a second-order tensor $\boldsymbol{P}$, namely

$$\boldsymbol{P}(\boldsymbol{v}) = (\boldsymbol{v} \cdot \boldsymbol{e})\boldsymbol{e}.$$

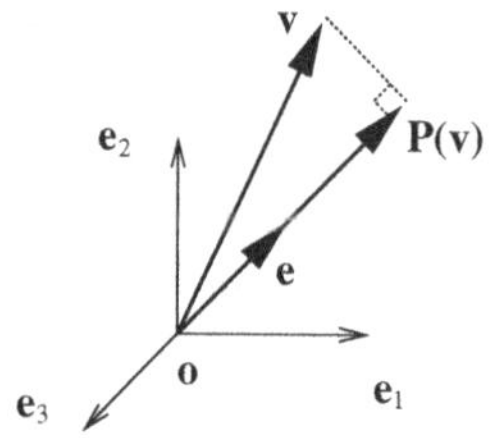

(2) **Rotations**. The action of rotating a vector $\boldsymbol{v}$ through a given angle θ about a fixed axis, say $\boldsymbol{e}_3$, can be expressed in terms of a second-order tensor $\boldsymbol{S}$. In particular, for each of the basis vectors $\boldsymbol{e}_i$ we have

$$\boldsymbol{S}(\boldsymbol{e}_1) = \cos\theta\, \boldsymbol{e}_1 + \sin\theta\, \boldsymbol{e}_2, \quad \boldsymbol{S}(\boldsymbol{e}_2) = \cos\theta\, \boldsymbol{e}_2 - \sin\theta\, \boldsymbol{e}_1, \quad \boldsymbol{S}(\boldsymbol{e}_3) = \boldsymbol{e}_3.$$

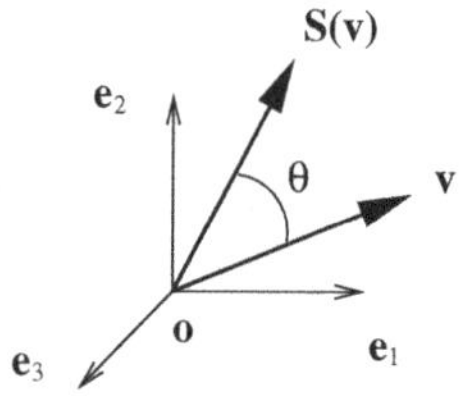

(3) **Reflections**. The action of reflecting a vector $\boldsymbol{v}$ with respect to a fixed plane, say with normal $\boldsymbol{e}_1$, can be expressed in terms of a second-order tensor $\boldsymbol{T}$, namely

$$\boldsymbol{T}(\boldsymbol{v}) = \boldsymbol{v} - 2(\boldsymbol{v} \cdot \boldsymbol{e}_1)\boldsymbol{e}_1.$$

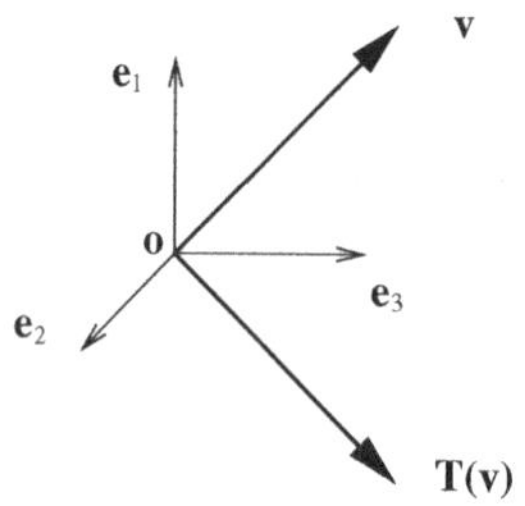

□

1.3.2 Second-Order Tensor Algebra

If we let $\mathcal{V}^2$ be the set of second-order tensors, then $\mathcal{V}^2$ has the structure of a real vector space since

$$\boldsymbol{S}+\boldsymbol{T} \in \mathcal{V}^2 \quad \forall \boldsymbol{S},\boldsymbol{T} \in \mathcal{V}^2 \quad \text{and} \quad \alpha\boldsymbol{T} \in \mathcal{V}^2 \quad \forall \alpha \in \mathbb{R},\ \boldsymbol{T} \in \mathcal{V}^2.$$

We can also compose second-order tensors in the sense that

$$\boldsymbol{S}\boldsymbol{T} \in \mathcal{V}^2 \quad \forall \boldsymbol{S},\boldsymbol{T} \in \mathcal{V}^2.$$

In particular, we define the sum $\boldsymbol{S}+\boldsymbol{T}$ via the relation $(\boldsymbol{S}+\boldsymbol{T})\boldsymbol{v} = \boldsymbol{S}\boldsymbol{v}+\boldsymbol{T}\boldsymbol{v}$ for all $\boldsymbol{v} \in \mathcal{V}$, and we define the composition $\boldsymbol{S}\boldsymbol{T}$ via the relation $(\boldsymbol{S}\boldsymbol{T})\boldsymbol{v} = \boldsymbol{S}(\boldsymbol{T}\boldsymbol{v})$ for all $\boldsymbol{v} \in \mathcal{V}$. A similar definition holds for $\alpha\boldsymbol{T}$.

1.3.3 Representation in a Coordinate Frame

By the **components** of a second-order tensor $\boldsymbol{S}$ in a coordinate frame $\{\boldsymbol{e}_i\}$ we mean the nine numbers S_{ij} $(1 \le i,j \le 3)$ defined by

$$S_{ij} = \boldsymbol{e}_i \cdot \boldsymbol{S}\boldsymbol{e}_j.$$

These components completely describe the transformation $\boldsymbol{S} : \mathcal{V} \to \mathcal{V}$. Given vectors $\boldsymbol{v} = v_i\boldsymbol{e}_i$ and $\boldsymbol{u} = u_j\boldsymbol{e}_j$ such that $\boldsymbol{v} = \boldsymbol{S}\boldsymbol{u}$ we have, by linearity

$$v_i = \boldsymbol{e}_i \cdot \boldsymbol{S}\boldsymbol{u} = (\boldsymbol{e}_i \cdot \boldsymbol{S}\boldsymbol{e}_j)u_j = S_{ij}u_j. \tag{1.6}$$

Thus the components of $\boldsymbol{S}$ are simply the coefficients in the linear relation between the components of $\boldsymbol{v}$ and $\boldsymbol{u}$.

Frequently, we arrange the nine components S_{ij} into a square matrix $[\boldsymbol{S}]$, in particular

$$[\boldsymbol{S}] = \begin{pmatrix} S_{11} & S_{12} & S_{13} \\ S_{21} & S_{22} & S_{23} \\ S_{31} & S_{32} & S_{33} \end{pmatrix} \in \mathbb{R}^{3\times 3},$$

where $\mathbb{R}^{3\times 3}$ denotes the set of all 3×3 matrices of real numbers. We call $[\boldsymbol{S}]$ the matrix **representation** of $\boldsymbol{S}$ in the given coordinate frame. We could also arrange the components into a 9×1 matrix, that is, an element of $\mathbb{R}^9$. However, the square matrix representation will be more useful for our developments.

If we denote the element of $[\boldsymbol{S}]$ at the ith-row and jth-column by $[\boldsymbol{S}]_{ij}$, then we have $[\boldsymbol{S}]_{ij} = S_{ij}$. The **transpose** of the matrix $[\boldsymbol{S}]$ is denoted by $[\boldsymbol{S}]^T$, and its elements are defined by the expression $[\boldsymbol{S}]^T_{ij} = [\boldsymbol{S}]_{ji}$. From the definitions of $[\boldsymbol{S}]$ and S_{ij} we deduce the convenient expression

$$[\boldsymbol{S}] = ([\boldsymbol{S}\boldsymbol{e}_1], [\boldsymbol{S}\boldsymbol{e}_2], [\boldsymbol{S}\boldsymbol{e}_3]),$$

where $[\boldsymbol{S}\boldsymbol{e}_j]$ is the 3×1 column matrix of components of $\boldsymbol{S}\boldsymbol{e}_j$.

In view of (1.6) and the rules of matrix multiplication, the tensor equation $\boldsymbol{v} = \boldsymbol{S}\boldsymbol{u}$ can be written in various equivalent ways

$$\boldsymbol{v} = \boldsymbol{S}\boldsymbol{u} \quad \Leftrightarrow \quad v_i = S_{ij}u_j \quad \Leftrightarrow \quad [\boldsymbol{v}] = [\boldsymbol{S}][\boldsymbol{u}].$$

Thus the matrix representations of vectors and second-order tensors have the property

$$[\boldsymbol{S}\boldsymbol{u}] = [\boldsymbol{S}][\boldsymbol{u}].$$

Examples:

(1) Let $\boldsymbol{S}$ and $\boldsymbol{T}$ be the rotation and reflection tensors introduced in the examples of Section 1.3.1. In the respective coordinate frame of each example, the matrix representation of $\boldsymbol{S}$ is

$$[\boldsymbol{S}] = \begin{pmatrix} \cos\theta & -\sin\theta & 0 \\ \sin\theta & \cos\theta & 0 \\ 0 & 0 & 1 \end{pmatrix},$$

and the matrix representation of $\boldsymbol{T}$ is

$$[\boldsymbol{T}] = \begin{pmatrix} -1 & 0 & 0 \\ 0 & 1 & 0 \\ 0 & 0 & 1 \end{pmatrix}.$$

(2) It is straightforward to show that, in any coordinate frame, the zero tensor $\boldsymbol{O}$ has components $O_{ij} = 0$ $(1 \le i, j \le 3)$, and the identity tensor $\boldsymbol{I}$ has components $I_{ij} = \delta_{ij}$ $(1 \le i, j \le 3)$.

□

1.3.4 Second-Order Dyadic Products, Bases

The **dyadic product** of two vectors $\boldsymbol{a}$ and $\boldsymbol{b}$ is the second-order tensor $\boldsymbol{a} \otimes \boldsymbol{b}$ defined by

$$(\boldsymbol{a} \otimes \boldsymbol{b})\boldsymbol{v} = (\boldsymbol{b} \cdot \boldsymbol{v})\boldsymbol{a}, \qquad \forall \boldsymbol{v} \in \mathcal{V}.$$

In terms of components $[\boldsymbol{a}\otimes\boldsymbol{b}]_{ij}$, the above equation is equivalent to

$$[\boldsymbol{a}\otimes\boldsymbol{b}]_{ij}v_j = (b_j v_j)a_i, \qquad \forall \boldsymbol{v}\in\mathcal{V},$$

which implies

$$[\boldsymbol{a}\otimes\boldsymbol{b}]_{ij} = a_i b_j.$$

To complete this last step simply take $\boldsymbol{v}=\boldsymbol{e}_1$, $\boldsymbol{e}_2$, $\boldsymbol{e}_3$ in turn. Throughout our developments we will move freely from similar expressions, which hold for all $\boldsymbol{v}$, to the statement with $\boldsymbol{v}$ removed. Notice that the matrix representation of $\boldsymbol{a}\otimes\boldsymbol{b}$ is

$$[\boldsymbol{a}\otimes\boldsymbol{b}] = \begin{pmatrix} a_1b_1 & a_1b_2 & a_1b_3 \\ a_2b_1 & a_2b_2 & a_2b_3 \\ a_3b_1 & a_3b_2 & a_3b_3 \end{pmatrix} = [\boldsymbol{a}][\boldsymbol{b}]^T.$$

Given any coordinate frame $\{\boldsymbol{e}_i\}$ the nine elementary dyadic products $\{\boldsymbol{e}_i\otimes\boldsymbol{e}_j\}$ form a **basis** for $\mathcal{V}^2$. In particular, each $\boldsymbol{S}\in\mathcal{V}^2$ can be uniquely represented as a linear combination

$$\boldsymbol{S} = S_{ij}\boldsymbol{e}_i\otimes\boldsymbol{e}_j, \tag{1.7}$$

where $S_{ij}=\boldsymbol{e}_i\cdot\boldsymbol{S}\boldsymbol{e}_j$. To see that the above representation is correct consider the expression $\boldsymbol{v}=\boldsymbol{S}\boldsymbol{u}$, where $\boldsymbol{v}=v_i\boldsymbol{e}_i$ and $\boldsymbol{u}=u_k\boldsymbol{e}_k$. Then

$$\begin{aligned} \boldsymbol{v} &= \boldsymbol{S}\boldsymbol{u} \\ &= S_{ij}\boldsymbol{e}_i\otimes\boldsymbol{e}_j\ u_k\boldsymbol{e}_k \\ &= S_{ij}\delta_{jk}u_k\boldsymbol{e}_i \qquad\qquad \text{(by dyadic property)} \\ &= S_{ij}u_j\boldsymbol{e}_i, \end{aligned}$$

which implies $v_i = S_{ij}u_j$. Thus (1.7) agrees with (1.6) and provides a unique representation of $\boldsymbol{S}$ in terms of its components S_{ij}.

1.3.5 Second-Order Tensor Algebra in Components

Let $\boldsymbol{S}=S_{ij}\boldsymbol{e}_i\otimes\boldsymbol{e}_j$ and $\boldsymbol{T}=T_{ij}\boldsymbol{e}_i\otimes\boldsymbol{e}_j$ be second-order tensors with matrix representations $[\boldsymbol{S}]$ and $[\boldsymbol{T}]$. Then from the definitions of $\boldsymbol{S}+\boldsymbol{T}$ and $\alpha\boldsymbol{T}$ it is straightforward to deduce

$$[\boldsymbol{S}+\boldsymbol{T}] = [\boldsymbol{S}]+[\boldsymbol{T}] \quad \text{and} \quad [\alpha\boldsymbol{T}] = \alpha[\boldsymbol{T}].$$

To establish a component expression for composition consider any two dyadic products $\boldsymbol{a} \otimes \boldsymbol{b}$ and $\boldsymbol{c} \otimes \boldsymbol{d}$. Then for any arbitrary vector $\boldsymbol{v}$ we have

$$\begin{aligned}((\boldsymbol{a} \otimes \boldsymbol{b})(\boldsymbol{c} \otimes \boldsymbol{d}))\,\boldsymbol{v} &= (\boldsymbol{a} \otimes \boldsymbol{b})\,((\boldsymbol{c} \otimes \boldsymbol{d})\boldsymbol{v}) \\ &= (\boldsymbol{a} \otimes \boldsymbol{b})\boldsymbol{c}(\boldsymbol{d} \cdot \boldsymbol{v}) \\ &= \boldsymbol{a}(\boldsymbol{b} \cdot \boldsymbol{c})(\boldsymbol{d} \cdot \boldsymbol{v}) \\ &= (\boldsymbol{b} \cdot \boldsymbol{c})(\boldsymbol{a} \otimes \boldsymbol{d})\boldsymbol{v},\end{aligned}$$

which, by the arbitrariness of $\boldsymbol{v}$, implies

$$(\boldsymbol{a} \otimes \boldsymbol{b})(\boldsymbol{c} \otimes \boldsymbol{d}) = (\boldsymbol{b} \cdot \boldsymbol{c})\boldsymbol{a} \otimes \boldsymbol{d}.$$

With this result we can express tensor composition in terms of components as follows. Let $\boldsymbol{S}$ and $\boldsymbol{T}$ be as above and let $\boldsymbol{U} = \boldsymbol{S}\boldsymbol{T}$. Then

$$\begin{aligned}\boldsymbol{U} &= (S_{ij}\boldsymbol{e}_i \otimes \boldsymbol{e}_j)(T_{kl}\boldsymbol{e}_k \otimes \boldsymbol{e}_l) \\ &= S_{ij}T_{kl}\delta_{jk}\boldsymbol{e}_i \otimes \boldsymbol{e}_l \\ &= S_{ij}T_{jl}\boldsymbol{e}_i \otimes \boldsymbol{e}_l,\end{aligned}$$

which implies that $\boldsymbol{U}$ has components $U_{il} = S_{ij}T_{jl}$. Notice that this corresponds to the standard notion of matrix multiplication for the representations of $\boldsymbol{S}$ and $\boldsymbol{T}$, that is

$$[\boldsymbol{U}] = [\boldsymbol{S}\boldsymbol{T}] = [\boldsymbol{S}][\boldsymbol{T}].$$

1.3.6 Special Classes of Tensors

To any tensor $\boldsymbol{S} \in \mathcal{V}^2$ we associate a **transpose** $\boldsymbol{S}^T \in \mathcal{V}^2$, which is the unique tensor with the property

$$\boldsymbol{S}\boldsymbol{u} \cdot \boldsymbol{v} = \boldsymbol{u} \cdot \boldsymbol{S}^T\boldsymbol{v} \quad \forall \boldsymbol{u}, \boldsymbol{v} \in \mathcal{V}.$$

We say $\boldsymbol{S}$ is **symmetric** if $\boldsymbol{S}^T = \boldsymbol{S}$ and **skew-symmetric** if $\boldsymbol{S}^T = -\boldsymbol{S}$.

A tensor $\boldsymbol{S} \in \mathcal{V}^2$ is said to be **positive-definite** if it satisfies

$$\boldsymbol{v} \cdot \boldsymbol{S}\boldsymbol{v} > 0 \quad \forall \boldsymbol{v} \neq \boldsymbol{0},$$

and is said to be **invertible** if there exists an **inverse** $\boldsymbol{S}^{-1} \in \mathcal{V}^2$ such that

$$\boldsymbol{S}\boldsymbol{S}^{-1} = \boldsymbol{S}^{-1}\boldsymbol{S} = \boldsymbol{I}.$$

The operations of inverse and transpose commute, that is, $(\boldsymbol{S}^{-1})^T = (\boldsymbol{S}^T)^{-1}$, and we denote the resulting tensor by $\boldsymbol{S}^{-T}$.

A tensor $\boldsymbol{Q} \in \mathcal{V}^2$ is said to be **orthogonal** if $\boldsymbol{Q}^T = \boldsymbol{Q}^{-1}$, that is

$$\boldsymbol{Q}\boldsymbol{Q}^T = \boldsymbol{Q}^T\boldsymbol{Q} = \boldsymbol{I}.$$

An orthogonal tensor is called a **rotation** if, given any right-handed orthonormal frame $\{\boldsymbol{e}_i\}$, the set of vectors $\{\boldsymbol{Q}\boldsymbol{e}_i\}$ is also a right-handed orthonormal frame. Later we will see that an orthogonal tensor is a rotation when it has positive determinant.

In any given coordinate frame the above ideas are just the standard notions of the transpose of a matrix, a symmetric matrix, a skew-symmetric matrix, a positive-definite matrix, the inverse of a matrix, an orthogonal matrix and a rotation matrix, respectively. Moreover, we have (see Exercise 11)

$$[\boldsymbol{S}^T] = [\boldsymbol{S}]^T, \qquad [\boldsymbol{S}^{-1}] = [\boldsymbol{S}]^{-1}, \qquad [\boldsymbol{S}^{-T}] = [\boldsymbol{S}]^{-T}.$$

The proof of the following result is straightforward.

Result 1.2 Symmetric-Skew Decomposition. *Every second-order tensor can be uniquely written as*

$$\boldsymbol{S} = \boldsymbol{E} + \boldsymbol{W},$$

where $\boldsymbol{E}$ is a symmetric tensor and $\boldsymbol{W}$ is a skew-symmetric tensor. Specifically, we have

$$\boldsymbol{E} = \tfrac{1}{2}(\boldsymbol{S} + \boldsymbol{S}^T) \quad \textit{and} \quad \boldsymbol{W} = \tfrac{1}{2}(\boldsymbol{S} - \boldsymbol{S}^T).$$

□

The above result motivates the two mappings sym : $\mathcal{V}^2 \to \mathcal{V}^2$ and skew : $\mathcal{V}^2 \to \mathcal{V}^2$ defined by

$$\mathrm{sym}(\boldsymbol{S}) = \tfrac{1}{2}(\boldsymbol{S} + \boldsymbol{S}^T) \quad \text{and} \quad \mathrm{skew}(\boldsymbol{S}) = \tfrac{1}{2}(\boldsymbol{S} - \boldsymbol{S}^T).$$

It is straightforward to show that, in any coordinate frame, a symmetric tensor $\boldsymbol{E}$ satisfies $E_{ij} = E_{ji}$, and a skew-symmetric tensor $\boldsymbol{W}$ satisfies $W_{ij} = -W_{ji}$.

Result 1.3 Skew Tensor as Vector Product. *Given any skew-symmetric tensor $\boldsymbol{W} \in \mathcal{V}^2$ there is a unique vector $\boldsymbol{w} \in \mathcal{V}$ such that*

$$\boldsymbol{W}\boldsymbol{v} = \boldsymbol{w} \times \boldsymbol{v}, \quad \forall \boldsymbol{v} \in \mathcal{V}.$$

We write $\boldsymbol{w} = \mathrm{vec}(\boldsymbol{W})$ and call it the **axial vector** *associated with $\boldsymbol{W}$. In any frame we have*

$$w_j = \frac{1}{2}\epsilon_{njm} W_{nm}.$$

Conversely, given any vector $\boldsymbol{w} \in \mathcal{V}$ *there is a unique skew-symmetric tensor* $\boldsymbol{W} \in \mathcal{V}^2$ *such that*

$$\boldsymbol{w} \times \boldsymbol{v} = \boldsymbol{W}\boldsymbol{v}, \quad \forall \boldsymbol{v} \in \mathcal{V}.$$

We write $\boldsymbol{W} = \mathrm{ten}(\boldsymbol{w})$ *and call it the* **axial tensor** *associated with* $\boldsymbol{w}$. *In any frame we have*

$$W_{ik} = \epsilon_{ijk} w_j.$$

The functions $\boldsymbol{W} = \mathrm{ten}(\boldsymbol{w})$ *and* $\boldsymbol{w} = \mathrm{vec}(\boldsymbol{W})$ *are inverses.* □

Proof To establish the first result let $\boldsymbol{W}$ be given. Then we seek $\boldsymbol{w}$ such that $\boldsymbol{W}\boldsymbol{v} = \boldsymbol{w} \times \boldsymbol{v}$ for all $\boldsymbol{v}$. In components this equation reads

$$W_{ik} v_k = \epsilon_{ijk} w_j v_k,$$

and by the arbitrariness of v_k we deduce

$$W_{ik} = \epsilon_{ijk} w_j. \tag{1.8}$$

We next show that there exists a vector w_j which satisfies this equation, and then show that the vector is unique. To establish existence let

$$w_j = \frac{1}{2} \epsilon_{njm} W_{nm}. \tag{1.9}$$

Substituting this expression into the right-hand side of (1.8), and using Result 1.1 and the fact that $W_{ki} = -W_{ik}$, we obtain

$$\epsilon_{ijk} w_j = \frac{1}{2} \epsilon_{ijk} \epsilon_{njm} W_{nm} = \frac{1}{2} (\delta_{in}\delta_{km} - \delta_{im}\delta_{kn}) W_{nm} = W_{ik}.$$

Thus the vector w_j defined in (1.9) satisfies (1.8). To establish uniqueness let w_j and u_j be any two solutions. Then from (1.8) we deduce $0 = \epsilon_{ijk}(w_j - u_j)$ for $i, k = 1, 2, 3$. Considering this equation with $i = 1$ and $k = 3$ we find $w_2 = u_2$, and by similar considerations we find $w_1 = u_1$ and $w_3 = u_3$. Thus the solution in (1.9) is unique.

To establish the second result let $\boldsymbol{w}$ be given. Then we seek a skew-symmetric tensor $\boldsymbol{W}$ such that $\boldsymbol{w} \times \boldsymbol{v} = \boldsymbol{W}\boldsymbol{v}$ for all $\boldsymbol{v}$. In components this equation reads

$$\epsilon_{ijk} w_j v_k = W_{ik} v_k,$$

and by the arbitrariness of v_k we deduce

$$W_{ik} = \epsilon_{ijk} w_j. \tag{1.10}$$

This equation provides an explicit expression for the tensor W_{ik}, which is skew-symmetric by properties of the permutation symbol. Uniqueness follows immediately since W_{ik} is explicitly determined by w_j.

To establish the final result let $\boldsymbol{w} = \text{vec}(\boldsymbol{W})$ be the map defined by (1.9) and let $\boldsymbol{W} = \text{ten}(\boldsymbol{w})$ be the map defined by (1.10). Then for any arbitrary skew-symmetric tensor $\boldsymbol{W}$ we have, using Result 1.1

$$\begin{aligned}
[\text{ten}(\text{vec}(\boldsymbol{W}))]_{ik} &= [\text{ten}(\boldsymbol{w})]_{ik} \\
&= \epsilon_{ijk} w_j \\
&= \frac{1}{2}\epsilon_{ijk}\epsilon_{njm} W_{nm} \\
&= \frac{1}{2}(\delta_{in}\delta_{km} - \delta_{im}\delta_{kn}) W_{nm} \\
&= W_{ik}.
\end{aligned}$$

Thus for any skew-symmetric tensor $\boldsymbol{W}$ we have $\text{ten}(\text{vec}(\boldsymbol{W})) = \boldsymbol{W}$. Similarly, for any vector $\boldsymbol{w}$ we find $\text{vec}(\text{ten}(\boldsymbol{w})) = \boldsymbol{w}$. Thus the functions $\boldsymbol{W} = \text{ten}(\boldsymbol{w})$ and $\boldsymbol{w} = \text{vec}(\boldsymbol{W})$ are inverses. □

1.3.7 Change of Basis

In any given coordinate frame the representation of a vector $\boldsymbol{v} \in \mathcal{V}$ is a triplet of components $[\boldsymbol{v}] \in \mathbb{R}^3$, and the representation of a second-order tensor $\boldsymbol{S} \in \mathcal{V}^2$ is a matrix of components $[\boldsymbol{S}] \in \mathbb{R}^{3\times 3}$. In this section we discuss how the representations $[\boldsymbol{v}]$ and $[\boldsymbol{S}]$ change when the coordinate frame is changed.

1.3.7.1 Change of Basis Tensors

Let $\{\boldsymbol{e}_i\}$ and $\{\boldsymbol{e}'_i\}$ be two coordinate frames for $\mathbb{E}^3$ as shown in Figure 1.4. By the **change of basis** tensor from $\{\boldsymbol{e}_i\}$ to $\{\boldsymbol{e}'_i\}$ we mean the tensor $\boldsymbol{A}$ defined by

$$\boldsymbol{A} = A_{ij}\, \boldsymbol{e}_i \otimes \boldsymbol{e}_j \quad \text{where} \quad A_{ij} = \boldsymbol{e}_i \cdot \boldsymbol{e}'_j. \tag{1.11}$$

We could also define a change of basis tensor $\boldsymbol{B}$ from $\{\boldsymbol{e}'_i\}$ to $\{\boldsymbol{e}_i\}$ by $\boldsymbol{B} = B_{ij}\, \boldsymbol{e}'_i \otimes \boldsymbol{e}'_j$ where $B_{ij} = \boldsymbol{e}'_i \cdot \boldsymbol{e}_j$. All that we say for $\boldsymbol{A}$ will also apply to $\boldsymbol{B}$. However, for convenience, we work only with $\boldsymbol{A}$.

Using the components of $\boldsymbol{A}$ we can express the basis vectors of one frame in terms of the other. For example, a basis vector $\boldsymbol{e}'_j$ may be expressed in the frame $\{\boldsymbol{e}_i\}$ as

$$\boldsymbol{e}'_j = (\boldsymbol{e}_1 \cdot \boldsymbol{e}'_j)\boldsymbol{e}_1 + (\boldsymbol{e}_2 \cdot \boldsymbol{e}'_j)\boldsymbol{e}_2 + (\boldsymbol{e}_3 \cdot \boldsymbol{e}'_j)\boldsymbol{e}_3 = (\boldsymbol{e}_i \cdot \boldsymbol{e}'_j)\boldsymbol{e}_i,$$

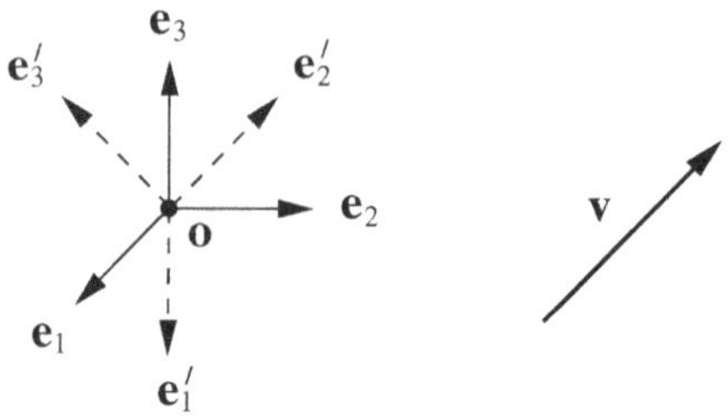

Fig. 1.4 Two different coordinate frames for describing vectors in space.

and by (1.11) this may be written as

$$\boldsymbol{e}_j' = A_{ij}\boldsymbol{e}_i. \tag{1.12}$$

Similarly, a basis vector $\boldsymbol{e}_i$ may be expressed in the frame $\{\boldsymbol{e}_k'\}$ as

$$\boldsymbol{e}_i = (\boldsymbol{e}_i \cdot \boldsymbol{e}_k')\boldsymbol{e}_k' = A_{ik}\boldsymbol{e}_k'. \tag{1.13}$$

The above relations imply the following:

Result 1.4 ***Orthogonality of Basis Change Tensor.*** *The components A_{ij} of the change of basis tensor $\boldsymbol{A}$ have the property that*

$$A_{ij}A_{ik} = \delta_{jk} \quad \textit{and} \quad A_{ik}A_{lk} = \delta_{il},$$

or in matrix notation

$$[\boldsymbol{A}]^T[\boldsymbol{A}] = [\boldsymbol{I}] \quad \textit{and} \quad [\boldsymbol{A}][\boldsymbol{A}]^T = [\boldsymbol{I}].$$

In particular, the change of basis tensor $\boldsymbol{A}$ is orthogonal. □

Proof Substituting (1.13) into (1.12) leads to

$$\boldsymbol{e}_j' = A_{ij}A_{ik}\boldsymbol{e}_k',$$

which implies $A_{ij}A_{ik} = \delta_{jk}$. Similarly, substituting (1.12) into (1.13) yields $A_{ik}A_{lk} = \delta_{il}$. □

1.3.7.2 Change of Representation

Consider an arbitrary vector $\boldsymbol{v} \in \mathcal{V}$ as illustrated in Figure 1.4. Let $[\boldsymbol{v}] = (v_1, v_2, v_3)^T$ be its representation in $\{\boldsymbol{e}_i\}$, let $[\boldsymbol{v}]' = (v_1', v_2', v_3')^T$ be its representation in $\{\boldsymbol{e}_i'\}$, and let $\boldsymbol{A}$ be the change of basis tensor from $\{\boldsymbol{e}_i\}$ to $\{\boldsymbol{e}_i'\}$. Because v_i and v_i' are components of the same physical vector they are related through A_{ij}. To see this, notice that by definition of components we have $\boldsymbol{v} = v_i \boldsymbol{e}_i = v'_j \boldsymbol{e}'_j$, and by definition of A_{ij} we have

$$\boldsymbol{v} = v_i \boldsymbol{e}_i = v'_j \boldsymbol{e}'_j = v'_j A_{ij} \boldsymbol{e}_i,$$

which implies $v_i = A_{ij} v'_j$. Similarly, $v'_k = A_{ik} v_i$. In matrix notation we write this as

$$[\boldsymbol{v}] = [\boldsymbol{A}][\boldsymbol{v}]' \quad \text{and} \quad [\boldsymbol{v}]' = [\boldsymbol{A}]^T [\boldsymbol{v}].$$

For second-order tensors we have

$$[\boldsymbol{S}] = [\boldsymbol{A}][\boldsymbol{S}]'[\boldsymbol{A}]^T \quad \text{and} \quad [\boldsymbol{S}]' = [\boldsymbol{A}]^T [\boldsymbol{S}][\boldsymbol{A}]. \tag{1.14}$$

The derivation of these results is straightforward (see Exercise 13).

1.3.8 Traces, Determinants and Exponentials

The **trace** function on second-order tensors is a linear mapping tr : $\mathcal{V}^2 \to \mathbb{R}$ defined by

$$\operatorname{tr} \boldsymbol{S} = \operatorname{tr}[\boldsymbol{S}] = [\boldsymbol{S}]_{ii},$$

where $[\boldsymbol{S}]$ denotes a matrix representation in any coordinate frame. Thus the trace of a second-order tensor $\boldsymbol{S}$ is just the usual trace (sum of diagonal elements) of a matrix representation $[\boldsymbol{S}]$. The next result shows that this definition does not depend on the coordinate frame.

Result 1.5 ***Invariance of the Trace.*** *Let $\boldsymbol{S}$ be a second-order tensor with matrix representations $[\boldsymbol{S}]$ and $[\boldsymbol{S}]'$ in the frames $\{\boldsymbol{e}_i\}$ and $\{\boldsymbol{e}'_i\}$, respectively. Then*

$$\operatorname{tr}[\boldsymbol{S}] = \operatorname{tr}[\boldsymbol{S}]'.$$

Thus the numerical value of the trace is independent of the coordinate frame in which it is computed. □

Proof Let $\boldsymbol{A}$ be the change of basis tensor from $\{\boldsymbol{e}_i\}$ to $\{\boldsymbol{e}'_i\}$. Then by (1.14) we have $[\boldsymbol{S}] = [\boldsymbol{A}][\boldsymbol{S}]'[\boldsymbol{A}]^T$ or $[\boldsymbol{S}]_{ij} = [\boldsymbol{A}]_{ik}[\boldsymbol{S}]'_{kl}[\boldsymbol{A}]_{jl}$, and

$$\begin{aligned}
\operatorname{tr}[\boldsymbol{S}] = [\boldsymbol{S}]_{ii} &= [\boldsymbol{S}]'_{kl}[\boldsymbol{A}]_{ik}[\boldsymbol{A}]_{il} \\
&= [\boldsymbol{S}]'_{kl}\delta_{kl} \qquad \text{(by Result 1.4)} \\
&= [\boldsymbol{S}]'_{kk} \\
&= \operatorname{tr}[\boldsymbol{S}]'.
\end{aligned}$$

□

The **determinant** function on second-order tensors is a mapping det : $\mathcal{V}^2 \to \mathbb{R}$ defined by

$$\det \boldsymbol{S} = \det[\boldsymbol{S}] = \epsilon_{ijk}[\boldsymbol{S}]_{i1}[\boldsymbol{S}]_{j2}[\boldsymbol{S}]_{k3},$$

where $[\boldsymbol{S}]$ denotes a matrix representation in any coordinate frame. Thus the determinant of a second-order tensor $\boldsymbol{S}$ is just the usual determinant of a matrix representation $[\boldsymbol{S}]$. The explicit expression in terms of the permutation symbol is a consequence of the relation in (1.4). The next result shows that this definition does not depend on the coordinate frame.

Result 1.6 ***Invariance of the Determinant.*** *Let $\boldsymbol{S}$ be a second-order tensor with matrix representations $[\boldsymbol{S}]$ and $[\boldsymbol{S}]'$ in the frames $\{\mathbf{e}_i\}$ and $\{\mathbf{e}'_i\}$, respectively. Then*

$$\det[\boldsymbol{S}] = \det[\boldsymbol{S}]'.$$

Thus the numerical value of the determinant is independent of the coordinate frame in which it is computed. □

Proof See Exercise 20. □

The quantity $|\det \boldsymbol{S}|$ has the geometric interpretation of being the volume of the parallelepiped defined by the three vectors $\boldsymbol{S}\mathbf{e}_1$, $\boldsymbol{S}\mathbf{e}_2$ and $\boldsymbol{S}\mathbf{e}_3$. The determinant also arises in the classification of certain types of tensors. For example, a tensor $\boldsymbol{S}$ is invertible if and only if $\det \boldsymbol{S} \neq 0$. Moreover, because $\det(\boldsymbol{AB}) = (\det \boldsymbol{A})(\det \boldsymbol{B})$, every orthogonal tensor $\boldsymbol{Q}$ has the property that $|\det \boldsymbol{Q}| = 1$, and rotation tensors are those orthogonal tensors which satisfy $\det \boldsymbol{Q} = 1$ (see Exercise 19).

The **exponential** function on second-order tensors is a mapping exp : $\mathcal{V}^2 \to \mathcal{V}^2$ defined by

$$\exp(\boldsymbol{S}) = \sum_{j=0}^{\infty} \frac{1}{j!}\boldsymbol{S}^j = \boldsymbol{I} + \boldsymbol{S} + \tfrac{1}{2}\boldsymbol{S}^2 + \cdots.$$

It can be shown that this series converges for any $\boldsymbol{S} \in \mathcal{V}^2$ and so the exponential function is well-defined. This function arises in the theory of linear, constant coefficient, ordinary differential equations. Our main interest in it is summarized in the following result, whose proof we omit for brevity.

Result 1.7 ***Exponential of a Skew Tensor.*** *Let* $\boldsymbol{W} \in \mathcal{V}^2$ *be an arbitrary skew-symmetric tensor. Then* $\exp(\boldsymbol{W}) \in \mathcal{V}^2$ *is a rotation in the sense that* $(\exp(\boldsymbol{W}))^T \exp(\boldsymbol{W}) = \boldsymbol{I}$ *and* $\det(\exp(\boldsymbol{W})) = 1$. *Moreover,* $(\exp(\boldsymbol{W}))^T = \exp(\boldsymbol{W}^T)$. □

1.3.9 Eigenvalues, Eigenvectors and Principal Invariants

By an **eigenpair** for a second-order tensor $\boldsymbol{S}$ we mean a scalar λ and a unit vector $\boldsymbol{e}$ satisfying

$$\boldsymbol{S}\boldsymbol{e} = \lambda \boldsymbol{e}.$$

Any such λ is called an **eigenvalue** and any such $\boldsymbol{e}$ an **eigenvector** of $\boldsymbol{S}$. From linear algebra we recall that λ is an eigenvalue if and only if it is a root of the characteristic (cubic) polynomial

$$p(\lambda) = \det(\boldsymbol{S} - \lambda \boldsymbol{I}). \tag{1.15}$$

Moreover, to any eigenvalue λ we associate one or more independent eigenvectors $\boldsymbol{e}$ by solving (in components) the linear, homogeneous equation

$$(\boldsymbol{S} - \lambda \boldsymbol{I})\boldsymbol{e} = \boldsymbol{0}. \tag{1.16}$$

When $\boldsymbol{S}$ is symmetric it can be shown that all three roots (possibly repeated) of the characteristic polynomial in (1.15) are real. Consequently, all solutions of the homogeneous equation in (1.16) are also real. Thus every symmetric tensor has exactly three eigenvalues, these eigenvalues are real, and each distinct eigenvalue has one or more independent eigenvectors, which are also real. Throughout our developments we consider eigenvalues and eigenvectors only for symmetric tensors.

Result 1.8 ***Eigenvalues, Eigenvectors of Symmetric Tensors.*** *(1) The eigenvalues of a symmetric, positive-definite tensor are strictly positive. (2) Any two eigenvectors corresponding to distinct eigenvalues of a symmetric tensor are orthogonal.* □

Proof

(1) Let $\boldsymbol{S}$ be symmetric, positive-definite and let $(\lambda, \boldsymbol{e})$ be an eigenpair. Then, since $\boldsymbol{S}\boldsymbol{e} = \lambda\boldsymbol{e}$ and $\boldsymbol{e}$ is a unit vector, we have

$$0 < \boldsymbol{e} \cdot \boldsymbol{S}\boldsymbol{e} = \lambda \boldsymbol{e} \cdot \boldsymbol{e} = \lambda.$$

(2) Let $\boldsymbol{S}$ be symmetric and let $(\lambda, \boldsymbol{e})$ and $(\omega, \boldsymbol{d})$ be two eigenpairs such that $\lambda \neq \omega$. Then, since $\boldsymbol{S}\boldsymbol{e} = \lambda\boldsymbol{e}$ and $\boldsymbol{S}\boldsymbol{d} = \omega\boldsymbol{d}$, we have

$$\lambda\boldsymbol{e}\cdot\boldsymbol{d} = \boldsymbol{S}\boldsymbol{e}\cdot\boldsymbol{d} = \boldsymbol{e}\cdot\boldsymbol{S}^T\boldsymbol{d} = \boldsymbol{e}\cdot\boldsymbol{S}\boldsymbol{d} = \omega\boldsymbol{e}\cdot\boldsymbol{d},$$

which implies $(\lambda-\omega)\boldsymbol{e}\cdot\boldsymbol{d} = 0$. Since $\lambda \neq \omega$ we must have $\boldsymbol{e}\cdot\boldsymbol{d} = 0$.

□

We state, without proof, the following result from linear algebra.

Result 1.9 ***Spectral Decomposition Theorem.*** *Let $\boldsymbol{S}$ be a symmetric second-order tensor. Then there exists a right-handed, orthonormal basis $\{\boldsymbol{e}_i\}$ for $\mathcal{V}$ consisting of eigenvectors of $\boldsymbol{S}$. The corresponding eigenvalues λ_i are the same (up to ordering) for any such basis and form the full set of eigenvalues of $\boldsymbol{S}$. The dyadic representation of $\boldsymbol{S}$ in any such basis is*

$$\boldsymbol{S} = \sum_{i=1}^{3} \lambda_i \boldsymbol{e}_i \otimes \boldsymbol{e}_i,$$

and the matrix representation is

$$[\boldsymbol{S}] = \begin{pmatrix} \lambda_1 & 0 & 0 \\ 0 & \lambda_2 & 0 \\ 0 & 0 & \lambda_3 \end{pmatrix}.$$

□

The **principal invariants** of a second-order tensor $\boldsymbol{S}$ are three scalars defined by

$$I_1(\boldsymbol{S}) = \operatorname{tr}\boldsymbol{S}, \qquad I_2(\boldsymbol{S}) = \tfrac{1}{2}\left[(\operatorname{tr}\boldsymbol{S})^2 - \operatorname{tr}(\boldsymbol{S}^2)\right], \qquad I_3(\boldsymbol{S}) = \det\boldsymbol{S}.$$

For any $\alpha \in \mathbb{R}$ the principal invariants satisfy the relation

$$\det(\boldsymbol{S} - \alpha\boldsymbol{I}) = -\alpha^3 + I_1(\boldsymbol{S})\alpha^2 - I_2(\boldsymbol{S})\alpha + I_3(\boldsymbol{S}),$$

which can be verified by working in any coordinate frame. If $\boldsymbol{S}$ is symmetric, then the principal invariants can be written as

$$I_1(\boldsymbol{S}) = \lambda_1 + \lambda_2 + \lambda_3,$$
$$I_2(\boldsymbol{S}) = \lambda_1\lambda_2 + \lambda_2\lambda_3 + \lambda_3\lambda_1, \qquad I_3(\boldsymbol{S}) = \lambda_1\lambda_2\lambda_3,$$

where λ_i are the eigenvalues of $\boldsymbol{S}$ (see Exercise 23). The quantities $I_i(\boldsymbol{S})$ are called invariants because their values are independent of any particular coordinate frame in which they are computed. The invariance properties of I_1 and I_2 follow from Result 1.5, and the invariance property of I_3 follows from Result 1.6 (see Exercise 24). For brevity, we will sometimes use the symbol $\mathcal{I}_S$ to denote the triplet $(I_1(\boldsymbol{S}), I_2(\boldsymbol{S}), I_3(\boldsymbol{S}))$.

Result 1.10 ***Cayley–Hamilton Theorem.*** *Any second-order tensor $\boldsymbol{S}$ satisfies the relation*

$$\boldsymbol{S}^3 - I_1(\boldsymbol{S})\boldsymbol{S}^2 + I_2(\boldsymbol{S})\boldsymbol{S} - I_3(\boldsymbol{S})\boldsymbol{I} = \boldsymbol{O}.$$

□

Proof See Exercise 25. □

1.3.10 Special Decompositions

Result 1.11 ***Tensor Square Root.*** *Let $\boldsymbol{C}$ be a symmetric, positive-definite tensor. Then there is a unique symmetric, positive-definite tensor $\boldsymbol{U}$ such that $\boldsymbol{U}^2 = \boldsymbol{C}$. In particular*

$$\boldsymbol{U} = \sum_{i=1}^{3} \sqrt{\lambda_i}\, \boldsymbol{e}_i \otimes \boldsymbol{e}_i,$$

where $\{\boldsymbol{e}_i\}$ is any frame consisting of eigenvectors of $\boldsymbol{C}$ and $\lambda_i > 0$ are the corresponding eigenvalues of $\boldsymbol{C}$. We usually write $\boldsymbol{U} = \sqrt{\boldsymbol{C}}$. □

Proof See Exercise 26. □

Result 1.12 ***Polar Decomposition Theorem.*** *Let $\boldsymbol{F}$ be a second-order tensor with positive determinant. Then there exist* **right** *and* **left polar decompositions** *of $\boldsymbol{F}$ taking the form, respectively*

$$\boldsymbol{F} = \boldsymbol{R}\boldsymbol{U} = \boldsymbol{V}\boldsymbol{R},$$

where $\boldsymbol{U} = \sqrt{\boldsymbol{F}^T\boldsymbol{F}}$ and $\boldsymbol{V} = \sqrt{\boldsymbol{F}\boldsymbol{F}^T}$ are symmetric, positive-definite tensors and $\boldsymbol{R}$ is a rotation. □

Proof Since $\boldsymbol{F}$ has positive determinant it is invertible. Thus $\boldsymbol{F}\boldsymbol{v} \neq \boldsymbol{0}$ for all $\boldsymbol{v} \neq \boldsymbol{0}$. Similarly, the transpose $\boldsymbol{F}^T$ is invertible and $\boldsymbol{F}^T\boldsymbol{v} \neq \boldsymbol{0}$ for all $\boldsymbol{v} \neq \boldsymbol{0}$. From these observations we deduce

$$\boldsymbol{v} \cdot \boldsymbol{F}^T\boldsymbol{F}\boldsymbol{v} = (\boldsymbol{F}\boldsymbol{v}) \cdot (\boldsymbol{F}\boldsymbol{v}) > 0 \quad \forall \boldsymbol{v} \neq \boldsymbol{0},$$
$$\boldsymbol{v} \cdot \boldsymbol{F}\boldsymbol{F}^T\boldsymbol{v} = (\boldsymbol{F}^T\boldsymbol{v}) \cdot (\boldsymbol{F}^T\boldsymbol{v}) > 0 \quad \forall \boldsymbol{v} \neq \boldsymbol{0}.$$

Thus $\boldsymbol{F}^T\boldsymbol{F}$ and $\boldsymbol{F}\boldsymbol{F}^T$ are positive-definite and clearly symmetric, and we can define $\boldsymbol{U}$ and $\boldsymbol{V}$ using Result 1.11.

Next, consider the tensor $\boldsymbol{R} = \boldsymbol{F}\boldsymbol{U}^{-1}$. Since $\boldsymbol{R}\boldsymbol{U} = \boldsymbol{F}$ and $\det(\boldsymbol{R}\boldsymbol{U}) = (\det \boldsymbol{R})(\det \boldsymbol{U})$ by properties of determinants (see Exercise 16), we have

$$\det \boldsymbol{R} = (\det \boldsymbol{F})/(\det \boldsymbol{U}).$$

From this we deduce $\det \boldsymbol{R} > 0$ since $\det \boldsymbol{F} > 0$ and $\det \boldsymbol{U} > 0$. Moreover

$$\boldsymbol{R}^T\boldsymbol{R} = \boldsymbol{U}^{-T}\boldsymbol{F}^T\boldsymbol{F}\boldsymbol{U}^{-1} = \boldsymbol{U}^{-1}\boldsymbol{U}^2\boldsymbol{U}^{-1} = \boldsymbol{I}.$$

Thus $\boldsymbol{R}$ is a rotation as claimed and the right polar decomposition is established.

To establish the left polar decomposition let $\boldsymbol{Q} = \boldsymbol{V}^{-1}\boldsymbol{F}$. Then by arguments similar to those above we deduce that $\boldsymbol{Q}$ is also a rotation. Thus we have $\boldsymbol{F} = \boldsymbol{R}\boldsymbol{U} = \boldsymbol{V}\boldsymbol{Q}$. To show that $\boldsymbol{Q} = \boldsymbol{R}$ let $\boldsymbol{S}$ be the symmetric, positive-definite tensor given by $\boldsymbol{S} = \boldsymbol{Q}^T\boldsymbol{V}\boldsymbol{Q}$. Then $\boldsymbol{R}\boldsymbol{U} = \boldsymbol{Q}\boldsymbol{S}$ and $\boldsymbol{R} = \boldsymbol{Q}\boldsymbol{S}\boldsymbol{U}^{-1}$. Since $\boldsymbol{R}$ is a rotation we have $\boldsymbol{R}^{-1} = \boldsymbol{R}^T$, which implies $\boldsymbol{S}^2 = \boldsymbol{U}^2 = \boldsymbol{F}^T\boldsymbol{F}$. From the uniqueness of the tensor square root we deduce $\boldsymbol{S} = \boldsymbol{U}$, which implies $\boldsymbol{R} = \boldsymbol{Q}$. Thus $\boldsymbol{F} = \boldsymbol{R}\boldsymbol{U} = \boldsymbol{V}\boldsymbol{R}$ as claimed. □

1.3.11 Scalar Product for Second-Order Tensors

Analogous to the scalar product for vectors, we define a scalar (inner) product for second-order tensors by

$$\boldsymbol{S} : \boldsymbol{D} = \operatorname{tr}(\boldsymbol{S}^T\boldsymbol{D}). \tag{1.17}$$

In any frame the dyadic basis $\{\boldsymbol{e}_i \otimes \boldsymbol{e}_j\}$ for $\mathcal{V}^2$ introduced in Section 1.3.4 is actually orthonormal in this scalar product (see Exercise 28).

Result 1.13 **Tensor Scalar Product.** *Let $\{\boldsymbol{e}_i\}$ be a coordinate frame and let $\boldsymbol{S} = S_{ij}\boldsymbol{e}_i \otimes \boldsymbol{e}_j$ and $\boldsymbol{D} = D_{ij}\boldsymbol{e}_i \otimes \boldsymbol{e}_j$ be any two second-order tensors. Then*

$$\boldsymbol{S} : \boldsymbol{D} = S_{ij}D_{ij}.$$

Furthermore, if $\boldsymbol{S}$ is symmetric, then

$$\boldsymbol{S} : \boldsymbol{D} = \boldsymbol{S} : \operatorname{sym}(\boldsymbol{D}) = \operatorname{sym}(\boldsymbol{S}) : \operatorname{sym}(\boldsymbol{D}).$$

□

Proof See Exercise 29. □

1.4 Fourth-Order Tensors

In the previous section we defined a second-order tensor as a linear transformation between vectors. In describing the behavior of material bodies we will also need the concept of a *linear transformation between second-order tensors.* For example, the relationship between stress and strain in a material body is sometimes modeled by such a transformation. Linear transformations between second-order tensors lead to the notion of a fourth-order tensor as defined below. As we will see, a fourth-order tensor will be described by eighty-one components in any Cartesian coordinate frame for $\boldsymbol{E}^3$.

1.4.1 Definition

By a **fourth-order tensor C** on the vector space $\mathcal{V}$ we mean a mapping $\mathsf{C} : \mathcal{V}^2 \to \mathcal{V}^2$ which is *linear* in the sense that:

(1) $\mathsf{C}(\boldsymbol{S} + \boldsymbol{T}) = \mathsf{C}\boldsymbol{S} + \mathsf{C}\boldsymbol{T}$ for all $\boldsymbol{S}, \boldsymbol{T} \in \mathcal{V}^2$,

(2) $\mathsf{C}(\alpha \boldsymbol{T}) = \alpha \mathsf{C}\boldsymbol{T}$ for all $\alpha \in \boldsymbol{R}$ and $\boldsymbol{T} \in \mathcal{V}^2$.

We denote the set of all fourth-order tensors on $\mathcal{V}$ by the symbol $\mathcal{V}^4$. We define a fourth-order **zero tensor O** with the property $\mathsf{O}\boldsymbol{T} = \boldsymbol{O}$ for all $\boldsymbol{T} \in \mathcal{V}^2$, and we define a fourth-order **identity tensor I** with the property $\mathsf{I}\boldsymbol{T} = \boldsymbol{T}$ for all $\boldsymbol{T} \in \mathcal{V}^2$. Two fourth-order tensors C and D are said to be equal if and only if $\mathsf{C}\boldsymbol{T} = \mathsf{D}\boldsymbol{T}$ for all $\boldsymbol{T} \in \mathcal{V}^2$.

Example: For any fixed $\boldsymbol{A} \in \mathcal{V}^2$ the mapping $\mathsf{C} : \mathcal{V}^2 \to \mathcal{V}^2$ given by

$$\mathsf{C}(\boldsymbol{T}) = \boldsymbol{A}\boldsymbol{T}$$

defines a fourth-order tensor. To see this we need only establish the two properties listed above, but from the algebra of second-order tensors we immediately have

$$\begin{aligned} \mathsf{C}(\alpha \boldsymbol{S} + \beta \boldsymbol{T}) &= \boldsymbol{A}(\alpha \boldsymbol{S} + \beta \boldsymbol{T}) \\ &= \alpha \boldsymbol{A}\boldsymbol{S} + \beta \boldsymbol{A}\boldsymbol{T} = \alpha \mathsf{C}(\boldsymbol{S}) + \beta \mathsf{C}(\boldsymbol{T}), \end{aligned}$$

for any $\alpha, \beta \in \boldsymbol{R}$ and $\boldsymbol{S}, \boldsymbol{T} \in \mathcal{V}^2$. □

1.4.2 Fourth-Order Tensor Algebra

If we let $\mathcal{V}^4$ be the set of fourth-order tensors, then $\mathcal{V}^4$ has the structure of a real vector space since

$$\mathsf{C} + \mathsf{D} \in \mathcal{V}^4 \quad \forall \mathsf{C}, \mathsf{D} \in \mathcal{V}^4 \quad \text{and} \quad \alpha \mathsf{C} \in \mathcal{V}^4 \quad \forall \alpha \in \mathbb{R}, \;\; \mathsf{C} \in \mathcal{V}^4.$$

We can also compose fourth-order tensors in the sense that

$$\mathsf{CD} \in \mathcal{V}^4 \quad \forall \mathsf{C}, \mathsf{D} \in \mathcal{V}^4.$$

Similar to the case of second-order tensors, we define the sum $\mathsf{C} + \mathsf{D}$ via the relation $(\mathsf{C} + \mathsf{D})\boldsymbol{T} = \mathsf{C}\boldsymbol{T} + \mathsf{D}\boldsymbol{T}$ for all $\boldsymbol{T} \in \mathcal{V}^2$, and we define the composition CD via the relation $(\mathsf{CD})\boldsymbol{T} = \mathsf{C}(\mathsf{D}\boldsymbol{T})$ for all $\boldsymbol{T} \in \mathcal{V}^2$. A similar definition holds for $\alpha\mathsf{C}$.

1.4.3 Representation in a Coordinate Frame

By the **components** of a fourth-order tensor C in a coordinate frame $\{\boldsymbol{e}_i\}$ we mean the eighty-one numbers C_{ijkl} $(1 \leq i, j, k, l \leq 3)$ defined by

$$\mathsf{C}_{ijkl} = \boldsymbol{e}_i \cdot \mathsf{C}(\boldsymbol{e}_k \otimes \boldsymbol{e}_l)\boldsymbol{e}_j.$$

That is, if we let $\boldsymbol{S}_{kl}$ be the second-order tensor given by $\mathsf{C}(\boldsymbol{e}_k \otimes \boldsymbol{e}_l)$, then

$$\mathsf{C}_{ijkl} = \boldsymbol{e}_i \cdot \boldsymbol{S}_{kl}\boldsymbol{e}_j.$$

These components completely describe the transformation $\mathsf{C} : \mathcal{V}^2 \to \mathcal{V}^2$. Given $\boldsymbol{U} = U_{ij}\boldsymbol{e}_i \otimes \boldsymbol{e}_j$ and $\boldsymbol{T} = T_{kl}\boldsymbol{e}_k \otimes \boldsymbol{e}_l$ related by $\boldsymbol{U} = \mathsf{C}(\boldsymbol{T})$, we have, by linearity

$$\begin{aligned} U_{ij} = \boldsymbol{e}_i \cdot \boldsymbol{U}\boldsymbol{e}_j &= \boldsymbol{e}_i \cdot \mathsf{C}(\boldsymbol{T})\boldsymbol{e}_j \\ &= \boldsymbol{e}_i \cdot \mathsf{C}(T_{kl}\boldsymbol{e}_k \otimes \boldsymbol{e}_l)\boldsymbol{e}_j \\ &= \boldsymbol{e}_i \cdot \mathsf{C}(\boldsymbol{e}_k \otimes \boldsymbol{e}_l)\boldsymbol{e}_j \, T_{kl} \\ &= \mathsf{C}_{ijkl}T_{kl}. \end{aligned} \tag{1.18}$$

Thus the components of C are simply the coefficients in the linear relation between the components of $\boldsymbol{U}$ and $\boldsymbol{T}$.

Example: Let $\boldsymbol{A} \in \mathcal{V}^2$ be given and consider the fourth-order tensor defined by $\mathsf{C}(\boldsymbol{T}) = \boldsymbol{A}\boldsymbol{T}$. Then the components of C are

$$\mathsf{C}_{ijkl} = \boldsymbol{e}_i \cdot \boldsymbol{A}(\boldsymbol{e}_k \otimes \boldsymbol{e}_l)\boldsymbol{e}_j = \boldsymbol{e}_i \cdot \boldsymbol{A}\boldsymbol{e}_k\delta_{lj} = A_{ik}\delta_{lj}.$$

□

An alternative way to compute the components C_{ijkl} is to use the scalar product for second-order tensors, namely

$$\mathsf{C}_{ijkl} = (\boldsymbol{e}_i \otimes \boldsymbol{e}_j) : \mathbf{C}(\boldsymbol{e}_k \otimes \boldsymbol{e}_l).$$

To see this, let $\boldsymbol{S}_{kl}$ be the second-order tensor given by $\mathbf{C}(\boldsymbol{e}_k \otimes \boldsymbol{e}_l)$ and notice that

$$\begin{aligned} \mathbf{C}(\boldsymbol{e}_k \otimes \boldsymbol{e}_l) &= \boldsymbol{S}_{kl} \\ &= (\boldsymbol{e}_i \cdot \boldsymbol{S}_{kl}\boldsymbol{e}_j)\boldsymbol{e}_i \otimes \boldsymbol{e}_j \\ &= \mathsf{C}_{ijkl}\boldsymbol{e}_i \otimes \boldsymbol{e}_j. \end{aligned}$$

Thus

$$\begin{aligned} (\boldsymbol{e}_m \otimes \boldsymbol{e}_n) : \mathbf{C}(\boldsymbol{e}_k \otimes \boldsymbol{e}_l) &= \mathsf{C}_{ijkl}(\boldsymbol{e}_m \otimes \boldsymbol{e}_n) : (\boldsymbol{e}_i \otimes \boldsymbol{e}_j) \\ &= \mathsf{C}_{ijkl}\delta_{mi}\delta_{nj} \\ &= \mathsf{C}_{mnkl}, \end{aligned}$$

as required.

Just as for second-order tensors, the components of a fourth-order tensor may be arranged into various matrix representations. For example, if we arrange the nine components of a second-order tensor $\boldsymbol{T}$ into a 9×1 column matrix $[[\boldsymbol{T}]] \in \mathbb{R}^9$, then the eighty-one components of a fourth-order tensor $\mathbf{C}$ could naturally be arranged into a 9×9 matrix $[[\mathbf{C}]] \in \mathbb{R}^{9\times 9}$. In this case the tensor equation $\boldsymbol{U} = \mathbf{C}\boldsymbol{T}$ would be equivalent to the matrix equation $[[\boldsymbol{U}]] = [[\mathbf{C}]]\,[[\boldsymbol{T}]]$. However, such matrix representations will not be exploited in our developments.

1.4.4 Fourth-Order Dyadic Products, Bases

The **dyadic product** of four vectors $\boldsymbol{a}$, $\boldsymbol{b}$, $\boldsymbol{c}$ and $\boldsymbol{d}$ is the fourth-order tensor $\boldsymbol{a} \otimes \boldsymbol{b} \otimes \boldsymbol{c} \otimes \boldsymbol{d}$ defined by

$$(\boldsymbol{a} \otimes \boldsymbol{b} \otimes \boldsymbol{c} \otimes \boldsymbol{d})\boldsymbol{T} = (\boldsymbol{c} \cdot \boldsymbol{T}\boldsymbol{d})\boldsymbol{a} \otimes \boldsymbol{b}, \qquad \forall \boldsymbol{T} \in \mathcal{V}^2.$$

Given any coordinate frame $\{\boldsymbol{e}_i\}$ the eighty-one elementary dyadic products $\{\boldsymbol{e}_i \otimes \boldsymbol{e}_j \otimes \boldsymbol{e}_k \otimes \boldsymbol{e}_l\}$ form a **basis** for $\mathcal{V}^4$. In particular, each $\mathbf{C} \in \mathcal{V}^4$ can be uniquely represented as a linear combination

$$\mathbf{C} = \mathsf{C}_{ijkl}\boldsymbol{e}_i \otimes \boldsymbol{e}_j \otimes \boldsymbol{e}_k \otimes \boldsymbol{e}_l, \tag{1.19}$$

where $\mathsf{C}_{ijkl} = \boldsymbol{e}_i \cdot \mathbf{C}(\boldsymbol{e}_k \otimes \boldsymbol{e}_l)\boldsymbol{e}_j$. To see that the above representation is correct consider the expression $\boldsymbol{U} = \mathbf{C}\boldsymbol{T}$, where $\boldsymbol{U} = U_{ij}\boldsymbol{e}_i \otimes \boldsymbol{e}_j$ and

$\boldsymbol{T} = T_{kl}\boldsymbol{e}_k \otimes \boldsymbol{e}_l$. Then

$$\begin{aligned} \boldsymbol{U} &= \mathsf{C}\boldsymbol{T} \\ &= (\mathsf{C}_{ijkl}\boldsymbol{e}_i \otimes \boldsymbol{e}_j \otimes \boldsymbol{e}_k \otimes \boldsymbol{e}_l)\boldsymbol{T} \\ &= \mathsf{C}_{ijkl}(\boldsymbol{e}_k \cdot \boldsymbol{T}\boldsymbol{e}_l)\boldsymbol{e}_i \otimes \boldsymbol{e}_j \\ &= \mathsf{C}_{ijkl}T_{kl}\boldsymbol{e}_i \otimes \boldsymbol{e}_j, \end{aligned}$$

which implies $U_{ij} = \mathsf{C}_{ijkl}T_{kl}$. Thus (1.19) agrees with (1.18) and provides a unique representation of C in terms of its components C_{ijkl}.

1.4.5 Symmetry Properties

A fourth-order tensor $\mathsf{C} \in \mathcal{V}^4$ is said to be **symmetric** or possess **major symmetry** if

$$\boldsymbol{A} : \mathsf{C}(\boldsymbol{B}) = \mathsf{C}(\boldsymbol{A}) : \boldsymbol{B}, \qquad \forall \boldsymbol{A}, \boldsymbol{B} \in \mathcal{V}^2.$$

It is said to possess a **right minor symmetry** if

$$\boldsymbol{A} : \mathsf{C}(\boldsymbol{B}) = \boldsymbol{A} : \mathsf{C}(\mathrm{sym}(\boldsymbol{B})), \qquad \forall \boldsymbol{A}, \boldsymbol{B} \in \mathcal{V}^2,$$

and a **left minor symmetry** if

$$\boldsymbol{A} : \mathsf{C}(\boldsymbol{B}) = \mathrm{sym}(\boldsymbol{A}) : \mathsf{C}(\boldsymbol{B}), \qquad \forall \boldsymbol{A}, \boldsymbol{B} \in \mathcal{V}^2.$$

The implications of the above conditions on the components of C are established in the following result.

Result 1.14 ***Fourth-order Tensor Symmetries.*** *If a fourth-order tensor* C *has major symmetry, then*

$$\mathsf{C}_{ijkl} = \mathsf{C}_{klij}.$$

If C *has right and left minor symmetries, then*

$$\mathsf{C}_{ijkl} = \mathsf{C}_{ijlk} \quad \textit{and} \quad \mathsf{C}_{ijkl} = \mathsf{C}_{jikl},$$

respectively. □

Proof See Exercise 33. □

1.5 Isotropic Tensor Functions

A function $\boldsymbol{G} : \mathcal{V}^2 \to \mathcal{V}^2$ is said to be **isotropic** if it satisfies

$$\boldsymbol{Q}\boldsymbol{G}(\boldsymbol{A})\boldsymbol{Q}^T = \boldsymbol{G}(\boldsymbol{Q}\boldsymbol{A}\boldsymbol{Q}^T),$$

for all $\boldsymbol{A} \in \mathcal{V}^2$ and all rotations $\boldsymbol{Q}$. Similarly, a function $g : \mathcal{V}^2 \to \mathbb{R}$ is said to be **isotropic** if it satisfies

$$g(\boldsymbol{A}) = g(\boldsymbol{Q}\boldsymbol{A}\boldsymbol{Q}^T),$$

for all $\boldsymbol{A} \in \mathcal{V}^2$ and all rotations $\boldsymbol{Q}$.

The condition of isotropy imposes severe restriction on functions. The next result, which we state without proof, shows that any isotropic function which maps the set of symmetric tensors into itself must have a particularly simple form. Recall the definition of, and notation for, principal invariants.

Result 1.15 ***Representation of Isotropic Tensor Functions.*** *Let $\boldsymbol{G} : \mathcal{V}^2 \to \mathcal{V}^2$ be an isotropic function which maps symmetric tensors to symmetric tensors, that is, $\boldsymbol{G}(\boldsymbol{A}) \in \mathcal{V}^2$ is symmetric if $\boldsymbol{A} \in \mathcal{V}^2$ is symmetric. Then there are functions $\alpha_0, \alpha_1, \alpha_2 : \mathbb{R}^3 \to \mathbb{R}$ such that*

$$\boldsymbol{G}(\boldsymbol{A}) = \alpha_0(\mathcal{I}_A)\boldsymbol{I} + \alpha_1(\mathcal{I}_A)\boldsymbol{A} + \alpha_2(\mathcal{I}_A)\boldsymbol{A}^2$$

for every symmetric $\boldsymbol{A} \in \mathcal{V}^2$. Consequently, by Result 1.10, there are functions $\beta_0, \beta_1, \beta_2 : \mathbb{R}^3 \to \mathbb{R}$ such that

$$\boldsymbol{G}(\boldsymbol{A}) = \beta_0(\mathcal{I}_A)\boldsymbol{I} + \beta_1(\mathcal{I}_A)\boldsymbol{A} + \beta_2(\mathcal{I}_A)\boldsymbol{A}^{-1}$$

for every symmetric $\boldsymbol{A} \in \mathcal{V}^2$ which is invertible. □

The above result may be strengthened when, in addition to the properties mentioned above, $\boldsymbol{G}$ is linear and satisfies $\boldsymbol{G}(\boldsymbol{W}) = \boldsymbol{O}$ for every skew-symmetric $\boldsymbol{W}$. This case, which we now study, will be of particular importance in Chapters 6 and 7 when discussing linear models for stress in fluid and solid bodies.

Result 1.16 ***Isotropic Fourth-Order Tensors.*** *Let $\mathsf{C} : \mathcal{V}^2 \to \mathcal{V}^2$ be an isotropic fourth-order tensor with the properties:*

(1) $\mathsf{C}(\boldsymbol{A}) \in \mathcal{V}^2$ is symmetric for every symmetric $\boldsymbol{A} \in \mathcal{V}^2$,

(2) $\mathsf{C}(\boldsymbol{W}) = \boldsymbol{O}$ for every skew-symmetric $\boldsymbol{W} \in \mathcal{V}^2$.

Then there are scalars μ and λ such that

$$\boldsymbol{C}(\boldsymbol{A}) = \lambda(\operatorname{tr}\boldsymbol{A})\boldsymbol{I} + 2\mu\,\operatorname{sym}(\boldsymbol{A}), \qquad \forall \boldsymbol{A} \in \mathcal{V}^2.$$

□

Proof Since $\mathsf{C} : \mathcal{V}^2 \to \mathcal{V}^2$ is isotropic and maps symmetric tensors to symmetric tensors it follows from Result 1.15 that there are functions $\alpha_0, \alpha_1, \alpha_2 : \mathbb{R}^3 \to \mathbb{R}$ such that

$$\mathsf{C}(\boldsymbol{H}) = \alpha_0(\mathcal{I}_H)\boldsymbol{I} + \alpha_1(\mathcal{I}_H)\boldsymbol{H} + \alpha_2(\mathcal{I}_H)\boldsymbol{H}^2$$

for every symmetric $\boldsymbol{H} \in \mathcal{V}^2$, where

$$\mathcal{I}_H = \left(\operatorname{tr}\boldsymbol{H}, \tfrac{1}{2}[(\operatorname{tr}\boldsymbol{H})^2 - \operatorname{tr}(\boldsymbol{H}^2)], \det\boldsymbol{H}\right).$$

Since $\mathsf{C}(\boldsymbol{H})$ is linear in $\boldsymbol{H}$ the only possibilities are

$$\begin{aligned}
\alpha_0(\mathcal{I}_H) &= c_0 \operatorname{tr}\boldsymbol{H} + c_1,\\
\alpha_1(\mathcal{I}_H) &= c_2,\\
\alpha_2(\mathcal{I}_H) &= 0,
\end{aligned}$$

where c_0, c_1 and c_2 are scalar constants. Since $\boldsymbol{C}(\boldsymbol{O}) = \boldsymbol{O}$ it follows that $c_1 = 0$. Thus, upon setting $\lambda = c_0$ and $\mu = c_2/2$, we have

$$\mathsf{C}(\boldsymbol{H}) = \lambda(\operatorname{tr}\boldsymbol{H})\boldsymbol{I} + 2\mu\boldsymbol{H}$$

for every symmetric $\boldsymbol{H}$. This together with the fact that $\mathsf{C}(\boldsymbol{W}) = \boldsymbol{O}$ for every skew-symmetric $\boldsymbol{W}$ leads to the expression

$$\mathsf{C}(\boldsymbol{H}) = \lambda(\operatorname{tr}\boldsymbol{H})\boldsymbol{I} + 2\mu\,\operatorname{sym}(\boldsymbol{H}), \qquad \forall \boldsymbol{H} \in \mathcal{V}^2.$$

□

Bibliographic Notes

Classic references for the material presented here are the texts by Bowen and Wang (1976) and Halmos (1974). A more modern treatment with applications to continuum mechanics is given in Knowles (1998). For simplicity of exposition we have chosen to work only in orthonormal coordinate frames. However, non-orthonormal coordinate frames arise naturally in many applications. A concise treatment of tensors in non-orthonormal frames can be found in Simmonds (1994), and a similar treatment from a more engineering point of view can be found in Malvern (1969).

The idea of a tensor can be defined in various different, but consistent ways. For example, many authors define an nth-order tensor on a vector space $\mathcal{V}$ to be a multilinear function from $\mathcal{V} \times \cdots \times \mathcal{V}$ (n copies) into $\mathbb{R}$. In this respect, a vector $\boldsymbol{v}$ is a first-order tensor because it can naturally (via the scalar product) be identified with a linear function $f : \mathcal{V} \to \mathbb{R}$, namely $f(\boldsymbol{w}) = \boldsymbol{v} \cdot \boldsymbol{w}$. Similarly, a second-order tensor $\boldsymbol{S}$ as defined here can naturally be identified with a bilinear function $f : \mathcal{V} \times \mathcal{V} \to \mathbb{R}$, namely $f(\boldsymbol{w}, \boldsymbol{u}) = \boldsymbol{w} \cdot \boldsymbol{S}\boldsymbol{u}$, and so on. The definitions of second- and fourth-order tensors given in this chapter as linear transformations are consistent with this more general definition. One reason for using the definition in terms of linear transformations is that it leads naturally to the idea of tensor components and matrix representations.

Vectors and tensors are of fundamental importance in mechanics and will be used throughout the remainder of our developments. In particular, vectors and tensors give us a way to state physical laws without reference to any particular coordinate frame. Various such laws for continuum bodies will be discussed in Chapter 5. A modern and more thorough discussion of tensors can be found in Bourne and Kendall (1992). For applications of tensors in differential geometry and analytical mechanics see Dodson and Poston (1991) and Abraham and Marsden (1978). For a general discussion of the history of vectors and tensors see Crowe (1967).

The theory of rotations, the exponential function and general isotropic functions are of special interest across various disciplines. A wealth of information on these subjects can be found, respectively, in Hughes (1986), Hirsch and Smale (1974) and Gurtin (1981). Proofs of our Results 1.7 and 1.15 can be found in Gurtin (1981), and proofs of our Results 1.9 and 1.10 can be found in both Bowen and Wang (1976) and Halmos (1974).

Exercises

1.1 Given the vectors $\boldsymbol{a} = 1\boldsymbol{i} + 2\boldsymbol{j} + 3\boldsymbol{k}$, $\boldsymbol{b} = 1\boldsymbol{i} + 3\boldsymbol{j} - 2\boldsymbol{k}$ and $\boldsymbol{c} = -2\boldsymbol{i} - 1\boldsymbol{j} + 0\boldsymbol{k}$, calculate:

(a) $\boldsymbol{a} \cdot \boldsymbol{b}$,

(b) $\boldsymbol{a} \times \boldsymbol{b}$,

(c) $\boldsymbol{a} \cdot \boldsymbol{b} \times \boldsymbol{c}$,

(d) $\boldsymbol{a} \times (\boldsymbol{b} \times \boldsymbol{c})$,

(e) $(\boldsymbol{a} \times \boldsymbol{b}) \times \boldsymbol{c}$.

1.2 Given a plane Π with normal $\boldsymbol{n} = 1\boldsymbol{i} - 2\boldsymbol{j} + 1\boldsymbol{k}$ and the vector $\boldsymbol{v} = 3\boldsymbol{i} + 4\boldsymbol{j} - 2\boldsymbol{k}$, calculate:

(a) the projection of $\boldsymbol{v}$ onto Π,
(b) the reflection of $\boldsymbol{v}$ with respect to Π.

1.3 Calculate $\delta_{ij}\delta_{ij}$ using the rules of index notation and the definition of the Kronecker delta.

1.4 Suppose a vector $\boldsymbol{v}$ satisfies the linear equation

$$\alpha\boldsymbol{v} + \boldsymbol{v} \times \boldsymbol{a} = \boldsymbol{b},$$

where $\alpha \neq 0$ is a given scalar, and $\boldsymbol{a}$ and $\boldsymbol{b}$ are given vectors. Use the dot and cross product operations to solve the above equation for $\boldsymbol{v}$. In particular, show that the unique solution is given by

$$\boldsymbol{v} = \frac{\alpha^2\boldsymbol{b} - \alpha(\boldsymbol{b} \times \boldsymbol{a}) + (\boldsymbol{b} \cdot \boldsymbol{a})\boldsymbol{a}}{\alpha(\alpha^2 + |\boldsymbol{a}|^2)}.$$

1.5 Let $\boldsymbol{a}$, $\boldsymbol{b}$ and $\boldsymbol{c}$ be vectors. Find scalars λ and μ such that

$$(\boldsymbol{a} \times \boldsymbol{b}) \times \boldsymbol{c} = \lambda\boldsymbol{b} - \mu\boldsymbol{a}.$$

1.6 Let $\boldsymbol{v}$ be an arbitrary vector and let $\boldsymbol{n}$ be an arbitrary unit vector. Show that:

(a) $\boldsymbol{v} = (\boldsymbol{v} \cdot \boldsymbol{n})\boldsymbol{n} - (\boldsymbol{v} \times \boldsymbol{n}) \times \boldsymbol{n}$,
(b) $\boldsymbol{v} = (\boldsymbol{v} \cdot \boldsymbol{n})\boldsymbol{n} + (\boldsymbol{v} \otimes \boldsymbol{n})\boldsymbol{n} - (\boldsymbol{n} \otimes \boldsymbol{v})\boldsymbol{n}$.

Remark: The identity in (a) shows that $\boldsymbol{v}$ can always be decomposed into parts parallel and perpendicular to $\boldsymbol{n}$.

1.7 Given two vectors $\boldsymbol{a}$ and $\boldsymbol{b}$, and a second-order tensor $\boldsymbol{S}$, prove:

(a) $\boldsymbol{S}(\boldsymbol{a} \otimes \boldsymbol{b}) = (\boldsymbol{S}\boldsymbol{a}) \otimes \boldsymbol{b}$,

(b) $(\boldsymbol{a} \otimes \boldsymbol{b})\boldsymbol{S} = \boldsymbol{a} \otimes (\boldsymbol{S}^T\boldsymbol{b})$,

(c) $(\boldsymbol{a} \otimes \boldsymbol{b})^T = (\boldsymbol{b} \otimes \boldsymbol{a})$.

Hint: Recall that two second-order tensors $\boldsymbol{A}$ and $\boldsymbol{B}$ are equal if and only if $\boldsymbol{A}\boldsymbol{v} = \boldsymbol{B}\boldsymbol{v}$ for all $\boldsymbol{v} \in \mathcal{V}$.

1.8 Consider any three vectors $\boldsymbol{a}$, $\boldsymbol{b}$ and $\boldsymbol{c}$ which are linearly independent, that is, $(\boldsymbol{a} \times \boldsymbol{b}) \cdot \boldsymbol{c} \neq 0$. Show that:

(a) $\boldsymbol{a} \times \boldsymbol{b}$, $\boldsymbol{b} \times \boldsymbol{c}$ and $\boldsymbol{c} \times \boldsymbol{a}$ are also linearly independent,
(b) $(\boldsymbol{a} \times \boldsymbol{b}) \otimes \boldsymbol{c} + (\boldsymbol{b} \times \boldsymbol{c}) \otimes \boldsymbol{a} + (\boldsymbol{c} \times \boldsymbol{a}) \otimes \boldsymbol{b} = (\boldsymbol{a} \times \boldsymbol{b} \cdot \boldsymbol{c})\boldsymbol{I}$.

1.9 A second-order tensor $\boldsymbol{P}$ is a **perpendicular projection** if $\boldsymbol{P}$ is symmetric and $\boldsymbol{P}^2 = \boldsymbol{P}$. Given two arbitrary unit vectors $\boldsymbol{n} \neq \boldsymbol{m}$, determine which of the following are perpendicular projections:

(a) $\boldsymbol{P} = \boldsymbol{I}$,

(b) $\boldsymbol{P} = \boldsymbol{n} \otimes \boldsymbol{m}$,

(c) $\boldsymbol{P} = \boldsymbol{n} \otimes \boldsymbol{n}$,

(d) $\boldsymbol{P} = \boldsymbol{I} - \boldsymbol{n} \otimes \boldsymbol{n}$,

(e) $\boldsymbol{P} = \boldsymbol{n} \otimes \boldsymbol{m} + \boldsymbol{m} \otimes \boldsymbol{n}$.

1.10 Let $\boldsymbol{Q}$ be a second-order tensor and let $\boldsymbol{I}$ be the identity tensor. Show that $\boldsymbol{Q}$ is orthogonal if $\boldsymbol{H} = \boldsymbol{Q} - \boldsymbol{I}$ satisfies

$$\boldsymbol{H} + \boldsymbol{H}^T + \boldsymbol{H}\boldsymbol{H}^T = \boldsymbol{O}.$$

1.11 Show that the transpose of a second-order tensor $\boldsymbol{S}$ is uniquely defined and that $[\boldsymbol{S}^T] = [\boldsymbol{S}]^T$.

1.12 Prove that a second-order tensor $\boldsymbol{S}$ cannot be both positive-definite and skew-symmetric.

1.13 Let $\boldsymbol{A}$ denote the change of basis tensor from a frame $\{\boldsymbol{e}_i\}$ to a frame $\{\boldsymbol{e}'_i\}$ with representation $[\boldsymbol{A}]$ in $\{\boldsymbol{e}_i\}$. Let $\boldsymbol{S}$ be a second-order tensor with representation $[\boldsymbol{S}]$ and $[\boldsymbol{S}]'$ in $\{\boldsymbol{e}_i\}$ and $\{\boldsymbol{e}'_i\}$, respectively. Show that

$$[\boldsymbol{S}]' = [\boldsymbol{A}]^T[\boldsymbol{S}][\boldsymbol{A}].$$

1.14 Consider a vector $\boldsymbol{a}$ with representation $[\boldsymbol{a}] = (1, 1, 1)^T$ in a coordinate frame $\{\boldsymbol{e}_i\}$. If $\boldsymbol{S}$ is an anti-clockwise rotation through an angle of $\pi/4$ about $\boldsymbol{a}$, find $[\boldsymbol{S}]$.

1.15 For an arbitrary second-order tensor $\boldsymbol{A} = A_{ij}\boldsymbol{e}_i \otimes \boldsymbol{e}_j$ show that

$$\det \boldsymbol{A} = \tfrac{1}{6}\epsilon_{ijk}\epsilon_{pqr}A_{ip}A_{jq}A_{kr},$$

and deduce that $\det \boldsymbol{A} = \det \boldsymbol{A}^T$.

1.16 For any two second-order tensors $\boldsymbol{A}$ and $\boldsymbol{B}$ show that

$$\det(\boldsymbol{A}\boldsymbol{B}) = (\det \boldsymbol{A})(\det \boldsymbol{B}).$$

Moreover, if $\boldsymbol{A}^{-1}$ exists, show that

$$\det \boldsymbol{A}^{-1} = 1/\det \boldsymbol{A}.$$

1.17 For any pair of vectors $\boldsymbol{u}$ and $\boldsymbol{v}$ and any invertible second-order tensor $\boldsymbol{F}$ show that

$$(\boldsymbol{F}\boldsymbol{u}) \times (\boldsymbol{F}\boldsymbol{v}) = (\det \boldsymbol{F})\boldsymbol{F}^{-T}(\boldsymbol{u} \times \boldsymbol{v}).$$

1.18 Prove Result 1.1.

1.19 Show that:

(a) $|\det \boldsymbol{Q}| = 1$ for any orthogonal tensor $\boldsymbol{Q}$,

(b) $\det \boldsymbol{Q} = 1$ for any rotation tensor $\boldsymbol{Q}$.

1.20 Prove Result 1.6.

1.21 Let $\boldsymbol{Q}$ be a rotation tensor and let $\boldsymbol{u}$, $\boldsymbol{v}$ be arbitrary vectors. Show that:

(a) $(\boldsymbol{Q}\boldsymbol{u}) \cdot (\boldsymbol{Q}\boldsymbol{v}) = \boldsymbol{u} \cdot \boldsymbol{v}$,
(b) $|\boldsymbol{Q}\boldsymbol{v}| = |\boldsymbol{v}|$,
(c) $(\boldsymbol{Q}\boldsymbol{u}) \times (\boldsymbol{Q}\boldsymbol{v}) = \boldsymbol{Q}(\boldsymbol{u} \times \boldsymbol{v})$.

Remark: The results in (a) and (b) together imply that the length of a vector and the angle between any two vectors are unchanged by a rotation. The result in (c) implies that rotations commute with the cross product operation; in particular, when two vectors are subject to a common rotation, the normal to their plane is subject to the same rotation.

1.22 Let $\boldsymbol{Q} \neq \boldsymbol{I}$ be a rotation tensor.

(a) Show that $\lambda = 1$ is always an eigenvalue of $\boldsymbol{Q}$. Hint: Use the characteristic polynomial and properties of determinants.

(b) Show that there is only one independent eigenvector $\boldsymbol{e}$ such that $\boldsymbol{Q}\boldsymbol{e} = \boldsymbol{e}$. Hint: Use part (c) of Exercise 21 to show that if there were more than one such independent eigenvector, then there must be three, which would imply $\boldsymbol{Q} = \boldsymbol{I}$.

(c) Let $\boldsymbol{n}$ be any unit vector orthogonal to $\boldsymbol{e}$. Show that $\boldsymbol{Q}\boldsymbol{n}$ is also a unit vector orthogonal to $\boldsymbol{e}$ and that the angle $\theta \in [0, \pi]$ between $\boldsymbol{n}$ and $\boldsymbol{Q}\boldsymbol{n}$ satisfies the relation

$$1 + 2\cos\theta = \operatorname{tr} \boldsymbol{Q}.$$

Hint: Express $\boldsymbol{Q}$ in the frame $\{\boldsymbol{e}, \boldsymbol{n}, \boldsymbol{e} \times \boldsymbol{n}\}$.

Remark: The vector $\boldsymbol{e}$ in part (b) is called the **rotation axis** and the angle θ in part (c) is called the **rotation angle** of $\boldsymbol{Q}$. The action of $\boldsymbol{Q}$ on any vector $\boldsymbol{v}$ can be understood by decomposing $\boldsymbol{v}$ into parts parallel and perpendicular to $\boldsymbol{e}$.

1.23 Show that the principal invariants of a symmetric second-order tensor $\boldsymbol{S}$ are given by

$$\begin{aligned} I_1(\boldsymbol{S}) &= \lambda_1 + \lambda_2 + \lambda_3, \\ I_2(\boldsymbol{S}) &= \lambda_1\lambda_2 + \lambda_1\lambda_3 + \lambda_2\lambda_3, \\ I_3(\boldsymbol{S}) &= \lambda_1\lambda_2\lambda_3, \end{aligned}$$

where λ_i are the eigenvalues of $\boldsymbol{S}$. Hint: Choose a simple frame in which to represent $\boldsymbol{S}$.

1.24 Let $\boldsymbol{S}$ be a second-order tensor and let $I_2(\boldsymbol{S})$ be its second principal invariant. Show that $I_2(\boldsymbol{S})$ has the same numerical value regardless of the coordinate frame in which it is computed.

1.25 Prove Result 1.10 for symmetric $\boldsymbol{S}$.

1.26 Prove Result 1.11.

1.27 For any vectors $\boldsymbol{a}$, $\boldsymbol{b}$, $\boldsymbol{c}$ and $\boldsymbol{d}$ show

$$(\boldsymbol{a} \otimes \boldsymbol{b}) : (\boldsymbol{c} \otimes \boldsymbol{d}) = (\boldsymbol{a} \cdot \boldsymbol{c})(\boldsymbol{b} \cdot \boldsymbol{d}).$$

1.28 For any given coordinate frame $\{\boldsymbol{e}_i\}$ show that the dyadic basis $\{\boldsymbol{e}_i \otimes \boldsymbol{e}_j\}$ for $\mathcal{V}^2$ is orthonormal in the inner product (1.17).

1.29 Prove Result 1.13.

1.30 Let $\boldsymbol{A}$, $\boldsymbol{B}$ and $\boldsymbol{C}$ be second-order tensors. Show that

$$\boldsymbol{A} : \boldsymbol{B}\boldsymbol{C} = \boldsymbol{A}\boldsymbol{C}^T : \boldsymbol{B} = \boldsymbol{B}^T\boldsymbol{A} : \boldsymbol{C}.$$

1.31 Let $\boldsymbol{a}$ and $\boldsymbol{b}$ be vectors and let $\boldsymbol{A}$ be a second-order tensor. Define C to be the fourth-order tensor given by $\mathsf{C}(\boldsymbol{S}) = \boldsymbol{A}\boldsymbol{S}$. Show that, if $\boldsymbol{a}$ is an eigenvector of $\boldsymbol{A}$ with eigenvalue α, then

$$(\mathsf{C}(\boldsymbol{a} \otimes \boldsymbol{b}))\boldsymbol{v} = \alpha(\boldsymbol{b} \cdot \boldsymbol{v})\boldsymbol{a}.$$

1.32 Find the components C_{ijkl} of the fourth-order tensor defined by

$$\mathsf{C}(\boldsymbol{S}) = \tfrac{1}{2}(\boldsymbol{S} - \boldsymbol{S}^T).$$

1.33 Prove Result 1.14.

1.34 Suppose two symmetric second-order tensors $\boldsymbol{S}$ and $\boldsymbol{E}$ satisfy $\boldsymbol{S} = \mathsf{C}\boldsymbol{E}$, where C is a fourth-order tensor with components $\mathsf{C}_{ijrs} = \lambda\delta_{ij}\delta_{rs} + \mu(\delta_{ir}\delta_{js} + \delta_{is}\delta_{jr})$ and λ and μ are scalar constants. Show that:

(a) $\boldsymbol{S} = \lambda(\operatorname{tr}\boldsymbol{E})\boldsymbol{I} + 2\mu\boldsymbol{E}$,

(b) $\boldsymbol{E} = \dfrac{1}{2\mu}\boldsymbol{S} - \dfrac{\lambda}{2\mu(3\lambda+2\mu)}(\operatorname{tr}\boldsymbol{S})\boldsymbol{I}$.

Answers to Selected Exercises

1.1 (a) $\boldsymbol{a}\cdot\boldsymbol{b} = 1$.
(b) $\boldsymbol{a}\times\boldsymbol{b} = -13\boldsymbol{i} + 5\boldsymbol{j} + \boldsymbol{k}$.
(c) $\boldsymbol{a}\cdot\boldsymbol{b}\times\boldsymbol{c} = \boldsymbol{c}\cdot\boldsymbol{a}\times\boldsymbol{b} = 21$.
(d) $\boldsymbol{a}\times(\boldsymbol{b}\times\boldsymbol{c}) = (\boldsymbol{a}\cdot\boldsymbol{c})\boldsymbol{b} - (\boldsymbol{a}\cdot\boldsymbol{b})\boldsymbol{c} = -2\boldsymbol{i} - 11\boldsymbol{j} + 8\boldsymbol{k}$.
(e) $(\boldsymbol{a}\times\boldsymbol{b})\times\boldsymbol{c} = (\boldsymbol{c}\cdot\boldsymbol{a})\boldsymbol{b} - (\boldsymbol{c}\cdot\boldsymbol{b})\boldsymbol{a} = \boldsymbol{i} - 2\boldsymbol{j} + 23\boldsymbol{k}$.

1.3 $\delta_{ij}\delta_{ij} = \delta_{ii} = \delta_{11} + \delta_{22} + \delta_{33} = 3$.

1.5 Using an epsilon-delta identity we have

$$\begin{aligned}(\boldsymbol{a}\times\boldsymbol{b})\times\boldsymbol{c} = (a_i\boldsymbol{e}_i\times b_j\boldsymbol{e}_j)\times c_k\boldsymbol{e}_k &= \varepsilon_{ijl}a_ib_jc_k\boldsymbol{e}_l\times\boldsymbol{e}_k\\ &= \varepsilon_{ijl}\varepsilon_{lkm}a_ib_jc_k\boldsymbol{e}_m = (\delta_{ik}\delta_{jm} - \delta_{im}\delta_{jk})a_ib_jc_k\boldsymbol{e}_m\\ &= (a_kc_k)b_j\boldsymbol{e}_j - (b_kc_k)a_m\boldsymbol{e}_m = (\boldsymbol{a}\cdot\boldsymbol{c})\boldsymbol{b} - (\boldsymbol{b}\cdot\boldsymbol{c})\boldsymbol{a}.\end{aligned}$$

Thus $\lambda = \boldsymbol{a}\cdot\boldsymbol{c}$ and $\mu = \boldsymbol{b}\cdot\boldsymbol{c}$.

1.7 (a) For arbitrary $\boldsymbol{v}$ we have

$$\begin{aligned}\boldsymbol{S}(\boldsymbol{a}\otimes\boldsymbol{b})\boldsymbol{v} &= \boldsymbol{S}(\boldsymbol{b}\cdot\boldsymbol{v})\boldsymbol{a}\\ &= (\boldsymbol{b}\cdot\boldsymbol{v})\boldsymbol{S}\boldsymbol{a}\\ &= ((\boldsymbol{S}\boldsymbol{a})\otimes\boldsymbol{b})\boldsymbol{v},\end{aligned}$$

which implies $\boldsymbol{S}(\boldsymbol{a}\otimes\boldsymbol{b}) = (\boldsymbol{S}\boldsymbol{a})\otimes\boldsymbol{b}$.

(b) For arbitrary $\boldsymbol{v}$ we have

$$\begin{aligned}(\boldsymbol{a}\otimes\boldsymbol{b})\boldsymbol{S}\boldsymbol{v} &= (\boldsymbol{b}\cdot\boldsymbol{S}\boldsymbol{v})\boldsymbol{a}\\ &= (\boldsymbol{S}^T\boldsymbol{b}\cdot\boldsymbol{v})\boldsymbol{a}\\ &= (\boldsymbol{a}\otimes\boldsymbol{S}^T\boldsymbol{b})\boldsymbol{v},\end{aligned}$$

which implies $(\boldsymbol{a}\otimes\boldsymbol{b})\boldsymbol{S} = \boldsymbol{a}\otimes(\boldsymbol{S}^T\boldsymbol{b})$.

(c) For arbitrary $\boldsymbol{u}$ and $\boldsymbol{v}$ we have

$$\begin{aligned}\boldsymbol{u}\cdot(\boldsymbol{a}\otimes\boldsymbol{b})^T\boldsymbol{v} &= (\boldsymbol{a}\otimes\boldsymbol{b})\boldsymbol{u}\cdot\boldsymbol{v}\\ &= (\boldsymbol{b}\cdot\boldsymbol{u})(\boldsymbol{a}\cdot\boldsymbol{v})\\ &= \boldsymbol{u}\cdot(\boldsymbol{b}\otimes\boldsymbol{a})\boldsymbol{v},\end{aligned}$$

which implies $(\boldsymbol{a}\otimes\boldsymbol{b})^T = (\boldsymbol{b}\otimes\boldsymbol{a})$.

1.9 (a) $\boldsymbol{P}^T = \boldsymbol{P}$ and $\boldsymbol{P}\boldsymbol{P} = \boldsymbol{P}$, thus perpendicular projection.
(b) $\boldsymbol{P}^T \neq \boldsymbol{P}$, thus not perpendicular projection.
(c) Perpendicular projection.
(d) Perpendicular projection.
(e) $\boldsymbol{P}^T = \boldsymbol{P}$ but $\boldsymbol{P}\boldsymbol{P} \neq \boldsymbol{P}$, thus not perpendicular projection.

1.11 For contradiction assume that there are two tensors $\boldsymbol{A}$ and $\boldsymbol{B}$ such that

$$\boldsymbol{S}\boldsymbol{u}\cdot\boldsymbol{v} = \boldsymbol{u}\cdot\boldsymbol{A}\boldsymbol{v} = \boldsymbol{u}\cdot\boldsymbol{B}\boldsymbol{v}, \quad \forall\boldsymbol{u},\boldsymbol{v}.$$

Then $\boldsymbol{u}\cdot(\boldsymbol{A}-\boldsymbol{B})\boldsymbol{v} = 0$ for all $\boldsymbol{u},\boldsymbol{v}$ and hence $(\boldsymbol{A}-\boldsymbol{B})\boldsymbol{v} = \boldsymbol{0}$ for all $\boldsymbol{v}$, which implies $\boldsymbol{A} = \boldsymbol{B}$. Thus there can be only one tensor $\boldsymbol{S}^T$ such that

$$\boldsymbol{S}\boldsymbol{u}\cdot\boldsymbol{v} = \boldsymbol{u}\cdot\boldsymbol{S}^T\boldsymbol{v}, \quad \forall\boldsymbol{u},\boldsymbol{v},$$

which proves uniqueness. For the second result notice that $[\boldsymbol{S}]^T_{ij} = [\boldsymbol{S}]_{ji}$. Furthermore

$$[\boldsymbol{S}]_{ji}u_iv_j = \boldsymbol{S}\boldsymbol{u}\cdot\boldsymbol{v} = \boldsymbol{u}\cdot\boldsymbol{S}^T\boldsymbol{v} = u_i[\boldsymbol{S}^T]_{ij}v_j.$$

Thus, by the arbitrariness of u_i and v_j, we have $[\boldsymbol{S}^T]_{ij} = [\boldsymbol{S}]_{ji} = [\boldsymbol{S}]^T_{ij}$ as required.

1.13 Consider any vector $\boldsymbol{v}$ and let $\boldsymbol{u} = \boldsymbol{S}\boldsymbol{v}$. Then in the two frames we have the representations $[\boldsymbol{u}] = [\boldsymbol{S}][\boldsymbol{v}]$ and $[\boldsymbol{u}]' = [\boldsymbol{S}]'[\boldsymbol{v}]'$. Moreover, by definition of $\boldsymbol{A}$ we have $[\boldsymbol{u}] = [\boldsymbol{A}][\boldsymbol{u}]'$ and $[\boldsymbol{v}] = [\boldsymbol{A}][\boldsymbol{v}]'$. Using the fact that $[\boldsymbol{A}]^{-1} = [\boldsymbol{A}]^T$ we have

$$[\boldsymbol{S}][\boldsymbol{v}] = [\boldsymbol{u}] = [\boldsymbol{A}][\boldsymbol{u}]' = [\boldsymbol{A}][\boldsymbol{S}]'[\boldsymbol{v}]' = [\boldsymbol{A}][\boldsymbol{S}]'[\boldsymbol{A}]^T[\boldsymbol{v}].$$

By the arbitrariness of $[\boldsymbol{v}]$, we obtain $[\boldsymbol{S}] = [\boldsymbol{A}][\boldsymbol{S}]'[\boldsymbol{A}]^T$, which implies $[\boldsymbol{S}]' = [\boldsymbol{A}]^T[\boldsymbol{S}][\boldsymbol{A}]$.

1.15 By definition of $\det\boldsymbol{A}$ we have

$$\det\boldsymbol{A} = \det[\boldsymbol{A}] = \varepsilon_{ijk}A_{i1}A_{j2}A_{k3}.$$

By properties of the permutation symbol and the determinant of a matrix under permutations of its columns, we find that $\det \boldsymbol{A}$ can be written in six equivalent ways, namely

$$\begin{aligned}\det \boldsymbol{A} &= \varepsilon_{ijk}\varepsilon_{123}A_{i1}A_{j2}A_{k3} = \varepsilon_{ijk}\varepsilon_{312}A_{i3}A_{j1}A_{k2} \\ &= \varepsilon_{ijk}\varepsilon_{231}A_{i2}A_{j3}A_{k1} = \varepsilon_{ijk}\varepsilon_{321}A_{i3}A_{j2}A_{k1} \\ &= \varepsilon_{ijk}\varepsilon_{132}A_{i1}A_{j3}A_{k2} = \varepsilon_{ijk}\varepsilon_{213}A_{i2}A_{j1}A_{k3}.\end{aligned}$$

Summing all six of the above expressions and making use of index notation we obtain

$$\det \boldsymbol{A} = \tfrac{1}{6}\epsilon_{ijk}\epsilon_{pqr}A_{ip}A_{jq}A_{kr},$$

which is the desired result. Since the above expression remains unchanged when A_{mn} is replaced by A_{nm} we deduce that $\det \boldsymbol{A} = \det \boldsymbol{A}^T$.

1.17 Consider any third vector $\boldsymbol{w}$. Then in any frame we have

$$\begin{aligned}(\boldsymbol{Fu} \times \boldsymbol{Fv}) \cdot \boldsymbol{w} &= \det([\boldsymbol{F}][\boldsymbol{u}], [\boldsymbol{F}][\boldsymbol{v}], [\boldsymbol{w}]) \\ &= \det\left([\boldsymbol{F}][\boldsymbol{u}], [\boldsymbol{F}][\boldsymbol{v}], [\boldsymbol{F}][\boldsymbol{F}]^{-1}[\boldsymbol{w}]\right) \\ &= (\det[\boldsymbol{F}])\ \det\left([\boldsymbol{u}], [\boldsymbol{v}], [\boldsymbol{F}]^{-1}[\boldsymbol{w}]\right) \\ &= (\det \boldsymbol{F})\ ((\boldsymbol{u} \times \boldsymbol{v}) \cdot \boldsymbol{F}^{-1}\boldsymbol{w}) \\ &= (\det \boldsymbol{F})\ \boldsymbol{F}^{-T}(\boldsymbol{u} \times \boldsymbol{v}) \cdot \boldsymbol{w}.\end{aligned}$$

The result follows by the arbitrariness of $\boldsymbol{w}$.

1.19 (a) Since $\boldsymbol{Q}^T\boldsymbol{Q} = \boldsymbol{I}$ and $\det \boldsymbol{Q}^T = \det \boldsymbol{Q}$ we have

$$1 = \det \boldsymbol{I} = \det(\boldsymbol{Q}^T\boldsymbol{Q}) = (\det \boldsymbol{Q}^T)(\det \boldsymbol{Q}) = (\det[\boldsymbol{Q}])^2,$$

which establishes the result.

(b) Given any right-handed frame $\{\boldsymbol{e}_i\}$ let $\boldsymbol{v}_i = \boldsymbol{Q}\boldsymbol{e}_i$ with corresponding matrix representation $[\boldsymbol{v}_i] = [\boldsymbol{Q}][\boldsymbol{e}_i]$ in the frame $\{\boldsymbol{e}_i\}$. Since $\boldsymbol{Q}$ is a rotation the vectors $\{\boldsymbol{v}_i\}$ form a right-handed frame and

$$\begin{aligned}1 = (\boldsymbol{v}_1 \times \boldsymbol{v}_2) \cdot \boldsymbol{v}_3 &= \det([\boldsymbol{v}_1], [\boldsymbol{v}_2], [\boldsymbol{v}_3]) \\ &= \det([\boldsymbol{Q}][\boldsymbol{e}_1], [\boldsymbol{Q}][\boldsymbol{e}_2], [\boldsymbol{Q}][\boldsymbol{e}_3]) \\ &= \det([\boldsymbol{Q}]([\boldsymbol{e}_1], [\boldsymbol{e}_2], [\boldsymbol{e}_3])) \\ &= (\det \boldsymbol{Q})\ (\boldsymbol{e}_1 \times \boldsymbol{e}_2) \cdot \boldsymbol{e}_3,\end{aligned}$$

from which we deduce $\det \boldsymbol{Q} = 1$.

1.21 (a) $\boldsymbol{Qu} \cdot \boldsymbol{Qv} = \boldsymbol{u} \cdot \boldsymbol{Q}^T \boldsymbol{Qv} = \boldsymbol{u} \cdot \boldsymbol{v}$.

(b) $|\boldsymbol{Qv}|^2 = \boldsymbol{Qv} \cdot \boldsymbol{Qv} = \boldsymbol{v} \cdot \boldsymbol{Q}^T \boldsymbol{Qv} = \boldsymbol{v} \cdot \boldsymbol{v} = |\boldsymbol{v}|^2 \geq 0$.

(c) Consider any third vector $\boldsymbol{w}$. Then in any frame we have

$$\begin{aligned}(\boldsymbol{Qu} \times \boldsymbol{Qv}) \cdot \boldsymbol{w} &= \det\left([\boldsymbol{Q}][\boldsymbol{u}], [\boldsymbol{Q}][\boldsymbol{v}], [\boldsymbol{w}]\right) \\ &= \det\left([\boldsymbol{Q}][\boldsymbol{u}], [\boldsymbol{Q}][\boldsymbol{v}], [\boldsymbol{Q}][\boldsymbol{Q}]^T[\boldsymbol{w}]\right) \\ &= (\det[\boldsymbol{Q}]) \ \det\left([\boldsymbol{u}], [\boldsymbol{v}], [\boldsymbol{Q}]^T[\boldsymbol{w}]\right) \\ &= (\det \boldsymbol{Q}) \ ((\boldsymbol{u} \times \boldsymbol{v}) \cdot \boldsymbol{Q}^T \boldsymbol{w}) \\ &= (\det \boldsymbol{Q}) \ \boldsymbol{Q}(\boldsymbol{u} \times \boldsymbol{v}) \cdot \boldsymbol{w}.\end{aligned}$$

The result follows by the arbitrariness of $\boldsymbol{w}$ and the fact that $\det \boldsymbol{Q} = 1$.

1.23 Since $\boldsymbol{S}$ is symmetric, the Spectral Theorem guarantees the existence of a frame $\{\boldsymbol{e}_i\}$ such that

$$\boldsymbol{S} = \sum_{i=1}^{3} \lambda_i \boldsymbol{e}_i \otimes \boldsymbol{e}_i,$$

where λ_i are the eigenvalues of $\boldsymbol{S}$. In this frame we have

$$[\boldsymbol{S}] = \begin{pmatrix} \lambda_1 & 0 & 0 \\ 0 & \lambda_2 & 0 \\ 0 & 0 & \lambda_3 \end{pmatrix}, \qquad [\boldsymbol{S}^2] = [\boldsymbol{S}]^2 = \begin{pmatrix} \lambda_1^2 & 0 & 0 \\ 0 & \lambda_2^2 & 0 \\ 0 & 0 & \lambda_3^2 \end{pmatrix}.$$

From the definition of the invariants we deduce

$$\begin{aligned}&I_1(\boldsymbol{S}) = \mathrm{tr}[\boldsymbol{S}] = \lambda_1 + \lambda_2 + \lambda_3, \\ &I_2(\boldsymbol{S}) = \tfrac{1}{2}\left((\mathrm{tr}[\boldsymbol{S}])^2 - \mathrm{tr}[\boldsymbol{S}]^2\right) = \lambda_1\lambda_2 + \lambda_1\lambda_3 + \lambda_2\lambda_3, \\ &I_3(\boldsymbol{S}) = \det[\boldsymbol{S}] = \lambda_1\lambda_2\lambda_3.\end{aligned}$$

1.25 The eigenvalues λ_i of $\boldsymbol{S}$ satisfy the characteristic equation

$$\det(\boldsymbol{S} - \lambda_i \boldsymbol{I}) = 0 \quad (i = 1, 2, 3),$$

which, in terms of the principal invariants, may be expressed as

$$\lambda_i^3 - I_1(\boldsymbol{S})\lambda_i^2 + I_2(\boldsymbol{S})\lambda_i - I_3(\boldsymbol{S}) = 0 \quad (i = 1, 2, 3). \qquad (1.20)$$

In the case that $\boldsymbol{S}$ is symmetric, the Spectral Theorem guarantees the existence of a frame $\{\boldsymbol{e}_i\}$ such that

$$\boldsymbol{S} = \sum_{i=1}^{3} \lambda_i \boldsymbol{e}_i \otimes \boldsymbol{e}_i,$$

where $\boldsymbol{e}_i$ is an eigenvector of $\boldsymbol{S}$ with eigenvalue λ_i. Moreover, we have

$$\boldsymbol{S}^2 = \sum_{i=1}^{3} \lambda_i^2 \boldsymbol{e}_i \otimes \boldsymbol{e}_i, \quad \boldsymbol{S}^3 = \sum_{i=1}^{3} \lambda_i^3 \boldsymbol{e}_i \otimes \boldsymbol{e}_i \quad \text{and} \quad \boldsymbol{I} = \sum_{i=1}^{3} \boldsymbol{e}_i \otimes \boldsymbol{e}_i.$$

The desired result now follows from (1.20), in particular

$$\begin{aligned} &\boldsymbol{S}^3 - I_1(\boldsymbol{S})\boldsymbol{S}^2 + I_2(\boldsymbol{S})\boldsymbol{S} - I_3(\boldsymbol{S})\boldsymbol{I} \\ &\qquad = \sum_{i=1}^{3} \left[\lambda_i^3 - I_1(\boldsymbol{S})\lambda_i^2 + I_2(\boldsymbol{S})\lambda_i - I_3(\boldsymbol{S})\right] \boldsymbol{e}_i \otimes \boldsymbol{e}_i = \boldsymbol{O}. \end{aligned}$$

1.27 Let $\boldsymbol{v}$ be an arbitrary vector. Since $(\boldsymbol{a} \otimes \boldsymbol{b})^T = (\boldsymbol{b} \otimes \boldsymbol{a})$ we have

$$\begin{aligned} (\boldsymbol{a} \otimes \boldsymbol{b})^T (\boldsymbol{c} \otimes \boldsymbol{d})\boldsymbol{v} &= (\boldsymbol{b} \otimes \boldsymbol{a})\boldsymbol{c}(\boldsymbol{d} \cdot \boldsymbol{v}) \\ &= \boldsymbol{b}(\boldsymbol{a} \cdot \boldsymbol{c})(\boldsymbol{d} \cdot \boldsymbol{v}) = (\boldsymbol{a} \cdot \boldsymbol{c})(\boldsymbol{b} \otimes \boldsymbol{d})\boldsymbol{v}, \end{aligned}$$

which implies

$$(\boldsymbol{a} \otimes \boldsymbol{b})^T (\boldsymbol{c} \otimes \boldsymbol{d}) = (\boldsymbol{a} \cdot \boldsymbol{c})(\boldsymbol{b} \otimes \boldsymbol{d}).$$

Using the fact that $[\boldsymbol{b} \otimes \boldsymbol{d}]_{ij} = b_i d_j$ then yields

$$\begin{aligned} (\boldsymbol{a} \otimes \boldsymbol{b}) : (\boldsymbol{c} \otimes \boldsymbol{d}) &= \operatorname{tr}((\boldsymbol{a} \otimes \boldsymbol{b})^T (\boldsymbol{c} \otimes \boldsymbol{d})) \\ &= (\boldsymbol{a} \cdot \boldsymbol{c}) \operatorname{tr}(\boldsymbol{b} \otimes \boldsymbol{d}) = (\boldsymbol{a} \cdot \boldsymbol{c})(\boldsymbol{b} \cdot \boldsymbol{d}). \end{aligned}$$

1.29 Since $[\boldsymbol{S}^T \boldsymbol{D}]_{ij} = S_{ki} D_{kj}$ we have

$$\boldsymbol{S} : \boldsymbol{D} = \operatorname{tr}(\boldsymbol{S}^T \boldsymbol{D}) = \operatorname{tr}[\boldsymbol{S}^T \boldsymbol{D}] = S_{ki} D_{ki}.$$

If $\boldsymbol{S}$ is symmetric, then $S_{ki} = S_{ik}$ and

$$\begin{aligned} \boldsymbol{S} : \boldsymbol{D} &= S_{ki} D_{ki} \\ &= \tfrac{1}{2}(S_{ki} + S_{ik}) D_{ki} \\ &= \tfrac{1}{2} S_{ki}(D_{ki} + D_{ik}) = \boldsymbol{S} : \operatorname{sym}(\boldsymbol{D}), \end{aligned}$$

where $\operatorname{sym}(\boldsymbol{D}) = \frac{1}{2}(\boldsymbol{D} + \boldsymbol{D}^T)$. Moreover, by symmetry of $\boldsymbol{S}$ we have $\boldsymbol{S} : \operatorname{sym}(\boldsymbol{D}) = \operatorname{sym}(\boldsymbol{S}) : \operatorname{sym}(\boldsymbol{D})$.

1.31 By definition of $\mathbf{C}$ we have

$$(\mathbf{C}(\boldsymbol{a} \otimes \boldsymbol{b}))\boldsymbol{v} = (\boldsymbol{A}(\boldsymbol{a} \otimes \boldsymbol{b}))\boldsymbol{v} = (\boldsymbol{b} \cdot \boldsymbol{v})\boldsymbol{A}\boldsymbol{a} = \alpha(\boldsymbol{b} \cdot \boldsymbol{v})\boldsymbol{a}.$$

1.33 Suppose C has major symmetry, that is, $\boldsymbol{A} : \mathsf{C}(\boldsymbol{B}) = \mathsf{C}(\boldsymbol{A}) : \boldsymbol{B}$ for all second-order tensors $\boldsymbol{A}$, $\boldsymbol{B}$. Since $\boldsymbol{A} : \mathsf{C}(\boldsymbol{B}) = A_{ij}\mathsf{C}_{ijkl}B_{kl}$ and $\mathsf{C}(\boldsymbol{A}) : \boldsymbol{B} = \mathsf{C}_{ijkl}A_{kl}B_{ij} = \mathsf{C}_{klij}A_{ij}B_{kl}$, we have

$$\boldsymbol{A} : \mathsf{C}(\boldsymbol{B}) - \mathsf{C}(\boldsymbol{A}) : \boldsymbol{B} = (\mathsf{C}_{ijkl} - \mathsf{C}_{klij})A_{ij}B_{kl} = 0, \quad \forall A_{ij}, B_{kl}.$$

This implies $\mathsf{C}_{ijkl} - \mathsf{C}_{klij} = 0$ and the result follows. Next, suppose C has left minor symmetry, that is, $\boldsymbol{A} : \mathsf{C}(\boldsymbol{B}) = \mathrm{sym}(\boldsymbol{A}) : \mathsf{C}(\boldsymbol{B})$ for all $\boldsymbol{A}$, $\boldsymbol{B}$. Since

$$\mathrm{sym}(\boldsymbol{A}) : \mathsf{C}(\boldsymbol{B}) = \tfrac{1}{2}(A_{ij} + A_{ji})\mathsf{C}_{ijkl}B_{kl} = \tfrac{1}{2}A_{ij}(\mathsf{C}_{ijkl} + \mathsf{C}_{jikl})B_{kl},$$

we get

$$\begin{aligned} &\boldsymbol{A} : \mathsf{C}(\boldsymbol{B}) - \mathrm{sym}(\boldsymbol{A}) : \mathsf{C}(\boldsymbol{B}) \\ &\quad = (\mathsf{C}_{ijkl} - \tfrac{1}{2}(\mathsf{C}_{ijkl} + \mathsf{C}_{jikl}))A_{ij}B_{kl} \\ &\quad = \tfrac{1}{2}(\mathsf{C}_{ijkl} - \mathsf{C}_{jikl})A_{ij}B_{kl} = 0, \quad \forall A_{ij}, B_{kl}. \end{aligned}$$

This implies $\mathsf{C}_{ijkl} - \mathsf{C}_{jikl} = 0$ and the result follows. The result for right major symmetry may be proved similarly.

2
Tensor Calculus

To describe the mechanical behavior of continuous media we will use scalars, vectors and second-order tensors which in general may vary from point to point in a material body. Since a body may be identified with a subset of Euclidean space we will be interested in functions of the form $\phi : \mathbb{E}^3 \to \mathbb{R}$, $\boldsymbol{v} : \mathbb{E}^3 \to \mathcal{V}$ and $\boldsymbol{S} : \mathbb{E}^3 \to \mathcal{V}^2$. Moreover, to describe the shapes of material bodies, and their reactions to changes in shape, we will be interested in functions of the form $\boldsymbol{\chi} : \mathbb{E}^3 \to \mathbb{E}^3$, $g : \mathcal{V}^2 \to \mathbb{R}$ and $\boldsymbol{G} : \mathcal{V}^2 \to \mathcal{V}^2$.

In this chapter we study various calculus concepts for different types of functions on $\mathbb{E}^3$ and $\mathcal{V}^2$. The important ideas that we introduce are: (i) the derivative or gradient, divergence, curl and Laplacian for different types of functions on $\mathbb{E}^3$; (ii) the Divergence, Stokes' and Localization Theorems pertaining to integrals over different types of subsets of $\mathbb{E}^3$; (iii) the derivative of different types of functions on $\mathcal{V}^2$.

For all the definitions and results in this chapter we assume, without further explicit statement, that the functions being differentiated are smooth in the sense that partial derivatives of all orders exist and are continuous. This is always stronger than what we need, but allows for a clean presentation.

2.1 Preliminaries

2.1.1 Points, Tensors and Representations

Throughout the remainder of our developments we assume that a single, fixed coordinate frame $\{\boldsymbol{e}_i\}$ has been specified for Euclidean space $\mathbb{E}^3$. Under this assumption we may identify points $\boldsymbol{x} \in \mathbb{E}^3$, vectors $\boldsymbol{v} \in \mathcal{V}$ and second-order tensors $\boldsymbol{S} \in \mathcal{V}^2$ with their respective matrix representations; in particular, we write

$$\boldsymbol{x} = x_i \boldsymbol{e}_i = \begin{Bmatrix} x_1 \\ x_2 \\ x_3 \end{Bmatrix} \in \mathbb{R}^3, \qquad \boldsymbol{v} = v_i \boldsymbol{e}_i = \begin{Bmatrix} v_1 \\ v_2 \\ v_3 \end{Bmatrix} \in \mathbb{R}^3,$$

and

$$\boldsymbol{S} = S_{ij} \boldsymbol{e}_i \otimes \boldsymbol{e}_j = \begin{pmatrix} S_{11} & S_{12} & S_{13} \\ S_{21} & S_{22} & S_{23} \\ S_{31} & S_{32} & S_{33} \end{pmatrix} \in \mathbb{R}^{3\times 3}.$$

Thus $\mathbb{E}^3$ and $\mathcal{V}$ are both identified with $\mathbb{R}^3$, and $\mathcal{V}^2$ is identified with $\mathbb{R}^{3\times 3}$. Any function of a point $\boldsymbol{x}$ is a function of the three real variables (x_1, x_2, x_3), and any function of a second-order tensor $\boldsymbol{S}$ is a function of the nine real variables $(S_{11}, S_{12}, \ldots, S_{33})$.

2.1.2 Standard Norms, Order Symbols

Here we exploit the above identifications to define norms on the spaces $\mathbb{E}^3$, $\mathcal{V}$ and $\mathcal{V}^2$ and introduce the order symbols $\mathcal{O}$ and o.

Definition 2.1 *By the standard (Euclidean)* **norms** *on the spaces $\mathbb{E}^3$, $\mathcal{V}$ and $\mathcal{V}^2$ we mean the scalar functions defined by*

$$\begin{aligned} |\boldsymbol{x}| &= \sqrt{\boldsymbol{x}\cdot\boldsymbol{x}}, && \forall \boldsymbol{x} \in \mathbb{E}^3, \\ |\boldsymbol{v}| &= \sqrt{\boldsymbol{v}\cdot\boldsymbol{v}}, && \forall \boldsymbol{v} \in \mathcal{V}, \\ |\boldsymbol{S}| &= \sqrt{\boldsymbol{S}:\boldsymbol{S}}, && \forall \boldsymbol{S} \in \mathcal{V}^2. \end{aligned}$$

□

In the same way that $|\boldsymbol{x}|$ and $|\boldsymbol{v}|$ provide a measure of the magnitude or size of the vectors $\boldsymbol{x}$ and $\boldsymbol{v}$, the quantity $|\boldsymbol{S}|$ provides a measure of the magnitude or size of $\boldsymbol{S}$. It has all the usual properties of a norm. Using these standard norms we can define limits in the spaces $\mathbb{E}^3$, $\mathcal{V}$ and $\mathcal{V}^2$, along with continuity of functions between them.

Definition 2.2 *Consider a function $\boldsymbol{f} : U \to W$, where U and W denote any of the spaces $\mathbb{E}^3$, $\mathcal{V}$, $\mathcal{V}^2$ or $\mathbb{R}$.*

(i) If there are constants $C > 0$ and $r > 0$ such that $|\boldsymbol{f}(\boldsymbol{u})| \le C|\boldsymbol{u}|^r$ as $\boldsymbol{u} \to \boldsymbol{0}$, then we write

$$\boldsymbol{f}(\boldsymbol{u}) = \mathcal{O}(|\boldsymbol{u}|^r) \quad \text{as} \quad \boldsymbol{u} \to \boldsymbol{0}.$$

(ii) If there is a constant $r > 0$ such that $|\boldsymbol{f}(\boldsymbol{u})|/|\boldsymbol{u}|^r \to 0$ as $\boldsymbol{u} \to \mathbf{0}$, then we write

$$\boldsymbol{f}(\boldsymbol{u}) = o(|\boldsymbol{u}|^r) \quad \text{as} \quad \boldsymbol{u} \to \mathbf{0}.$$

The symbols $\mathcal{O}$ and o are called the standard **order symbols.** □

The order symbols provide a way to describe the behavior of a function $\boldsymbol{f}(\boldsymbol{u})$ as $\boldsymbol{u}$ approaches zero, or, by translation, any other given value. The statement $\boldsymbol{f}(\boldsymbol{u}) = \mathcal{O}(|\boldsymbol{u}|^r)$ as $\boldsymbol{u} \to \mathbf{0}$ means that $|\boldsymbol{f}(\boldsymbol{u})|$ goes to zero at least as fast as $|\boldsymbol{u}|^r$. The statement $\boldsymbol{f}(\boldsymbol{u}) = o(|\boldsymbol{u}|^r)$ as $\boldsymbol{u} \to \mathbf{0}$ means that $|\boldsymbol{f}(\boldsymbol{u})|$ goes to zero faster than $|\boldsymbol{u}|^r$. Thus $\boldsymbol{f}(\boldsymbol{u}) = o(|\boldsymbol{u}|^r)$ implies $\boldsymbol{f}(\boldsymbol{u}) = \mathcal{O}(|\boldsymbol{u}|^r)$, but the converse need not be true. However, when $p > r$, we find that $\boldsymbol{f}(\boldsymbol{u}) = \mathcal{O}(|\boldsymbol{u}|^p)$ implies $\boldsymbol{f}(\boldsymbol{u}) = o(|\boldsymbol{u}|^r)$.

Given two functions $\boldsymbol{f}, \boldsymbol{g} : U \to W$ we use the notation

$$\boldsymbol{f}(\boldsymbol{u}) = \boldsymbol{g}(\boldsymbol{u}) + \mathcal{O}(|\boldsymbol{u}|^r) \quad \text{as} \quad \boldsymbol{u} \to \mathbf{0}$$

to signify that

$$\boldsymbol{f}(\boldsymbol{u}) - \boldsymbol{g}(\boldsymbol{u}) = \mathcal{O}(|\boldsymbol{u}|^r) \quad \text{as} \quad \boldsymbol{u} \to \mathbf{0}.$$

Similar notation is also used for the order symbol o. We sometimes indicate the limit $\boldsymbol{u} \to \mathbf{0}$ by saying that $\boldsymbol{u}$ is small. When there is no cause for confusion we omit reference to this limit altogether.

2.2 Differentiation of Tensor Fields

We use the term **field** to denote a function which is defined in a region of Euclidean space $\mathbb{E}^3$. By a **scalar field** we mean a function of the form $\phi : \mathbb{E}^3 \to \mathbb{R}$, by a **vector field** we mean $\boldsymbol{v} : \mathbb{E}^3 \to \mathcal{V}$ and by a **second-order tensor field** we mean $\boldsymbol{S} : \mathbb{E}^3 \to \mathcal{V}^2$. In this section we define various differential operations on such tensor fields and discuss some of their properties.

2.2.1 Derivatives, Gradients

Here we define the derivative or gradient of scalar and vector fields. All other differential operations that we perform on tensor fields will be based on these definitions.

Definition 2.3 *A scalar field $\phi : \boldsymbol{E}^3 \to \boldsymbol{R}$ is said to be* **differentiable** *at $\boldsymbol{x} \in \boldsymbol{E}^3$ if there exists a vector $\nabla\phi(\boldsymbol{x}) \in \mathcal{V}$ such that*

$$\phi(\boldsymbol{x} + \boldsymbol{h}) = \phi(\boldsymbol{x}) + \nabla\phi(\boldsymbol{x}) \cdot \boldsymbol{h} + o(|\boldsymbol{h}|),$$

or equivalently

$$\nabla\phi(\boldsymbol{x}) \cdot \boldsymbol{a} = \frac{d}{d\alpha}\phi(\boldsymbol{x} + \alpha\boldsymbol{a})\Big|_{\alpha=0}, \quad \forall \boldsymbol{a} \in \mathcal{V},$$

where $\alpha \in \boldsymbol{R}$. The vector $\nabla\phi(\boldsymbol{x})$ is called the **derivative** *or* **gradient** *of ϕ at $\boldsymbol{x}$.* □

If ϕ is differentiable at $\boldsymbol{x}$ it can be shown that the vector $\nabla\phi(\boldsymbol{x})$ is necessarily unique. The second characterization follows from the first upon setting $\boldsymbol{h} = \alpha\boldsymbol{a}$, dividing through by α, and taking the limit $\alpha \to 0$. The following result provides an explicit characterization of the derivative $\nabla\phi(\boldsymbol{x})$ in any coordinate frame.

Result 2.4 ***Scalar Gradient in Coordinates.*** *Let $\{\boldsymbol{e}_i\}$ be an arbitrary frame. Then*

$$\nabla\phi(\boldsymbol{x}) = \frac{\partial\phi}{\partial x_i}(\boldsymbol{x})\,\boldsymbol{e}_i,$$

where (x_1, x_2, x_3) are the coordinates of $\boldsymbol{x}$ in $\{\boldsymbol{e}_i\}$. □

Proof Writing ϕ as a function of the coordinates x_i we have, by a slight abuse of notation, $\phi(\boldsymbol{x}) = \phi(x_1, x_2, x_3)$. For any scalar α and vector $\boldsymbol{a} = a_k\boldsymbol{e}_k$ this gives

$$\phi(\boldsymbol{x} + \alpha\boldsymbol{a}) = \phi(x_1 + \alpha a_1, x_2 + \alpha a_2, x_3 + \alpha a_3).$$

By Definition 2.3 and the chain rule we find

$$\begin{aligned}\nabla\phi(\boldsymbol{x}) \cdot \boldsymbol{a} &= \frac{d}{d\alpha}\phi(\boldsymbol{x} + \alpha\boldsymbol{a})\Big|_{\alpha=0} \\ &= \frac{\partial\phi}{\partial x_1}(\boldsymbol{x})a_1 + \frac{\partial\phi}{\partial x_2}(\boldsymbol{x})a_2 + \frac{\partial\phi}{\partial x_3}(\boldsymbol{x})a_3 \\ &= \frac{\partial\phi}{\partial x_i}(\boldsymbol{x})\boldsymbol{e}_i \cdot a_k\boldsymbol{e}_k,\end{aligned}$$

which implies $\nabla\phi(\boldsymbol{x}) = \frac{\partial\phi}{\partial x_i}(\boldsymbol{x})\boldsymbol{e}_i$. □

Remarks:

(1) The derivative of a scalar field $\phi : \mathbb{E}^3 \to \mathbb{R}$ is a vector field $\nabla\phi : \mathbb{E}^3 \to \mathcal{V}$.

(2) We sometimes denote $\partial\phi/\partial x_i$ by $\phi_{,i}$.

(3) Since all partial derivatives are assumed to exist and be continuous, we can use Taylor's Theorem to deduce, for any small vector $\boldsymbol{h} \in \mathcal{V}$

$$\phi(\boldsymbol{x}+\boldsymbol{h}) = \phi(\boldsymbol{x}) + \frac{\partial\phi}{\partial x_i}(\boldsymbol{x})h_i + \mathcal{O}(|\boldsymbol{h}|^2),$$

or equivalently

$$\phi(\boldsymbol{x}+\boldsymbol{h}) = \phi(\boldsymbol{x}) + \nabla\phi(\boldsymbol{x})\cdot\boldsymbol{h} + \mathcal{O}(|\boldsymbol{h}|^2).$$

Thus, in the smooth case, the difference between $\phi(\boldsymbol{x}+\boldsymbol{h})$ and $\phi(\boldsymbol{x}) + \nabla\phi(\boldsymbol{x})\cdot\boldsymbol{h}$ is order $\mathcal{O}(|\boldsymbol{h}|^2)$, not just $o(|\boldsymbol{h}|)$ as indicated in the definition of $\nabla\phi(\boldsymbol{x})$. □

Definition 2.5 *A vector field $\boldsymbol{v} : \mathbb{E}^3 \to \mathcal{V}$ is said to be* **differentiable** *at $\boldsymbol{x} \in \mathbb{E}^3$ if there exists a second-order tensor $\nabla\boldsymbol{v}(\boldsymbol{x}) \in \mathcal{V}^2$ such that*

$$\boldsymbol{v}(\boldsymbol{x}+\boldsymbol{h}) = \boldsymbol{v}(\boldsymbol{x}) + \nabla\boldsymbol{v}(\boldsymbol{x})\boldsymbol{h} + o(|\boldsymbol{h}|),$$

or equivalently

$$\nabla\boldsymbol{v}(\boldsymbol{x})\boldsymbol{a} = \frac{d}{d\alpha}\boldsymbol{v}(\boldsymbol{x}+\alpha\boldsymbol{a})\Big|_{\alpha=0}, \quad \forall\boldsymbol{a} \in \mathcal{V},$$

where $\alpha \in \mathbb{R}$. The tensor $\nabla\boldsymbol{v}(\boldsymbol{x})$ is called the **derivative** *or* **gradient** *of $\boldsymbol{v}$ at $\boldsymbol{x}$.* □

If $\boldsymbol{v}$ is differentiable at $\boldsymbol{x}$ it can be shown that the tensor $\nabla\boldsymbol{v}(\boldsymbol{x})$ is necessarily unique. As before, the second characterization follows from the first upon setting $\boldsymbol{h} = \alpha\boldsymbol{a}$, dividing through by α, and taking the limit $\alpha \to 0$. The following result provides an explicit characterization of the derivative $\nabla\boldsymbol{v}(\boldsymbol{x})$ in any coordinate frame.

Result 2.6 **Vector Gradient in Coordinates.** *Let $\{\boldsymbol{e}_i\}$ be an arbitrary frame and let $\boldsymbol{v}(\boldsymbol{x}) = v_i(\boldsymbol{x})\boldsymbol{e}_i$. Then*

$$\nabla\boldsymbol{v}(\boldsymbol{x}) = \frac{\partial v_i}{\partial x_j}(\boldsymbol{x})\,\boldsymbol{e}_i \otimes \boldsymbol{e}_j,$$

where (x_1, x_2, x_3) are the coordinates of $\boldsymbol{x}$ in $\{\boldsymbol{e}_i\}$. □

Proof Writing the components v_i as functions of the coordinates x_j we have, by a slight abuse of notation, $v_i(\boldsymbol{x}) = v_i(x_1, x_2, x_3)$. For any scalar α and vector $\boldsymbol{a} = a_k \boldsymbol{e}_k$ this gives

$$v_i(\boldsymbol{x} + \alpha \boldsymbol{a}) = v_i(x_1 + \alpha a_1, x_2 + \alpha a_2, x_3 + \alpha a_3),$$

and by the chain rule we find

$$\begin{aligned} \frac{d}{d\alpha} v_i(\boldsymbol{x} + \alpha \boldsymbol{a})\Big|_{\alpha=0} &= \frac{\partial v_i}{\partial x_1}(\boldsymbol{x}) a_1 + \frac{\partial v_i}{\partial x_2}(\boldsymbol{x}) a_2 + \frac{\partial v_i}{\partial x_3}(\boldsymbol{x}) a_3 \\ &= \frac{\partial v_i}{\partial x_j}(\boldsymbol{x}) a_j. \end{aligned}$$

Using this result, together with Definition 2.5 and properties of dyadic products (see Chapter 1), we obtain

$$\begin{aligned} \nabla \boldsymbol{v}(\boldsymbol{x}) \boldsymbol{a} = \frac{d}{d\alpha} \boldsymbol{v}(\boldsymbol{x} + \alpha \boldsymbol{a})\Big|_{\alpha=0} &= \frac{d}{d\alpha} v_i(\boldsymbol{x} + \alpha \boldsymbol{a})\Big|_{\alpha=0} \boldsymbol{e}_i \\ &= \frac{\partial v_i}{\partial x_j}(\boldsymbol{x}) a_j \boldsymbol{e}_i \\ &= \left(\frac{\partial v_i}{\partial x_j}(\boldsymbol{x})\, \boldsymbol{e}_i \otimes \boldsymbol{e}_j \right) a_k \boldsymbol{e}_k, \end{aligned}$$

which implies $\nabla \boldsymbol{v}(\boldsymbol{x}) = \frac{\partial v_i}{\partial x_j}(\boldsymbol{x})\, \boldsymbol{e}_i \otimes \boldsymbol{e}_j$. □

Remarks:

(1) The derivative of a vector field $\boldsymbol{v} : \mathbb{E}^3 \to \mathcal{V}$ is a second-order tensor field $\nabla \boldsymbol{v} : \mathbb{E}^3 \to \mathcal{V}^2$.

(2) As for scalar fields, we often denote $\partial v_i / \partial x_j$ by $v_{i,j}$.

(3) Since all partial derivatives are assumed to exist and be continuous, we can use Taylor's Theorem on each component v_i to deduce, for any small vector $\boldsymbol{h} \in \mathcal{V}$

$$v_i(\boldsymbol{x} + \boldsymbol{h}) = v_i(\boldsymbol{x}) + \frac{\partial v_i}{\partial x_j}(\boldsymbol{x}) h_j + \mathcal{O}(|\boldsymbol{h}|^2),$$

or in tensor notation

$$\boldsymbol{v}(\boldsymbol{x} + \boldsymbol{h}) = \boldsymbol{v}(\boldsymbol{x}) + \nabla \boldsymbol{v}(\boldsymbol{x}) \boldsymbol{h} + \mathcal{O}(|\boldsymbol{h}|^2).$$

Thus, in the smooth case, the difference between $\boldsymbol{v}(\boldsymbol{x} + \boldsymbol{h})$ and $\boldsymbol{v}(\boldsymbol{x}) + \nabla \boldsymbol{v}(\boldsymbol{x}) \boldsymbol{h}$ is order $\mathcal{O}(|\boldsymbol{h}|^2)$, not just $o(|\boldsymbol{h}|)$ as indicated in the definition of $\nabla \boldsymbol{v}(\boldsymbol{x})$.

(4) Since we identify points $\boldsymbol{x} \in \mathbb{E}^3$ with their position vectors $\boldsymbol{x}-\boldsymbol{o} \in \mathcal{V}$ relative to a fixed origin, any result that we state for vector fields $\boldsymbol{v} : \mathbb{E}^3 \to \mathcal{V}$ will also apply to **point mappings**, which are functions of the form $\boldsymbol{\chi} : \mathbb{E}^3 \to \mathbb{E}^3$. Thus the derivative or gradient of a point mapping $\boldsymbol{\chi}$ is a second-order tensor field $\nabla\boldsymbol{\chi} : \mathbb{E}^3 \to \mathcal{V}^2$, where

$$\nabla\boldsymbol{\chi}(\boldsymbol{x}) = \frac{\partial \chi_i}{\partial x_j}(\boldsymbol{x})\, \boldsymbol{e}_i \otimes \boldsymbol{e}_j.$$

□

2.2.2 Divergence

Here we introduce the divergence of vector and second-order tensor fields. These quantities arise in the statements of various integral theorems introduced later.

Definition 2.7 *To any vector field $\boldsymbol{v} : \mathbb{E}^3 \to \mathcal{V}$ we associate a scalar field $\nabla \cdot \boldsymbol{v} : \mathbb{E}^3 \to \mathbb{R}$ defined by*

$$\nabla \cdot \boldsymbol{v} = \operatorname{tr} \nabla \boldsymbol{v}.$$

We call $\nabla \cdot \boldsymbol{v}$ the **divergence** *of $\boldsymbol{v}$.* □

If we interpret the vector field $\boldsymbol{v}$ as the velocity field in a flowing fluid or gas, then the scalar $\nabla \cdot \boldsymbol{v}$ at a point $\boldsymbol{x}$ can be interpreted as the rate of volume expansion at $\boldsymbol{x}$. (See the discussion following the Divergence Theorem in Section 2.3.) The next result, which follows directly from Result 2.6 and the definition of the trace function, provides an explicit characterization of the divergence $\nabla \cdot \boldsymbol{v}$ in any coordinate frame.

Result 2.8 ***Vector Divergence in Coordinates.*** *Let $\{\boldsymbol{e}_i\}$ be an arbitrary frame and let $\boldsymbol{v}(\boldsymbol{x}) = v_i(\boldsymbol{x})\boldsymbol{e}_i$. Then*

$$(\nabla \cdot \boldsymbol{v})(\boldsymbol{x}) = \frac{\partial v_i}{\partial x_i}(\boldsymbol{x}) = v_{i,i}(\boldsymbol{x}),$$

where (x_1, x_2, x_3) are the coordinates of $\boldsymbol{x}$ in $\{\boldsymbol{e}_i\}$. □

To formulate balance laws for material bodies it will prove useful to extend the above definition of divergence to second-order tensor fields. While different extensions are possible, we adopt one which leads to a convenient form of the Tensor Divergence Theorem (Result 2.17).

Definition 2.9 *To any second-order tensor field $\boldsymbol{S} : \mathbb{E}^3 \to \mathcal{V}^2$ we associate a vector field $\nabla \cdot \boldsymbol{S} : \mathbb{E}^3 \to \mathcal{V}$ defined by the relation*

$$(\nabla \cdot \boldsymbol{S}) \cdot \boldsymbol{a} = \nabla \cdot (\boldsymbol{S}^T \boldsymbol{a}),$$

for all constant vectors $\boldsymbol{a}$. We call $\nabla \cdot \boldsymbol{S}$ the **divergence** *of $\boldsymbol{S}$.* □

Thus the divergence of a second-order tensor field is defined using the corresponding notion for vector fields. The following result provides an explicit characterization of $\nabla \cdot \boldsymbol{S}$ in any coordinate frame.

Result 2.10 ***Tensor Divergence in Coordinates.*** *Let $\{\boldsymbol{e}_i\}$ be an arbitrary frame and let $\boldsymbol{S}(\boldsymbol{x}) = S_{ij}(\boldsymbol{x})\boldsymbol{e}_i \otimes \boldsymbol{e}_j$. Then*

$$(\nabla \cdot \boldsymbol{S})(\boldsymbol{x}) = \frac{\partial S_{ij}}{\partial x_j}(\boldsymbol{x})\boldsymbol{e}_i = S_{ij,j}(\boldsymbol{x})\boldsymbol{e}_i,$$

where (x_1, x_2, x_3) are the coordinates of $\boldsymbol{x}$ in $\{\boldsymbol{e}_i\}$. □

Proof Consider an arbitrary, constant vector $\boldsymbol{a} = a_k \boldsymbol{e}_k$ and let $\boldsymbol{q} = \boldsymbol{S}^T \boldsymbol{a}$. Then we have $\boldsymbol{q} = q_j \boldsymbol{e}_j$, where $q_j = S_{ij} a_i$. Using Definition 2.9 together with Result 2.8 we get

$$(\nabla \cdot \boldsymbol{S}) \cdot \boldsymbol{a} = \nabla \cdot \boldsymbol{q} = q_{j,j} = S_{ij,j} a_i = \Big(S_{ij,j} \boldsymbol{e}_i\Big) \cdot a_k \boldsymbol{e}_k,$$

which implies $\nabla \cdot \boldsymbol{S} = S_{ij,j}\, \boldsymbol{e}_i$. □

The next result summarizes some useful product rules for the gradient and divergence operations on scalar, vector and second-order tensor fields. Many other similar rules are established in the exercises.

Result 2.11 ***Gradient and Divergence Product Rules.*** *Let ϕ, $\boldsymbol{v}$ and $\boldsymbol{S}$ be scalar, vector and second-order tensor fields, respectively. Then*

$$\nabla(\phi \boldsymbol{v}) = \boldsymbol{v} \otimes \nabla\phi + \phi \nabla \boldsymbol{v},$$
$$\nabla \cdot (\phi \boldsymbol{S}) = \phi \nabla \cdot \boldsymbol{S} + \boldsymbol{S} \nabla \phi,$$
$$\nabla \cdot (\boldsymbol{S}^T \boldsymbol{v}) = (\nabla \cdot \boldsymbol{S}) \cdot \boldsymbol{v} + \boldsymbol{S} : \nabla \boldsymbol{v}.$$

□

Proof Let $\boldsymbol{v} = v_i \boldsymbol{e}_i$ and $\boldsymbol{S} = S_{ij} \boldsymbol{e}_i \otimes \boldsymbol{e}_j$. To establish the first result we notice that $\phi \boldsymbol{v}$ is a vector field and use Result 2.6 to obtain

$$\begin{aligned}\nabla(\phi \boldsymbol{v}) &= \frac{\partial(\phi v_i)}{\partial x_j} \boldsymbol{e}_i \otimes \boldsymbol{e}_j \\ &= (v_i \phi,_j + \phi v_{i,j}) \boldsymbol{e}_i \otimes \boldsymbol{e}_j \\ &= \boldsymbol{v} \otimes \nabla \phi + \phi \nabla \boldsymbol{v}.\end{aligned}$$

To establish the second result we notice that $\phi \boldsymbol{S}$ is a second-order tensor field and use Result 2.10 to obtain

$$\begin{aligned}\nabla \cdot (\phi \boldsymbol{S}) &= \frac{\partial(\phi S_{ij})}{\partial x_j} \boldsymbol{e}_i \\ &= (S_{ij} \phi,_j + \phi S_{ij,j}) \boldsymbol{e}_i \\ &= \boldsymbol{S} \nabla \phi + \phi \nabla \cdot \boldsymbol{S}.\end{aligned}$$

To establish the third result we notice that $\boldsymbol{S}^T \boldsymbol{v}$ is a vector field, in particular, $\boldsymbol{S}^T \boldsymbol{v} = S_{ij} v_i \boldsymbol{e}_j$. Using Result 2.8 we obtain

$$\begin{aligned}\nabla \cdot (\boldsymbol{S}^T \boldsymbol{v}) &= \frac{\partial(S_{ij} v_i)}{\partial x_j} \\ &= S_{ij,j} v_i + S_{ij} v_{i,j} \\ &= (\nabla \cdot \boldsymbol{S}) \cdot \boldsymbol{v} + \boldsymbol{S} : \nabla \boldsymbol{v}.\end{aligned}$$

□

2.2.3 Curl

Here we introduce the curl of a vector field. This quantity arises in the statement of Stokes' Theorem introduced later and plays an important role in describing the motions of fluids and gases.

Definition 2.12 *To any vector field $\boldsymbol{v} : \boldsymbol{E}^3 \to \mathcal{V}$ we associate another vector field $\nabla \times \boldsymbol{v} : \boldsymbol{E}^3 \to \mathcal{V}$ defined by*

$$(\nabla \times \boldsymbol{v}) \times \boldsymbol{a} = (\nabla \boldsymbol{v} - \nabla \boldsymbol{v}^T) \boldsymbol{a},$$

for all constant vectors $\boldsymbol{a}$. We call $\nabla \times \boldsymbol{v}$ the **curl** *of $\boldsymbol{v}$.* □

If we interpret the vector field $\boldsymbol{v}$ as the velocity field in a flowing fluid or gas, then the vector $\nabla \times \boldsymbol{v}$ at a point $\boldsymbol{x}$ provides information on the rate and direction of rotation at $\boldsymbol{x}$. (See the discussion following Stokes' Theorem in Section 2.3.) The following result provides an explicit characterization of the curl $\nabla \times \boldsymbol{v}$ in any coordinate frame.

Result 2.13 **Vector Curl in Coordinates.** *Let $\{e_i\}$ be an arbitrary frame and let $\boldsymbol{v}(\boldsymbol{x}) = v_i(\boldsymbol{x})e_i$. Then*

$$(\nabla \times \boldsymbol{v})(\boldsymbol{x}) = \epsilon_{ijk} v_{i,k}(\boldsymbol{x})\, \boldsymbol{e}_j,$$

where (x_1, x_2, x_3) are the coordinates of $\boldsymbol{x}$ in $\{e_i\}$. Equivalently, we have

$$\nabla \times \boldsymbol{v} = \left(\frac{\partial v_3}{\partial x_2} - \frac{\partial v_2}{\partial x_3}\right) e_1 + \left(\frac{\partial v_1}{\partial x_3} - \frac{\partial v_3}{\partial x_1}\right) e_2 + \left(\frac{\partial v_2}{\partial x_1} - \frac{\partial v_1}{\partial x_2}\right) e_3.$$

□

Proof In view of Result 1.3 we notice that $\nabla \times \boldsymbol{v}$ is the axial vector associated with the skew-symmetric tensor $\nabla \boldsymbol{v} - \nabla \boldsymbol{v}^T$. If we let $\boldsymbol{w} = \nabla \times \boldsymbol{v}$ and $\boldsymbol{T} = \nabla \boldsymbol{v} - \nabla \boldsymbol{v}^T$, then in any frame we have, by Result 1.3

$$w_j = \frac{1}{2}\epsilon_{ijk} T_{ik} = \frac{1}{2}\epsilon_{ijk}(v_{i,k} - v_{k,i}).$$

Carrying out the multiplication on the right-hand side of the last equation and using the fact that $\epsilon_{ijk} = -\epsilon_{kji}$ we get

$$w_j = \frac{1}{2}(\epsilon_{ijk} v_{i,k} - \epsilon_{ijk} v_{k,i}) = \frac{1}{2}(\epsilon_{ijk} v_{i,k} + \epsilon_{kji} v_{k,i}) = \epsilon_{ijk} v_{i,k},$$

where the last equality follows from the fact that i and k are dummy indices. Since $\boldsymbol{w} = \nabla \times \boldsymbol{v}$ we obtain

$$\nabla \times \boldsymbol{v} = \epsilon_{ijk} v_{i,k}\, \boldsymbol{e}_j.$$

The second result is obtained by carrying out the indicated sums using the definition of the permutation symbol. □

2.2.4 Laplacian

Here we introduce the Laplacian of scalar and vector fields. These quantities are second-order differential operations that arise frequently in applications.

Definition 2.14 *To any scalar field $\phi : \boldsymbol{E}^3 \to \boldsymbol{R}$ we associate another scalar field $\Delta\phi : \boldsymbol{E}^3 \to \boldsymbol{R}$ defined by*

$$\Delta\phi = \nabla \cdot (\nabla\phi),$$

and to any vector field $\boldsymbol{v} : \boldsymbol{E}^3 \to \mathcal{V}$ we associate another vector field $\Delta\boldsymbol{v} : \boldsymbol{E}^3 \to \mathcal{V}$ defined by

$$\Delta\boldsymbol{v} = \nabla \cdot (\nabla\boldsymbol{v}).$$

We call $\Delta\phi$ *and* $\Delta\boldsymbol{v}$ *the* **Laplacians** *of* ϕ *and* $\boldsymbol{v}$*, respectively.* □

Thus the Laplacian of a scalar or vector field is the divergence of the corresponding gradient. The following result provides an explicit characterization of $\Delta\phi$ and $\Delta\boldsymbol{v}$ in any coordinate frame.

Result 2.15 ***Scalar and Vector Laplacian in Coordinates.*** *Let* $\{\boldsymbol{e}_i\}$ *be an arbitrary frame and let* $\boldsymbol{v}(\boldsymbol{x}) = v_i(\boldsymbol{x})\boldsymbol{e}_i$*. Then*

$$\Delta\phi(\boldsymbol{x}) = \phi_{,ii}(\boldsymbol{x}) \quad \text{and} \quad \Delta\boldsymbol{v}(\boldsymbol{x}) = v_{i,jj}(\boldsymbol{x})\boldsymbol{e}_i,$$

where (x_1, x_2, x_3) *are the coordinates of* $\boldsymbol{x}$ *in* $\{\boldsymbol{e}_i\}$*.* □

Proof By Result 2.4 we have $\nabla\phi = \phi_{,i}\boldsymbol{e}_i$ and using Result 2.8 we obtain

$$\Delta\phi = \nabla\cdot(\nabla\phi) = (\phi_{,i})_{,i} = \phi_{,ii}.$$

Similarly, by Result 2.6 we have $\nabla\boldsymbol{v} = v_{i,j}\,\boldsymbol{e}_i\otimes\boldsymbol{e}_j$ and using Result 2.10 we obtain

$$\Delta\boldsymbol{v} = \nabla\cdot(\nabla\boldsymbol{v}) = (v_{i,j})_{,j}\,\boldsymbol{e}_i = v_{i,jj}\,\boldsymbol{e}_i.$$

□

2.3 Integral Theorems

In this section we state some integral theorems that will be useful later when we derive local statements of balance laws for continuum bodies.

2.3.1 Divergence Theorem

Here we state, without proof, a result that provides a relationship between the volume integral of the divergence of a vector field in a region B of $\mathbb{E}^3$, and the surface integral of an associated field over the bounding surface ∂B of B; we then generalize the result to second-order tensor fields. The validity of these results requires certain assumptions on the region B. For our purposes it will be sufficient to only consider those regions that are **regular** in the following sense: (i) B consists of a finite number of open, disjoint and bounded components; (ii) the bounding surface ∂B is piecewise smooth and consists of a finite number of disjoint components; (iii) each component of ∂B is orientable in the sense

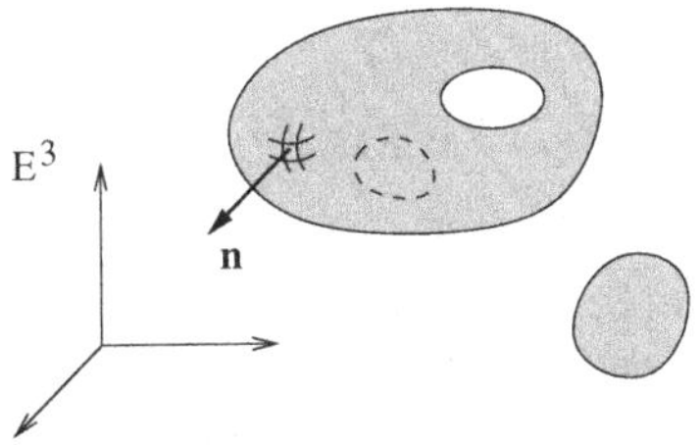

Fig. 2.1 Graphical illustration of a regular region consisting of two disjoint components. One component has an internal void and a hole.

that it clearly has two sides. The notion of a regular region of $\mathbb{E}^3$ encompasses many bodies of interest in mechanics, including those with a finite number of holes and voids as depicted in Figure 2.1.

In the following, dV denotes an infinitesimal volume element in B and dA denotes an infinitesimal area element in ∂B.

Result 2.16 ***Divergence Theorem.*** *Let B denote a regular region in $\mathbb{E}^3$ with piecewise smooth boundary ∂B, and consider an arbitrary vector field $\boldsymbol{v} : B \to \mathcal{V}$. Then*

$$\int_{\partial B} \boldsymbol{v} \cdot \boldsymbol{n} \, dA = \int_B \nabla \cdot \boldsymbol{v} \, dV,$$

or in components

$$\int_{\partial B} v_i n_i \, dA = \int_B v_{i,i} \, dV,$$

where $\boldsymbol{n}$ is the outward unit normal field on ∂B. The quantity appearing on the left-hand side of the above relation is called the **flux** *of $\boldsymbol{v}$ across the oriented surface ∂B.* □

The Divergence Theorem can be used to gain some insight into the physical meaning of the divergence of a vector field. To this end, consider an arbitrary point $\boldsymbol{y} \in \mathbb{E}^3$ and let Ω_δ denote a ball centered at $\boldsymbol{y}$ with radius δ and boundary $\partial\Omega_\delta$. Then by the Divergence Theorem we have

$$\begin{aligned}
\int_{\partial\Omega_\delta} \boldsymbol{v}(\boldsymbol{x}) \cdot \boldsymbol{n}(\boldsymbol{x}) \, dA &= \int_{\Omega_\delta} (\nabla \cdot \boldsymbol{v})(\boldsymbol{x}) \, dV \\
&= \int_{\Omega_\delta} [\,(\nabla \cdot \boldsymbol{v})(\boldsymbol{y}) + \mathcal{O}(\delta)\,] \, dV \\
&= \mathrm{vol}(\Omega_\delta)[\,(\nabla \cdot \boldsymbol{v})(\boldsymbol{y}) + \mathcal{O}(\delta)\,],
\end{aligned}$$

which implies

$$(\nabla \cdot \boldsymbol{v})(\boldsymbol{y}) = \lim_{\delta \to 0} \frac{1}{\mathrm{vol}(\Omega_\delta)} \int_{\partial \Omega_\delta} \boldsymbol{v} \cdot \boldsymbol{n} \, dA.$$

Thus, for small δ, we see that $(\nabla \cdot \boldsymbol{v})(\boldsymbol{y})$ is the flux of $\boldsymbol{v}$ across $\partial\Omega_\delta$ divided by the volume of Ω_δ. If we interpret $\boldsymbol{v}$ as the velocity field in a fluid, then the flux of $\boldsymbol{v}$ across $\partial\Omega_\delta$ is equal to the net rate (volume per unit time) at which fluid is exiting Ω_δ. In this sense $(\nabla \cdot \boldsymbol{v})(\boldsymbol{y})$ provides a measure of volume expansion at $\boldsymbol{y}$.

The following generalization of the Divergence Theorem to second-order tensor fields will be useful.

Result 2.17 ***Tensor Divergence Theorem.*** *Let B denote a regular region in $\boldsymbol{E}^3$ with piecewise smooth boundary ∂B, and consider an arbitrary second-order tensor field $\boldsymbol{S} : B \to \mathcal{V}^2$. Then*

$$\int_{\partial B} \boldsymbol{S}\boldsymbol{n} \, dA = \int_B \nabla \cdot \boldsymbol{S} \, dV,$$

or in components

$$\int_{\partial B} S_{ij} n_j \, dA = \int_B S_{ij,j} \, dV,$$

where $\boldsymbol{n}$ is the outward unit normal field on ∂B. □

Proof Let $\boldsymbol{a}$ be an arbitrary, constant vector. Then

$$\boldsymbol{a} \cdot \int_{\partial B} \boldsymbol{S}\boldsymbol{n} \, dA = \int_{\partial B} \boldsymbol{a} \cdot \boldsymbol{S}\boldsymbol{n} \, dA = \int_{\partial B} (\boldsymbol{S}^T \boldsymbol{a}) \cdot \boldsymbol{n} \, dA.$$

Since $\boldsymbol{S}^T \boldsymbol{a}$ is a vector field we can apply the Divergence Theorem in Result 2.16 to get

$$\int_{\partial B} (\boldsymbol{S}^T \boldsymbol{a}) \cdot \boldsymbol{n} \, dA = \int_B \nabla \cdot (\boldsymbol{S}^T \boldsymbol{a}) \, dV,$$

and by Definition 2.9 we obtain

$$\int_B \nabla \cdot (\boldsymbol{S}^T \boldsymbol{a}) \, dV = \int_B \boldsymbol{a} \cdot (\nabla \cdot \boldsymbol{S}) \, dV = \boldsymbol{a} \cdot \int_B \nabla \cdot \boldsymbol{S} \, dV.$$

Thus from the arbitrariness of $\boldsymbol{a}$ we deduce

$$\int_{\partial B} \boldsymbol{S}\boldsymbol{n} \, dA = \int_B \nabla \cdot \boldsymbol{S} \, dV.$$

□

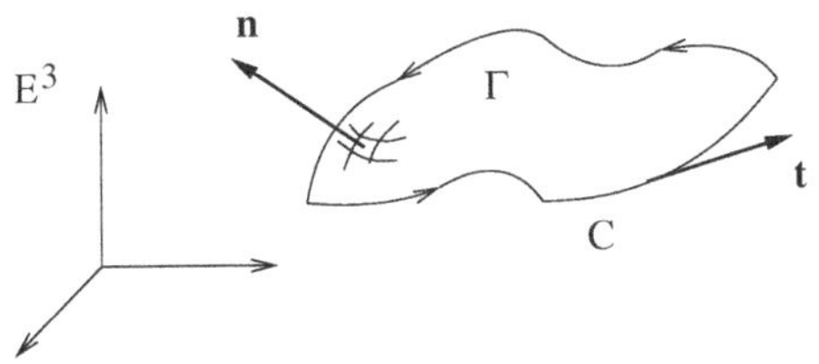

Fig. 2.2 Graphical illustration of a surface Γ with a bounding curve C.

2.3.2 Stokes' Theorem

Here we state, without proof, a result that provides a relationship between the line integral of a vector field along a simple closed curve C and the surface integral of an associated field over a surface Γ whose boundary is $C = \partial\Gamma$ as shown in Figure 2.2. The validity of this result requires certain assumptions on C and Γ. For our purposes it will be sufficient to only consider curves C which do not intersect themselves, and which are bounded and piecewise smooth. Moreover, we assume that the surface Γ is orientable, bounded and piecewise smooth. The orientations for the unit tangent field $\boldsymbol{t}$ on C and the unit normal field $\boldsymbol{n}$ on Γ are assumed to be chosen as follows: a person walking around C in the direction $\boldsymbol{t}$ with their head in the direction $\boldsymbol{n}$ has the surface Γ to their left.

In the following, dA is an infinitesimal area element in Γ and ds is an infinitesimal length element in C.

Result 2.18 ***Stokes' Theorem.*** *Let Γ be a surface in $\boldsymbol{E}^3$ with a piecewise smooth boundary curve $C = \partial\Gamma$, and consider an arbitrary vector field $\boldsymbol{v} : \boldsymbol{E}^3 \to \mathcal{V}$. Then*

$$\int_\Gamma (\nabla \times \boldsymbol{v}) \cdot \boldsymbol{n}\, dA = \int_C \boldsymbol{v} \cdot \boldsymbol{t}\, ds,$$

where $\boldsymbol{n}$ is a unit normal field on Γ and $\boldsymbol{t}$ is a unit tangent field on C oriented as discussed above. The quantity appearing on the right-hand side of the above relation is called the **circulation** *of $\boldsymbol{v}$ around the oriented curve C.* □

Stokes' Theorem can be used to gain some insight into the physical meaning of the curl of a vector field. To this end, consider an arbitrary point $\boldsymbol{y} \in \boldsymbol{E}^3$ and let Γ_δ be a disc centered at $\boldsymbol{y}$ with radius δ and

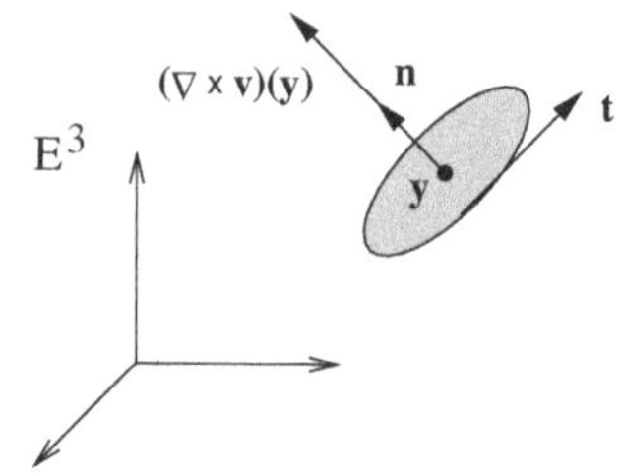

Fig. 2.3 Interpretation of the curl of a vector field in terms of circulation.

boundary $\partial\Gamma_\delta$. Then by Stokes' Theorem we have

$$\begin{aligned}\int_{\partial\Gamma_\delta} \boldsymbol{v}(\boldsymbol{x}) \cdot \boldsymbol{t}(\boldsymbol{x})\, ds &= \int_{\Gamma_\delta} (\nabla \times \boldsymbol{v})(\boldsymbol{x}) \cdot \boldsymbol{n}(\boldsymbol{x})\, dA \\ &= \int_{\Gamma_\delta} [\, (\nabla \times \boldsymbol{v})(\boldsymbol{y}) \cdot \boldsymbol{n}(\boldsymbol{y}) + \mathcal{O}(\delta)\,]\, dA \\ &= \text{area}(\Gamma_\delta)[\, (\nabla \times \boldsymbol{v})(\boldsymbol{y}) \cdot \boldsymbol{n}(\boldsymbol{y}) + \mathcal{O}(\delta)\,],\end{aligned}$$

which implies

$$\boldsymbol{n}(\boldsymbol{y}) \cdot (\nabla \times \boldsymbol{v})(\boldsymbol{y}) = \lim_{\delta\to 0} \frac{1}{\text{area}(\Gamma_\delta)} \int_{\partial\Gamma_\delta} \boldsymbol{v} \cdot \boldsymbol{t}\, ds.$$

At the point $\boldsymbol{y}$ in question suppose $(\nabla \times \boldsymbol{v})(\boldsymbol{y}) \neq \boldsymbol{0}$ and choose the disc Γ_δ such that its normal is given by $\boldsymbol{n}(\boldsymbol{y}) = (\nabla \times \boldsymbol{v})(\boldsymbol{y})/|(\nabla \times \boldsymbol{v})(\boldsymbol{y})|$ as shown in Figure 2.3. For this choice of Γ_δ we obtain

$$|(\nabla \times \boldsymbol{v})(\boldsymbol{y})| = \lim_{\delta\to 0} \frac{1}{\text{area}(\Gamma_\delta)} \int_{\partial\Gamma_\delta} \boldsymbol{v} \cdot \boldsymbol{t}\, ds.$$

Thus, for small δ, we see that $|(\nabla \times \boldsymbol{v})(\boldsymbol{y})|$ is the circulation of $\boldsymbol{v}$ around $\partial\Gamma_\delta$ divided by the area of Γ_δ. Here Γ_δ is a disc of radius δ centered at $\boldsymbol{y}$ with normal $\boldsymbol{n}$ in the direction of $(\nabla \times \boldsymbol{v})(\boldsymbol{y})$. If we interpret $\boldsymbol{v}$ as the velocity field in a fluid, then $(\nabla \times \boldsymbol{v})(\boldsymbol{y})$ can be interpreted as twice the local angular velocity of the fluid at $\boldsymbol{y}$.

2.3.3 Localization Theorem

Here we establish a result which states that, if the integral of a function over arbitrary subsets of its domain is zero, then the function itself must be zero. This result will be used frequently in later chapters.

Result 2.19 Localization Theorem. *Consider an open set B in $\boldsymbol{E}^3$ and a continuous function $\phi : B \to \boldsymbol{R}$, and let Ω denote an arbitrary open subset of B. If*

$$\int_\Omega \phi(\boldsymbol{x})\, dV = 0, \quad \forall \Omega \subseteq B, \tag{2.1}$$

then

$$\phi(\boldsymbol{x}) = 0, \quad \forall \boldsymbol{x} \in B. \tag{2.2}$$

□

Proof The result follows by contradiction. To begin, assume the hypothesis (2.1) is true and assume the statement (2.2) is false; that is, there exists some point $\boldsymbol{y} \in B$ for which $\phi(\boldsymbol{y}) \neq 0$, say $\phi(\boldsymbol{y}) = 2\delta > 0$; the case $\delta < 0$ is dealt with similarly. Then, by continuity of ϕ and the fact that B is open, there exists an open neighborhood Ω of $\boldsymbol{y}$ such that $\Omega \subset B$ and $\phi(\boldsymbol{x}) > \delta$ for all $\boldsymbol{x} \in \Omega$. For this subset we have

$$\int_\Omega \phi(\boldsymbol{x})\, dV > \delta \operatorname{vol}[\Omega] > 0,$$

which contradicts (2.1). Hence the statement (2.2) must be true. □

Remarks:

(1) Result 2.19 also holds when the scalar field ϕ is replaced by a vector field $\boldsymbol{v}$ or second-order tensor field $\boldsymbol{S}$. In particular, the result for scalar fields can be applied to each component of a vector or second-order tensor field.

(2) Result 2.19 can also be generalized to integrals other than volume integrals. For example, if Γ is an open disc and $\int_\Omega \phi(\boldsymbol{x})\, dA = 0$ for all open subsets $\Omega \subseteq \Gamma$, then a similar argument shows that $\phi(\boldsymbol{x}) = 0$ for all $\boldsymbol{x} \in \Gamma$. □

2.4 Functions of Second-Order Tensors

Here we define the derivative of functions of the form $\psi : \mathcal{V}^2 \to \boldsymbol{R}$, such as the determinant and trace functions, and functions of the form $\boldsymbol{\Sigma} : \mathcal{V}^2 \to \mathcal{V}^2$, such as the exponential function and the stress response functions to be considered in Chapters 7 and 9.

2.4.1 Scalar-Valued Functions

Definition 2.20 *A function* $\psi : \mathcal{V}^2 \to \mathbb{R}$ *is said to be* **differentiable** *at* $\boldsymbol{A} \in \mathcal{V}^2$ *if there exists a second-order tensor* $D\psi(\boldsymbol{A}) \in \mathcal{V}^2$ *such that*

$$\psi(\boldsymbol{A}+\boldsymbol{H}) = \psi(\boldsymbol{A}) + D\psi(\boldsymbol{A}) : \boldsymbol{H} + o(|\boldsymbol{H}|),$$

or equivalently

$$D\psi(\boldsymbol{A}) : \boldsymbol{B} = \frac{d}{d\alpha}\psi(\boldsymbol{A}+\alpha\boldsymbol{B})\Big|_{\alpha=0}, \quad \forall \boldsymbol{B} \in \mathcal{V}^2,$$

where $\alpha \in \mathbb{R}$. *The tensor* $D\psi(\boldsymbol{A})$ *is called the* **derivative** *of* ψ *at* $\boldsymbol{A}$. □

If ψ is differentiable at $\boldsymbol{A}$ it can be shown that the tensor $D\psi(\boldsymbol{A})$ is necessarily unique. The second characterization follows from the first upon setting $\boldsymbol{H} = \alpha\boldsymbol{B}$, dividing through by α, and taking the limit $\alpha \to 0$. The following result provides an explicit characterization of the derivative $D\psi(\boldsymbol{A})$ in any coordinate frame.

Result 2.21 ***Derivative of Scalar Function in Components.*** *Let* $\{\boldsymbol{e}_i\}$ *be an arbitrary frame. Then*

$$D\psi(\boldsymbol{A}) = \frac{\partial\psi}{\partial A_{ij}}(\boldsymbol{A})\, \boldsymbol{e}_i \otimes \boldsymbol{e}_j,$$

where $(A_{11}, A_{12}, \ldots, A_{33})$ *are the components of* $\boldsymbol{A}$ *in the frame* $\{\boldsymbol{e}_i\}$. □

Proof Writing ψ as a function of the components A_{ij} we have, by a slight abuse of notation, $\psi(\boldsymbol{A}) = \psi(A_{11}, A_{12}, \ldots, A_{33})$. For any scalar α and tensor $\boldsymbol{B} = B_{kl}\boldsymbol{e}_k \otimes \boldsymbol{e}_l$ this gives

$$\psi(\boldsymbol{A}+\alpha\boldsymbol{B}) = \psi(A_{11}+\alpha B_{11}, \ldots, A_{33}+\alpha B_{33}).$$

By Definition 2.20 and the chain rule we find

$$\begin{aligned} D\psi(\boldsymbol{A}) : \boldsymbol{B} &= \frac{d}{d\alpha}\psi(\boldsymbol{A}+\alpha\boldsymbol{B})\Big|_{\alpha=0} \\ &= \frac{\partial\psi}{\partial A_{11}}(\boldsymbol{A})B_{11} + \cdots + \frac{\partial\psi}{\partial A_{33}}(\boldsymbol{A})B_{33} \\ &= \left(\frac{\partial\psi}{\partial A_{ij}}(\boldsymbol{A})\boldsymbol{e}_i \otimes \boldsymbol{e}_j\right) : B_{kl}\boldsymbol{e}_k \otimes \boldsymbol{e}_l, \end{aligned}$$

which implies $D\psi(\boldsymbol{A}) = \frac{\partial\psi}{\partial A_{ij}}(\boldsymbol{A})\, \boldsymbol{e}_i \otimes \boldsymbol{e}_j$. □

Remarks:

(1) The derivative of a scalar-valued function $\psi : \mathcal{V}^2 \to I\!R$ is a second-order tensor-valued function $D\psi : \mathcal{V}^2 \to \mathcal{V}^2$.

(2) Since all partial derivatives are assumed to exist and be continuous, we can use Taylor's Theorem to deduce, for any small tensor $\boldsymbol{H} \in \mathcal{V}^2$

$$\psi(\boldsymbol{A}+\boldsymbol{H}) = \psi(\boldsymbol{A}) + \frac{\partial \psi}{\partial A_{ij}}(\boldsymbol{A})H_{ij} + \mathcal{O}(|\boldsymbol{H}|^2),$$

or equivalently

$$\psi(\boldsymbol{A}+\boldsymbol{H}) = \psi(\boldsymbol{A}) + D\psi(\boldsymbol{A}) : \boldsymbol{H} + \mathcal{O}(|\boldsymbol{H}|^2).$$

Thus, in the smooth case, the difference between $\psi(\boldsymbol{A}+\boldsymbol{H})$ and $\psi(\boldsymbol{A})+D\psi(\boldsymbol{A}) : \boldsymbol{H}$ is order $\mathcal{O}(|\boldsymbol{H}|^2)$, not just $o(|\boldsymbol{H}|)$ as indicated in the definition of $D\psi(\boldsymbol{A})$. □

Example: Consider the function $\psi : \mathcal{V}^2 \to I\!R$ defined by

$$\psi(\boldsymbol{A}) = \tfrac{1}{2}\boldsymbol{A} : \boldsymbol{A} = \tfrac{1}{2}A_{kl}A_{kl}.$$

To compute $D\psi(\boldsymbol{A})$ we first take the partial derivative of ψ with respect to a general component A_{ij} to obtain

$$\frac{\partial \psi}{\partial A_{ij}}(\boldsymbol{A}) = \tfrac{1}{2}\frac{\partial A_{kl}}{\partial A_{ij}}A_{kl} + \tfrac{1}{2}A_{kl}\frac{\partial A_{kl}}{\partial A_{ij}} = \frac{\partial A_{kl}}{\partial A_{ij}}A_{kl}.$$

Since $\partial A_{kl}/\partial A_{ij} = \delta_{ki}\delta_{lj}$ we find

$$\frac{\partial \psi}{\partial A_{ij}}(\boldsymbol{A}) = \delta_{ki}\delta_{lj}A_{kl} = A_{ij},$$

from which we deduce $D\psi(\boldsymbol{A}) = A_{ij}\,\boldsymbol{e}_i \otimes \boldsymbol{e}_j = \boldsymbol{A}$. □

Result 2.22 Derivative of Determinant. *Let $\psi(\boldsymbol{S}) = \det \boldsymbol{S}$. If $\boldsymbol{A}$ is invertible, then the derivative of ψ at $\boldsymbol{A}$ is $D\psi(\boldsymbol{A}) = \det(\boldsymbol{A})\boldsymbol{A}^{-T}$.* □

Proof Let $\boldsymbol{B} \in \mathcal{V}^2$ be arbitrary. Then

$$\begin{aligned}\psi(\boldsymbol{A}+\alpha\boldsymbol{B}) &= \det(\boldsymbol{A}+\alpha\boldsymbol{B}) \\ &= \det\left(\alpha\boldsymbol{A}(\boldsymbol{A}^{-1}\boldsymbol{B} - \lambda\boldsymbol{I})\right) \\ &= \det(\alpha\boldsymbol{A})\det(\boldsymbol{A}^{-1}\boldsymbol{B} - \lambda\boldsymbol{I}),\end{aligned}$$

where $\lambda = -1/\alpha$. By definition of the determinant of $\alpha \boldsymbol{A} \in \mathcal{V}^2$ we have

$$\det(\alpha \boldsymbol{A}) = \alpha^3 \det \boldsymbol{A},$$

and by definition of the principal invariants of the second-order tensor $\boldsymbol{A}^{-1}\boldsymbol{B}$ we find

$$\begin{aligned}\det(\boldsymbol{A}^{-1}\boldsymbol{B} - \lambda \boldsymbol{I}) &= -\lambda^3 + \lambda^2 I_1(\boldsymbol{A}^{-1}\boldsymbol{B}) - \lambda I_2(\boldsymbol{A}^{-1}\boldsymbol{B}) + I_3(\boldsymbol{A}^{-1}\boldsymbol{B}) \\ &= \frac{1}{\alpha^3} + \frac{1}{\alpha^2} I_1(\boldsymbol{A}^{-1}\boldsymbol{B}) + \frac{1}{\alpha} I_2(\boldsymbol{A}^{-1}\boldsymbol{B}) + I_3(\boldsymbol{A}^{-1}\boldsymbol{B}),\end{aligned}$$

which implies

$$\begin{aligned}\psi(\boldsymbol{A} + \alpha \boldsymbol{B}) = \det(\boldsymbol{A}) &+ \alpha \det(\boldsymbol{A}) I_1(\boldsymbol{A}^{-1}\boldsymbol{B}) \\ &+ \alpha^2 \det(\boldsymbol{A}) I_2(\boldsymbol{A}^{-1}\boldsymbol{B}) + \alpha^3 \det(\boldsymbol{A}) I_3(\boldsymbol{A}^{-1}\boldsymbol{B}).\end{aligned}$$

From this expression and Definition 2.20 we deduce

$$D\psi(\boldsymbol{A}) : \boldsymbol{B} = \det(\boldsymbol{A}) I_1(\boldsymbol{A}^{-1}\boldsymbol{B}),$$

and since $I_1(\boldsymbol{A}^{-1}\boldsymbol{B}) = \mathrm{tr}(\boldsymbol{A}^{-1}\boldsymbol{B}) = \boldsymbol{A}^{-T} : \boldsymbol{B}$ we find

$$D\psi(\boldsymbol{A}) : \boldsymbol{B} = \det(\boldsymbol{A}) \boldsymbol{A}^{-T} : \boldsymbol{B}.$$

The desired result $D\psi(\boldsymbol{A}) = \det(\boldsymbol{A})\boldsymbol{A}^{-T}$ follows from the arbitrariness of $\boldsymbol{B}$. □

Result 2.23 ***Time Derivative of Determinant.*** *Let* $\boldsymbol{S}$ *be a time-dependent second-order tensor, that is,* $\boldsymbol{S} : \mathbb{R} \to \mathcal{V}^2$*. Then*

$$\frac{d}{dt}(\det \boldsymbol{S}) = (\det \boldsymbol{S})\, \mathrm{tr}(\boldsymbol{S}^{-1}\dot{\boldsymbol{S}}) = \det(\boldsymbol{S})\boldsymbol{S}^{-T} : \dot{\boldsymbol{S}},$$

where

$$\dot{\boldsymbol{S}} = \frac{d\boldsymbol{S}}{dt} = \frac{dS_{ij}}{dt} \boldsymbol{e}_i \otimes \boldsymbol{e}_j.$$

□

Proof Let $\psi(\boldsymbol{S}) = \det \boldsymbol{S}$. Then

$$\frac{d}{dt}\psi(\boldsymbol{S}) = \frac{d}{dt}\psi(S_{11}, S_{12}, \ldots, S_{33}),$$

and by the chain rule we have

$$\frac{d}{dt}\psi(\boldsymbol{S}) = \frac{\partial \psi}{\partial S_{11}}(\boldsymbol{S})\frac{dS_{11}}{dt} + \cdots + \frac{\partial \psi}{\partial S_{33}}(\boldsymbol{S})\frac{dS_{33}}{dt} = D\psi(\boldsymbol{S}) : \dot{\boldsymbol{S}}.$$

Since $D\psi(\boldsymbol{S}) = \det(\boldsymbol{S})\boldsymbol{S}^{-T}$ we obtain

$$\frac{d}{dt}\psi(\boldsymbol{S}) = \det(\boldsymbol{S})\boldsymbol{S}^{-T} : \dot{\boldsymbol{S}} = \det(\boldsymbol{S})\,\mathrm{tr}(\boldsymbol{S}^{-1}\dot{\boldsymbol{S}}).$$

□

2.4.2 Tensor-Valued Functions

Definition 2.24 *A function $\boldsymbol{\Sigma} : \mathcal{V}^2 \to \mathcal{V}^2$ is said to be* **differentiable** *at $\boldsymbol{A} \in \mathcal{V}^2$ if there exists a fourth-order tensor $D\boldsymbol{\Sigma}(\boldsymbol{A}) \in \mathcal{V}^4$ such that*

$$\boldsymbol{\Sigma}(\boldsymbol{A}+\boldsymbol{H}) = \boldsymbol{\Sigma}(\boldsymbol{A}) + D\boldsymbol{\Sigma}(\boldsymbol{A})\boldsymbol{H} + o(|\boldsymbol{H}|),$$

or equivalently

$$D\boldsymbol{\Sigma}(\boldsymbol{A})\boldsymbol{B} = \frac{d}{d\alpha}\boldsymbol{\Sigma}(\boldsymbol{A}+\alpha\boldsymbol{B})\Big|_{\alpha=0}, \quad \forall \boldsymbol{B} \in \mathcal{V}^2,$$

where $\alpha \in \mathbb{R}$. The tensor $D\boldsymbol{\Sigma}(\boldsymbol{A})$ is called the **derivative** *of $\boldsymbol{\Sigma}$ at $\boldsymbol{A}$.*

□

If $\boldsymbol{\Sigma}$ is differentiable at $\boldsymbol{A}$ it can be shown that the tensor $D\boldsymbol{\Sigma}(\boldsymbol{A})$ is necessarily unique. The second characterization follows from the first upon setting $\boldsymbol{H} = \alpha\boldsymbol{B}$, dividing through by α, and taking the limit $\alpha \to 0$. The following result provides an explicit characterization of the derivative $D\boldsymbol{\Sigma}(\boldsymbol{A})$ in any coordinate frame.

Result 2.25 **Derivative of Tensor Function in Components.** *Let $\{\boldsymbol{e}_i\}$ be an arbitrary frame and let $\boldsymbol{\Sigma}(\boldsymbol{A}) = \Sigma_{ij}(\boldsymbol{A})\,\boldsymbol{e}_i \otimes \boldsymbol{e}_j$. Then*

$$D\boldsymbol{\Sigma}(\boldsymbol{A}) = \frac{\partial \Sigma_{ij}}{\partial A_{kl}}(\boldsymbol{A})\,\boldsymbol{e}_i \otimes \boldsymbol{e}_j \otimes \boldsymbol{e}_k \otimes \boldsymbol{e}_l,$$

where $(A_{11}, A_{12}, \ldots, A_{33})$ are the components of $\boldsymbol{A}$ in the frame $\{\boldsymbol{e}_i\}$. □

Proof Writing the components Σ_{ij} as functions of the components A_{kl} we have, by a slight abuse of notation, $\Sigma_{ij}(\boldsymbol{A}) = \Sigma_{ij}(A_{11}, A_{12}, \ldots, A_{33})$. For any scalar α and tensor $\boldsymbol{B} = B_{kl}\boldsymbol{e}_k \otimes \boldsymbol{e}_l$ this gives

$$\Sigma_{ij}(\boldsymbol{A}+\alpha\boldsymbol{B}) = \Sigma_{ij}(A_{11}+\alpha B_{11}, \ldots, A_{33}+\alpha B_{33}),$$

and by the chain rule we find

$$\begin{aligned}\frac{d}{d\alpha}\Sigma_{ij}(\boldsymbol{A}+\alpha\boldsymbol{B})\Big|_{\alpha=0} &= \frac{\partial\Sigma_{ij}}{\partial A_{11}}(\boldsymbol{A})B_{11}+\cdots+\frac{\partial\Sigma_{ij}}{\partial A_{33}}(\boldsymbol{A})B_{33}\\ &= \frac{\partial\Sigma_{ij}}{\partial A_{kl}}(\boldsymbol{A})B_{kl}.\end{aligned}$$

Using this result, together with Definition 2.24 and properties of dyadic products (see Chapter 1), we obtain

$$\begin{aligned}D\boldsymbol{\Sigma}(\boldsymbol{A})\boldsymbol{B} &= \frac{d}{d\alpha}\boldsymbol{\Sigma}(\boldsymbol{A}+\alpha\boldsymbol{B})\Big|_{\alpha=0}\\ &= \frac{d}{d\alpha}\Sigma_{ij}(\boldsymbol{A}+\alpha\boldsymbol{B})\Big|_{\alpha=0}\boldsymbol{e}_i\otimes\boldsymbol{e}_j\\ &= \frac{\partial\Sigma_{ij}}{\partial A_{kl}}(\boldsymbol{A})B_{kl}\,\boldsymbol{e}_i\otimes\boldsymbol{e}_j\\ &= \left(\frac{\partial\Sigma_{ij}}{\partial A_{kl}}(\boldsymbol{A})\,\boldsymbol{e}_i\otimes\boldsymbol{e}_j\otimes\boldsymbol{e}_k\otimes\boldsymbol{e}_l\right)\boldsymbol{B},\end{aligned}$$

which implies $D\boldsymbol{\Sigma}(\boldsymbol{A})=\frac{\partial\Sigma_{ij}}{\partial A_{kl}}(\boldsymbol{A})\,\boldsymbol{e}_i\otimes\boldsymbol{e}_j\otimes\boldsymbol{e}_k\otimes\boldsymbol{e}_l$. □

Remarks:

(1) The derivative of a second-order tensor-valued function $\boldsymbol{\Sigma}:\mathcal{V}^2\to\mathcal{V}^2$ is a fourth-order tensor-valued function $D\boldsymbol{\Sigma}:\mathcal{V}^2\to\mathcal{V}^4$.

(2) Since all partial derivatives are assumed to exist and be continuous, we can use Taylor's Theorem to deduce, for any small tensor $\boldsymbol{H}\in\mathcal{V}^2$

$$\Sigma_{ij}(\boldsymbol{A}+\boldsymbol{H})=\Sigma_{ij}(\boldsymbol{A})+\frac{\partial\Sigma_{ij}}{\partial A_{kl}}(\boldsymbol{A})H_{kl}+\mathcal{O}(|\boldsymbol{H}|^2),$$

or in tensor notation

$$\boldsymbol{\Sigma}(\boldsymbol{A}+\boldsymbol{H})=\boldsymbol{\Sigma}(\boldsymbol{A})+D\boldsymbol{\Sigma}(\boldsymbol{A})\boldsymbol{H}+\mathcal{O}(|\boldsymbol{H}|^2).$$

Thus, in the smooth case, the difference between $\boldsymbol{\Sigma}(\boldsymbol{A}+\boldsymbol{H})$ and $\boldsymbol{\Sigma}(\boldsymbol{A})+D\boldsymbol{\Sigma}(\boldsymbol{A})\boldsymbol{H}$ is order $\mathcal{O}(|\boldsymbol{H}|^2)$, not just $o(|\boldsymbol{H}|)$ as indicated in the definition of $D\boldsymbol{\Sigma}(\boldsymbol{A})$. □

Example: Consider the function $\boldsymbol{\Sigma}:\mathcal{V}^2\to\mathcal{V}^2$ defined by

$$\boldsymbol{\Sigma}(\boldsymbol{A})=\mathrm{tr}(\boldsymbol{A})\boldsymbol{A}\quad\text{or in components}\quad\Sigma_{ij}(\boldsymbol{A})=A_{mm}A_{ij}.$$

The components $[D\boldsymbol{\Sigma}(\boldsymbol{A})]_{ijkl}$ of the derivative $D\boldsymbol{\Sigma}(\boldsymbol{A})$ are

$$\begin{aligned}[D\boldsymbol{\Sigma}(\boldsymbol{A})]_{ijkl} &= \frac{\partial \Sigma_{ij}}{\partial A_{kl}}(\boldsymbol{A}) \\ &= \frac{\partial A_{mm}}{\partial A_{kl}} A_{ij} + A_{mm} \frac{\partial A_{ij}}{\partial A_{kl}} \\ &= \delta_{mk}\delta_{ml}A_{ij} + A_{mm}\delta_{ik}\delta_{jl} \\ &= \delta_{kl}A_{ij} + A_{mm}\delta_{ik}\delta_{jl}.\end{aligned}$$

For any $\boldsymbol{B} \in \mathcal{V}^2$ we obtain

$$\begin{aligned}D\boldsymbol{\Sigma}(\boldsymbol{A})\boldsymbol{B} &= [D\boldsymbol{\Sigma}(\boldsymbol{A})]_{ijkl}B_{kl}\, \boldsymbol{e}_i \otimes \boldsymbol{e}_j \\ &= (\delta_{kl}A_{ij}B_{kl} + A_{mm}\delta_{ik}\delta_{jl}B_{kl})\, \boldsymbol{e}_i \otimes \boldsymbol{e}_j \\ &= (A_{ij}B_{kk} + A_{mm}B_{ij})\, \boldsymbol{e}_i \otimes \boldsymbol{e}_j \\ &= \mathrm{tr}(\boldsymbol{B})\boldsymbol{A} + \mathrm{tr}(\boldsymbol{A})\boldsymbol{B}.\end{aligned}$$

□

Bibliographic Notes

Much of the material on vector and tensor analysis in Cartesian coordinates can be found in standard texts on advanced multivariable calculus, for example Bourne and Kendall (1992) and Marsden and Tromba (1988). The book by Chadwick (1976) also contains a concise and readable treatment of this subject, oriented towards applications in continuum mechanics. For classic treatments in general curvilinear coordinates the reader is referred to Brand (1947), Sokolnikoff (1951) and Brillouin (1964). More modern treatments with applications can be found in Ogden (1984) and Simmonds (1994).

Proofs of the Divergence and Stokes' Theorems in general dimensions can be found in Munkres (1991). For an extensive analysis of the types of regions and surfaces for which these theorems are valid the reader is referred to Kellogg (1967).

Exercises

2.1 Consider the scalar field $\phi(\boldsymbol{x}) = (x_1)^2 x_3 + x_2(x_3)^2$ and the vector field $\boldsymbol{v}(\boldsymbol{x}) = x_3\boldsymbol{e}_1 + x_2\sin(x_1)\boldsymbol{e}_3$. Find the components of $\nabla\phi(\boldsymbol{x})$ and $\nabla\boldsymbol{v}(\boldsymbol{x})$.

2.2 Consider the vector and second-order tensor fields

$$\boldsymbol{v}(\boldsymbol{x}) = x_1\boldsymbol{e}_1 + x_2x_1\boldsymbol{e}_2 + x_3x_1\boldsymbol{e}_3,$$
$$\boldsymbol{S}(\boldsymbol{x}) = x_2\boldsymbol{e}_1 \otimes \boldsymbol{e}_2 + x_1x_3\boldsymbol{e}_3 \otimes \boldsymbol{e}_3.$$

Find: (a) $\nabla \cdot \boldsymbol{v}$, (b) $\nabla \times \boldsymbol{v}$, (c) $\nabla \cdot \boldsymbol{S}$.

2.3 Use index notation to prove the following, where $\boldsymbol{A}$ is a constant second-order tensor:

(a) $\nabla\boldsymbol{x} = \boldsymbol{I}$,

(b) $\nabla \cdot \boldsymbol{x} = 3$,

(c) $\nabla(\boldsymbol{x} \cdot \boldsymbol{A}\boldsymbol{x}) = (\boldsymbol{A} + \boldsymbol{A}^T)\boldsymbol{x}$.

2.4 Let ϕ be a scalar field, $\boldsymbol{v}$ and $\boldsymbol{w}$ vector fields, and $\boldsymbol{S}$ a second-order tensor field. Use index notation to prove the following:

(a) $\nabla \cdot (\phi\boldsymbol{v}) = (\nabla\phi) \cdot \boldsymbol{v} + \phi(\nabla \cdot \boldsymbol{v})$,

(b) $\nabla(\boldsymbol{v} \cdot \boldsymbol{w}) = (\nabla\boldsymbol{v})^T\boldsymbol{w} + (\nabla\boldsymbol{w})^T\boldsymbol{v}$,

(c) $\nabla \cdot (\boldsymbol{v} \otimes \boldsymbol{w}) = (\nabla\boldsymbol{v})\boldsymbol{w} + (\nabla \cdot \boldsymbol{w})\boldsymbol{v}$.

2.5 Consider the scalar field $\phi(\boldsymbol{x}) = 1/|\boldsymbol{x}|$, $\boldsymbol{x} \neq \boldsymbol{0}$, and the vector field $\boldsymbol{v}(\boldsymbol{x}) = \phi(\boldsymbol{x})\boldsymbol{n}$, where $\boldsymbol{n}$ is a constant vector. Show that:

(a) $\nabla\phi(\boldsymbol{x}) = -\boldsymbol{x}/|\boldsymbol{x}|^3$ for all $\boldsymbol{x} \neq \boldsymbol{0}$,

(b) $\Delta\phi(\boldsymbol{x}) = 0$ for all $\boldsymbol{x} \neq \boldsymbol{0}$,

(c) $\Delta\boldsymbol{v}(\boldsymbol{x}) = \boldsymbol{0}$ for all $\boldsymbol{x} \neq \boldsymbol{0}$.

Remark: Scalar and vector fields satisfying $\Delta\phi = 0$ and $\Delta\boldsymbol{v} = \boldsymbol{0}$ in a region B are said to be **harmonic** in B. In this sense the above fields are harmonic in any region which excludes the origin.

2.6 Consider a vector field $\boldsymbol{v} : \mathbb{E}^3 \to \mathcal{V}$. Show that:

(a) $\nabla \cdot (\nabla\boldsymbol{v}^T) = \nabla(\nabla \cdot \boldsymbol{v})$,

(b) $\nabla \times \boldsymbol{v} = \boldsymbol{0}$ if and only if $\nabla\boldsymbol{v} = \nabla\boldsymbol{v}^T$,

(c) if $\nabla \cdot \boldsymbol{v} = 0$ and $\nabla \times \boldsymbol{v} = \boldsymbol{0}$, then $\boldsymbol{v}$ is harmonic.

2.7 Let $\boldsymbol{v}$ and $\boldsymbol{w}$ be vector fields and ϕ a scalar field. Prove the following identities:

(a) $\nabla \cdot (\nabla \times \boldsymbol{v}) = 0$,

(b) $\nabla \cdot (\boldsymbol{v} \times \boldsymbol{w}) = \boldsymbol{w} \cdot (\nabla \times \boldsymbol{v}) - \boldsymbol{v} \cdot (\nabla \times \boldsymbol{w})$,

(c) $\nabla \times (\nabla \times \boldsymbol{v}) = \nabla(\nabla \cdot \boldsymbol{v}) - \Delta \boldsymbol{v}$,

(d) $\nabla \times (\nabla \phi) = \boldsymbol{0}$.

2.8 Let B be a region in $\mathbb{E}^3$ with boundary ∂B, let $\boldsymbol{n}$ be the outward unit normal field on ∂B, and let $\boldsymbol{v}$ be a vector field in B. Use the divergence theorem for second-order tensors to show

$$\int_B \nabla \boldsymbol{v} \, dV = \int_{\partial B} \boldsymbol{v} \otimes \boldsymbol{n} \, dA.$$

Hint: Two second-order tensors $\boldsymbol{A}$ and $\boldsymbol{B}$ are equal if and only if $\boldsymbol{A}\boldsymbol{a} = \boldsymbol{B}\boldsymbol{a}$ for all vectors $\boldsymbol{a}$.

2.9 Let $\boldsymbol{S}$ be an arbitrary second-order tensor field and consider the vector field $\boldsymbol{x}$. Show that

$$\nabla \cdot (\boldsymbol{S}\boldsymbol{x}) = (\nabla \cdot \boldsymbol{S}^T) \cdot \boldsymbol{x} + \mathrm{tr}(\boldsymbol{S}).$$

2.10 Let B be a region in $\mathbb{E}^3$ with boundary ∂B, and let $\boldsymbol{n}$ be the unit outward normal field on ∂B. Let $\boldsymbol{v}$ and $\boldsymbol{w}$ be vector fields and $\boldsymbol{S}$ a second-order tensor field in B. Use the divergence theorems to show:

(a) $\int_{\partial B} (\boldsymbol{S}\boldsymbol{n}) \otimes \boldsymbol{v} \, dA = \int_B (\nabla \cdot \boldsymbol{S}) \otimes \boldsymbol{v} + \boldsymbol{S} \nabla \boldsymbol{v}^T \, dV$,

(b) $\int_{\partial B} \boldsymbol{v} \cdot (\boldsymbol{S}\boldsymbol{n}) \, dA = \int_B (\nabla \cdot \boldsymbol{S}) \cdot \boldsymbol{v} + \boldsymbol{S} : \nabla \boldsymbol{v} \, dV$,

(c) $\int_{\partial B} \boldsymbol{v}(\boldsymbol{w} \cdot \boldsymbol{n}) \, dA = \int_B (\nabla \cdot \boldsymbol{w})\boldsymbol{v} + (\nabla \boldsymbol{v})\boldsymbol{w} \, dV$.

2.11 Let ϕ, $\boldsymbol{v}$ and $\boldsymbol{S}$ be scalar, vector and second-order tensor fields, respectively. Prove the following identities:

(a) $\nabla \cdot (\phi \boldsymbol{S}\boldsymbol{v}) = \phi(\nabla \cdot \boldsymbol{S}^T) \cdot \boldsymbol{v} + \nabla\phi \cdot (\boldsymbol{S}\boldsymbol{v}) + \phi \boldsymbol{S} : (\nabla \boldsymbol{v})^T$,

(b) $\nabla \cdot (\boldsymbol{S}\boldsymbol{v}) = \boldsymbol{S}^T : \nabla \boldsymbol{v} + \boldsymbol{v} \cdot (\nabla \cdot \boldsymbol{S}^T)$.

2.12 Let $\boldsymbol{v}$ and $\boldsymbol{w}$ be time-dependent vector fields and consider the derivative operation defined by

$$\frac{D_{\boldsymbol{v}}\boldsymbol{w}}{Dt} = \frac{\partial \boldsymbol{w}}{\partial t} + (\nabla \cdot \boldsymbol{w})\boldsymbol{v} + \nabla \times (\boldsymbol{w} \times \boldsymbol{v}).$$

(a) Show that

$$\frac{D_{\boldsymbol{v}}\boldsymbol{w}}{Dt} = \frac{\partial \boldsymbol{w}}{\partial t} + (\nabla \boldsymbol{w})\boldsymbol{v} - (\nabla \boldsymbol{v})\boldsymbol{w} + (\nabla \cdot \boldsymbol{v})\boldsymbol{w}.$$

(b) For any vector $\boldsymbol{b}$, show that

$$\boldsymbol{b}\cdot\frac{D_{\boldsymbol{v}}\boldsymbol{w}}{Dt} = \boldsymbol{b}\cdot\left[\frac{\partial\boldsymbol{w}}{\partial t} + (\nabla\boldsymbol{w})\boldsymbol{v}\right] + \left[(\boldsymbol{b}\cdot\boldsymbol{w})\boldsymbol{I} - \boldsymbol{b}\otimes\boldsymbol{w}\right] : \nabla\boldsymbol{v}.$$

2.13 Let $\{\boldsymbol{e}_i\}$ and $\{\widehat{\boldsymbol{e}}_i(t)\}$ be two coordinate frames for $\mathbb{E}^3$, where the first is fixed (time-independent) and the second is moving (time-dependent). Moreover, let $\boldsymbol{Q}(t)$ be the change of basis tensor from $\{\boldsymbol{e}_i\}$ to $\{\widehat{\boldsymbol{e}}_i(t)\}$ so that (omitting arguments t for brevity)

$$\widehat{\boldsymbol{e}}_i = \boldsymbol{Q}\boldsymbol{e}_i.$$

(a) Show that the time derivative $d\widehat{\boldsymbol{e}}_i/dt$, as measured by an observer in the fixed frame, may be expressed as

$$\frac{d\widehat{\boldsymbol{e}}_i}{dt} = \boldsymbol{\Omega}\widehat{\boldsymbol{e}}_i = \boldsymbol{\omega}\times\widehat{\boldsymbol{e}}_i,$$

where $\boldsymbol{\Omega}(t)$ is the skew-symmetric tensor defined by

$$\frac{d\boldsymbol{Q}}{dt} = \boldsymbol{\Omega}\boldsymbol{Q} \qquad \text{or equivalently} \qquad \boldsymbol{\Omega} = \frac{d\boldsymbol{Q}}{dt}\boldsymbol{Q}^T,$$

and $\boldsymbol{\omega}(t)$ is the axial vector of $\boldsymbol{\Omega}(t)$.

(b) For any vector $\boldsymbol{v} = \widehat{v}_i\widehat{\boldsymbol{e}}_i = v_i\boldsymbol{e}_i$, show that

$$\frac{d\widehat{v}_i}{dt} = Q_{ji}\left[\frac{dv_j}{dt} - \Omega_{jk}v_k\right], \tag{2.3}$$

where Q_{ij} and Ω_{jk} are the components of $\boldsymbol{Q}$ and $\boldsymbol{\Omega}$ in the fixed frame.

Remark: An observer in the fixed frame $\{\boldsymbol{e}_i\}$ sees the time derivative of $\boldsymbol{v}$ as

$$\dot{\boldsymbol{v}} = \frac{dv_i}{dt}\boldsymbol{e}_i,$$

while an observer moving with the frame $\{\widehat{\boldsymbol{e}}_i(t)\}$ sees the time derivative of $\boldsymbol{v}$ as

$$\overset{\circ}{\boldsymbol{v}} = \frac{d\widehat{v}_j}{dt}\widehat{\boldsymbol{e}}_j.$$

The components of $\dot{\boldsymbol{v}}$ and $\overset{\circ}{\boldsymbol{v}}$ in the fixed frame $\{\boldsymbol{e}_i\}$ are

$$\dot{v}_i = \dot{\boldsymbol{v}}\cdot\boldsymbol{e}_i = \frac{dv_i}{dt},$$

$$\overset{\circ}{v}_i = \overset{\circ}{\boldsymbol{v}}\cdot\boldsymbol{e}_i = \frac{d\widehat{v}_j}{dt}\widehat{\boldsymbol{e}}_j\cdot\boldsymbol{e}_i = Q_{ij}\frac{d\widehat{v}_j}{dt}.$$

In view of (2.3) the vector $\overset{\circ}{\boldsymbol{v}}$ can be expressed in the fixed frame as

$$\overset{\circ}{\boldsymbol{v}} = \overset{\circ}{v}_i \boldsymbol{e}_i = \left[\frac{dv_i}{dt} - \Omega_{ik} v_k \right] \boldsymbol{e}_i.$$

The vector $\overset{\circ}{\boldsymbol{v}}$ is called the **co-rotational** or **Jaumann derivative** of $\boldsymbol{v}$ with respect to $\boldsymbol{\Omega}$. It represents the rate of change of $\boldsymbol{v}$ with respect to time as seen by an observer in a moving frame. In tensor notation we have

$$\overset{\circ}{\boldsymbol{v}} = \dot{\boldsymbol{v}} - \boldsymbol{\Omega} \boldsymbol{v}.$$

2.14 Let $\{\boldsymbol{e}_i\}$, $\{\widehat{\boldsymbol{e}}_i(t)\}$, $\boldsymbol{Q}(t)$ and $\boldsymbol{\Omega}(t)$ be as in Exercise 13, and consider a second-order tensor $\boldsymbol{S} = S_{ij} \boldsymbol{e}_i \otimes \boldsymbol{e}_j = \widehat{S}_{ij} \widehat{\boldsymbol{e}}_i \otimes \widehat{\boldsymbol{e}}_j$. Show that

$$\frac{d\widehat{S}_{ij}}{dt} = Q_{ki} Q_{lj} \left[\frac{dS_{kl}}{dt} - \Omega_{km} S_{ml} - \Omega_{lm} S_{km} \right],$$

where Q_{ij} and Ω_{jk} are the components of $\boldsymbol{Q}$ and $\boldsymbol{\Omega}$ in the fixed frame $\{\boldsymbol{e}_i\}$.

Remark: The tensor $\overset{\circ}{\boldsymbol{S}}$ given by

$$\overset{\circ}{\boldsymbol{S}} = \left[\frac{dS_{kl}}{dt} - \Omega_{km} S_{ml} - \Omega_{lm} S_{km} \right] \boldsymbol{e}_k \otimes \boldsymbol{e}_l$$

is called the **co-rotational** or **Jaumann derivative** of $\boldsymbol{S}$ with respect to $\boldsymbol{\Omega}$. It represents the rate of change of $\boldsymbol{S}$ as seen by an observer in a moving frame. In tensor notation, noting that $\boldsymbol{\Omega}^T = -\boldsymbol{\Omega}$, we have

$$\overset{\circ}{\boldsymbol{S}} = \dot{\boldsymbol{S}} - \boldsymbol{\Omega} \boldsymbol{S} + \boldsymbol{S} \boldsymbol{\Omega}.$$

2.15 Consider the scalar-valued function $\psi(\boldsymbol{A}) = \mathrm{tr}(\boldsymbol{A})\, \boldsymbol{A} : \boldsymbol{B}$, where $\boldsymbol{B}$ is a constant second-order tensor. Show that

$$D\psi(\boldsymbol{A}) = (\boldsymbol{A} : \boldsymbol{B}) \boldsymbol{I} + \mathrm{tr}(\boldsymbol{A}) \boldsymbol{B}.$$

2.16 Consider the tensor-valued function $\boldsymbol{\Sigma}(\boldsymbol{A}) = \boldsymbol{A}^2$. Show that

$$D\boldsymbol{\Sigma}(\boldsymbol{A})\boldsymbol{B} = \boldsymbol{B}\boldsymbol{A} + \boldsymbol{A}\boldsymbol{B}, \quad \forall \boldsymbol{B} \in \mathcal{V}^2.$$

2.17 Let $\mathbf{C} : \mathcal{V}^2 \to \mathcal{V}^2$ be a fourth-order tensor defined by $\mathbf{C}(\boldsymbol{A}) = \mathrm{tr}(\boldsymbol{A})\boldsymbol{I} + \boldsymbol{A}$, and let $\psi : \mathcal{V}^2 \to I\!R$ be a function defined by $\psi(\boldsymbol{A}) = \frac{1}{2}\boldsymbol{A} : \mathbf{C}(\boldsymbol{A})$. Show that:

(a) $\boldsymbol{A} : \mathbf{C}(\boldsymbol{B}) = \mathbf{C}(\boldsymbol{A}) : \boldsymbol{B}, \quad \forall \boldsymbol{A}, \boldsymbol{B} \in \mathcal{V}^2$,

(b) $D\psi(\boldsymbol{A}) = \mathbf{C}(\boldsymbol{A})$.

Answers to Selected Exercises

2.1

$$[\nabla\phi] = \begin{Bmatrix} \phi_{,1} \\ \phi_{,2} \\ \phi_{,3} \end{Bmatrix} = \begin{Bmatrix} 2x_1x_3 \\ (x_3)^2 \\ (x_1)^2 + 2x_2x_3 \end{Bmatrix}.$$

$$[\nabla\boldsymbol{v}] = \begin{pmatrix} v_{1,1} & v_{1,2} & v_{1,3} \\ v_{2,1} & v_{2,2} & v_{2,3} \\ v_{3,1} & v_{3,2} & v_{3,3} \end{pmatrix} = \begin{pmatrix} 0 & 0 & 1 \\ 0 & 0 & 0 \\ x_2\cos(x_1) & \sin(x_1) & 0 \end{pmatrix}.$$

2.3 (a) $\nabla\boldsymbol{x} = \frac{\partial x_i}{\partial x_j}\boldsymbol{e}_i \otimes \boldsymbol{e}_j = \delta_{ij}\boldsymbol{e}_i \otimes \boldsymbol{e}_j = \boldsymbol{I}$.

(b) $\nabla \cdot \boldsymbol{x} = \frac{\partial x_i}{\partial x_i} = \delta_{ii} = 3$.

(c)

$$\begin{aligned} \nabla(\boldsymbol{x}\cdot\boldsymbol{A}\boldsymbol{x}) &= \frac{\partial}{\partial x_i}(x_kA_{kl}x_l)\boldsymbol{e}_i \\ &= (\delta_{ki}A_{kl}x_l + x_kA_{kl}\delta_{li})\boldsymbol{e}_i \\ &= (A_{il}x_l + A_{ki}x_k)\boldsymbol{e}_i \\ &= (\boldsymbol{A} + \boldsymbol{A}^T)\boldsymbol{x}. \end{aligned}$$

2.5 (a) Since $|\boldsymbol{x}| = (x_kx_k)^{1/2}$ and $\partial x_k/\partial x_i = \delta_{ki}$, we find by the chain rule that $\partial|\boldsymbol{x}|/\partial x_i = x_i/|\boldsymbol{x}|$. Thus

$$\nabla\phi = \frac{\partial}{\partial x_i}\left(\frac{1}{|\boldsymbol{x}|}\right)\boldsymbol{e}_i = -\frac{1}{|\boldsymbol{x}|^2}\frac{\partial|\boldsymbol{x}|}{\partial x_i}\boldsymbol{e}_i = -\frac{1}{|\boldsymbol{x}|^2}\frac{x_i}{|\boldsymbol{x}|}\boldsymbol{e}_i = -\frac{\boldsymbol{x}}{|\boldsymbol{x}|^3}.$$

(b) Using the facts that $\Delta\phi = \nabla\cdot(\nabla\phi)$ and $[\nabla\phi]_i = -x_i/|\boldsymbol{x}|^3$, we have

$$\begin{aligned}\Delta\phi = [\nabla\phi]_{i,i} &= \frac{\partial}{\partial x_i}\Big(-\frac{x_i}{|\boldsymbol{x}|^3}\Big)\\ &= -\frac{1}{|\boldsymbol{x}|^3}\frac{\partial x_i}{\partial x_i} + \frac{3x_i}{|\boldsymbol{x}|^4}\frac{\partial|\boldsymbol{x}|}{\partial x_i}\\ &= -\frac{1}{|\boldsymbol{x}|^3}\delta_{ii} + \frac{3x_i}{|\boldsymbol{x}|^4}\frac{x_i}{|\boldsymbol{x}|}\\ &= 0,\end{aligned}$$

where the last line follows from the facts that $\delta_{ii} = 3$ and $x_i x_i = |\boldsymbol{x}|^2$.

(c) We have $\boldsymbol{v} = v_i\boldsymbol{e}_i$, where $v_i = \phi n_i$ and n_i are the constant components of $\boldsymbol{n}$. Using the facts that $\Delta\boldsymbol{v} = v_{i,jj}\boldsymbol{e}_i$ and $\Delta\phi = \phi_{,jj}$, we have

$$\Delta\boldsymbol{v} = (\phi n_i)_{,jj}\boldsymbol{e}_i = \phi_{,jj}n_i\boldsymbol{e}_i = \Delta\phi\boldsymbol{n}.$$

The result $\Delta\boldsymbol{v} = \boldsymbol{0}$ follows from part (b).

2.7 (a)

$$\nabla\cdot(\nabla\times\boldsymbol{v}) = \frac{\partial}{\partial x_i}\left(\varepsilon_{ijk}\frac{\partial v_k}{\partial x_j}\right) = \varepsilon_{ijk}v_{k,ji}.$$

By the commutativity property of mixed partial derivatives and the permutation properties of ε_{ijk}, we have

$$\varepsilon_{ijk}v_{k,ji} = \varepsilon_{ijk}v_{k,ij} = -\varepsilon_{jik}v_{k,ij} = -\varepsilon_{ijk}v_{k,ji},$$

which implies that $\varepsilon_{ijk}v_{k,ji} = 0$ so that $\nabla\cdot(\nabla\times\boldsymbol{v}) = 0$.

(b)

$$\begin{aligned}\nabla\cdot(\boldsymbol{v}\times\boldsymbol{w}) &= \frac{\partial}{\partial x_i}(\varepsilon_{ijk}v_j w_k)\\ &= \varepsilon_{ijk}\frac{\partial v_j}{\partial x_i}w_k + \varepsilon_{ijk}v_j\frac{\partial w_k}{\partial x_i}\\ &= w_k\varepsilon_{kij}\frac{\partial v_j}{\partial x_i} - v_j\varepsilon_{jik}\frac{\partial w_k}{\partial x_i}\\ &= \boldsymbol{w}\cdot(\nabla\times\boldsymbol{v}) - \boldsymbol{v}\cdot(\nabla\times\boldsymbol{w}).\end{aligned}$$

(c)

$$
\begin{aligned}
\nabla \times (\nabla \times \boldsymbol{v}) &= \varepsilon_{ijk}(\varepsilon_{klm} v_{m,l})_{,j} \boldsymbol{e}_i \\
&= \varepsilon_{ijk}\varepsilon_{klm} v_{m,lj} \boldsymbol{e}_i \\
&= [\delta_{il}\delta_{jm} - \delta_{im}\delta_{jl}] v_{m,lj} \boldsymbol{e}_i \\
&= v_{j,ij} \boldsymbol{e}_i - v_{i,jj} \boldsymbol{e}_i \\
&= \nabla(\nabla \cdot \boldsymbol{v}) - \Delta \boldsymbol{v}.
\end{aligned}
$$

(d)

$$
\begin{aligned}
\nabla \times (\nabla \phi) &= \varepsilon_{ijk}(\phi_{,k})_{,j} \boldsymbol{e}_i \\
&= \varepsilon_{ijk}\phi_{,kj} \boldsymbol{e}_i \\
&= \tfrac{1}{2}[\varepsilon_{ijk}\phi_{,kj} + \varepsilon_{ikj}\phi_{,jk}] \boldsymbol{e}_i \\
&= \tfrac{1}{2}[\varepsilon_{ijk}\phi_{,kj} - \varepsilon_{ijk}\phi_{,jk}] \boldsymbol{e}_i \\
&= \boldsymbol{0}.
\end{aligned}
$$

2.9

$$
\begin{aligned}
\nabla \cdot (\boldsymbol{S}\boldsymbol{x}) &= \frac{\partial}{\partial x_i}(S_{ij}x_j) \\
&= \frac{\partial S_{ij}}{\partial x_i} x_j + S_{ij}\frac{\partial x_j}{\partial x_i} \\
&= \frac{\partial S_{ij}}{\partial x_i} x_j + S_{ij}\delta_{ij} \\
&= (\nabla \cdot \boldsymbol{S}^T) \cdot \boldsymbol{x} + \mathrm{tr}(\boldsymbol{S}).
\end{aligned}
$$

2.11 (a)

$$
\begin{aligned}
\nabla \cdot (\boldsymbol{S}(\phi \boldsymbol{v})) &= \nabla \cdot (S_{ij}\phi v_j \boldsymbol{e}_i) \\
&= \frac{\partial}{\partial x_i}(S_{ij}\phi v_j) \\
&= \frac{\partial S_{ij}}{\partial x_i}\phi v_j + S_{ij}\frac{\partial \phi}{\partial x_i} v_j + S_{ij}\phi\frac{\partial v_j}{\partial x_i} \\
&= \phi(\nabla \cdot \boldsymbol{S}^T) \cdot \boldsymbol{v} + \nabla\phi \cdot (\boldsymbol{S}\boldsymbol{v}) + \phi \boldsymbol{S} : (\nabla \boldsymbol{v})^T.
\end{aligned}
$$

(b)

$$
\begin{aligned}
\nabla \cdot (\boldsymbol{S}\boldsymbol{v}) &= \frac{\partial}{\partial x_i}(S_{ik}v_k) \\
&= \frac{\partial S_{ik}}{\partial x_i} v_k + S_{ik}\frac{\partial v_k}{\partial x_i} \\
&= (\nabla \cdot \boldsymbol{S}^T) \cdot \boldsymbol{v} + \boldsymbol{S}^T : \nabla \boldsymbol{v}.
\end{aligned}
$$

2.13 (a) Let $\boldsymbol{\Omega} = \dot{\boldsymbol{Q}}\boldsymbol{Q}^T$ so that $\dot{\boldsymbol{Q}} = \boldsymbol{\Omega}\boldsymbol{Q}$. Then

$$\frac{d\hat{\boldsymbol{e}}_i}{dt} = \frac{d}{dt}(\boldsymbol{Q}\boldsymbol{e}_i) = \frac{d\boldsymbol{Q}}{dt}\boldsymbol{e}_i = \boldsymbol{\Omega}\boldsymbol{Q}\boldsymbol{e}_i = \boldsymbol{\Omega}\hat{\boldsymbol{e}}_i.$$

Moreover, differentiation of the relation $\boldsymbol{Q}\boldsymbol{Q}^T = \boldsymbol{I}$ gives

$$\boldsymbol{O} = \frac{d\boldsymbol{Q}}{dt}\boldsymbol{Q}^T + \boldsymbol{Q}\frac{d\boldsymbol{Q}^T}{dt} = \boldsymbol{\Omega} + \boldsymbol{\Omega}^T,$$

which shows that $\boldsymbol{\Omega}$ is skew-symmetric. If we let $\boldsymbol{\omega}$ denote the axial vector of $\boldsymbol{\Omega}$, then by definition (see Chapter 1)

$$\frac{d\hat{\boldsymbol{e}}_i}{dt} = \boldsymbol{\Omega}\hat{\boldsymbol{e}}_i = \boldsymbol{\omega} \times \hat{\boldsymbol{e}}_i.$$

(b) By definition, the components of the change of basis tensor $\boldsymbol{Q}$ are given by $Q_{mn} = \boldsymbol{e}_m \cdot \widehat{\boldsymbol{e}}_n$. Thus

$$\widehat{v}_i = \boldsymbol{v} \cdot \widehat{\boldsymbol{e}}_i = (v_k \boldsymbol{e}_k) \cdot \widehat{\boldsymbol{e}}_i = Q_{ki} v_k,$$

and differentiation with respect to time yields

$$\frac{d\widehat{v}_i}{dt} = \frac{dQ_{ki}}{dt} v_k + Q_{ki}\frac{dv_k}{dt}.$$

Using the fact that $\dot{\boldsymbol{Q}} = \boldsymbol{\Omega}\boldsymbol{Q}$, or equivalently $\dot{Q}_{ki} = \Omega_{kj}Q_{ji}$, we get

$$\begin{aligned}\frac{d\widehat{v}_i}{dt} &= \Omega_{kj}Q_{ji}v_k + Q_{ji}\frac{dv_j}{dt} \\ &= Q_{ji}\left[\frac{dv_j}{dt} + \Omega_{kj}v_k\right] \\ &= Q_{ji}\left[\frac{dv_j}{dt} - \Omega_{jk}v_k\right],\end{aligned}$$

where the last line follows from the skew-symmetry of $\boldsymbol{\Omega}$.

2.15 In components, we have $\psi(\boldsymbol{A}) = A_{mm}A_{kl}B_{kl}$ and

$$\begin{aligned}\frac{\partial \psi}{\partial A_{ij}} &= \frac{\partial A_{mm}}{\partial A_{ij}} A_{kl}B_{kl} + A_{mm}\frac{\partial A_{kl}}{\partial A_{ij}} B_{kl} \\ &= \delta_{mi}\delta_{mj} A_{kl}B_{kl} + A_{mm}\delta_{ki}\delta_{lj}B_{kl} \\ &= \delta_{ij}A_{kl}B_{kl} + A_{mm}B_{ij}.\end{aligned}$$

Thus

$$\begin{aligned}D\psi(\boldsymbol{A}) &= \frac{\partial \psi}{\partial A_{ij}}\boldsymbol{e}_i \otimes \boldsymbol{e}_j \\ &= A_{kl}B_{kl}\delta_{ij}\boldsymbol{e}_i \otimes \boldsymbol{e}_j + A_{mm}B_{ij}\boldsymbol{e}_i \otimes \boldsymbol{e}_j \\ &= (\boldsymbol{A} : \boldsymbol{B})\boldsymbol{I} + \mathrm{tr}(\boldsymbol{A})\boldsymbol{B}.\end{aligned}$$

2.17 (a) Let $\boldsymbol{A}$ and $\boldsymbol{B}$ be arbitrary second-order tensors. Then, by definition of $\mathbf{C}$, we have

$$\begin{aligned}\boldsymbol{A} : \mathbf{C}(\boldsymbol{B}) &= \boldsymbol{A} : (\mathrm{tr}(\boldsymbol{B})\boldsymbol{I} + \boldsymbol{B}) \\ &= \mathrm{tr}(\boldsymbol{B})\,\mathrm{tr}(\boldsymbol{A}) + \boldsymbol{A} : \boldsymbol{B} \\ &= \boldsymbol{B} : (\mathrm{tr}(\boldsymbol{A})\boldsymbol{I} + \boldsymbol{A}) \\ &= \boldsymbol{B} : \mathbf{C}(\boldsymbol{A}),\end{aligned}$$

which establishes the result.

(b) Let $\boldsymbol{A} \in \mathcal{V}^2$ be given and consider the function $\boldsymbol{A}(\alpha) = \boldsymbol{A} + \alpha\boldsymbol{B}$, where $\alpha \in \mathbb{R}$ and $\boldsymbol{B} \in \mathcal{V}^2$ is arbitrary. Then the derivative $D\psi(\boldsymbol{A})$ is defined by the relation

$$D\psi(\boldsymbol{A}) : \boldsymbol{B} = \frac{d}{d\alpha}\psi(\boldsymbol{A}(\alpha))\Big|_{\alpha=0}, \quad \forall \boldsymbol{B} \in \mathcal{V}^2. \tag{2.4}$$

From the definition of $\psi(\boldsymbol{A})$, we have

$$\psi(\boldsymbol{A}(\alpha)) = \frac{1}{2}\boldsymbol{A}(\alpha) : \mathbf{C}(\boldsymbol{A}(\alpha)),$$

and differentiation with respect to α (denoted by a prime) gives

$$\begin{aligned}\frac{d}{d\alpha}\psi(\boldsymbol{A}(\alpha)) &= \frac{1}{2}\boldsymbol{A}'(\alpha) : \mathbf{C}(\boldsymbol{A}(\alpha)) + \frac{1}{2}\boldsymbol{A}(\alpha) : \mathbf{C}(\boldsymbol{A}'(\alpha)) \\ &= \frac{1}{2}\boldsymbol{A}'(\alpha) : \mathbf{C}(\boldsymbol{A}(\alpha)) + \frac{1}{2}\mathbf{C}(\boldsymbol{A}(\alpha)) : \boldsymbol{A}'(\alpha) \\ &= \mathbf{C}(\boldsymbol{A}(\alpha)) : \boldsymbol{A}'(\alpha),\end{aligned} \tag{2.5}$$

where we have used the linearity and major symmetry of $\mathbf{C}$. Combining (2.4) and (2.5) then gives

$$D\psi(\boldsymbol{A}) : \boldsymbol{B} = \mathbf{C}(\boldsymbol{A}) : \boldsymbol{B}, \quad \forall \boldsymbol{B} \in \mathcal{V}^2,$$

which implies $D\psi(\boldsymbol{A}) = \mathbf{C}(\boldsymbol{A})$.

3
Continuum Mass and Force Concepts

Here we introduce the notion of a continuum body and discuss how to describe its mass properties, and the various types of forces that may act on it. As we will see, the discussion of internal forces in a continuum body will lead to the notion of a stress tensor field – our first example of a second-order tensor field arising in a physical context. We introduce the basic conditions necessary for the mechanical equilibrium of a continuum body and then derive a corresponding statement in terms of differential equations.

The important ideas in this chapter are: (i) the notion of a mass density field, which enables us to define the mass of an arbitrary part of a body; (ii) the notion of body and surface force fields, which enable us to define the resultant force and torque on an arbitrary part of a body; (iii) the notion of a Cauchy stress field and its relation to surface force fields; (iv) the equations of equilibrium for a body.

3.1 Continuum Bodies

The most basic assumption we make in our study of any material body, whether it be a solid, liquid or gas, is that the material involved can be modeled as a **continuum**: we ignore the atomic nature of the material and assume it is infinitely divisible. This assumption can lead to very effective material models at length scales much longer than typical interatomic spacings. However, at length scales comparable to or smaller than this, it is no longer expected to be valid.

The continuum assumption allows us, at any fixed instant of time, to identify a material body with an open subset B of Euclidean space $\mathbb{E}^3$. In particular, each material particle is identified with a point $\boldsymbol{x} \in B$. We call B a **placement** or **configuration** of the body in $\mathbb{E}^3$. Unless mentioned otherwise, we will always assume that B is a regular region of $\mathbb{E}^3$ in the sense defined in Section 2.3. We choose to identify material bodies with open subsets of space purely for mathematical convenience. It would also be reasonable to identify bodies with the closure of such sets.

Physical experience tells us that a body will tend to move and generally change shape under the action of external influences. Thus a given material body will generally assume various configurations as time progresses. For the purposes of this chapter, however, we focus attention on a fixed time and identify a material body with a single configuration B.

3.2 Mass

Mass is a physical property of matter which quantifies its resistance to acceleration. Here we introduce the idea of a mass density field for a continuum body along with the idea of a center of mass. These concepts will appear throughout our developments.

3.2.1 Mass Density

In accordance with the continuum assumption we assume that the mass of a body B is distributed continuously throughout its volume. Moreover, we assume that any subset of B with positive volume has positive mass, and that this mass tends to zero as the volume tends to zero.

To make the above ideas precise, let mass$[\Omega]$ denote the mass of an arbitrary open subset Ω of B. Then we assume there exists a **mass density field** per unit volume $\rho : B \to \mathbb{R}$ such that

$$\mathrm{mass}[\Omega] = \int_\Omega \rho(\boldsymbol{x}) \, dV_{\boldsymbol{x}}.$$

Moreover, we assume $\rho(\boldsymbol{x}) > 0$ for all $\boldsymbol{x} \in B$. Here we use $dV_{\boldsymbol{x}}$ to denote an infinitesimal volume element at $\boldsymbol{x} \in \Omega$. In particular, the volume of Ω is

$$\mathrm{vol}[\Omega] = \int_\Omega dV_{\boldsymbol{x}}.$$

(In later chapters we will integrate over both $\boldsymbol{x}$ and $\boldsymbol{X}$; hence we introduce the good practice of explicitly denoting the variable of integration in surface and volume integrals.)

The mass density field ρ can be formally defined as follows. Let $\boldsymbol{x}$ be an arbitrary point in B and let $\Omega_\delta(\boldsymbol{x})$ denote a family of volumes with the properties that $\mathrm{vol}[\Omega_\delta(\boldsymbol{x})] \to 0$ as $\delta \to 0$ and $\boldsymbol{x} \in \Omega_\delta(\boldsymbol{x})$ for every $\delta > 0$. Then

$$\rho(\boldsymbol{x}) = \lim_{\delta \to 0} \frac{\mathrm{mass}[\Omega_\delta(\boldsymbol{x})]}{\mathrm{vol}[\Omega_\delta(\boldsymbol{x})]}.$$

Our basic assumption on the mass distribution in a continuum is that this limit exists and is positive for each $\boldsymbol{x}$ in B. Furthermore, we assume this limit is the same for any family $\Omega_\delta(\boldsymbol{x})$ with the properties described.

3.2.2 Center of Mass

By the **center of mass** of an open subset Ω of B we mean the point

$$\boldsymbol{x}_{\mathrm{com}}[\Omega] = \frac{1}{\mathrm{mass}[\Omega]} \int_\Omega \boldsymbol{x} \rho(\boldsymbol{x}) \, dV_{\boldsymbol{x}}.$$

This is related to, but in general different from, the **center of volume** or centroid of Ω, which is defined as

$$\boldsymbol{x}_{\mathrm{cov}}[\Omega] = \frac{1}{\mathrm{vol}[\Omega]} \int_\Omega \boldsymbol{x} \, dV_{\boldsymbol{x}}.$$

Notice that $\boldsymbol{x}_{\mathrm{com}}[\Omega]$ and $\boldsymbol{x}_{\mathrm{cov}}[\Omega]$ need not be points in B (see Exercises 1 and 2). As discussed below, these points play a special role in determining the resultant torque on a body due to body forces that are uniform in an appropriate sense. The center of mass also plays a special role in the laws of inertia for bodies as discussed later in Chapter 5.

3.3 Force

Mechanical interactions between parts of a body or between a body and its environment are described by forces. Here we describe two basic types of forces: body forces, which are exerted at the interior points of a body, and surface forces, which are exerted on internal surfaces between separate parts of a body, or on external surfaces between the body and its environment.

3.3.1 Body Forces

We use the term **body force** to denote any force that does not arise due to physical contact between bodies. Such a force is the result of action at a distance. A prototypical example is a gravitational force. The body force, per unit volume, exerted by an external influence on a body B is assumed to be given by a function $\widehat{\boldsymbol{b}} : B \to \mathcal{V}$, which we call a **body force field** on B.

Let Ω be an arbitrary open subset of B. Then the resultant force on Ω due to a body force field per unit volume is defined as

$$\boldsymbol{r}_b[\Omega] = \int_\Omega \widehat{\boldsymbol{b}}(\boldsymbol{x})\, dV_{\boldsymbol{x}},$$

and the resultant torque on Ω, about a point $\boldsymbol{z}$, is defined as

$$\boldsymbol{\tau}_b[\Omega] = \int_\Omega (\boldsymbol{x} - \boldsymbol{z}) \times \widehat{\boldsymbol{b}}(\boldsymbol{x})\, dV_{\boldsymbol{x}}.$$

Sometimes it is convenient to introduce a body force field per unit mass defined by

$$\boldsymbol{b}(\boldsymbol{x}) = \rho(\boldsymbol{x})^{-1}\widehat{\boldsymbol{b}}(\boldsymbol{x}).$$

In this case, the resultant force becomes

$$\boldsymbol{r}_b[\Omega] = \int_\Omega \rho(\boldsymbol{x})\boldsymbol{b}(\boldsymbol{x})\, dV_{\boldsymbol{x}},$$

and the resultant torque, about a point $\boldsymbol{z}$, becomes

$$\boldsymbol{\tau}_b[\Omega] = \int_\Omega (\boldsymbol{x} - \boldsymbol{z}) \times \rho(\boldsymbol{x})\boldsymbol{b}(\boldsymbol{x})\, dV_{\boldsymbol{x}}.$$

The resultant torque $\boldsymbol{\tau}_b[\Omega]$ can be made to vanish through an appropriate choice of the reference point $\boldsymbol{z}$. For example, when the body force per unit mass $\boldsymbol{b}$ is uniform (constant), the resultant torque about the center of mass $\boldsymbol{x}_{\mathrm{com}}[\Omega]$ vanishes. Similarly, when the body force per unit volume $\widehat{\boldsymbol{b}}$ is uniform, the resultant torque about the center of volume $\boldsymbol{x}_{\mathrm{cov}}[\Omega]$ vanishes (see Exercises 4 and 5).

3.3.2 Surface Forces

We use the term **surface force** to denote any force that arises due to physical contact between bodies. When referring to a surface force along an imaginary surface within the interior of a body, we say that the force is **internal**. In contrast, when referring to a surface force along the bounding surface of a body, we say that the force is **external**. Roughly speaking, internal surface forces are what hold a loaded body together: they resist the tendency of one part of a body from being pulled away from another part. External surface forces are contact forces applied to a body by its environment.

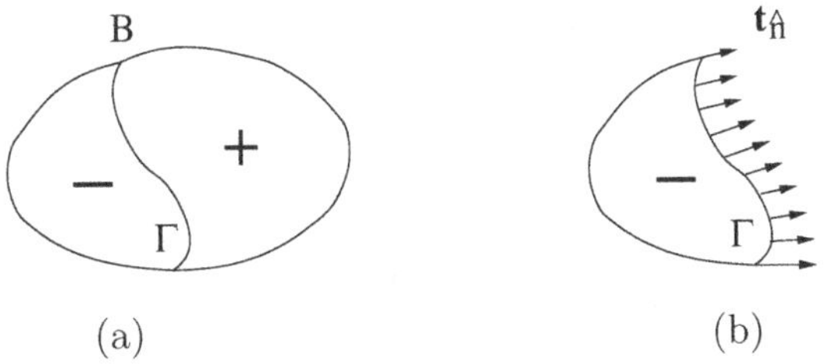

Fig. 3.1 Graphical illustration of internal surfaces forces. (a) a body B separated into positive and negative parts by an oriented surface Γ (unit normal field $\widehat{\boldsymbol{n}}$ not shown). (b) the action of the positive material upon the negative material is described by a traction field $\boldsymbol{t}_{\widehat{\boldsymbol{n}}}$ on Γ.

3.3.2.1 Concept of a Traction Field

Let Γ be an arbitrary, oriented surface in B with unit normal field $\widehat{\boldsymbol{n}} : \Gamma \to \mathcal{V}$. At each point $\boldsymbol{x}$ the unit normal $\widehat{\boldsymbol{n}}(\boldsymbol{x})$ defines a positive side and a negative side of Γ as shown in Figure 3.1. The force, per unit area, exerted by material on the positive side upon material on the negative side is assumed to be given by a function $\boldsymbol{t}_{\widehat{\boldsymbol{n}}} : \Gamma \to \mathcal{V}$, which we call the **traction** or **surface force field** for Γ. When Γ is part of the bounding surface of B we always choose $\widehat{\boldsymbol{n}}$ to be the outward unit normal field. In this case, the traction field $\boldsymbol{t}_{\widehat{\boldsymbol{n}}}$ represents the force, per unit area, applied to the surface of B by external means.

The resultant force due to a traction field on an oriented surface Γ is defined as

$$\boldsymbol{r}_s[\Gamma] = \int_\Gamma \boldsymbol{t}_{\widehat{\boldsymbol{n}}}(\boldsymbol{x})\, dA_{\boldsymbol{x}},$$

where $dA_{\boldsymbol{x}}$ represents an infinitesimal surface area element at $\boldsymbol{x} \in \Gamma$. Similarly, the resultant torque, about a point $\boldsymbol{z}$, due to a traction field on Γ is defined as

$$\boldsymbol{\tau}_s[\Gamma] = \int_\Gamma (\boldsymbol{x} - \boldsymbol{z}) \times \boldsymbol{t}_{\widehat{\boldsymbol{n}}}(\boldsymbol{x})\, dA_{\boldsymbol{x}}.$$

For examples and applications of these ideas see Exercises 6 and 7.

3.3.2.2 Cauchy's Postulate, Law of Action and Reaction

The theory of surface forces in classical continuum mechanics is based on the following assumption typically referred to as **Cauchy's Postulate**.

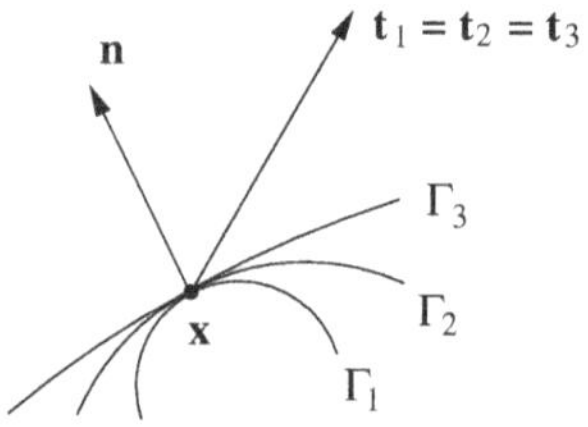

Fig. 3.2 Graphical illustration of Cauchy's postulate: surfaces through a point $\boldsymbol{x}$ with normal $\boldsymbol{n}$ at $\boldsymbol{x}$ share the same traction vector at $\boldsymbol{x}$.

Axiom 3.1 Dependence of Traction on Surface Geometry. *The traction field $\boldsymbol{t}_{\widehat{\boldsymbol{n}}}$ on a surface Γ in B depends only pointwise on the unit normal field $\widehat{\boldsymbol{n}}$. In particular, there is a function $\boldsymbol{t} : \mathcal{N} \times B \to \mathcal{V}$, where $\mathcal{N} \subset \mathcal{V}$ denotes the set of all unit vectors, such that*

$$\boldsymbol{t}_{\widehat{\boldsymbol{n}}}(\boldsymbol{x}) = \boldsymbol{t}(\widehat{\boldsymbol{n}}(\boldsymbol{x}), \boldsymbol{x}).$$

The function $\boldsymbol{t} : \mathcal{N} \times B \to \mathcal{V}$ is called the **traction function** *for B.* □

Thus the traction vector $\boldsymbol{t}_{\widehat{\boldsymbol{n}}}$ at a point $\boldsymbol{x}$ depends only on the values of $\boldsymbol{x}$ and $\widehat{\boldsymbol{n}}(\boldsymbol{x})$. In particular, it does not depend on the value of $\nabla\widehat{\boldsymbol{n}}(\boldsymbol{x})$, which describes the curvature of the surface. To see a consequence of this, consider a collection of surfaces Γ_1, Γ_2 and Γ_3 in B. Suppose that these surfaces all pass through a point $\boldsymbol{x}$, and that they share a common unit normal $\boldsymbol{n}$ at $\boldsymbol{x}$ as shown in Figure 3.2. In particular, $\widehat{\boldsymbol{n}}_i(\boldsymbol{x}) = \boldsymbol{n}$ where $\widehat{\boldsymbol{n}}_i$ is the unit normal field for the surface Γ_i. If we let $\boldsymbol{t}_{\widehat{\boldsymbol{n}}_i}$ be the traction field for surface Γ_i, then

$$\boldsymbol{t}_{\widehat{\boldsymbol{n}}_1}(\boldsymbol{x}) = \boldsymbol{t}_{\widehat{\boldsymbol{n}}_2}(\boldsymbol{x}) = \boldsymbol{t}_{\widehat{\boldsymbol{n}}_3}(\boldsymbol{x}),$$

as depicted in Figure 3.2. Since $\widehat{\boldsymbol{n}}_i(\boldsymbol{x}) = \boldsymbol{n}$ the common value of these three traction vectors is $\boldsymbol{t}(\boldsymbol{n}, \boldsymbol{x})$.

The next result shows that the traction function satisfies a certain law of action and reaction. In the following, Ω denotes a regular, open subset of B with boundary $\partial\Omega$ and outward unit normal field $\widehat{\boldsymbol{n}}$.

Result 3.2 Law of Action and Reaction. *Let $\boldsymbol{t} : \mathcal{N} \times B \to \mathcal{V}$ be the traction function for a body B. Suppose that $\boldsymbol{t}(\boldsymbol{n}, \boldsymbol{x})$ is continuous and that for any sequence of subsets Ω whose volumes tend to zero*

$$\frac{1}{\text{area}(\partial\Omega)} \int_{\partial\Omega} \boldsymbol{t}(\widehat{\boldsymbol{n}}(\boldsymbol{x}), \boldsymbol{x})\, dA_{\boldsymbol{x}} \to \boldsymbol{0} \quad \textit{as} \quad \text{vol}(\Omega) \to 0. \tag{3.1}$$

Then

$$\boldsymbol{t}(-\boldsymbol{n},\boldsymbol{x}) = -\boldsymbol{t}(\boldsymbol{n},\boldsymbol{x}), \quad \forall \boldsymbol{n} \in \mathcal{N},\ \boldsymbol{x} \in B.$$

□

Proof Let $\boldsymbol{x} \in B$ and $\boldsymbol{n} \in \mathcal{N}$ be arbitrary, and let $D \subset B$ be a disc of arbitrary, fixed radius centered at $\boldsymbol{x}$ with normal $\boldsymbol{n}$. For $\delta > 0$ let $\Omega_\delta \subset B$ be the cylinder with center $\boldsymbol{x}$, height δ, axis $\boldsymbol{n}$, end faces $\Gamma_\pm$ parallel to D, and lateral surface Γ_δ as depicted in Figure 3.3(a). Notice that $\text{area}(\Gamma_\delta) \to 0$ and $\Gamma_\pm \to D$ as $\delta \to 0$. Let $\widehat{\boldsymbol{n}}$ be the outward unit normal field on $\partial\Omega_\delta$ and notice that $\widehat{\boldsymbol{n}}(\boldsymbol{y}) = \pm\boldsymbol{n}$ (constant) on $\Gamma_\pm$. Considering only the second factor in (3.1), we have

$$\lim_{\delta\to 0} \int_{\partial\Omega_\delta} \boldsymbol{t}(\widehat{\boldsymbol{n}}(\boldsymbol{y}),\boldsymbol{y})\, dA_{\boldsymbol{y}} = \boldsymbol{0},$$

and since $\partial\Omega_\delta = \Gamma_\delta \cup \Gamma_+ \cup \Gamma_-$, we have

$$\lim_{\delta\to 0} \left[\int_{\Gamma_\delta} \boldsymbol{t}(\widehat{\boldsymbol{n}}(\boldsymbol{y}),\boldsymbol{y})\, dA_{\boldsymbol{y}} + \int_{\Gamma_+} \boldsymbol{t}(\boldsymbol{n},\boldsymbol{y})\, dA_{\boldsymbol{y}} + \int_{\Gamma_-} \boldsymbol{t}(-\boldsymbol{n},\boldsymbol{y})\, dA_{\boldsymbol{y}} \right] = \boldsymbol{0}.$$

The first term vanishes by continuity (hence boundedness) of $\boldsymbol{t}(\widehat{\boldsymbol{n}}(\boldsymbol{y}),\boldsymbol{y})$ and the fact that $\text{area}(\Gamma_\delta) \to 0$. Using the fact that $\Gamma_\pm \to D$ the above limit implies

$$\int_D [\boldsymbol{t}(\boldsymbol{n},\boldsymbol{y}) + \boldsymbol{t}(-\boldsymbol{n},\boldsymbol{y})]\, dA_{\boldsymbol{y}} = \boldsymbol{0}.$$

Since the radius of D is arbitrary we deduce, by a slight generalization of the Localization Theorem (Result 2.19), that the integrand must vanish at the center of D. Thus

$$\boldsymbol{t}(\boldsymbol{n},\boldsymbol{x}) + \boldsymbol{t}(-\boldsymbol{n},\boldsymbol{x}) = \boldsymbol{0},$$

which establishes the result. □

Remarks:

(1) The above result implies that the traction exerted by material on the positive side upon material on the negative side of an oriented surface is equal and opposite, at each point, to the traction exerted by material on the negative side upon material on the positive side.

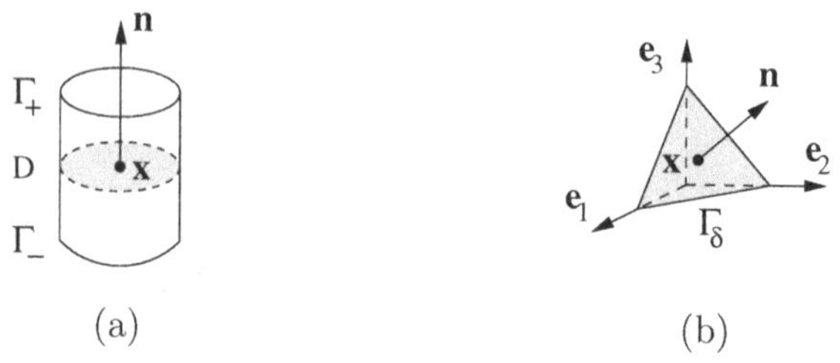

Fig. 3.3 Illustration of the regions used in the proofs of Results 3.2 and 3.3. (a) cylinder defined by a disc D. (b) tetrahedron defined by a triangular region Γ_δ.

(2) The condition (3.1) states that the resultant surface force on a body, divided by the surface area, tends to zero as the body volume tends to zero. This condition is consistent with the axiom on balance of linear momentum (Axiom 5.2) under the assumption that the body force and acceleration fields are bounded. Throughout our developments we always assume this to be the case.

(3) The proof of Result 3.2 does not make complete use of (3.1). In particular, the proof makes use of only the second factor in (3.1), which implies that the resultant surface force tends to zero as the body volume tends to zero. Complete use of (3.1) will be made in Result 3.3 below. □

3.3.3 The Stress Tensor

Using Result 3.2 we can determine the specific way in which the traction function $\boldsymbol{t}(\boldsymbol{n}, \boldsymbol{x})$ depends on $\boldsymbol{n}$. The following result is typically referred to as **Cauchy's Theorem** and will be fundamental throughout our developments.

Result 3.3 Existence of Stress Tensor. *Let* $\mathbf{t} : \mathcal{N} \times B \to \mathcal{V}$ *be the traction function for a body* B *and suppose it satisfies the conditions of Result 3.2. Then* $\boldsymbol{t}(\boldsymbol{n}, \boldsymbol{x})$ *is linear in* $\boldsymbol{n}$, *that is, for each* $\boldsymbol{x} \in B$ *there is a second-order tensor* $\boldsymbol{S}(\boldsymbol{x}) \in \mathcal{V}^2$ *such that*

$$\boldsymbol{t}(\boldsymbol{n}, \boldsymbol{x}) = \boldsymbol{S}(\boldsymbol{x})\boldsymbol{n}.$$

The field $\boldsymbol{S} : B \to \mathcal{V}^2$ *is called the* **Cauchy stress field** *for* B. □

Proof Let $\{\boldsymbol{e}_j\}$ be an arbitrary frame, $\boldsymbol{x} \in B$ an arbitrary point and consider any $\boldsymbol{n} \in \mathcal{N}$ such that $n_j = \boldsymbol{n} \cdot \boldsymbol{e}_j > 0$ for each j. For $\delta > 0$ let $\Gamma_\delta \subset B$ denote a triangular region with center $\boldsymbol{x}$, normal $\boldsymbol{n}$, maximum edge length δ, and each edge in a different coordinate plane as depicted in Figure 3.3(b). Moreover, let $\Omega_\delta \subset B$ be the tetrahedron bounded by Γ_δ and the three coordinate planes. In particular, Ω_δ has three faces Γ_j with outward normals $-\boldsymbol{e}_j$, one face Γ_δ with outward normal $\boldsymbol{n}$, and $\mathrm{vol}(\Omega_\delta) \to 0$ as $\delta \to 0$. Let $\widehat{\boldsymbol{n}}$ be the outward unit normal field on $\partial\Omega_\delta$ and notice that $\widehat{\boldsymbol{n}}(\boldsymbol{y}) = -\boldsymbol{e}_j$ (constant) on Γ_j and $\widehat{\boldsymbol{n}}(\boldsymbol{y}) = \boldsymbol{n}$ (constant) on Γ_δ. By (3.1), we have

$$\lim_{\delta\to 0} \frac{1}{\mathrm{area}(\partial\Omega_\delta)} \int_{\partial\Omega_\delta} \boldsymbol{t}(\widehat{\boldsymbol{n}}(\boldsymbol{y}), \boldsymbol{y})\, dA_{\boldsymbol{y}} = \boldsymbol{0},$$

and since $\partial\Omega_\delta = \Gamma_\delta \cup \Gamma_1 \cup \Gamma_2 \cup \Gamma_3$, we have

$$\lim_{\delta\to 0} \frac{1}{\mathrm{area}(\partial\Omega_\delta)} \left[\int_{\Gamma_\delta} \boldsymbol{t}(\boldsymbol{n}, \boldsymbol{y})\, dA_{\boldsymbol{y}} + \sum_{j=1}^{3} \int_{\Gamma_j} \boldsymbol{t}(-\boldsymbol{e}_j, \boldsymbol{y})\, dA_{\boldsymbol{y}} \right] = \boldsymbol{0}. \quad (3.2)$$

Since each face Γ_j can be linearly mapped onto Γ_δ with constant Jacobian $n_j > 0$, we have $\mathrm{area}(\Gamma_j) = n_j\, \mathrm{area}(\Gamma_\delta)$ and

$$\mathrm{area}(\partial\Omega_\delta) = \mathrm{area}(\Gamma_\delta) + \sum_{j=1}^{3} \mathrm{area}(\Gamma_j) = \lambda\, \mathrm{area}(\Gamma_\delta),$$

where $\lambda = 1 + \sum_{j=1}^{3} n_j$ is independent of δ. Substituting this result into (3.2), multiplying by λ, and changing variables between Γ_j and Γ_δ, we obtain

$$\lim_{\delta\to 0} \frac{1}{\mathrm{area}(\Gamma_\delta)} \int_{\Gamma_\delta} \left[\boldsymbol{t}(\boldsymbol{n}, \boldsymbol{y}) + \sum_{j=1}^{3} \boldsymbol{t}(-\boldsymbol{e}_j, \boldsymbol{y}) n_j \right] dA_{\boldsymbol{y}} = \boldsymbol{0}.$$

By the Mean Value Theorem for integrals, since the integrand is continuous and Γ_δ is shrinking to the point $\boldsymbol{x}$, the limit on the left-hand side is the integrand at $\boldsymbol{x}$. Thus we have

$$\boldsymbol{t}(\boldsymbol{n}, \boldsymbol{x}) + \sum_{j=1}^{3} \boldsymbol{t}(-\boldsymbol{e}_j, \boldsymbol{x}) n_j = \boldsymbol{0}.$$

Using Result 3.2, and employing the summation convention, this can be written as

$$\boldsymbol{t}(\boldsymbol{n}, \boldsymbol{x}) = \boldsymbol{t}(\boldsymbol{e}_j, \boldsymbol{x}) n_j = \Big(\boldsymbol{t}(\boldsymbol{e}_j, \boldsymbol{x}) \otimes \boldsymbol{e}_j \Big) \boldsymbol{n} = \boldsymbol{S}(\boldsymbol{x}) \boldsymbol{n}, \quad (3.3)$$

where

$$\boldsymbol{S}(\boldsymbol{x}) = \boldsymbol{t}(\boldsymbol{e}_j, \boldsymbol{x}) \otimes \boldsymbol{e}_j. \tag{3.4}$$

Thus for arbitrary $\boldsymbol{x}$ there exists a second-order tensor $\boldsymbol{S}(\boldsymbol{x})$ such that $\boldsymbol{t}(\boldsymbol{n}, \boldsymbol{x}) = \boldsymbol{S}(\boldsymbol{x})\boldsymbol{n}$ for all $\boldsymbol{n}$ such that $\boldsymbol{n} \cdot \boldsymbol{e}_j > 0$.

The above result can be extended to other unit vectors. For example, if $\boldsymbol{n}$ satisfies $\boldsymbol{n} \cdot \boldsymbol{e}_1 < 0$, with the other components being positive, then we can introduce a new frame $\{\boldsymbol{e}'_j\}$ such that $\boldsymbol{e}'_1 = -\boldsymbol{e}_1$, $\boldsymbol{e}'_2 = \boldsymbol{e}_3$ and $\boldsymbol{e}'_3 = \boldsymbol{e}_2$. In this frame $\boldsymbol{n}$ satisfies $\boldsymbol{n} \cdot \boldsymbol{e}'_j > 0$ and the above arguments apply giving

$$\boldsymbol{t}(\boldsymbol{n}, \boldsymbol{x}) = \Big(\boldsymbol{t}(\boldsymbol{e}'_j, \boldsymbol{x}) \otimes \boldsymbol{e}'_j\Big)\boldsymbol{n}.$$

However, by definition of $\{\boldsymbol{e}'_j\}$ and Result 3.2, we have, still employing the summation convention

$$\boldsymbol{t}(\boldsymbol{e}'_j, \boldsymbol{x}) \otimes \boldsymbol{e}'_j = \boldsymbol{t}(\boldsymbol{e}_j, \boldsymbol{x}) \otimes \boldsymbol{e}_j.$$

Thus (3.3) also holds for all $\boldsymbol{n}$ with $\boldsymbol{n} \cdot \boldsymbol{e}_1 < 0$ and $\boldsymbol{n} \cdot \boldsymbol{e}_j > 0$ $(j = 2, 3)$. Continuing in this way we find that $\boldsymbol{t}(\boldsymbol{n}, \boldsymbol{x}) = \boldsymbol{S}(\boldsymbol{x})\boldsymbol{n}$ for all $\boldsymbol{n}$ such that $\boldsymbol{n} \cdot \boldsymbol{e}_j \neq 0$ $(j = 1, 2, 3)$. By the assumed continuity of $\boldsymbol{t}(\boldsymbol{n}, \boldsymbol{x})$ this result then extends to all $\boldsymbol{n}$. □

Remarks:

(1) Throughout the remainder of our developments we will denote normal fields by $\boldsymbol{n}(\boldsymbol{x})$ rather than $\widehat{\boldsymbol{n}}(\boldsymbol{x})$. Thus, by Axiom 3.1 and Result 3.3, the traction field on a surface with normal $\boldsymbol{n}(\boldsymbol{x})$ is

$$\boldsymbol{t}(\boldsymbol{n}(\boldsymbol{x}), \boldsymbol{x}) = \boldsymbol{S}(\boldsymbol{x})\boldsymbol{n}(\boldsymbol{x}).$$

When there is no cause for confusion, we will abbreviate this relation by $\boldsymbol{t}(\boldsymbol{x}) = \boldsymbol{S}(\boldsymbol{x})\boldsymbol{n}$, or more simply

$$\boldsymbol{t} = \boldsymbol{S}\boldsymbol{n}.$$

In components in any frame $\{\boldsymbol{e}_i\}$ this reads $t_i = S_{ij}n_j$.

(2) The nine components of the stress tensor $\boldsymbol{S}(\boldsymbol{x})$ can be understood as the components of the three traction vectors $\boldsymbol{t}(\boldsymbol{e}_j, \boldsymbol{x})$ on the coordinate planes at $\boldsymbol{x}$. In particular, substituting $\boldsymbol{t}(\boldsymbol{e}_j, \boldsymbol{x}) = t_i(\boldsymbol{e}_j, \boldsymbol{x})\boldsymbol{e}_i$ into (3.4) we get

$$\boldsymbol{S}(\boldsymbol{x}) = S_{ij}(\boldsymbol{x})\boldsymbol{e}_i \otimes \boldsymbol{e}_j \quad \text{where} \quad S_{ij}(\boldsymbol{x}) = t_i(\boldsymbol{e}_j, \boldsymbol{x}).$$

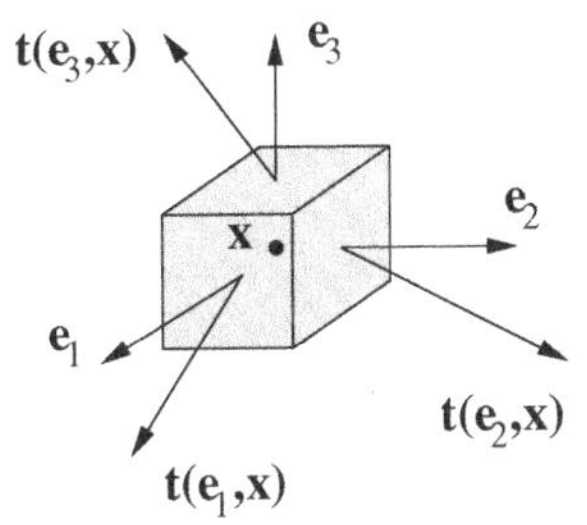

Fig. 3.4 Illustration of the traction vectors on the coordinate planes with normals $\boldsymbol{e}_1$, $\boldsymbol{e}_2$ and $\boldsymbol{e}_3$ at $\boldsymbol{x}$. These traction vectors can be interpreted as the surface forces (per unit area) on an infinitesimal cube centered at $\boldsymbol{x}$.

Thus the traction vectors on the coordinate planes with normals $\boldsymbol{e}_1$, $\boldsymbol{e}_2$ and $\boldsymbol{e}_3$ at $\boldsymbol{x}$ are (see Figure 3.4)

$$\begin{aligned}
\boldsymbol{t}(\boldsymbol{e}_1,\boldsymbol{x}) &= t_i(\boldsymbol{e}_1,\boldsymbol{x})\boldsymbol{e}_i = S_{i1}(\boldsymbol{x})\boldsymbol{e}_i,\\
\boldsymbol{t}(\boldsymbol{e}_2,\boldsymbol{x}) &= t_i(\boldsymbol{e}_2,\boldsymbol{x})\boldsymbol{e}_i = S_{i2}(\boldsymbol{x})\boldsymbol{e}_i,\\
\boldsymbol{t}(\boldsymbol{e}_3,\boldsymbol{x}) &= t_i(\boldsymbol{e}_3,\boldsymbol{x})\boldsymbol{e}_i = S_{i3}(\boldsymbol{x})\boldsymbol{e}_i.
\end{aligned}$$

□

3.4 Equilibrium

Here we state an axiom which provides necessary conditions for a continuum body to be in mechanical equilibrium; that is, at rest in a fixed frame for Euclidean space $\mathbb{E}^3$. We then derive a local form of the necessary conditions in terms of differential equations.

3.4.1 Preliminaries

Consider a continuum body in Euclidean space $\mathbb{E}^3$. Suppose that at a given instant of time it is motionless and has a configuration B_0 as shown in Figure 3.5(a). Suppose next that the body is subject to an external traction field and a body force field such that the body changes shape and comes to rest in a configuration B as shown in Figure 3.5(b). In the configuration B we denote the mass density field per unit volume by $\rho : B \to \mathbb{R}$, the applied external traction field per unit area by $\boldsymbol{h} : \partial B \to \mathcal{V}$ and the body force field per unit mass by $\boldsymbol{b} : B \to \mathcal{V}$. We implicitly assume that ρ, $\boldsymbol{h}$ and $\boldsymbol{b}$ do not depend on time.

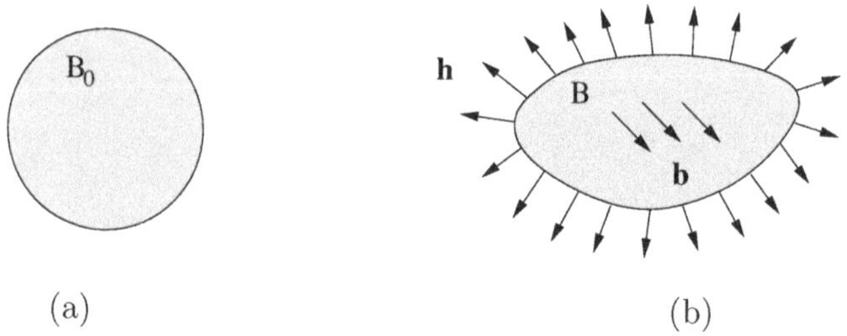

Fig. 3.5 Two equilibrium configurations of a material body. (a) equilibrium in the absence of external loads. (b) equilibrium in the presence of a traction $\boldsymbol{h}$ and a body force $\boldsymbol{b}$.

3.4.2 Necessary Conditions

Let Ω be any open subset of B and let $\boldsymbol{t} : \partial\Omega \to \mathcal{V}$ be the traction field acting on its bounding surface, with orientation determined by the unit outward normal field. (Notice that $\boldsymbol{t} = \boldsymbol{h}$ in the special case when $\Omega = B$.) The resultant force on Ω due to body and surface forces is

$$\boldsymbol{r}[\Omega] = \boldsymbol{r}_b[\Omega] + \boldsymbol{r}_s[\partial\Omega] = \int_\Omega \rho(\boldsymbol{x})\boldsymbol{b}(\boldsymbol{x})\, dV_{\boldsymbol{x}} + \int_{\partial\Omega} \boldsymbol{t}(\boldsymbol{x})\, dA_{\boldsymbol{x}}, \tag{3.5}$$

and the resultant torque on Ω, about a point $\boldsymbol{z}$, due to body and surface forces is

$$\begin{aligned}\boldsymbol{\tau}[\Omega] &= \boldsymbol{\tau}_b[\Omega] + \boldsymbol{\tau}_s[\partial\Omega] \\ &= \int_\Omega (\boldsymbol{x} - \boldsymbol{z}) \times \rho(\boldsymbol{x})\boldsymbol{b}(\boldsymbol{x})\, dV_{\boldsymbol{x}} + \int_{\partial\Omega} (\boldsymbol{x} - \boldsymbol{z}) \times \boldsymbol{t}(\boldsymbol{x})\, dA_{\boldsymbol{x}}.\end{aligned} \tag{3.6}$$

Necessary conditions for equilibrium, which actually follow from more general axioms discussed later (see Axiom 5.2), can be stated as follows.

Axiom 3.4 ***Conditions for Equilibrium.*** *If a body in a configuration B is in mechanical equilibrium, then the resultant force and resultant torque about any fixed point, say the origin, must vanish for every open subset Ω of B. That is*

$$\left.\begin{aligned}\boldsymbol{r}[\Omega] &= \int_\Omega \rho(\boldsymbol{x})\boldsymbol{b}(\boldsymbol{x})\, dV_{\boldsymbol{x}} + \int_{\partial\Omega} \boldsymbol{t}(\boldsymbol{x})\, dA_{\boldsymbol{x}} = \boldsymbol{0} \\ \boldsymbol{\tau}[\Omega] &= \int_\Omega \boldsymbol{x} \times \rho(\boldsymbol{x})\boldsymbol{b}(\boldsymbol{x})\, dV_{\boldsymbol{x}} + \int_{\partial\Omega} \boldsymbol{x} \times \boldsymbol{t}(\boldsymbol{x})\, dA_{\boldsymbol{x}} = \boldsymbol{0}\end{aligned}\right\} \quad \forall \Omega \subseteq B. \tag{3.7}$$

□

The freedom in the choice of reference point for the resultant torque is a consequence of $(3.7)_1$. In particular, if $\boldsymbol{r}[\Omega]$ vanishes, then $\boldsymbol{\tau}[\Omega]$ as defined in (3.6) is independent of $\boldsymbol{z}$ (see Exercise 8).

3.4.3 Local Equations

Here we derive a set of local or pointwise equations corresponding to the necessary conditions in Axiom 3.4. In view of Result 3.3, these equations naturally involve the Cauchy stress field $\boldsymbol{S}$.

Result 3.5 ***Local Equilibrium Equations.*** *If the Cauchy stress field* $\boldsymbol{S}$ *is continuously differentiable, and the density field* ρ *and body force field* $\boldsymbol{b}$ *are continuous, then the equilibrium conditions in* (3.7) *are equivalent to*

$$\left.\begin{array}{l}(\nabla\cdot\boldsymbol{S})(\boldsymbol{x})+\rho(\boldsymbol{x})\boldsymbol{b}(\boldsymbol{x})=\boldsymbol{0}\\ \boldsymbol{S}^T(\boldsymbol{x})=\boldsymbol{S}(\boldsymbol{x})\end{array}\right\}\quad\forall\boldsymbol{x}\in B, \tag{3.8}$$

or in components

$$\left.\begin{array}{l}S_{ij,j}(\boldsymbol{x})+\rho(\boldsymbol{x})b_i(\boldsymbol{x})=0\\ S_{ij}(\boldsymbol{x})=S_{ji}(\boldsymbol{x})\end{array}\right\}\quad\forall\boldsymbol{x}\in B.$$

□

Proof To establish $(3.8)_1$ we use the definition of the Cauchy stress field (Result 3.3) to rewrite $(3.7)_1$ as

$$\int_{\partial\Omega}\boldsymbol{S}\boldsymbol{n}\,dA_{\boldsymbol{x}}+\int_{\Omega}\rho\boldsymbol{b}\,dV_{\boldsymbol{x}}=\boldsymbol{0},$$

where for convenience we omit the explicit dependence on $\boldsymbol{x}$ of the integrands, and $\boldsymbol{n}$ denotes the outward unit normal field on $\partial\Omega$. Using the Tensor Divergence Theorem (Result 2.17) we obtain

$$\int_{\Omega}(\nabla\cdot\boldsymbol{S}+\rho\boldsymbol{b})\,dV_{\boldsymbol{x}}=\boldsymbol{0}.$$

Since the above equation must hold for an arbitrary region Ω in B, and the integrand is continuous by assumption, we apply the Localization Theorem (Result 2.19) to conclude the expression in $(3.8)_1$ (see Exercise 9).

To establish $(3.8)_2$ we use the definition of the Cauchy stress field to rewrite $(3.7)_2$ as

$$\int_{\partial\Omega} \boldsymbol{x} \times (\boldsymbol{S}\boldsymbol{n})\, dA_{\boldsymbol{x}} + \int_{\Omega} \boldsymbol{x} \times \rho \boldsymbol{b}\, dV_{\boldsymbol{x}} = \boldsymbol{0},$$

and then use the result $\rho\boldsymbol{b} = -\nabla \cdot \boldsymbol{S}$ from $(3.8)_1$ to rewrite the above equation as

$$\int_{\partial\Omega} \boldsymbol{x} \times (\boldsymbol{S}\boldsymbol{n})\, dA_{\boldsymbol{x}} - \int_{\Omega} \boldsymbol{x} \times (\nabla \cdot \boldsymbol{S})\, dV_{\boldsymbol{x}} = \boldsymbol{0}. \tag{3.9}$$

To simplify (3.9) we define a second-order tensor field $\boldsymbol{R} = R_{il}\boldsymbol{e}_i \otimes \boldsymbol{e}_l$ by

$$R_{il} = \epsilon_{ijk} x_j S_{kl}.$$

This tensor field has the property that $\boldsymbol{R}\boldsymbol{n} = \boldsymbol{x} \times (\boldsymbol{S}\boldsymbol{n})$, and allows (3.9) to be written as

$$\int_{\partial\Omega} \boldsymbol{R}\boldsymbol{n}\, dA_{\boldsymbol{x}} - \int_{\Omega} \boldsymbol{x} \times (\nabla \cdot \boldsymbol{S})\, dV_{\boldsymbol{x}} = \boldsymbol{0}. \tag{3.10}$$

An application of the Tensor Divergence Theorem (Result 2.17) on the first term in (3.10) yields

$$\int_{\Omega} \nabla \cdot \boldsymbol{R} - \boldsymbol{x} \times (\nabla \cdot \boldsymbol{S})\ \ dV_{\boldsymbol{x}} = \boldsymbol{0}.$$

Since this equation must hold for any subset Ω of B, and the integrand is continuous by assumption, we deduce by the Localization Theorem (Result 2.19) that

$$\nabla \cdot \boldsymbol{R} - \boldsymbol{x} \times (\nabla \cdot \boldsymbol{S}) = \boldsymbol{0}, \quad \forall \boldsymbol{x} \in B,$$

which in components becomes

$$(\epsilon_{ijk} x_j S_{kl})_{,l} - \epsilon_{ijk} x_j S_{kl,l} = 0, \quad \forall \boldsymbol{x} \in B. \tag{3.11}$$

The left-hand side of this equation may be simplified

$$\begin{aligned}
(\epsilon_{ijk} x_j S_{kl})_{,l} - \epsilon_{ijk} x_j S_{kl,l} &= \epsilon_{ijk} x_{j,l} S_{kl} + \epsilon_{ijk} x_j S_{kl,l} - \epsilon_{ijk} x_j S_{kl,l} \\
&= \epsilon_{ijk} x_{j,l} S_{kl} \\
&= \epsilon_{ijk} \delta_{jl} S_{kl} \\
&= \epsilon_{ijk} S_{kj}.
\end{aligned}$$

Thus (3.11) gives $\epsilon_{ijk} S_{kj} = 0$, which implies $\epsilon_{ijk} S_{kj} + \epsilon_{ikj} S_{jk} = 0$, since j and k are dummy indices. By properties of the permutation symbol we can rewrite this last expression as

$$\epsilon_{ijk}[S_{kj} - S_{jk}] = 0.$$

Taking (i, j, k) to be distinct, and allowing $i = 1, 2, 3$, we conclude that $S_{kj} = S_{jk}$ $(1 \le j, k \le 3)$ for each point $\boldsymbol{x} \in B$, which is the statement in $(3.8)_2$.

The integral equations (3.7) can similarly be derived from the local equations (3.8). In particular, we need only integrate (3.8) over an arbitrary open subset $\Omega \subset B$ and reverse the order of the arguments above. □

Remarks:

(1) The symmetry property of the Cauchy stress field holds even when a body is not in equilibrium (see Chapter 5).

(2) Some texts define a stress tensor $\boldsymbol{T}$ which is the transpose of the Cauchy stress field $\boldsymbol{S}$ as defined here; that is, $\boldsymbol{T} = \boldsymbol{S}^T$. However, because of the symmetry of the Cauchy stress field this distinction is immaterial in practice.

(3) The local equations (3.8) do not completely determine the Cauchy stress field for a body in equilibrium. In particular, we have three partial differential equations in $(3.8)_1$ and three independent algebraic equations in $(3.8)_2$ with which to determine the nine unknown components of $\boldsymbol{S}$. This shortage of equations will arise frequently in later chapters. It is addressed by adding **constitutive equations** which characterize the specific material properties of a body.

(4) The traction field $\boldsymbol{h}$ on the bounding surface ∂B represents the surface force, per unit area, exerted by the environment on B. Thus by Result 3.3 we obtain $\boldsymbol{S}\boldsymbol{n} = \boldsymbol{h}$ for all $\boldsymbol{x} \in \partial B$, where $\boldsymbol{n}$ is the outward unit normal field on ∂B. This provides **boundary conditions** for the partial differential equations in $(3.8)_1$ (see Exercise 12).

(5) In deriving the equilibrium equations we have assumed continuity of ρ and $\boldsymbol{b}$, and continuous differentiability of $\boldsymbol{S}$. In practice, establishing such regularity properties for the field equations of continuum mechanics is an important part of the subject. Throughout our developments will we assume, without explicit reference, that the fields in question are sufficiently regular to pass from integral laws to differential equations via localization. □

3.5 Basic Stress Concepts

Here we describe various states of stress that may exist at a point in a continuum body and define principal stress values and stress directions. We introduce the concepts of normal and shear stresses at a point on a surface with a prescribed normal vector, and then determine those surfaces that experience maximum normal and shear stresses. We also introduce a decomposition of the stress field into spherical and deviatoric parts and discuss their physical interpretations.

3.5.1 Simple States of Stress

When the stress tensor $\boldsymbol{S}$ is of a particular form at a point $\boldsymbol{x}$ in a continuum body, we often say that the stress field is in a particular **state** at the point $\boldsymbol{x}$. For example, if at a point $\boldsymbol{x}$ the stress tensor $\boldsymbol{S}$ is of the form

$$\boldsymbol{S} = -\pi \boldsymbol{I},$$

where π is a scalar, we say that a **spherical** or **Eulerian** state of stress exists at $\boldsymbol{x}$. For this stress state the traction on any surface with normal $\boldsymbol{n}$ at $\boldsymbol{x}$ is given by

$$\boldsymbol{t} = \boldsymbol{S}\boldsymbol{n} = -\pi \boldsymbol{n},$$

which shows that the traction is always normal to the surface. For example, assuming $\boldsymbol{S}$ is constant in a neighborhood of $\boldsymbol{x}$, the traction field on the surface of a small sphere centered at $\boldsymbol{x}$ would be everywhere normal to the sphere, as illustrated in Figure 3.6(a) (see Exercise 13).

The state of stress at a point $\boldsymbol{x}$ is said to be **uniaxial** if there exists a unit vector $\boldsymbol{\gamma}$ and a scalar σ such that

$$\boldsymbol{S} = \sigma\, \boldsymbol{\gamma} \otimes \boldsymbol{\gamma}.$$

Such a state is called a pure tension if $\sigma > 0$, and a pure compression if $\sigma < 0$. For a uniaxial stress state the traction on any surface with normal $\boldsymbol{n}$ at $\boldsymbol{x}$ is given by

$$\boldsymbol{t} = \boldsymbol{S}\boldsymbol{n} = (\boldsymbol{\gamma} \cdot \boldsymbol{n})\sigma\, \boldsymbol{\gamma}.$$

Notice that the traction is always parallel to $\boldsymbol{\gamma}$ and vanishes when $\boldsymbol{n}$ is orthogonal to $\boldsymbol{\gamma}$. For example, if we assume $\boldsymbol{S}$ is constant in a neighborhood of $\boldsymbol{x}$ and consider a small box centered at $\boldsymbol{x}$ having one axis aligned with $\boldsymbol{\gamma}$, then the traction field on the surface of the box would

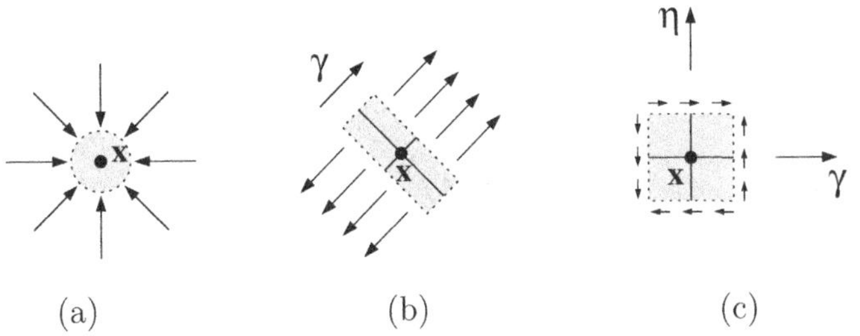

Fig. 3.6 Graphical illustration of three simple states of stress at a point in a continuum body: (a) spherical, (b) uniaxial and (c) pure shear.

be non-zero only on those faces with normal $\boldsymbol{n} = \pm\boldsymbol{\gamma}$, as illustrated in Figure 3.6(b).

If at a point $\boldsymbol{x}$ there exists a pair of orthogonal unit vectors $\boldsymbol{\gamma}$ and $\boldsymbol{\eta}$ and a scalar τ such that

$$\boldsymbol{S} = \tau\,(\boldsymbol{\gamma} \otimes \boldsymbol{\eta} + \boldsymbol{\eta} \otimes \boldsymbol{\gamma}),$$

we say that a state of **pure shear** exists at $\boldsymbol{x}$. For this stress state the traction on any surface with normal $\boldsymbol{n}$ at $\boldsymbol{x}$ is given by

$$\boldsymbol{t} = \boldsymbol{S}\boldsymbol{n} = (\boldsymbol{\eta} \cdot \boldsymbol{n})\tau\,\boldsymbol{\gamma} + (\boldsymbol{\gamma} \cdot \boldsymbol{n})\tau\,\boldsymbol{\eta}.$$

Notice that $\boldsymbol{t} = \tau\boldsymbol{\gamma}$ when $\boldsymbol{n} = \boldsymbol{\eta}$, and $\boldsymbol{t} = \tau\boldsymbol{\eta}$ when $\boldsymbol{n} = \boldsymbol{\gamma}$. For example, if we assume $\boldsymbol{S}$ is constant in a neighborhood of $\boldsymbol{x}$ and consider a small box centered at $\boldsymbol{x}$ having one axis aligned with $\boldsymbol{\gamma}$ and another with $\boldsymbol{\eta}$, then the traction field on the surface of the box would be non-zero only on those faces with normal $\boldsymbol{n} = \pm\boldsymbol{\gamma}$ and $\pm\boldsymbol{\eta}$, as illustrated in Figure 3.6(c).

If at a point $\boldsymbol{x}$ there exists a pair of orthogonal unit vectors $\boldsymbol{\gamma}$ and $\boldsymbol{\eta}$ such that the matrix representation of $\boldsymbol{S}$ in the basis $\boldsymbol{e}_1 = \boldsymbol{\gamma}$, $\boldsymbol{e}_2 = \boldsymbol{\eta}$ and $\boldsymbol{e}_3 = \boldsymbol{\gamma} \times \boldsymbol{\eta}$ is of the form

$$[\boldsymbol{S}] = \begin{pmatrix} S_{11} & S_{12} & 0 \\ S_{21} & S_{22} & 0 \\ 0 & 0 & 0 \end{pmatrix},$$

we say that a state of **plane stress** exists at $\boldsymbol{x}$. It is straightforward to verify that uniaxial and pure shear stress states are special cases of a plane stress state (see Exercise 14).

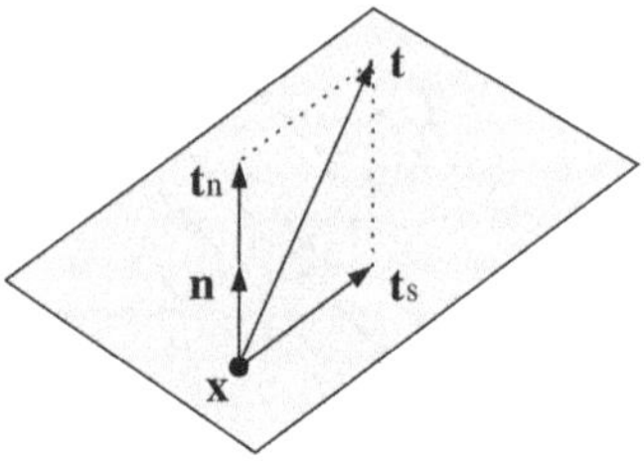

Fig. 3.7 Decomposition of the traction vector into normal and shear tractions at a point $\boldsymbol{x}$. The shear traction is tangent to the surface and the normal traction is normal to the surface.

3.5.2 Principal, Normal and Shear Stresses

Given a point $\boldsymbol{x}$ in a continuum body we call the eigenvalues of $\boldsymbol{S}$ the **principal stresses** and the corresponding eigenvectors the **principal stress directions** at $\boldsymbol{x}$. Because the stress tensor is symmetric, we note that there exist three mutually perpendicular principal directions and three corresponding principal stresses for each point $\boldsymbol{x}$ (see Result 1.9).

If we let $\boldsymbol{e}$ be a principal direction with corresponding principal stress σ, then the traction vector on a surface with normal $\boldsymbol{e}$ at $\boldsymbol{x}$ is given by

$$\boldsymbol{t} = \boldsymbol{S}\boldsymbol{e} = \sigma\boldsymbol{e}.$$

Hence the traction vector is itself in the direction $\boldsymbol{e}$.

Consider an arbitrary surface with unit normal $\boldsymbol{n}$ at $\boldsymbol{x}$. Then the traction vector can be decomposed into the sum of two parts: a **normal traction**

$$\boldsymbol{t}_n = (\boldsymbol{t} \cdot \boldsymbol{n})\boldsymbol{n},$$

and a **shear traction**

$$\boldsymbol{t}_s = \boldsymbol{t} - (\boldsymbol{t} \cdot \boldsymbol{n})\boldsymbol{n}.$$

In particular, we have $\boldsymbol{t} = \boldsymbol{t}_n + \boldsymbol{t}_s$, as illustrated in Figure 3.7. We call $\sigma_n = |\boldsymbol{t}_n|$ the **normal stress** and $\sigma_s = |\boldsymbol{t}_s|$ the **shear stress** on the surface with normal $\boldsymbol{n}$ at $\boldsymbol{x}$.

If we consider a surface with normal $\boldsymbol{n} = \boldsymbol{e}$, where $\boldsymbol{e}$ is a principal direction with corresponding principal stress σ, then we find that $\boldsymbol{t}_n = \sigma\boldsymbol{e}$ and $\boldsymbol{t}_s = \boldsymbol{0}$. Hence the normal stress on such a surface is $\sigma_n = |\sigma|$ and the shear stress is zero (see Exercises 15 and 16). In particular, the shear stress on a surface is zero if and only if $\boldsymbol{n}$ is a principal direction.

3.5.3 Maximum Normal and Shear Stresses

Given a point $\boldsymbol{x}$ in a continuum body it is of interest to know what families of surfaces through $\boldsymbol{x}$ experience a maximum normal stress or a maximum shear stress. That is, we would like to know for what normal direction $\boldsymbol{n}$ at $\boldsymbol{x}$ is σ_n a maximum; similarly, for what normal direction is σ_s a maximum.

To begin, we denote the principal stresses at the point $\boldsymbol{x}$ by σ_i and we denote the corresponding principal directions by $\boldsymbol{e}_i$. For simplicity, we assume that the principal stresses are distinct and ordered in the sense that

$$\sigma_1 > \sigma_2 > \sigma_3. \tag{3.12}$$

When the principal stresses are not distinct as above, there may not be isolated directions $\boldsymbol{n}$ for which σ_n or σ_s achieve maximum values. For example, when $\sigma_i = \alpha$ $(i = 1, 2, 3)$, we can verify that $\sigma_n = |\alpha|$ and $\sigma_s = 0$ for all $\boldsymbol{n}$. Thus there is no isolated direction $\boldsymbol{n}$ for which σ_n or σ_s achieves a maximum value. The next result shows that, when the principal stresses are distinct as in (3.12), there are isolated direction pairs $(\boldsymbol{n}, -\boldsymbol{n})$ for which σ_n and σ_s achieve maximum values.

Result 3.6 ***Maximal Normal and Shear Stresses.*** *Consider a point $\boldsymbol{x}$ in a continuum body and assume the principal stresses are distinct and ordered as in* (3.12). *Then:*

(1) the maximum value of σ_n is $|\sigma_k|$ and this value is achieved for pairs $\boldsymbol{n} = \pm\boldsymbol{e}_k$, where $k = 1$ and/or $k = 3$ depending on which value of k achieves the maximum in the definition

$$|\sigma_k| = \max\{|\sigma_1|, |\sigma_3|\},$$

(2) the maximum value of σ_s is $\frac{1}{2}|\sigma_1 - \sigma_3|$ and this value is achieved for the two pairs

$$\boldsymbol{n} = \pm\tfrac{1}{\sqrt{2}}(\boldsymbol{e}_1 + \boldsymbol{e}_3), \quad \pm\tfrac{1}{\sqrt{2}}(\boldsymbol{e}_1 - \boldsymbol{e}_3).$$

□

Proof See Exercises 17, 18. □

3.5.4 Spherical and Deviatoric Stress Tensors

At any point $\boldsymbol{x}$ in a continuum body the Cauchy stress tensor can be decomposed into the sum of two parts: a **spherical stress tensor**

$$\boldsymbol{S}_S = -p\boldsymbol{I},$$

and a **deviatoric stress tensor**

$$\boldsymbol{S}_D = \boldsymbol{S} + p\boldsymbol{I},$$

where $p = -\frac{1}{3}\operatorname{tr}\boldsymbol{S}$ is called the **pressure**. In particular, we have $\boldsymbol{S} = \boldsymbol{S}_S + \boldsymbol{S}_D$.

The spherical stress tensor $\boldsymbol{S}_S$ is completely determined by the pressure p. Noting that

$$-p = \tfrac{1}{3}\operatorname{tr}\boldsymbol{S} = \tfrac{1}{3}(\sigma_1 + \sigma_2 + \sigma_3),$$

we see that $-p$ is the arithmetic average of the three principal stresses at a point $\boldsymbol{x}$. Equivalently, $-p$ can be interpreted as the average of the normal component of traction $\boldsymbol{t}\cdot\boldsymbol{n}$, where the average is taken over all possible directions $\boldsymbol{n}$ at $\boldsymbol{x}$ (see Exercise 19). Roughly speaking, the spherical stress is that part of the overall stress that tends to change the volume of a body without changing its shape: a round ball of uniform material subject to a uniform pressure should experience a change of volume while retaining its round shape. Notice that the sign convention for pressure is such that $p > 0$ corresponds to a compression.

The deviatoric stress is that part of the overall stress which is complementary to the spherical stress. In particular, the deviatoric stress tensor contains information about shear stresses, whereas the spherical stress tensor contains information about normal stresses. Roughly speaking, the deviatoric stress is that part of the overall stress that tends to change the shape of a body while producing little or no change in volume: a cube of uniform material subject to a state of pure shear should lose its cubical shape while experiencing only a small change in volume.

By virtue of its definition the deviatoric stress tensor has the property that

$$\operatorname{tr}\boldsymbol{S}_D = 0. \tag{3.13}$$

Hence the first principal invariant $I_1(\boldsymbol{S}_D)$ is zero (see Section 1.3.9). The remaining principal invariants $I_2(\boldsymbol{S}_D)$ and $I_3(\boldsymbol{S}_D)$ are usually denoted by $-J_2(\boldsymbol{S})$ and $J_3(\boldsymbol{S})$. In particular, by (3.13), the symmetry of $\boldsymbol{S}_D$, and the definition of the principal invariants, we obtain

$$J_2(\boldsymbol{S}) = -I_2(\boldsymbol{S}_D) = \tfrac{1}{2}\boldsymbol{S}_D : \boldsymbol{S}_D \quad \text{and} \quad J_3(\boldsymbol{S}) = I_3(\boldsymbol{S}_D) = \det \boldsymbol{S}_D.$$

These invariants play a central role in the formulation of so-called *yield conditions* in theories describing the plastic deformation of isotropic materials at small strains (see Exercise 20).

Bibliographic Notes

Most of the material presented here on bodies, mass and force can be found in standard texts in continuum mechanics, for example Gurtin (1981), Chadwick (1976), Mase (1970) and Malvern (1969). A rigorous, axiomatic treatment of these primitive concepts can be found in Truesdell (1991). In some applications it is useful to introduce the additional primitive concept of a couple, which can be described as a torque that is not the moment of a force. For a discussion of couples in continuum mechanics see Malvern (1969).

The theory of internal surface forces in a continuum body as described here consists of four main parts: (i) the existence of a traction field on an oriented surface, which is referred to as the Euler–Cauchy Cut Principle; (ii) Cauchy's Postulate, which describes the manner in which the traction field is dependent on the surface geometry; (iii) the law of action and reaction for traction fields, which is typically referred to as Cauchy's Lemma; (iv) Cauchy's Theorem, which establishes the existence of the stress tensor. A rigorous account of all these aspects of the theory is contained in Truesdell (1991), and a more concise account is given in Gurtin (1981).

The equilibrium equations presented here are special cases of the balance laws which govern the rate of change of total linear and angular momentum for a continuum body. These laws are typically referred to as Euler's or Cauchy's laws of motion depending on whether they are in integral or local form, respectively. In particular, the integral form of the equilibrium equations given in Axiom 3.4 arises when the rates of change of linear and angular momentum in Euler's laws of motion in Axiom 5.2 are set to zero. A discussion of the momentum balance laws will be given in Chapter 5. For a detailed treatment and historical overview of Euler's and Cauchy's laws see Truesdell (1991).

The analysis of normal and shear stresses on families of surfaces at a point in a body is important in many engineering applications. Such an analysis can be carried out graphically using the technique of Mohr's circles. For a discussion of this technique see Mase (1970) and Malvern (1969). In this respect we remark that, while we have defined the normal stress as the magnitude of the normal traction vector, many authors define it as the normal component of the traction vector. These two definitions are equivalent up to sign.

Exercises

3.1 Consider a body with configuration $B = \{\boldsymbol{x} \in \mathbb{E}^3 \mid |x_i| < 1\}$ and mass density field $\rho = \exp(x_3)$. Find:

(a) $\text{vol}[B]$, (b) $\boldsymbol{x}_{\text{cov}}[B]$, (c) $\text{mass}[B]$, (d) $\boldsymbol{x}_{\text{com}}[B]$.

3.2 Provide an example of a body B and corresponding mass density field ρ for which $\boldsymbol{x}_{\text{cov}}[B] \notin B$ and $\boldsymbol{x}_{\text{com}}[B] \notin B$.

3.3 Consider a body B with density field ρ and volume $\text{vol}[B] > 0$. The **inertia tensor** of B with respect to a point $\boldsymbol{y}$ is a second-order tensor $\mathbf{I}_{\boldsymbol{y}}$ defined by

$$\mathbf{I}_{\boldsymbol{y}} = \int_B \rho(\boldsymbol{x}) \left[|\boldsymbol{x} - \boldsymbol{y}|^2 \boldsymbol{I} - (\boldsymbol{x} - \boldsymbol{y}) \otimes (\boldsymbol{x} - \boldsymbol{y}) \right] \, dV_{\boldsymbol{x}}.$$

(a) Show that $\mathbf{I}_{\boldsymbol{y}}$ is symmetric, positive-definite for any point $\boldsymbol{y}$.

(b) Let M be the total mass and $\overline{\boldsymbol{x}}$ the center of mass of B. Show that $\mathbf{I}_{\boldsymbol{y}}$ can be decomposed as

$$\mathbf{I}_{\boldsymbol{y}} = \mathbf{I}_{\overline{\boldsymbol{x}}} + M \left[|\overline{\boldsymbol{x}} - \boldsymbol{y}|^2 \boldsymbol{I} - (\overline{\boldsymbol{x}} - \boldsymbol{y}) \otimes (\overline{\boldsymbol{x}} - \boldsymbol{y}) \right]$$

for any point $\boldsymbol{y}$.

3.4 Suppose a body with configuration $B = \{\boldsymbol{x} \in \mathbb{E}^3 \mid 0 < x_i < 1\}$ and constant mass density $\rho > 0$ is subject to a gravitational force field per unit mass $\boldsymbol{b} = -g\boldsymbol{e}_3$, where g is the gravitational acceleration constant. Find:

(a) the resultant force $\boldsymbol{r}_b[B]$ (weight of B),

(b) the resultant torque $\boldsymbol{\tau}_b[B]$ about the origin,

(c) the resultant torque $\boldsymbol{\tau}_b[B]$ about the mass center $\boldsymbol{x}_{\text{com}}[B]$.

3.5 Consider an arbitrary body B subject to a body force.

(a) Show that the resultant torque about the center of mass vanishes when the body is subject to a uniform (constant) body force field per unit mass. Does the resultant torque about any other reference point also vanish in this case?

(b) Show that the resultant torque about the center of volume vanishes when the body is subject to a uniform body force field per unit volume. Does the resultant torque about any other reference point also vanish in this case?

3.6 Suppose a body with configuration $B = \{\boldsymbol{x} \in \boldsymbol{E}^3 \mid 0 < x_i < 1\}$ is subject to a body force field per unit volume $\widehat{\boldsymbol{b}} = \alpha x_3 \boldsymbol{e}_3$ and a traction field on its bounding surface ∂B given by

$$\boldsymbol{t} = \begin{cases} x_1 x_2 (1 - x_1)(1 - x_2)\boldsymbol{e}_3, & \text{on face } x_3 = 0, \\ \boldsymbol{0}, & \text{on all other faces.} \end{cases}$$

Find the value of α (constant) for which the resultant body and surface forces are balanced, that is, $\boldsymbol{r}_b[B] + \boldsymbol{r}_s[\partial B] = \boldsymbol{0}$.

3.7 Consider a solid body B immersed in a liquid of constant mass density ρ_* and subject to a uniform gravitational force field per unit mass. Suppose that the free surface of the liquid coincides with the plane $x_3 = 0$, and that the downward direction (into the liquid) coincides with $\boldsymbol{e}_3$. In this case, the liquid exerts a hydrostatic surface force field on the bounding surface of B

$$\boldsymbol{t} = -p\boldsymbol{n},$$

where $\boldsymbol{n}$ is the outward unit normal on the surface of B, $p = \rho_* g x_3$ is the hydrostatic pressure in the liquid, and g is the gravitational acceleration constant.

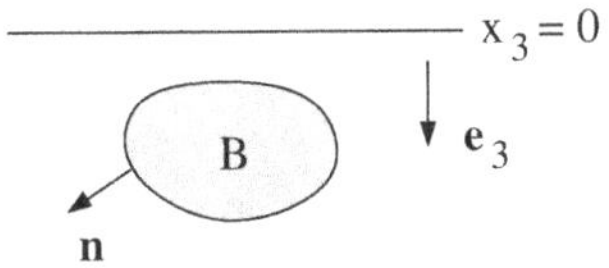

(a) Use the Divergence Theorem to show that the resultant hydrostatic surface force on B (the buoyant force) is given by $\boldsymbol{r}_s[\partial B] = -W\boldsymbol{e}_3$, where W is the weight of the liquid displaced by B.

(b) Show that the hydrostatic surface force has a zero resultant torque about the center of volume of B, that is, $\boldsymbol{\tau}_s[\partial B] = \boldsymbol{0}$.

Remark: The result in (a) is known as **Archimedes' Principle**. It states that the buoyant force on an object equals the weight of the displaced liquid. The result in (b) shows that the buoyant force acts at the center of volume of the object.

3.8 Prove that if $\boldsymbol{r}[\Omega]$ as defined in (3.5) vanishes, then $\boldsymbol{\tau}[\Omega]$ as defined in (3.6) is independent of $\boldsymbol{z}$.

3.9 Consider a body $B = \{\boldsymbol{x} \in \mathbb{E}^3 \mid 0 < x_i < 1\}$ with constant mass density $\rho > 0$ subject to a constant body force per unit mass $[\boldsymbol{b}] = (0, 0, -g)^T$. Suppose the Cauchy stress field in B is given by

$$[\boldsymbol{S}] = \begin{pmatrix} x_2 & x_3 & 0 \\ x_3 & x_1 & 0 \\ 0 & 0 & \rho g x_3 \end{pmatrix}.$$

(a) Show that $\boldsymbol{S}$ and $\boldsymbol{b}$ satisfy the local equilibrium equations in Result 3.5.

(b) Find the traction field on each of the six faces of the bounding surface ∂B.

(c) Find by direct calculation the resultant surface force $\boldsymbol{r}_s[\partial B]$ and the resultant body force $\boldsymbol{r}_b[B]$ and verify that these forces are balanced, that is, $\boldsymbol{r}_s[\partial B] + \boldsymbol{r}_b[B] = \boldsymbol{0}$. Briefly explain how this result is consistent with part (a).

3.10 Consider a body B with uniform mass density field $\rho > 0$ (constant) under the influence of a body force per unit mass $\boldsymbol{b} = -4\boldsymbol{C}\boldsymbol{x}$, where $\boldsymbol{C}$ is a given second-order tensor (constant). Moreover, suppose the Cauchy stress field in B is of the form

$$\boldsymbol{S} = \rho\,(\boldsymbol{C}\boldsymbol{x}) \otimes \boldsymbol{x}.$$

(a) Show that $\boldsymbol{S}$ and $\boldsymbol{b}$ satisfy the local equilibrium equation $\nabla \cdot \boldsymbol{S} + \rho \boldsymbol{b} = \boldsymbol{0}$ for balance of forces.

(b) Find conditions on $\boldsymbol{C}$ for which the local equilibrium equation $\boldsymbol{S}^T = \boldsymbol{S}$ for balance of torques will be satisfied.

3.11 Consider a continuum body B with mass density ρ subject to a body force per unit mass $\boldsymbol{b}$ and a traction $\boldsymbol{h}$ on its bounding surface. Assume the Cauchy stress field $\boldsymbol{S}$ in B is related to a vector field $\boldsymbol{u} : B \to \mathcal{V}$ by the expression

$$\boldsymbol{S} = \mathsf{C}\boldsymbol{E} \quad \text{or} \quad S_{ij} = \mathsf{C}_{ijkl}E_{kl},$$

where C is a constant fourth-order tensor and $\boldsymbol{E} : B \to \mathcal{V}^2$ is defined by

$$\boldsymbol{E} = \mathrm{sym}(\nabla \boldsymbol{u}) = \tfrac{1}{2}(\nabla \boldsymbol{u} + \nabla \boldsymbol{u}^T).$$

Here $\boldsymbol{u}$ is the **displacement field** of the body from an unstressed state and $\boldsymbol{E}$ is the **infinitesimal strain tensor**. Moreover, let W denote the **strain energy** in B, defined by

$$W = \frac{1}{2}\int_B \boldsymbol{E}(\boldsymbol{x}) : \mathbf{C}\boldsymbol{E}(\boldsymbol{x})\, dV_{\boldsymbol{x}}.$$

Assuming B is in equilibrium show that

$$W = \frac{1}{2}\left(\int_B \rho(\boldsymbol{x})\boldsymbol{b}(\boldsymbol{x}) \cdot \boldsymbol{u}(\boldsymbol{x})\, dV_{\boldsymbol{x}} + \int_{\partial B} \boldsymbol{h}(\boldsymbol{x}) \cdot \boldsymbol{u}(\boldsymbol{x})\, dA_{\boldsymbol{x}}\right).$$

3.12 Consider a straight bar B of uniform cross-section whose axis is parallel to the z-axis of an xyz-coordinate system. Let Ω denote a typical cross-section of B and assume the boundary $\partial\Omega$ is described by a smooth curve C in the xy-plane as illustrated in the figure below.

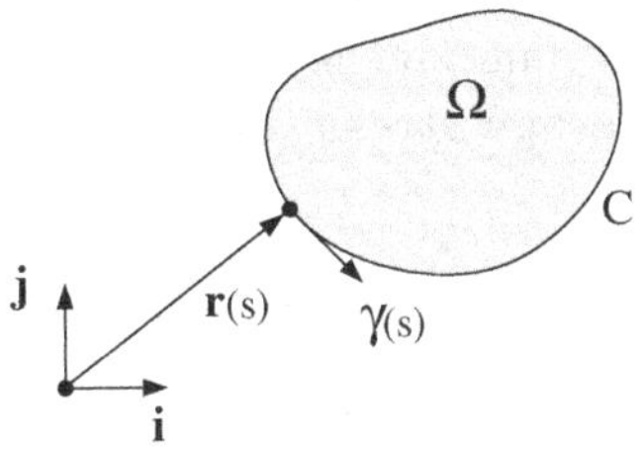

(a) Let $\boldsymbol{r}(s) = x(s)\boldsymbol{e}_1 + y(s)\boldsymbol{e}_2$, $0 \leq s \leq L$, be an arclength parametrization of C, so that

$$\boldsymbol{\gamma} = \frac{d\boldsymbol{r}}{ds} = \frac{dx}{ds}\boldsymbol{e}_1 + \frac{dy}{ds}\boldsymbol{e}_2$$

is a unit tangent vector field on C in the direction of increasing arclength parameter s. Show that the vector field

$$\boldsymbol{n} = \frac{dy}{ds}\boldsymbol{i} - \frac{dx}{ds}\boldsymbol{j}$$

is a unit vector field on C, is everywhere orthogonal to $\boldsymbol{\gamma}$, and is oriented such that $\boldsymbol{n} \times \boldsymbol{\gamma} = \boldsymbol{e}_3$.

(b) Suppose the ends of the bar B are twisted relative to each other by an amount so small that the configuration of B remains essentially unchanged; in particular, cross-sections do not warp and remain perpendicular to the z-axis. For such twisting, we may assume that the Cauchy stress in B is of the form

$$[\boldsymbol{S}] = \begin{pmatrix} 0 & 0 & \tau_x \\ 0 & 0 & \tau_y \\ \tau_x & \tau_y & 0 \end{pmatrix},$$

where τ_x and τ_y are each functions of x and y only, and that these two functions are related to a single scalar function $\phi(x, y)$ by

$$\tau_x = \frac{\partial \phi}{\partial y}, \qquad \tau_y = -\frac{\partial \phi}{\partial x}.$$

Determine the boundary condition imposed on the function ϕ by the requirement that $\boldsymbol{S}\boldsymbol{n} = \boldsymbol{0}$ on C. That is, what conditions must ϕ satisfy on C in order that the lateral surface of the bar be traction-free?

3.13 Consider a continuum body B subject to a body force per unit volume $\widehat{\boldsymbol{b}}$ and a traction field $\boldsymbol{h}$ on its bounding surface. Suppose B is in equilibrium and let $\boldsymbol{S}$ denote its Cauchy stress field. Define the **average stress tensor** $\overline{\boldsymbol{S}}$ (a constant) in B by

$$\overline{\boldsymbol{S}} = \frac{1}{\text{vol}[B]} \int_B \boldsymbol{S} \, dV_{\boldsymbol{x}}.$$

(a) Use the local equilibrium equations for B to show that

$$\overline{\boldsymbol{S}} = \frac{1}{\text{vol}[B]} \left[\int_B \boldsymbol{x} \otimes \widehat{\boldsymbol{b}} \, dV_{\boldsymbol{x}} + \int_{\partial B} \boldsymbol{x} \otimes \boldsymbol{h} \, dA_{\boldsymbol{x}} \right].$$

(b) Suppose that $\widehat{\boldsymbol{b}} = \boldsymbol{0}$ and that the applied traction $\boldsymbol{h}$ is a uniform pressure, so that

$$\boldsymbol{h} = -p\boldsymbol{n},$$

where p is a constant and $\boldsymbol{n}$ is the outward unit normal field on ∂B. Use the result in (a) to show that the average stress tensor is spherical, namely

$$\overline{\boldsymbol{S}} = -p\boldsymbol{I}.$$

(c) Under the same conditions as in (b) show that the uniform, spherical stress field $\boldsymbol{S} = -p\boldsymbol{I}$ satisfies the equations of equilibrium in B and the boundary condition $\boldsymbol{S}\boldsymbol{n} = \boldsymbol{h}$ on ∂B. Thus in this case we have $\boldsymbol{S}(\boldsymbol{x}) = \overline{\boldsymbol{S}}$ for all $\boldsymbol{x} \in B$.

Remark: The result in (a) is known as **Signorini's Theorem**. It states that the average value of the Cauchy stress tensor in a body in equilibrium is completely determined by the external surface traction $\boldsymbol{h}$ and the body force $\widehat{\boldsymbol{b}}$.

3.14 Show that uniaxial and pure shear stress states are examples of plane stress states. In particular, for each of these states find a basis in which the matrix representation $[\boldsymbol{S}]$ has the required form.

3.15 Suppose the Cauchy stress tensor at a point $\boldsymbol{x}$ in a body has the form

$$[\boldsymbol{S}] = \begin{pmatrix} 5 & 3 & -3 \\ 3 & 0 & 2 \\ -3 & 2 & 0 \end{pmatrix},$$

and consider a surface Γ with normal $[\boldsymbol{n}] = (0, 1/\sqrt{2}, 1/\sqrt{2})^T$ and a surface Γ' with normal $[\boldsymbol{n}'] = (1,0,0)^T$ at $\boldsymbol{x}$.

(a) Find the normal and shear tractions $[\boldsymbol{t}_n]$ and $[\boldsymbol{t}_s]$ on each surface at $\boldsymbol{x}$. In particular, show that Γ experiences no shear traction at $\boldsymbol{x}$, whereas Γ' does.

(b) Find the principal stresses and stress directions at $\boldsymbol{x}$ and verify that $[\boldsymbol{n}]$ is a principal direction.

3.16 Suppose the Cauchy stress field in a body $B = \{\boldsymbol{x} \in \mathbb{E}^3 \mid |x_i| < 1\}$ is uniaxial of the form

$$[\boldsymbol{S}] = \begin{pmatrix} 0 & 0 & 0 \\ 0 & \sigma & 0 \\ 0 & 0 & 0 \end{pmatrix},$$

where $\sigma \neq 0$ is constant. In this case notice that the traction field $\boldsymbol{t}$ on any plane through B will be constant because $\boldsymbol{S}$ and $\boldsymbol{n}$ are constant.

(a) Consider the family of planes Γ_θ through the origin which contain the x_1-axis and have unit normal $[\boldsymbol{n}] = (0, \cos\theta, \sin\theta)^T$, $\theta \in [0, \pi/2]$. Find the normal and shear stresses σ_n and σ_s on these planes as a function of θ.

(b) Show that the maximum normal stress is $\sigma_n = |\sigma|$ and that this value occurs on the plane with $\theta = 0$. Similarly, show that the maximum shear stress is $\sigma_s = \frac{1}{2}|\sigma|$ and that this value occurs on the plane with $\theta = \pi/4$.

Remark: The result in (b) illustrates the principle that planes of maximum shear stress occur at 45-degree angles to planes of maximum normal stress.

3.17 Prove Result 3.6 (1).

3.18 Prove Result 3.6 (2).

3.19 Let p be the pressure at a point $\boldsymbol{x}$ in a continuum body and let $\mathcal{N}$ be the set of all unit vectors $\boldsymbol{n}$. By identifying $\mathcal{N}$ with the standard unit sphere show

$$p = -\frac{\int_{\mathcal{N}} \boldsymbol{n} \cdot \boldsymbol{t} \, dA_{\boldsymbol{n}}}{\int_{\mathcal{N}} dA_{\boldsymbol{n}}},$$

where $\boldsymbol{t}$ is the traction vector on a surface with normal $\boldsymbol{n}$ at $\boldsymbol{x}$.

3.20 Two different measures of stress intensity in a body are provided by the functions $f(\boldsymbol{S}) = |p|$ and $g(\boldsymbol{S}) = \sqrt{J_2}$, where p is the pressure and J_2 is the second deviatoric invariant associated with the Cauchy stress field $\boldsymbol{S}$. Find f and g for the following stress states $[\boldsymbol{S}]$:

$$\text{(a)} \begin{pmatrix} \sigma & 0 & 0 \\ 0 & \sigma & 0 \\ 0 & 0 & \sigma \end{pmatrix}, \quad \text{(b)} \begin{pmatrix} 0 & \tau & 0 \\ \tau & 0 & 0 \\ 0 & 0 & 0 \end{pmatrix}, \quad \text{(c)} \begin{pmatrix} \sigma & 0 & 0 \\ 0 & -\sigma & 0 \\ 0 & 0 & 0 \end{pmatrix}.$$

Remark: The function $g(\boldsymbol{S}) = \sqrt{J_2}$ is typically called the **Mises yield function** and is employed in the modeling of rigid-plastic materials. Such materials are considered to be rigid, and deform only when $g(\boldsymbol{S})$ exceeds a threshold value.

Answers to Selected Exercises

3.1 Let $e = \exp(1)$ be the base of the natural logarithm. Then:

(a) $\text{vol}[B] = \int_{-1}^{1} \int_{-1}^{1} \int_{-1}^{1} dx_1 dx_2 dx_3 = 8.$

(b) $\boldsymbol{x}_{\text{cov}}[B] = \frac{1}{\text{vol}[B]} \left(\int_{-1}^{1} \int_{-1}^{1} \int_{-1}^{1} x_i \, dx_1 dx_2 dx_3 \right) \boldsymbol{e}_i = \boldsymbol{0}.$

(c) $\text{mass}[B] = \int_{-1}^{1} \int_{-1}^{1} \int_{-1}^{1} e^{x_3} \, dx_1 dx_2 dx_3 = 4e - 4e^{-1}.$

(d) $\boldsymbol{x}_{\text{com}}[B] = \frac{1}{\text{mass}[B]} \left(\int_{-1}^{1} \int_{-1}^{1} \int_{-1}^{1} e^{x_3} x_i \, dx_1 dx_2 dx_3 \right) \boldsymbol{e}_i = (\frac{2}{e^2 - 1}) \boldsymbol{e}_3.$

3.3 (a) In components we have

$$\begin{aligned}[\mathbf{I}_{\boldsymbol{y}}]_{ij} &= \boldsymbol{e}_i \cdot \mathbf{I}_{\boldsymbol{y}} \boldsymbol{e}_j \\ &= \int_B \rho(\boldsymbol{x}) \left[|\boldsymbol{x}-\boldsymbol{y}|^2 \delta_{ij} - ((\boldsymbol{x}-\boldsymbol{y}) \cdot \boldsymbol{e}_i)((\boldsymbol{x}-\boldsymbol{y}) \cdot \boldsymbol{e}_j) \right] \, dV_{\boldsymbol{x}}.\end{aligned}$$

Thus $\mathbf{I}_{\boldsymbol{y}}$ is symmetric since $[\mathbf{I}_{\boldsymbol{y}}]_{ij} = [\mathbf{I}_{\boldsymbol{y}}]_{ji}$. Next, for any vector $\boldsymbol{a} \neq \boldsymbol{0}$ we notice that

$$\boldsymbol{a} \cdot \mathbf{I}_{\boldsymbol{y}} \boldsymbol{a} = \int_B \rho(\boldsymbol{x}) \left[|\boldsymbol{x}-\boldsymbol{y}|^2 |\boldsymbol{a}|^2 - |(\boldsymbol{x}-\boldsymbol{y}) \cdot \boldsymbol{a}|^2 \right] \, dV_{\boldsymbol{x}} \geq 0. \quad (3.14)$$

This follows from the fact that $\rho > 0$ in B and $|\boldsymbol{u} \cdot \boldsymbol{v}| \leq |\boldsymbol{u}||\boldsymbol{v}|$ for any two vectors $\boldsymbol{u}$ and $\boldsymbol{v}$. Moreover, equality in (3.14) can be achieved only if

$$|\boldsymbol{x}-\boldsymbol{y}|^2 |\boldsymbol{a}|^2 - |(\boldsymbol{x}-\boldsymbol{y}) \cdot \boldsymbol{a}|^2 = 0, \quad \forall \boldsymbol{x} \in B,$$

which is possible only if $\boldsymbol{x}-\boldsymbol{y}$ is parallel to $\boldsymbol{a}$ for all $\boldsymbol{x} \in B$. This implies that all points in B must lie on a straight line, which contradicts the assumption that B has non-zero volume. Thus (3.14) must hold with strict inequality for all vectors $\boldsymbol{a} \neq \boldsymbol{0}$ and we conclude that $\mathbf{I}_{\boldsymbol{y}}$ is positive-definite.

(b) By definition of the center of mass $\overline{\boldsymbol{x}}$ we have

$$\int_B \rho(\boldsymbol{x})(\boldsymbol{x}-\overline{\boldsymbol{x}}) \, dV_{\boldsymbol{x}} = \boldsymbol{0}.$$

Thus for any constant vector $\boldsymbol{u}$ we deduce

$$\begin{aligned}&\int_B \rho(\boldsymbol{x}) \boldsymbol{u} \cdot (\boldsymbol{x}-\overline{\boldsymbol{x}}) \, dV_{\boldsymbol{x}} = 0, \\ &\int_B \rho(\boldsymbol{x}) \boldsymbol{u} \otimes (\boldsymbol{x}-\overline{\boldsymbol{x}}) \, dV_{\boldsymbol{x}} = \boldsymbol{O}.\end{aligned} \quad (3.15)$$

Next, let $\boldsymbol{v} = \boldsymbol{x} - \overline{\boldsymbol{x}}$ and $\boldsymbol{u} = \overline{\boldsymbol{x}} - \boldsymbol{y}$. Then using the substitution $\boldsymbol{x}-\boldsymbol{y} = \boldsymbol{v}+\boldsymbol{u}$ in the definition of $\mathbf{I}_{\boldsymbol{y}}$ yields

$$\begin{aligned}\mathbf{I}_{\boldsymbol{y}} &= \mathbf{I}_{\overline{\boldsymbol{x}}} + \int_B \rho(\boldsymbol{x}) \left[|\boldsymbol{u}|^2 \boldsymbol{I} - \boldsymbol{u} \otimes \boldsymbol{u} \right] \, dV_{\boldsymbol{x}} \\ &\quad - \int_B \rho(\boldsymbol{x}) \left[2(\boldsymbol{u} \cdot \boldsymbol{v}) \boldsymbol{I} + \boldsymbol{u} \otimes \boldsymbol{v} + \boldsymbol{v} \otimes \boldsymbol{u} \right] \, dV_{\boldsymbol{x}} \\ &= \mathbf{I}_{\overline{\boldsymbol{x}}} + \left(\int_B \rho(\boldsymbol{x}) \, dV_{\boldsymbol{x}} \right) \left[|\boldsymbol{u}|^2 \boldsymbol{I} - \boldsymbol{u} \otimes \boldsymbol{u} \right],\end{aligned}$$

where the last line follows from (3.15) and the fact that $\boldsymbol{u}$ is constant. This establishes the result.

3.5 (a) If the body force per unit mass $\boldsymbol{b}$ is constant, then

$$\begin{aligned}\mathcal{T}_b[B] &= \int_B (\boldsymbol{x} - \boldsymbol{z}) \times \rho(\boldsymbol{x})\boldsymbol{b}\, dV_{\boldsymbol{x}} \\ &= \left(\int_B \boldsymbol{x}\rho(\boldsymbol{x})\, dV_{\boldsymbol{x}}\right) \times \boldsymbol{b} - \text{mass}[B](\boldsymbol{z} \times \boldsymbol{b}).\end{aligned}$$

Thus $\mathcal{T}_b[B] = \boldsymbol{0}$ if and only if

$$\boldsymbol{z} \times \boldsymbol{b} = \left(\frac{1}{\text{mass}[B]} \int_B \boldsymbol{x}\rho(\boldsymbol{x})\, dV_{\boldsymbol{x}}\right) \times \boldsymbol{b} = \boldsymbol{x}_{\text{com}}[B] \times \boldsymbol{b}.$$

By inspection, a solution to this equation is given by $\boldsymbol{z} = \boldsymbol{x}_{\text{com}}[B]$. However, this solution is not unique. In particular

$$\boldsymbol{z} = \boldsymbol{x}_{\text{com}}[B] + \alpha\boldsymbol{b}$$

is a solution for any real α. Thus $\mathcal{T}_b[B] = \boldsymbol{0}$ about any reference point $\boldsymbol{z}$ on the line through $\boldsymbol{x}_{\text{com}}[B]$ parallel to $\boldsymbol{b}$.

(b) If the body force per unit volume $\widehat{\boldsymbol{b}}$ is constant, then an argument similar to that in (a) shows that $\mathcal{T}_b[B] = \boldsymbol{0}$ about any reference point $\boldsymbol{z}$ on the line through $\boldsymbol{x}_{\text{cov}}[B]$ parallel to $\widehat{\boldsymbol{b}}$.

3.7 (a) Since $\boldsymbol{t} = -p\boldsymbol{n}$ the resultant surface force on B is

$$\boldsymbol{r}_s[\partial B] = \int_{\partial B} \boldsymbol{t}\, dA_{\boldsymbol{x}} = \int_{\partial B} -p\boldsymbol{n}\, dA_{\boldsymbol{x}}.$$

Let $\boldsymbol{c}$ be an arbitrary constant vector. Then

$$\begin{aligned}\boldsymbol{c} \cdot \boldsymbol{r}_s[\partial B] &= \int_{\partial B} -p\boldsymbol{c} \cdot \boldsymbol{n}\, dA_{\boldsymbol{x}} \\ &= \int_B -\nabla \cdot (p\boldsymbol{c})\, dV_{\boldsymbol{x}} = \int_B -\boldsymbol{c} \cdot \nabla p\, dV_{\boldsymbol{x}},\end{aligned}$$

where the second equality follows from the Divergence Theorem and the third follows from the constancy of $\boldsymbol{c}$. Using the fact that $\nabla p = \rho_* g\boldsymbol{e}_3$ we obtain

$$\boldsymbol{r}_s[\partial B] = \int_B -\nabla p\, dV_{\boldsymbol{x}} = -\left(\int_B \rho_*\, dV_{\boldsymbol{x}}\right) g\boldsymbol{e}_3 = -Mg\boldsymbol{e}_3,$$

where M is the mass of the fluid displaced by B. The result follows from the fact that $W = Mg$.

(b) Let $\overline{\boldsymbol{x}}$ denote the center of volume of B. Then the resultant torque on B about $\overline{\boldsymbol{x}}$ due to the hydrostatic surface force is

$$\mathcal{T}_s[\partial B] = \int_{\partial B} (\boldsymbol{x} - \overline{\boldsymbol{x}}) \times \boldsymbol{t}\, dA_{\boldsymbol{x}} = \int_{\partial B} -(\boldsymbol{x} - \overline{\boldsymbol{x}}) \times p\boldsymbol{n}\, dA_{\boldsymbol{x}}.$$

Let $\boldsymbol{c}$ be an arbitrary constant vector and let $\boldsymbol{v} = \boldsymbol{x} - \overline{\boldsymbol{x}}$. Then

$$\begin{aligned}\boldsymbol{c} \cdot \boldsymbol{\tau}_s[\partial B] &= \int_{\partial B} -p\boldsymbol{c} \cdot \boldsymbol{v} \times \boldsymbol{n}\, dA_{\boldsymbol{x}} \\ &= \int_{\partial B} -p\boldsymbol{c} \times \boldsymbol{v} \cdot \boldsymbol{n}\, dA_{\boldsymbol{x}} \\ &= \int_B -\nabla \cdot (p\boldsymbol{c} \times \boldsymbol{v})\, dV_{\boldsymbol{x}} = \int_B -\boldsymbol{c} \cdot \boldsymbol{v} \times \nabla p\, dV_{\boldsymbol{x}},\end{aligned}$$

where the third equality follows from the Divergence Theorem and the fourth follows from the constancy of $\boldsymbol{c}$ and the definition of $\boldsymbol{v}$. Using the fact that ∇p is constant we obtain

$$\boldsymbol{\tau}_s[\partial B] = -\int_B \boldsymbol{v} \times \nabla p\, dV_{\boldsymbol{x}} = -\left(\int_B \boldsymbol{v}\, dV_{\boldsymbol{x}}\right) \times \nabla p = \boldsymbol{0},$$

where the last equality follows from the definition of $\boldsymbol{v}$ and $\overline{\boldsymbol{x}}$.

3.9 (a) From the given form of $[\boldsymbol{S}]$ and $[\boldsymbol{b}]$ we find

$$\begin{aligned}S_{1j,j} + \rho b_1 &= \frac{\partial x_2}{\partial x_1} + \frac{\partial x_3}{\partial x_2} + 0 + 0 = 0, \\ S_{2j,j} + \rho b_2 &= \frac{\partial x_3}{\partial x_1} + \frac{\partial x_1}{\partial x_2} + 0 + 0 = 0, \\ S_{3j,j} + \rho b_3 &= 0 + 0 + \rho g - \rho g = 0,\end{aligned}$$

which shows that $S_{ij,j} + \rho b_i = 0$ for all $\boldsymbol{x} \in B$. Moreover, from the given form of $[\boldsymbol{S}]$ we see that $S_{ij} = S_{ji}$ for all $\boldsymbol{x} \in B$.

(b) Let Γ_i^0 and Γ_i^1 denote the faces corresponding to $x_i = 0$ and $x_i = 1$, respectively. Then

$$\begin{aligned}&\Gamma_1^0: \quad [\boldsymbol{n}] = (-1,0,0)^T, \quad [\boldsymbol{t}] = [\boldsymbol{S}][\boldsymbol{n}] = (-x_2, -x_3, 0)^T, \\ &\Gamma_1^1: \quad [\boldsymbol{n}] = (1,0,0)^T, \quad [\boldsymbol{t}] = [\boldsymbol{S}][\boldsymbol{n}] = (x_2, x_3, 0)^T, \\ &\Gamma_2^0: \quad [\boldsymbol{n}] = (0,-1,0)^T, \quad [\boldsymbol{t}] = [\boldsymbol{S}][\boldsymbol{n}] = (-x_3, -x_1, 0)^T, \\ &\Gamma_2^1: \quad [\boldsymbol{n}] = (0,1,0)^T, \quad [\boldsymbol{t}] = [\boldsymbol{S}][\boldsymbol{n}] = (x_3, x_1, 0)^T, \\ &\Gamma_3^0: \quad [\boldsymbol{n}] = (0,0,-1)^T, \quad [\boldsymbol{t}] = [\boldsymbol{S}][\boldsymbol{n}] = (0,0,0)^T, \\ &\Gamma_3^1: \quad [\boldsymbol{n}] = (0,0,1)^T, \quad [\boldsymbol{t}] = [\boldsymbol{S}][\boldsymbol{n}] = (0,0,\rho g)^T.\end{aligned}$$

(c) Since $\partial B = \Gamma_1^0 \cup \Gamma_1^1 \cup \cdots \cup \Gamma_3^1$ we have

$$\boldsymbol{r}_s[\partial B] = \int_{\partial B} \boldsymbol{t}\, dA_{\boldsymbol{x}} = \int_{\Gamma_1^0} \boldsymbol{t}\, dA_{\boldsymbol{x}} + \cdots + \int_{\Gamma_3^1} \boldsymbol{t}\, dA_{\boldsymbol{x}},$$

and using the results from part (b) we find

$$
\begin{aligned}
\int_{\Gamma_1^0} \boldsymbol{t}\, dA_{\boldsymbol{x}} &= \int_0^1 \int_0^1 -x_2\boldsymbol{e}_1 - x_3\boldsymbol{e}_2\, dx_2 dx_3 &&= -\tfrac{1}{2}\boldsymbol{e}_1 - \tfrac{1}{2}\boldsymbol{e}_2,\\
\int_{\Gamma_1^1} \boldsymbol{t}\, dA_{\boldsymbol{x}} &= \int_0^1 \int_0^1 x_2\boldsymbol{e}_1 + x_3\boldsymbol{e}_2\, dx_2 dx_3 &&= \tfrac{1}{2}\boldsymbol{e}_1 + \tfrac{1}{2}\boldsymbol{e}_2,\\
\int_{\Gamma_2^0} \boldsymbol{t}\, dA_{\boldsymbol{x}} &= \int_0^1 \int_0^1 -x_3\boldsymbol{e}_1 - x_1\boldsymbol{e}_2\, dx_1 dx_3 &&= -\tfrac{1}{2}\boldsymbol{e}_1 - \tfrac{1}{2}\boldsymbol{e}_2,\\
\int_{\Gamma_2^1} \boldsymbol{t}\, dA_{\boldsymbol{x}} &= \int_0^1 \int_0^1 x_3\boldsymbol{e}_1 + x_1\boldsymbol{e}_2\, dx_1 dx_3 &&= \tfrac{1}{2}\boldsymbol{e}_1 + \tfrac{1}{2}\boldsymbol{e}_2,\\
\int_{\Gamma_3^0} \boldsymbol{t}\, dA_{\boldsymbol{x}} &= \int_0^1 \int_0^1 \boldsymbol{0}\, dx_1 dx_2 &&= \boldsymbol{0},\\
\int_{\Gamma_3^1} \boldsymbol{t}\, dA_{\boldsymbol{x}} &= \int_0^1 \int_0^1 \rho g\boldsymbol{e}_3\, dx_1 dx_2 &&= \rho g\boldsymbol{e}_3.
\end{aligned}
$$

This gives a resultant surface force of $\boldsymbol{r}_s[\partial B] = \rho g\boldsymbol{e}_3$. For the resultant body force we have

$$
\boldsymbol{r}_b[B] = \int_B \rho\boldsymbol{b}\, dV_{\boldsymbol{x}} = \int_0^1 \int_0^1 \int_0^1 -\rho g\boldsymbol{e}_3\, dx_1 dx_2 dx_3 = -\rho g\boldsymbol{e}_3,
$$

and we see that the resultants are balanced in the sense that $\boldsymbol{r}_s[\partial B] + \boldsymbol{r}_b[B] = \boldsymbol{0}$. This result is consistent with the local equilibrium equation $\nabla \cdot \boldsymbol{S} + \rho\boldsymbol{b} = \boldsymbol{0}$ verified in part (a). In particular, the local equilibrium equation implies the force balance $\boldsymbol{r}_s[\partial\Omega] + \boldsymbol{r}_b[\Omega] = \boldsymbol{0}$ for all open subsets $\Omega \subset B$, including the case when $\Omega = B$.

3.11 Substituting $\boldsymbol{S} = \mathbf{C}\boldsymbol{E}$ and $\boldsymbol{E} = \mathrm{sym}(\nabla\boldsymbol{u})$ into the definition of W gives

$$
\begin{aligned}
W = \tfrac{1}{2}\int_B \mathbf{C}\boldsymbol{E} : \boldsymbol{E}\, dV_{\boldsymbol{x}} &= \tfrac{1}{2}\int_B \boldsymbol{S} : \mathrm{sym}(\nabla\boldsymbol{u})\, dV_{\boldsymbol{x}}\\
&= \tfrac{1}{2}\int_B \boldsymbol{S} : \nabla\boldsymbol{u}\, dV_{\boldsymbol{x}},
\end{aligned}
\tag{3.16}
$$

where the last line follows from the symmetry of $\boldsymbol{S}$, which is implied by the assumption that B is in equilibrium. From Chapter 2 we recall the Divergence Product Rule

$$
(\nabla \cdot \boldsymbol{S}) \cdot \boldsymbol{u} = \nabla \cdot (\boldsymbol{S}^T\boldsymbol{u}) - \boldsymbol{S} : \nabla\boldsymbol{u},
$$

which, since $\boldsymbol{S}$ is symmetric, can be written as

$$
\boldsymbol{S} : \nabla\boldsymbol{u} = \nabla \cdot (\boldsymbol{S}\boldsymbol{u}) - (\nabla \cdot \boldsymbol{S}) \cdot \boldsymbol{u}.
$$

Substituting this expression into (3.16) and using the Divergence

Theorem we obtain

$$\begin{aligned} W &= \tfrac{1}{2}\int_B \nabla\cdot(\boldsymbol{Su})\,dV_{\boldsymbol{x}} - \tfrac{1}{2}\int_B (\nabla\cdot\boldsymbol{S})\cdot\boldsymbol{u}\,dV_{\boldsymbol{x}} \\ &= \tfrac{1}{2}\int_{\partial B} \boldsymbol{Sn}\cdot\boldsymbol{u}\,dA_{\boldsymbol{x}} - \tfrac{1}{2}\int_B (\nabla\cdot\boldsymbol{S})\cdot\boldsymbol{u}\,dV_{\boldsymbol{x}}. \end{aligned} \tag{3.17}$$

Since B is in equilibrium we have $\nabla\cdot\boldsymbol{S}+\rho\boldsymbol{b}=\boldsymbol{0}$ in B. Moreover, by definition $\boldsymbol{h} = \boldsymbol{Sn}$ on ∂B. Substituting these results into (3.17) leads to the desired result

$$W = \tfrac{1}{2}\left(\int_{\partial B} \boldsymbol{h}\cdot\boldsymbol{u}\,dA_{\boldsymbol{x}} + \int_B \rho\boldsymbol{b}\cdot\boldsymbol{u}\,dV_{\boldsymbol{x}}\right).$$

3.13 (a) Using the identity

$$S_{ij} = (x_j S_{ik})_{,k} - x_j S_{ik,k},$$

and the equilibrium equation $S_{ik,k} + \widehat{b}_i = 0$, we get

$$S_{ij} = (x_j S_{ik})_{,k} + x_j \widehat{b}_i.$$

Moreover, by interchanging the indices i and j and using the equilibrium equation $S_{ji} = S_{ij}$, we find

$$S_{ij} = (x_i S_{jk})_{,k} + x_i \widehat{b}_j.$$

Integrating this result over B and using the Divergence Theorem we obtain

$$\begin{aligned} \int_B S_{ij}\,dV_{\boldsymbol{x}} &= \int_B (x_i S_{jk})_{,k}\,dV_{\boldsymbol{x}} + \int_B x_i\widehat{b}_j\,dV_{\boldsymbol{x}} \\ &= \int_{\partial B} x_i S_{jk} n_k\,dA_{\boldsymbol{x}} + \int_B x_i\widehat{b}_j\,dV_{\boldsymbol{x}} \\ &= \int_{\partial B} x_i h_j\,dA_{\boldsymbol{x}} + \int_B x_i\widehat{b}_j\,dV_{\boldsymbol{x}}, \end{aligned}$$

where the last line follows from the fact that $h_j = S_{jk}n_k$. From the above result we deduce

$$\begin{aligned} \overline{\boldsymbol{S}} &= \frac{1}{\mathrm{vol}[B]}\int_B \boldsymbol{S}\,dV_{\boldsymbol{x}} \\ &= \frac{1}{\mathrm{vol}[B]}\left[\int_{\partial B} \boldsymbol{x}\otimes\boldsymbol{h}\,dA_{\boldsymbol{x}} + \int_B \boldsymbol{x}\otimes\widehat{\boldsymbol{b}}\,dV_{\boldsymbol{x}}\right], \end{aligned}$$

which is the desired result.

(b) Substituting $\widehat{\boldsymbol{b}} = \boldsymbol{0}$ and $\boldsymbol{h} = -p\boldsymbol{n}$ into the result for $\overline{\boldsymbol{S}}$ gives

$$\overline{S}_{ij} = \frac{1}{\text{vol}[B]} \int_{\partial B} -p x_i n_j \, dA_{\boldsymbol{x}}.$$

Using the Divergence Theorem together with the fact that p is constant we obtain

$$\overline{S}_{ij} = \frac{1}{\text{vol}[B]} \int_{B} -(p x_i)_{,j} \, dV_{\boldsymbol{x}} = \frac{1}{\text{vol}[B]} \int_{B} -p\delta_{ij} \, dV_{\boldsymbol{x}},$$

which leads to the result $\overline{\boldsymbol{S}} = -p\boldsymbol{I}$.

(c) The equilibrium equations are $\nabla \cdot \boldsymbol{S} + \widehat{\boldsymbol{b}} = \boldsymbol{0}$ and $\boldsymbol{S}^T = \boldsymbol{S}$ in B, and the boundary condition is $\boldsymbol{S}\boldsymbol{n} = \boldsymbol{h}$ on ∂B. Assuming $\widehat{\boldsymbol{b}} = \boldsymbol{0}$ and $\boldsymbol{h} = -p\boldsymbol{n}$ we see that the equilibrium equations and boundary condition are both satisfied by $\boldsymbol{S} = -p\boldsymbol{I}$ (constant).

3.15 (a) The traction vector on Γ at $\boldsymbol{x}$ is $[\boldsymbol{t}] = [\boldsymbol{S}][\boldsymbol{n}] = (0, \sqrt{2}, \sqrt{2})^T$. Since $\boldsymbol{t}_n = (\boldsymbol{t} \cdot \boldsymbol{n})\boldsymbol{n}$ and $\boldsymbol{t}_s = \boldsymbol{t} - (\boldsymbol{t} \cdot \boldsymbol{n})\boldsymbol{n}$ we obtain

$$[\boldsymbol{t}_n] = (0, \sqrt{2}, \sqrt{2})^T \quad \text{and} \quad [\boldsymbol{t}_s] = (0, 0, 0)^T.$$

The traction vector on Γ' at $\boldsymbol{x}$ is $[\boldsymbol{t}] = [\boldsymbol{S}][\boldsymbol{n}'] = (5, 3, -3)^T$. Since $\boldsymbol{t}_n = (\boldsymbol{t} \cdot \boldsymbol{n}')\boldsymbol{n}'$ and $\boldsymbol{t}_s = \boldsymbol{t} - (\boldsymbol{t} \cdot \boldsymbol{n}')\boldsymbol{n}'$ we obtain

$$[\boldsymbol{t}_n] = (5, 0, 0)^T \quad \text{and} \quad [\boldsymbol{t}_s] = (0, 3, -3)^T.$$

(b) By definition, the principal stresses (eigenvalues) are roots of the characteristic polynomial

$$p(\sigma) = \det(\boldsymbol{S} - \sigma \boldsymbol{I}) = \det([\boldsymbol{S}] - \sigma[\boldsymbol{I}]) = -\sigma^3 + 5\sigma^2 + 22\sigma - 56.$$

In particular, we find that $p(\sigma)$ has three distinct roots: $\sigma_1 = 7$, $\sigma_2 = 2$, $\sigma_3 = -4$. The principal direction $\boldsymbol{e}_i$ (unit vector) associated with each σ_i is found (up to sign) by solving the linear homogeneous equation

$$([\boldsymbol{S}] - \sigma_i[\boldsymbol{I}])[\boldsymbol{e}_i] = [\boldsymbol{0}], \qquad \text{(no sum)}.$$

In particular, for each σ_i we find one independent direction

$$[\boldsymbol{e}_1] = \frac{1}{\sqrt{11}}(-3, -1, 1)^T,$$

$$[\boldsymbol{e}_2] = \frac{1}{\sqrt{2}}(0, 1, 1)^T, \quad [\boldsymbol{e}_3] = \frac{1}{\sqrt{22}}(2, -3, 3)^T.$$

Notice that $[\boldsymbol{n}] = [\boldsymbol{e}_2]$ is indeed a principal direction at $\boldsymbol{x}$.

3.17 Using the three principal stress directions $\boldsymbol{e}_i$ at $\boldsymbol{x}$ as a basis we can express any unit vector $\boldsymbol{n}$ as $\boldsymbol{n} = \sum_{i=1}^{3} n_i \boldsymbol{e}_i$, where $n_i = \boldsymbol{n}\cdot\boldsymbol{e}_i$. Because each $\boldsymbol{e}_i$ is a principal stress direction we have

$$\boldsymbol{S}\boldsymbol{n} = \sum_{i=1}^{3} n_i \sigma_i \boldsymbol{e}_i,$$

and hence

$$\boldsymbol{t}\cdot\boldsymbol{n} = (\boldsymbol{S}\boldsymbol{n})\cdot\boldsymbol{n} = \left(\sum_{i=1}^{3} n_i\sigma_i\boldsymbol{e}_i\right)\cdot\left(\sum_{j=1}^{3} n_j\boldsymbol{e}_j\right) = \sum_{i=1}^{3} n_i^2\sigma_i.$$

Since the maximum of $\sigma_n = |\boldsymbol{t}\cdot\boldsymbol{n}|$ coincides with the maximum of $\sigma_n^2 = (\boldsymbol{t}\cdot\boldsymbol{n})^2$, we consider the problem of maximizing the smooth function

$$\Phi(n_1, n_2, n_3) = \sigma_n^2 = \left(\sum_{i=1}^{3} n_i^2\sigma_i\right)^2,$$

subject to the condition that $\boldsymbol{n}$ be a unit vector, that is

$$g(n_1, n_2, n_3) = 0 \quad \text{where} \quad g(n_1, n_2, n_3) = \sum_{i=1}^{3} n_i^2 - 1.$$

To maximize $\Phi(\boldsymbol{n})$ subject to the condition $g(\boldsymbol{n}) = 0$ we employ the method of Lagrange multipliers. According to this method, the maxima occur among those directions $\boldsymbol{n}$ satisfying

$$\begin{aligned} \frac{\partial\Phi}{\partial n_i}(\boldsymbol{n}) &= \lambda\frac{\partial g}{\partial n_i}(\boldsymbol{n}) \quad (i = 1, 2, 3), \\ g(\boldsymbol{n}) &= 0, \end{aligned}$$

where λ is the Lagrange multiplier. Writing the above equations in explicit form, we seek n_i $(i = 1, 2, 3)$ and λ such that

$$\left.\begin{aligned} 4\left(\sum_{j=1}^{3} n_j^2\sigma_j\right) n_i\sigma_i = 2\lambda n_i \quad (i = 1, 2, 3), \\ \sum_{i=1}^{3} n_i^2 - 1 = 0. \end{aligned}\right\} \tag{3.18}$$

To solve (3.18) we first rewrite $(3.18)_1$ as

$$\left[2\left(\sum_{j=1}^{3} n_j^2\sigma_j\right)\sigma_i - \lambda\right] n_i = 0 \qquad (i = 1, 2, 3). \tag{3.19}$$

One set of solutions occurs when $\lambda = 0$ and $\boldsymbol{n}$ is an arbitrary unit vector with $\Phi(\boldsymbol{n}) = 0$. Next we consider solutions with $\lambda \neq 0$. Note that if

$$\lambda \neq 2\left(\sum_{j=1}^{3} n_j^2 \sigma_j\right)\sigma_i, \qquad \forall i = 1,2,3,$$

then (3.19) implies $n_i = 0$ $(i = 1,2,3)$, which contradicts $(3.18)_2$. Thus, any solution of (3.18) must have the property that

$$\lambda = 2\left(\sum_{j=1}^{3} n_j^2 \sigma_j\right)\sigma_i = 2\sqrt{\Phi(\boldsymbol{n})}\,\sigma_i \tag{3.20}$$

for some $i = 1$, 2, or 3. Assuming (3.20) holds with $i = 1$, we find the following solutions to (3.18)

$$\left.\begin{array}{ll} \lambda = 0, & \boldsymbol{n} \text{ any unit vector such that } \Phi(\boldsymbol{n}) = 0, \\ \lambda = 2\sigma_1^2, & \boldsymbol{n} = \pm\boldsymbol{e}_1. \end{array}\right\} \tag{3.21}$$

Assuming (3.20) holds with $i = 2$, we find the solutions

$$\left.\begin{array}{ll} \lambda = 0, & \boldsymbol{n} \text{ any unit vector such that } \Phi(\boldsymbol{n}) = 0, \\ \lambda = 2\sigma_2^2, & \boldsymbol{n} = \pm\boldsymbol{e}_2, \end{array}\right\} \tag{3.22}$$

and assuming (3.20) holds with $i = 3$, we find the solutions

$$\left.\begin{array}{ll} \lambda = 0, & \boldsymbol{n} \text{ any unit vector such that } \Phi(\boldsymbol{n}) = 0, \\ \lambda = 2\sigma_3^2, & \boldsymbol{n} = \pm\boldsymbol{e}_3. \end{array}\right\} \tag{3.23}$$

The maximum of $\Phi(\boldsymbol{n})$ subject to $g(\boldsymbol{n}) = 0$ must occur among those directions $\boldsymbol{n}$ in (3.21), (3.22) and (3.23). Using this fact, together with the assumption that the principal stresses are distinct, we deduce that the maximum value of $\Phi(\boldsymbol{n})$ is σ_k^2, and this value is achieved for $\boldsymbol{n} = \pm\boldsymbol{e}_k$, where k is any index such that

$$|\sigma_k| = \max\{|\sigma_1|, |\sigma_2|, |\sigma_3|\} = \max\{|\sigma_1|, |\sigma_3|\}.$$

3.19 By identifying $\mathcal{N}$ with the standard unit sphere the components of any vector $\boldsymbol{n} \in \mathcal{N}$ can be written as

$$[\boldsymbol{n}] = (\cos\phi\cos\theta, \cos\phi\sin\theta, \sin\phi)^T, \; \phi \in \left[-\frac{\pi}{2}, \frac{\pi}{2}\right], \; \theta \in [-\pi, \pi].$$

In terms of this parametrization we have $dA_{\boldsymbol{n}} = \cos\phi \; d\phi \; d\theta$ and

$$\int_{\mathcal{N}} dA_{\boldsymbol{n}} = \int_{-\pi}^{\pi}\int_{-\pi/2}^{\pi/2} \cos\phi \; d\phi \; d\theta = 4\pi.$$

Moreover, since $\boldsymbol{t} = \boldsymbol{Sn}$ we have

$$\int_{\mathcal{N}} \boldsymbol{n} \cdot \boldsymbol{t} \, dA_{\boldsymbol{n}} = \int_{\mathcal{N}} \boldsymbol{n} \cdot \boldsymbol{Sn} \, dA_{\boldsymbol{n}} = \int_{\mathcal{N}} n_i S_{ij} n_j \, dA_{\boldsymbol{n}}.$$

Integrating the first term in the sum we find

$$\int_{\mathcal{N}} n_1 S_{11} n_1 \, dA_{\boldsymbol{n}} \\ = \int_{-\pi}^{\pi} \int_{-\pi/2}^{\pi/2} S_{11} (\cos\phi \cos\theta)^2 \; \cos\phi \, d\phi \, d\theta = \frac{4\pi S_{11}}{3}.$$

Similar results are obtained for the other diagonal terms

$$\int_{\mathcal{N}} n_2 S_{22} n_2 \, dA_{\boldsymbol{n}} = \frac{4\pi S_{22}}{3}, \quad \int_{\mathcal{N}} n_3 S_{33} n_3 \, dA_{\boldsymbol{n}} = \frac{4\pi S_{33}}{3}.$$

Moreover, by direct calculation we find that each off-diagonal term $(i \neq j)$ in the sum is zero. Combining the above results yields

$$\frac{\int_{\mathcal{N}} \boldsymbol{n} \cdot \boldsymbol{t} \, dA_{\boldsymbol{n}}}{\int_{\mathcal{N}} dA_{\boldsymbol{n}}} = \tfrac{1}{3}(S_{11} + S_{22} + S_{33}) = -p,$$

which is the desired result.

4
Kinematics

Kinematics is the study of motion exclusive of the influences of mass, force and stress: it is the study of the geometry of motion. In this chapter we study the kinematics of continuum bodies. We focus on how to quantify strain and rate of strain in a body whose shape changes with time. In subsequent chapters we present various types of constitutive relations between stress and strain that characterize different types of materials. The contents of this chapter provide the basis for understanding the physical reasoning behind those relations, as well as the mathematical language for their precise statement.

The important ideas in this chapter are: (i) the notion of a deformation and a deformation map; (ii) the distinction between material and spatial coordinates; (iii) the notion of strain and its quantification; (iv) the idea of a motion and the material time derivative; (v) the notion of rate of strain and its quantification; (vi) the change of variable formulae for integration over material and spatial coordinates; (vii) special types of deformations and motions which preserve the shape and/or volume of a body.

4.1 Configurations and Deformations

At any instant of time a material body occupies an open subset B of Euclidean space $\mathbb{E}^3$ as described in Section 3.1. The identification of material particles with points of B defines the **configuration** of the body at that instant. We usually refer to the subset B as the configuration, and consider the identification of particles with points to be implicitly understood.

By a **deformation** we mean a change of configuration between some

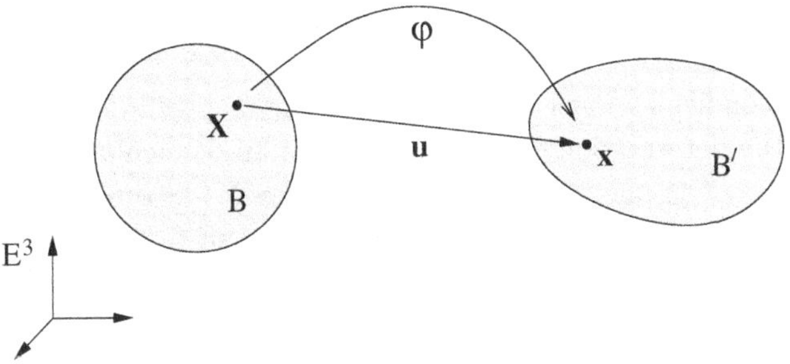

Fig. 4.1 Illustration of a deformation: B denotes the reference configuration and B' denotes the deformed configuration. The black dot in B and B' represents one and the same material particle.

initial (undeformed) configuration B and a subsequent (deformed) configuration B' (see Figure 4.1). By convention, we call B the **reference configuration** and B' the **deformed configuration**. We denote points in B by $\boldsymbol{X} = (X_1, X_2, X_3)^T$ and points in B' by $\boldsymbol{x} = (x_1, x_2, x_3)^T$. Because B and B' are two configurations of a single material body, each particle in the body has two sets of coordinates: a set of **material coordinates** X_i for its location in B, and a set of **spatial coordinates** x_i for its location in B'.

We usually think of a deformation as changing the shape of a body. However, this need not be the case. For example, the rotation of a solid ball about its center is a deformation that does not change the shape of the ball: it occupies the same region of space, but its material particles have changed location within that region. Thus a deformation can take place even when $B' = B$. It will sometimes be useful to assume that a material body occupies all of space. In this case we always have $B' = B = \mathbb{E}^3$.

4.2 The Deformation Map

The deformation of a body from a configuration B onto another configuration B' is described by a function $\boldsymbol{\varphi} : B \to B'$, which maps each point $\boldsymbol{X} \in B$ to a point $\boldsymbol{x} = \boldsymbol{\varphi}(\boldsymbol{X}) \in B'$ (see Figure 4.1). We call $\boldsymbol{\varphi}$ the **deformation map** relative to the reference configuration B. The displacement of a material particle from its initial location $\boldsymbol{X}$ to its final location $\boldsymbol{x}$ is given by

$$\boldsymbol{u}(\boldsymbol{X}) = \boldsymbol{\varphi}(\boldsymbol{X}) - \boldsymbol{X}. \tag{4.1}$$

We call $\boldsymbol{u} : B \to \mathcal{V}$ the **displacement field** associated with $\boldsymbol{\varphi}$ (see Exercises 1 and 2).

Certain conditions must be imposed on the map $\boldsymbol{\varphi}$ in order for it to represent the deformation of a material body. In particular, we assume: (i) $\boldsymbol{\varphi} : B \to B'$ is one-to-one, and (ii) $\det \nabla\boldsymbol{\varphi}(\boldsymbol{X}) > 0$ for all $\boldsymbol{X} \in B$. Deformations satisfying these assumptions are called **admissible**. Assumption (i) implies that two or more distinct points from B cannot simultaneously occupy the same position in B'. Assumption (ii) implies that deformations should preserve the orientation of a body; in particular, a body cannot be continuously deformed onto its mirror image (see Exercises 3 and 4). Since it is both one-to-one and onto, an admissible deformation $\boldsymbol{\varphi}$ is a bijection between B and B'.

Throughout the remainder of our developments we assume that all deformations are admissible. Furthermore, we assume that all deformations are smooth in the sense that partial derivatives of all orders exist and are continuous. This is always stronger than what we need, but allows for a clean presentation.

4.3 Measures of Strain

Consider an open ball Ω of radius $\alpha > 0$ centered at a point $\boldsymbol{X}_0$ in B as shown in Figure 4.2. Under a deformation $\boldsymbol{\varphi}$, the point $\boldsymbol{X}_0 \in B$ is mapped to a point $\boldsymbol{x}_0 \in B'$, and the ball $\Omega \subset B$ is mapped onto a region $\Omega' \subset B'$. In particular, we have $\boldsymbol{x}_0 = \boldsymbol{\varphi}(\boldsymbol{X}_0)$ and

$$\Omega' = \{\boldsymbol{x} \in B' \mid \boldsymbol{x} = \boldsymbol{\varphi}(\boldsymbol{X}),\ \boldsymbol{X} \in \Omega\},$$

which, for brevity, we abbreviate as $\Omega' = \boldsymbol{\varphi}(\Omega)$. Any relative difference in shape between Ω' and Ω in the limit $\alpha \to 0$ is called **strain** at $\boldsymbol{X}_0$. Intuitively, strain refers to the local stretching of a body caused by a deformation $\boldsymbol{\varphi}$. The concept of strain plays a central role in the study of solids, and in later chapters we will see various generalizations of Hooke's Law for elastic solids which relate stress to strain. In the following sections we introduce various measures of strain and study each in some detail.

4.3.1 The Deformation Gradient F

A natural way to quantify strain is through the **deformation gradient**, which is a second-order tensor field $\boldsymbol{F} : B \to \mathcal{V}^2$ defined by

$$\boldsymbol{F}(\boldsymbol{X}) = \nabla\boldsymbol{\varphi}(\boldsymbol{X}).$$

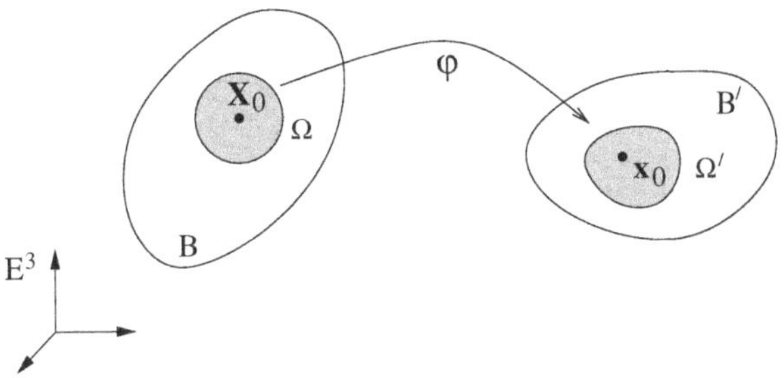

Fig. 4.2 A small region Ω in B is mapped onto a small region Ω' in B' by a deformation φ. The difference in shape between Ω' and Ω leads to the concept of strain.

The field $\boldsymbol{F}$ provides information on the local behavior of a deformation φ. In particular, given $\boldsymbol{X}_0 \in B$ we can use a Taylor expansion to write

$$\varphi(\boldsymbol{X}) = \varphi(\boldsymbol{X}_0) + \boldsymbol{F}(\boldsymbol{X}_0)(\boldsymbol{X} - \boldsymbol{X}_0) + \mathcal{O}(|\boldsymbol{X} - \boldsymbol{X}_0|^2),$$

or equivalently

$$\varphi(\boldsymbol{X}) = \boldsymbol{c} + \boldsymbol{F}(\boldsymbol{X}_0)\boldsymbol{X} + \mathcal{O}(|\boldsymbol{X} - \boldsymbol{X}_0|^2), \qquad (4.2)$$

where $\boldsymbol{c} = \varphi(\boldsymbol{X}_0) - \boldsymbol{F}(\boldsymbol{X}_0)\boldsymbol{X}_0$. Thus $\boldsymbol{F}(\boldsymbol{X}_0)$ characterizes the local behavior of $\varphi(\boldsymbol{X})$ for all points $\boldsymbol{X}$ in a neighborhood of $\boldsymbol{X}_0$.

4.3.2 Interpretation of F, Homogeneous Deformations

Here we describe the way in which $\boldsymbol{F}$ provides a measure of strain. For simplicity, we initially assume that the deformation φ is **homogeneous**, which means that the deformation gradient field $\boldsymbol{F}$ is constant. In this case, the higher-order terms in (4.2) vanish and φ can be written as

$$\varphi(\boldsymbol{X}) = \boldsymbol{c} + \boldsymbol{F}\boldsymbol{X}. \qquad (4.3)$$

The case when the deformation φ is not homogeneous will be discussed at the end of this section (see Exercises 5 and 6).

By studying the homogeneous case we will reveal how $\boldsymbol{F}$ quantifies the amount of stretch and rotation experienced by a body when it is deformed from a configuration B onto a configuration B'. Due to the linear nature of (4.3), we notice that any line segment in B is mapped to a corresponding line segment in B'. This fact is exploited in the figures below (see Exercises 7 and 8).

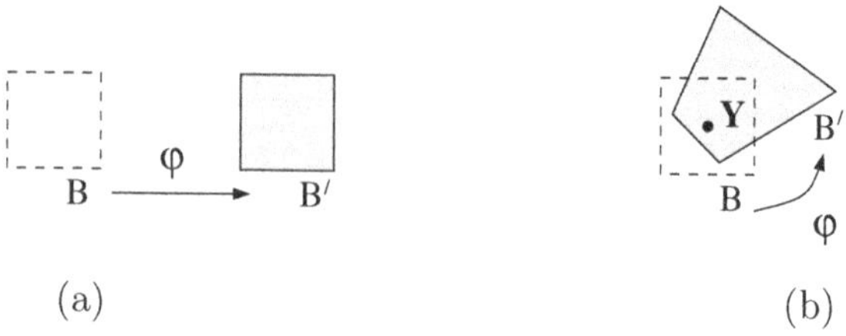

Fig. 4.3 Illustration of two homogeneous deformations. (a) translation. (b) deformation with a fixed point $\boldsymbol{Y}$.

4.3.2.1 Translations and Fixed Points

A homogeneous deformation φ is called a **translation** if

$$\varphi(\boldsymbol{X}) = \boldsymbol{X} + \boldsymbol{c}$$

for some constant vector $\boldsymbol{c}$. In such a deformation, each point in the body is translated or displaced along the vector $\boldsymbol{c}$, and there is no change in the shape and orientation of the body, as illustrated in Figure 4.3(a).

A homogeneous deformation φ is said to have a **fixed point** at $\boldsymbol{Y}$ if

$$\varphi(\boldsymbol{X}) = \boldsymbol{Y} + \boldsymbol{F}(\boldsymbol{X} - \boldsymbol{Y}). \tag{4.4}$$

The point $\boldsymbol{Y}$ is fixed in the sense that $\varphi(\boldsymbol{Y}) = \boldsymbol{Y}$. Any other point $\boldsymbol{X}$ is displaced by an amount determined by $\boldsymbol{F}$ and the relative position of $\boldsymbol{X}$ with respect to $\boldsymbol{Y}$. In such a deformation, the body may change both shape and orientation, as illustrated in Figure 4.3(b). We typically assume that $\boldsymbol{Y}$ is a point in B. However, this need not be the case.

The following result shows that an arbitrary homogeneous deformation can always be expressed as the composition of a translation and a deformation with a given fixed point. Below and throughout we use the symbol $\circ$ to denote the composition of maps.

Result 4.1 ***Translation-Fixed Point Decomposition.*** *Let φ be an arbitrary homogeneous deformation with constant deformation gradient $\boldsymbol{F}$. Then, given any point $\boldsymbol{Y}$, we can decompose φ as*

$$\varphi = \boldsymbol{d}_1 \circ \boldsymbol{g} = \boldsymbol{g} \circ \boldsymbol{d}_2,$$

where $\boldsymbol{d}_1$ and $\boldsymbol{d}_2$ are translations and $\boldsymbol{g}$ is a homogeneous deformation with a fixed point at $\boldsymbol{Y}$, in particular

$$\boldsymbol{g}(\boldsymbol{X}) = \boldsymbol{Y} + \boldsymbol{F}(\boldsymbol{X} - \boldsymbol{Y}).$$

□

(a) (b)

Fig. 4.4 Illustration of two homogeneous deformations with fixed points. (a) rotation. (b) stretch.

Proof See Exercise 9. □

Since translations of a body are easy to understand, we concentrate on homogeneous deformations which exclude translations, that is, deformations of the form (4.4) with a fixed point $\boldsymbol{Y}$. Our goal is to show that (4.4) may be completely understood in simple geometric terms provided the polar decomposition of $\boldsymbol{F}$ is known (see Result 1.12).

4.3.2.2 Rotations and Stretches

A homogeneous deformation φ is called a **rotation** about a fixed point $\boldsymbol{Y}$ if

$$\varphi(\boldsymbol{X}) = \boldsymbol{Y} + \boldsymbol{Q}(\boldsymbol{X} - \boldsymbol{Y})$$

for some rotation tensor $\boldsymbol{Q}$. In such a deformation, the orientation of the body is changed holding $\boldsymbol{Y}$ fixed, but there is no change in the shape of the body, as illustrated in Figure 4.4(a).

A homogeneous deformation φ is called a **stretch** from $\boldsymbol{Y}$ if

$$\varphi(\boldsymbol{X}) = \boldsymbol{Y} + \boldsymbol{S}(\boldsymbol{X} - \boldsymbol{Y})$$

for some symmetric, positive-definite tensor $\boldsymbol{S}$. In such a deformation, the body is extended by different amounts in different directions holding $\boldsymbol{Y}$ fixed, but there is no net change in the overall orientation of the body, as illustrated in Figure 4.4(b).

The following result shows that an arbitrary homogeneous deformation with a given fixed point can always be decomposed into a rotation and a stretch about the same fixed point.

Result 4.2 ***Stretch-Rotation Decomposition.*** *Let φ be an arbitrary homogeneous deformation with a fixed point $\boldsymbol{Y}$ and deformation gradient $\boldsymbol{F}$ as in (4.4), and let $\boldsymbol{F} = \boldsymbol{RU} = \boldsymbol{VR}$ be the right and left*

Fig. 4.5 Illustration of homogeneous extensions along a direction $\boldsymbol{e}$. (a) $\lambda > 1$. (b) $0 < \lambda < 1$.

polar decompositions given in Result 1.12. Then φ *can be decomposed as*

$$\varphi = \boldsymbol{r} \circ \boldsymbol{s}_1 = \boldsymbol{s}_2 \circ \boldsymbol{r},$$

where $\boldsymbol{r} = \boldsymbol{Y} + \boldsymbol{R}(\boldsymbol{X} - \boldsymbol{Y})$ *is a rotation about* $\boldsymbol{Y}$, *and* $\boldsymbol{s}_1 = \boldsymbol{Y} + \boldsymbol{U}(\boldsymbol{X} - \boldsymbol{Y})$ *and* $\boldsymbol{s}_2 = \boldsymbol{Y} + \boldsymbol{V}(\boldsymbol{X} - \boldsymbol{Y})$ *are stretches from* $\boldsymbol{Y}$. □

Proof See Exercise 10. □

4.3.2.3 Extensions

A homogeneous deformation φ is called an **extension** at $\boldsymbol{Y}$ in the direction of a unit vector $\boldsymbol{e}$ if

$$\varphi(\boldsymbol{X}) = \boldsymbol{Y} + \boldsymbol{F}(\boldsymbol{X} - \boldsymbol{Y}) \quad \text{and} \quad \boldsymbol{F} = \boldsymbol{I} + (\lambda - 1)\boldsymbol{e} \otimes \boldsymbol{e}$$

for some number $\lambda > 0$. This terminology is based on the observation that $\boldsymbol{F}$ extends any vector parallel to $\boldsymbol{e}$ by a factor λ, that is

$$\boldsymbol{F}\boldsymbol{e} = \boldsymbol{e} + (\lambda - 1)(\boldsymbol{e} \cdot \boldsymbol{e})\boldsymbol{e} = \lambda \boldsymbol{e}.$$

This type of deformation is a special case of a stretch deformation introduced earlier. In this case, the body is extended by a factor λ along the direction $\boldsymbol{e}$ holding $\boldsymbol{Y}$ fixed, but no deformation occurs in directions orthogonal to $\boldsymbol{e}$. Moreover, there is no change in the overall orientation of the body, as illustrated in Figure 4.5. Notice that when $\lambda = 1$ we have $\boldsymbol{F} = \boldsymbol{I}$, which yields $\varphi(\boldsymbol{X}) = \boldsymbol{X}$. In this case, each point of the body is fixed and the body is unchanged.

The following result shows that the two stretches $\boldsymbol{s}_1$ and $\boldsymbol{s}_2$ appearing in Result 4.2 can be expressed as the composition of three extensions defined by the eigenvalues and eigenvectors of $\boldsymbol{U}$ and $\boldsymbol{V}$.

Result 4.3 Stretches as Extensions. *Let $\boldsymbol{s}_1$ and $\boldsymbol{s}_2$ be the stretches defined in Result 4.2 and let $\{\lambda_i, \boldsymbol{u}_i\}$ and $\{\lambda_i, \boldsymbol{v}_i\}$ $(i = 1, 2, 3)$ be the eigenpairs associated with $\boldsymbol{U}$ and $\boldsymbol{V}$, respectively. Then*

$$\boldsymbol{s}_1 = \boldsymbol{f}_1 \circ \boldsymbol{f}_2 \circ \boldsymbol{f}_3 \quad \text{and} \quad \boldsymbol{s}_2 = \boldsymbol{h}_1 \circ \boldsymbol{h}_2 \circ \boldsymbol{h}_3,$$

where $\boldsymbol{f}_i$ is an extension at $\boldsymbol{Y}$ by λ_i in the direction $\boldsymbol{u}_i$, and $\boldsymbol{h}_i$ is an extension at $\boldsymbol{Y}$ by λ_i in the direction $\boldsymbol{v}_i$. □

Proof See Exercise 11. □

Remarks:

(1) The tensors $\boldsymbol{U}$ and $\boldsymbol{V}$ appearing in the right and left polar decompositions of $\boldsymbol{F}$ have the same eigenvalues, but different eigenvectors in general (see Exercise 12). Using this fact it follows that the stretches $\boldsymbol{s}_1$ and $\boldsymbol{s}_2$ give rise to extensions by the same amounts, but in different directions.

(2) Motivated by the above result, we call the eigenvalues λ_i the **principal stretches** associated with a deformation gradient $\boldsymbol{F}$, and call the eigenvectors $\boldsymbol{u}_i$ and $\boldsymbol{v}_i$ the right and left **principal directions**. Moreover, we call $\boldsymbol{U}$ and $\boldsymbol{V}$ the right and left **stretch tensors**, respectively.

(3) When a principal stretch $\lambda_i = 1$ for some i, the body is not extended along the corresponding principal direction. Whether it is the right or left principal direction depends on whether the rotation $\boldsymbol{r}$ is applied last or first, respectively (see Result 4.2 and summary below). When the principal stretches are all unity we have $\boldsymbol{U} = \boldsymbol{V} = \boldsymbol{I}$. In this case, $\boldsymbol{s}_1(\boldsymbol{X}) = \boldsymbol{s}_2(\boldsymbol{X}) = \boldsymbol{X}$ and the body is not extended in any direction. □

4.3.2.4 Summary

We have shown that, given any point $\boldsymbol{Y}$, an arbitrary homogeneous deformation $\boldsymbol{\varphi}$ can be decomposed as

$$\boldsymbol{\varphi} = \boldsymbol{d}_1 \circ \boldsymbol{g} = \boldsymbol{g} \circ \boldsymbol{d}_2,$$

where $\boldsymbol{d}_1$ and $\boldsymbol{d}_2$ are translations, and $\boldsymbol{g}$ has a fixed point at $\boldsymbol{Y}$. The deformation $\boldsymbol{g}$ can itself be decomposed as

$$\boldsymbol{g} = \boldsymbol{r} \circ \boldsymbol{s}_1 = \boldsymbol{s}_2 \circ \boldsymbol{r},$$

where $\boldsymbol{r}$ is a rotation about $\boldsymbol{Y}$, and $\boldsymbol{s}_1$ and $\boldsymbol{s}_2$ are stretches from $\boldsymbol{Y}$. Furthermore, these two stretches can be decomposed as

$$\boldsymbol{s}_1 = \boldsymbol{f}_1 \circ \boldsymbol{f}_2 \circ \boldsymbol{f}_3 \quad \text{and} \quad \boldsymbol{s}_2 = \boldsymbol{h}_1 \circ \boldsymbol{h}_2 \circ \boldsymbol{h}_3,$$

where $\boldsymbol{f}_i$ and $\boldsymbol{h}_i$ are extensions from $\boldsymbol{Y}$ along the right and left principal directions, respectively. The amount of each extension is determined by the corresponding principal stretch. The principal stretches, principal directions and rotation are entirely determined by the polar decomposition of the deformation gradient $\boldsymbol{F}$.

From the above results we deduce that an arbitrary homogeneous deformation $\boldsymbol{\varphi}$ can be understood in several different, but equivalent ways. For example, we may write $\boldsymbol{\varphi} = \boldsymbol{s}_2 \circ \boldsymbol{r} \circ \boldsymbol{d}_2$, which implies that $\boldsymbol{\varphi}$ can be decomposed into a sequence of three elementary operations: a translation, followed by a rotation, followed by extensions along the left principal directions. Equivalently, we may write $\boldsymbol{\varphi} = \boldsymbol{r} \circ \boldsymbol{s}_1 \circ \boldsymbol{d}_2$, which implies that $\boldsymbol{\varphi}$ can be decomposed into a different sequence: a translation, followed by extensions along the right principal directions, followed by a rotation (see Exercise 13). Other decompositions are also possible.

The above results about homogeneous deformations can also be applied to general deformations for which the deformation gradient field is not constant. In particular, the Taylor expansion in (4.2) shows that a general deformation is approximately homogeneous in a small neighborhood of each point $\boldsymbol{X}_0 \in B$. Thus the polar decomposition of the deformation gradient $\boldsymbol{F}(\boldsymbol{X}_0)$ completely determines how material in a neighborhood of $\boldsymbol{X}_0$ is stretched and rotated. For this reason the deformation gradient plays a central role in the modeling of stress in various types of material bodies.

4.3.3 The Cauchy–Green Strain Tensor C

Consider a general deformation $\boldsymbol{\varphi} : B \to B'$ with associated deformation gradient $\boldsymbol{F} = \nabla\boldsymbol{\varphi}$. Then another measure of strain is provided by

$$\boldsymbol{C} = \boldsymbol{F}^T \boldsymbol{F}.$$

We call $\boldsymbol{C} : B \to \mathcal{V}^2$ the **(right) Cauchy–Green strain tensor** field associated with $\boldsymbol{\varphi}$. Notice that $\boldsymbol{C}$ is symmetric and positive-definite at each point in B.

While $\boldsymbol{F}$ contains a mixture of information on both rotations and stretches, $\boldsymbol{C}$ contains information on stretches only. This observation follows from the right polar decomposition $\boldsymbol{F} = \boldsymbol{R}\boldsymbol{U}$, which implies

$$\boldsymbol{C} = \boldsymbol{F}^T\boldsymbol{F} = \boldsymbol{U}^2.$$

Notice that the rotation tensor $\boldsymbol{R}$, which is implicit in $\boldsymbol{F}$, does not appear in $\boldsymbol{C}$. For this reason, $\boldsymbol{C}$ is typically a more useful measure of strain than $\boldsymbol{F}$ (see Exercise 14). The tensor $\boldsymbol{C}$ will be employed in Chapters 7 and 9 when we study models of stress in elastic and thermoelastic solids at large strains.

Remarks:

(1) The right stretch tensor $\boldsymbol{U}$ is also independent of $\boldsymbol{R}$ and contains information on stretches only. However, we introduce $\boldsymbol{C} = \boldsymbol{F}^T\boldsymbol{F}$ because it is often easier to compute than $\boldsymbol{U} = \sqrt{\boldsymbol{F}^T\boldsymbol{F}}$. In particular, we avoid the tensor square root (see Exercise 15).

(2) By Result 1.9, the right stretch tensor can be written as $\boldsymbol{U} = \sum_{i=1}^{3} \lambda_i \boldsymbol{u}_i \otimes \boldsymbol{u}_i$, where $\lambda_i > 0$ are the principal stretches and $\boldsymbol{u}_i$ are the right principal directions. From the relation $\boldsymbol{C} = \boldsymbol{U}^2$ we deduce $\boldsymbol{C} = \sum_{i=1}^{3} \lambda_i^2 \boldsymbol{u}_i \otimes \boldsymbol{u}_i$, which implies that the eigenvalues of $\boldsymbol{C}$ are the squares of the principal stretches, and the eigenvectors are the right principal directions. Thus the principal stretches and directions can be determined from $\boldsymbol{C}$ without calculating $\boldsymbol{U}$.

(3) Another measure of strain related to $\boldsymbol{C}$ is the **left Cauchy–Green strain tensor** $\boldsymbol{B} = \boldsymbol{F}\boldsymbol{F}^T$. However, this strain tensor will not be employed in any of our developments. □

4.3.4 Interpretation of C

Here we show how changes in the relative position and orientation of points in a material body are quantified by $\boldsymbol{C}$. To begin, consider an arbitrary point $\boldsymbol{X}$ in the reference configuration B, and let Ω be the open ball of radius $\alpha > 0$ centered at $\boldsymbol{X}$. For any two unit vectors $\boldsymbol{e}$ and $\boldsymbol{d}$ consider the points $\boldsymbol{Y} = \boldsymbol{X} + \alpha\boldsymbol{e}$ and $\boldsymbol{Z} = \boldsymbol{X} + \alpha\boldsymbol{d}$ as shown in Figure 4.6. Let $\boldsymbol{x}$, $\boldsymbol{y}$ and $\boldsymbol{z}$ denote the corresponding points in Ω', and let $\phi \in [0, \pi]$ denote the angle between the vectors $\boldsymbol{v} = \boldsymbol{y} - \boldsymbol{x}$ and $\boldsymbol{w} = \boldsymbol{z} - \boldsymbol{x}$.

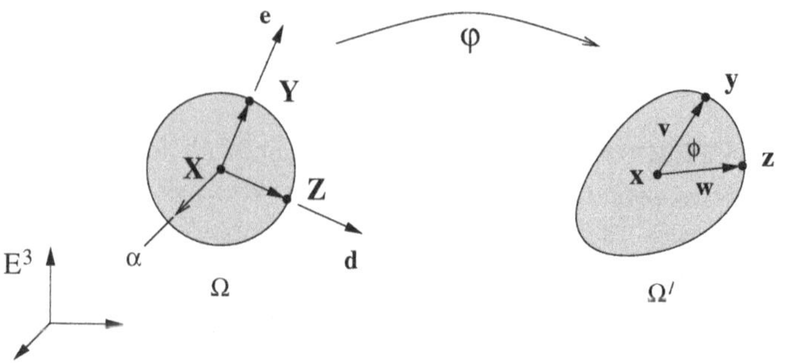

Fig. 4.6 Setup for interpreting the strain tensor $\boldsymbol{C}$. Three points $\boldsymbol{X}$, $\boldsymbol{Y}$, $\boldsymbol{Z}$ in Ω are mapped to corresponding points $\boldsymbol{x}$, $\boldsymbol{y}$, $\boldsymbol{z}$ in Ω'. The change in the relative position and orientation of the points can be quantified in terms of $\boldsymbol{C}$.

Result 4.4 ***Cauchy–Green Strain Relations.*** *For any point $\boldsymbol{X} \in B$ and unit vectors $\boldsymbol{e}$ and $\boldsymbol{d}$ define $\lambda(\boldsymbol{e}) > 0$ and $\theta(\boldsymbol{e}, \boldsymbol{d}) \in [0, \pi]$ by*

$$\lambda(\boldsymbol{e}) = \sqrt{\boldsymbol{e} \cdot \boldsymbol{C}\boldsymbol{e}} \quad \text{and} \quad \cos\theta(\boldsymbol{e}, \boldsymbol{d}) = \frac{\boldsymbol{e} \cdot \boldsymbol{C}\boldsymbol{d}}{\sqrt{\boldsymbol{e} \cdot \boldsymbol{C}\boldsymbol{e}}\sqrt{\boldsymbol{d} \cdot \boldsymbol{C}\boldsymbol{d}}}.$$

Then as $\alpha \to 0$ we have

$$\frac{|\boldsymbol{y} - \boldsymbol{x}|}{|\boldsymbol{Y} - \boldsymbol{X}|} \to \lambda(\boldsymbol{e}), \qquad \frac{|\boldsymbol{z} - \boldsymbol{x}|}{|\boldsymbol{Z} - \boldsymbol{X}|} \to \lambda(\boldsymbol{d}) \tag{4.5}$$

and

$$\cos\phi \to \cos\theta(\boldsymbol{e}, \boldsymbol{d}). \tag{4.6}$$

□

Proof To establish $(4.5)_1$ we notice that $\boldsymbol{v} = \boldsymbol{y} - \boldsymbol{x}$ can be written as

$$\boldsymbol{v} = \boldsymbol{\varphi}(\boldsymbol{Y}) - \boldsymbol{\varphi}(\boldsymbol{X}) = \boldsymbol{\varphi}(\boldsymbol{X} + \alpha\boldsymbol{e}) - \boldsymbol{\varphi}(\boldsymbol{X}).$$

Using a Taylor expansion about $\alpha = 0$ we obtain

$$\boldsymbol{v} = \alpha\boldsymbol{F}(\boldsymbol{X})\boldsymbol{e} + \mathcal{O}(\alpha^2), \tag{4.7}$$

and using the relations $|\boldsymbol{v}|^2 = \boldsymbol{v} \cdot \boldsymbol{v}$ and $\boldsymbol{C} = \boldsymbol{F}^T\boldsymbol{F}$ we find

$$|\boldsymbol{v}|^2 = \alpha^2\boldsymbol{F}(\boldsymbol{X})\boldsymbol{e} \cdot \boldsymbol{F}(\boldsymbol{X})\boldsymbol{e} + \mathcal{O}(\alpha^3) = \alpha^2\boldsymbol{e} \cdot \boldsymbol{C}(\boldsymbol{X})\boldsymbol{e} + \mathcal{O}(\alpha^3).$$

Dividing through by α^2 and noting that $|\boldsymbol{y}-\boldsymbol{x}| = |\boldsymbol{v}|$ and $|\boldsymbol{Y}-\boldsymbol{X}| = \alpha$ we get

$$\frac{|\boldsymbol{y}-\boldsymbol{x}|^2}{|\boldsymbol{Y}-\boldsymbol{X}|^2} = \frac{|\boldsymbol{v}|^2}{\alpha^2} = \boldsymbol{e}\cdot\boldsymbol{C}(\boldsymbol{X})\boldsymbol{e} + \mathcal{O}(\alpha), \tag{4.8}$$

which leads to $(4.5)_1$. To establish $(4.5)_2$ we notice that $\boldsymbol{w} = \boldsymbol{z} - \boldsymbol{x}$ can be written as

$$\boldsymbol{w} = \boldsymbol{\varphi}(\boldsymbol{Z}) - \boldsymbol{\varphi}(\boldsymbol{X}) = \boldsymbol{\varphi}(\boldsymbol{X}+\alpha\boldsymbol{d}) - \boldsymbol{\varphi}(\boldsymbol{X}).$$

Using a Taylor expansion as above we obtain

$$\boldsymbol{w} = \alpha\boldsymbol{F}(\boldsymbol{X})\boldsymbol{d} + \mathcal{O}(\alpha^2), \tag{4.9}$$

which implies

$$|\boldsymbol{w}|^2 = \alpha^2\boldsymbol{d}\cdot\boldsymbol{C}(\boldsymbol{X})\boldsymbol{d} + \mathcal{O}(\alpha^3).$$

Dividing through by α^2 and noting that $|\boldsymbol{z}-\boldsymbol{x}| = |\boldsymbol{w}|$ and $|\boldsymbol{Z}-\boldsymbol{X}| = \alpha$ gives

$$\frac{|\boldsymbol{z}-\boldsymbol{x}|^2}{|\boldsymbol{Z}-\boldsymbol{X}|^2} = \frac{|\boldsymbol{w}|^2}{\alpha^2} = \boldsymbol{d}\cdot\boldsymbol{C}(\boldsymbol{X})\boldsymbol{d} + \mathcal{O}(\alpha), \tag{4.10}$$

which leads to $(4.5)_2$. To establish (4.6) we use the relation

$$\cos\phi = \frac{\boldsymbol{v}\cdot\boldsymbol{w}}{|\boldsymbol{v}||\boldsymbol{w}|}. \tag{4.11}$$

From (4.7) and (4.9) we get

$$\boldsymbol{v}\cdot\boldsymbol{w} = \alpha^2\boldsymbol{F}(\boldsymbol{X})\boldsymbol{e}\cdot\boldsymbol{F}(\boldsymbol{X})\boldsymbol{d} + \mathcal{O}(\alpha^3) = \alpha^2\boldsymbol{e}\cdot\boldsymbol{C}(\boldsymbol{X})\boldsymbol{d} + \mathcal{O}(\alpha^3),$$

which implies

$$\frac{\boldsymbol{v}\cdot\boldsymbol{w}}{\alpha^2} = \boldsymbol{e}\cdot\boldsymbol{C}(\boldsymbol{X})\boldsymbol{d} + \mathcal{O}(\alpha). \tag{4.12}$$

When (4.12), (4.10) and (4.8) are combined with (4.11) we obtain the result in (4.6). □

Remarks:

(1) The above result shows that $\lambda(\boldsymbol{e})$ is the limiting value of the ratio $|\boldsymbol{y}-\boldsymbol{x}|/|\boldsymbol{Y}-\boldsymbol{X}|$ as $\boldsymbol{Y}$ tends to $\boldsymbol{X}$ along a direction $\boldsymbol{e}$. For this reason $\lambda(\boldsymbol{e})$ is called the **stretch** in the direction $\boldsymbol{e}$ at $\boldsymbol{X}$. It is the ratio of deformed length to initial length of an infinitesimal line segment that, prior to deformation, was in the direction $\boldsymbol{e}$ at $\boldsymbol{X}$ (see Exercise 16).

(2) Let $\boldsymbol{u}_i$ be a right principal direction at $\boldsymbol{X}$ with corresponding principal stretch λ_i so that $\boldsymbol{C}\boldsymbol{u}_i = \lambda_i^2 \boldsymbol{u}_i$ (no sum). Then a straightforward calculation shows that $\lambda(\boldsymbol{u}_i) = \lambda_i$, which provides further justification for referring to λ_i $(i = 1, 2, 3)$ as the principal stretches.

(3) An analysis similar to that in Section 3.5.3 reveals that, at any point $\boldsymbol{X}$, the extreme values of $\lambda(\boldsymbol{e})$ occur when $\boldsymbol{e}$ is an eigenvector of $\boldsymbol{C}$ (right principal direction). From this we deduce that the extreme values of $\lambda(\boldsymbol{e})$ at a point are given by the maximum and minimum principle stretches at that point.

(4) The angle $\theta(\boldsymbol{e}, \boldsymbol{d})$ is the limiting value of ϕ as $\boldsymbol{Y}, \boldsymbol{Z}$ tend to $\boldsymbol{X}$ along directions $\boldsymbol{e}, \boldsymbol{d}$. If we denote the angle between $\boldsymbol{e}$ and $\boldsymbol{d}$ by $\Theta(\boldsymbol{e}, \boldsymbol{d})$, then the quantity $\gamma(\boldsymbol{e}, \boldsymbol{d}) = \Theta(\boldsymbol{e}, \boldsymbol{d}) - \theta(\boldsymbol{e}, \boldsymbol{d})$ is called the **shear** between the directions $\boldsymbol{e}$ and $\boldsymbol{d}$ at $\boldsymbol{X}$. Physically, shear is the change in angle between two infinitesimal line segments that, prior to deformation, were aligned with $\boldsymbol{e}$ and $\boldsymbol{d}$ at $\boldsymbol{X}$. A straightforward analysis shows that the shear between any two eigenvectors of $\boldsymbol{C}$ (right principal directions) is zero (see Exercises 17 and 18).

□

The following result shows that, in contrast to $\boldsymbol{F}$, the components of $\boldsymbol{C}$ explicitly quantify the stretch and shear caused by a deformation $\boldsymbol{\varphi}$.

Result 4.5 ***Components of C.*** *Let C_{ij} be the components of $\boldsymbol{C}$ in an arbitrary coordinate frame $\{\boldsymbol{e}_i\}$. Then for any point $\boldsymbol{X} \in B$ we have*

$$C_{ii} = \lambda^2(\boldsymbol{e}_i) \quad \text{and} \quad C_{ij} = \lambda(\boldsymbol{e}_i)\lambda(\boldsymbol{e}_j)\sin\gamma(\boldsymbol{e}_i, \boldsymbol{e}_j) \quad (\text{no sum, } i \neq j)\,.$$

Thus the diagonal components of $\boldsymbol{C}$ are the squares of the stretches along the coordinate directions, and the off-diagonal components are related to the shear between the corresponding pairs of coordinate directions. □

Proof See Exercise 19. □

4.3.5 Rigid Deformations

A (homogeneous) deformation $\boldsymbol{\varphi} : B \to B'$ is called **rigid** if

$$\boldsymbol{\varphi}(\boldsymbol{X}) = \boldsymbol{c} + \boldsymbol{Q}\boldsymbol{X}$$

for some vector $\boldsymbol{c}$ and rotation tensor $\boldsymbol{Q}$. Notice that this class of deformations contains both translations and rotations as considered in Section 4.3.2. For a rigid deformation we have $\boldsymbol{F} = \boldsymbol{Q}$, which yields

$$\boldsymbol{C}(\boldsymbol{X}) = \boldsymbol{F}(\boldsymbol{X})^T \boldsymbol{F}(\boldsymbol{X}) = \boldsymbol{Q}^T \boldsymbol{Q} = \boldsymbol{I}, \quad \forall \boldsymbol{X} \in B.$$

In view of Results 4.4 and 4.5 we see that rigid deformations produce no strain as measured by $\boldsymbol{C}$. In particular, the stretch in every direction is unity and the shear between any two directions is zero at each point in B. Indeed, an inspection of the proof of Result 4.4 reveals that the relative position and orientation between *any* three points in B is unchanged by a rigid deformation. Rigid deformations can be completely characterized in terms of the tensor field $\boldsymbol{C}$. In particular, it can be shown that a deformation φ is rigid if and only if $\boldsymbol{C}(\boldsymbol{X}) = \boldsymbol{I}$ for all $\boldsymbol{X} \in B$.

4.3.6 The Infinitesimal Strain Tensor E

Consider a deformation $\varphi : B \to B'$ with associated displacement field $\boldsymbol{u}$ and displacement gradient $\nabla \boldsymbol{u}$. Then another measure of strain is provided by

$$\boldsymbol{E} = \mathrm{sym}(\nabla \boldsymbol{u}) = \tfrac{1}{2}(\nabla \boldsymbol{u} + \nabla \boldsymbol{u}^T).$$

We call $\boldsymbol{E} : B \to \mathcal{V}^2$ the **infinitesimal strain tensor** field associated with φ. Notice that, by definition, $\boldsymbol{E}$ is symmetric at each point in B.

The tensor $\boldsymbol{E}$ is related to the deformation gradient $\boldsymbol{F}$ and the Cauchy-Green tensor $\boldsymbol{C}$. In particular, from the definition of $\boldsymbol{u}$ in (4.1) we deduce that $\nabla \boldsymbol{u} = \boldsymbol{F} - \boldsymbol{I}$, which leads to the result

$$\boldsymbol{E} = \mathrm{sym}(\boldsymbol{F} - \boldsymbol{I}).$$

Moreover, since $\boldsymbol{C} = \boldsymbol{F}^T \boldsymbol{F}$ we also find

$$\boldsymbol{E} = \tfrac{1}{2}(\boldsymbol{C} - \boldsymbol{I}) - \tfrac{1}{2}\nabla \boldsymbol{u}^T \nabla \boldsymbol{u}. \tag{4.13}$$

The tensor $\boldsymbol{E}$ is particularly useful in the case of small deformations. We say that a deformation φ is **small** if there is a number $0 \leq \epsilon \ll 1$ such that $|\nabla \boldsymbol{u}| = \mathcal{O}(\epsilon)$, or equivalently $\partial u_i / \partial X_j = \mathcal{O}(\epsilon)$, for all points $\boldsymbol{X} \in B$. Thus a deformation is small if the norm of the displacement gradient is small at all points in the body. In this case we deduce from (4.13) that

$$\boldsymbol{E} = \tfrac{1}{2}(\boldsymbol{C} - \boldsymbol{I}) + \mathcal{O}(\epsilon^2). \tag{4.14}$$

Notice that, if terms of order $\mathcal{O}(\epsilon^2)$ are neglected, then $\boldsymbol{E}$ is equivalent to $\boldsymbol{C}$ up to a constant multiplicative factor and offset. The tensor $\boldsymbol{E}$ will arise naturally in Chapters 7 and 9 when we study linearized models of stress in elastic and thermoelastic solids.

Remarks:

(1) When $\nabla \boldsymbol{u} = \boldsymbol{O}$ for all $\boldsymbol{X} \in B$ we have $\boldsymbol{F} = \boldsymbol{I}$ for all $\boldsymbol{X} \in B$, which implies that φ is a translation. Thus φ is a small deformation when it deviates only slightly from a pure translation (see Exercise 20).

(2) For small deformations the tensor $\boldsymbol{E}$ contains essentially the same information as $\boldsymbol{C}$. However, there is one important difference: $\boldsymbol{E}$ depends linearly on $\boldsymbol{u}$ (hence φ), whereas $\boldsymbol{C}$ depends non-linearly on $\boldsymbol{u}$. □

4.3.7 Interpretation of E

Here we describe the way in which $\boldsymbol{E}$ provides a measure of strain for a small deformation. In particular, we consider a deformation φ and assume it is small in the sense that $\partial u_i/\partial X_j = \mathcal{O}(\epsilon)$ for all $\boldsymbol{X} \in B$ where $0 \leq \epsilon \ll 1$.

Result 4.6 ***Components of E.*** *Let E_{ij} be the components of $\boldsymbol{E}$ in an arbitrary coordinate frame $\{\mathbf{e}_i\}$. If we neglect terms of order $\mathcal{O}(\epsilon^2)$, then for any point $\boldsymbol{X} \in B$ we have*

$$E_{ii} \approx \lambda(\mathbf{e}_i) - 1 \quad \textit{and} \quad E_{ij} \approx \tfrac{1}{2}\sin\gamma(\mathbf{e}_i, \mathbf{e}_j) \quad (\textit{no sum, } i \neq j),$$

where $\lambda(\mathbf{e}_i)$ is the stretch in the direction $\mathbf{e}_i$ and $\gamma(\mathbf{e}_i, \mathbf{e}_j)$ is the shear between the directions $\mathbf{e}_i$ and $\mathbf{e}_j$. □

Proof To establish the result for the diagonal components we consider the case $i = 1$. From (4.14) we have

$$C_{11} = 1 + 2E_{11} + \mathcal{O}(\epsilon^2),$$

and since $E_{11} = \mathcal{O}(\epsilon)$ it follows by properties of the square root function that

$$\sqrt{C_{11}} = 1 + E_{11} + \mathcal{O}(\epsilon^2).$$

Neglecting the terms of order $\mathcal{O}(\epsilon^2)$ and using Result 4.5 gives

$$E_{11} \approx \sqrt{C_{11}} - 1 = \lambda(\mathbf{e}_1) - 1,$$

which is the desired result. The other diagonal cases follow similarly. To establish the result for the off-diagonal components we consider the case $i = 1$ and $j = 2$. Then from Result 4.5 we have

$$\sin\gamma(\boldsymbol{e}_1, \boldsymbol{e}_2) = \frac{C_{12}}{\sqrt{C_{11}}\sqrt{C_{22}}}. \qquad (4.15)$$

From (4.14) and the fact that E_{11}, E_{22} and E_{12} are all order $\mathcal{O}(\epsilon)$ we obtain

$$C_{12} = 2E_{12} + \mathcal{O}(\epsilon^2), \quad C_{11} = 1 + \mathcal{O}(\epsilon), \quad C_{22} = 1 + \mathcal{O}(\epsilon),$$

which together with (4.15) implies

$$\sin\gamma(\boldsymbol{e}_1, \boldsymbol{e}_2) = 2E_{12} + \mathcal{O}(\epsilon^2).$$

Neglecting the terms of order $\mathcal{O}(\epsilon^2)$ gives

$$E_{12} \approx \tfrac{1}{2}\sin\gamma(\boldsymbol{e}_1, \boldsymbol{e}_2),$$

which is the desired result. The other off-diagonal cases follow similarly.
□

Remarks:

(1) As shown in Result 4.4, the stretch $\lambda(\boldsymbol{e}_i)$ is the limiting value of the ratio $|\boldsymbol{y} - \boldsymbol{x}|/|\boldsymbol{Y} - \boldsymbol{X}|$ as $\boldsymbol{Y}$ tends to $\boldsymbol{X}$ along a direction $\boldsymbol{e}_i$ (see Figure 4.6). From this we deduce that $\lambda(\boldsymbol{e}_i) - 1$ is the limiting value of the quantity

$$\frac{|\boldsymbol{y} - \boldsymbol{x}|}{|\boldsymbol{Y} - \boldsymbol{X}|} - 1 = \frac{|\boldsymbol{y} - \boldsymbol{x}| - |\boldsymbol{Y} - \boldsymbol{X}|}{|\boldsymbol{Y} - \boldsymbol{X}|}.$$

Thus, for a small deformation, the diagonal component E_{ii} (no sum) is approximately equal to the relative change in length of an infinitesimal line segment that, prior to deformation, was in the direction $\boldsymbol{e}_i$ at $\boldsymbol{X}$.

(2) When the shear angle $\gamma(\boldsymbol{e}_i, \boldsymbol{e}_j)$ is small we obtain

$$E_{ij} \approx \tfrac{1}{2}\sin\gamma(\boldsymbol{e}_i, \boldsymbol{e}_j) \approx \tfrac{1}{2}\gamma(\boldsymbol{e}_i, \boldsymbol{e}_j).$$

Thus, for a small deformation, the off-diagonal component E_{ij} is approximately equal to half of the shear angle between two infinitesimal line segments that, prior to deformation, were in the directions $\boldsymbol{e}_i$ and $\boldsymbol{e}_j$ at $\boldsymbol{X}$ (see Exercise 21). □

4.3.8 Infinitesimally Rigid Deformations

A (homogeneous) deformation $\varphi : B \to B'$ is called **infinitesimally rigid** if the associated displacement field $\boldsymbol{u}$ is of the form

$$\boldsymbol{u}(\boldsymbol{X}) = \boldsymbol{c} + \boldsymbol{W}\boldsymbol{X}, \quad \forall \boldsymbol{X} \in B$$

for some vector $\boldsymbol{c}$ and skew-symmetric tensor $\boldsymbol{W}$. Equivalently, by Result 1.3, the displacement field for an infinitesimally rigid deformation may be written as

$$\boldsymbol{u}(\boldsymbol{X}) = \boldsymbol{c} + \boldsymbol{w} \times \boldsymbol{X},$$

where $\boldsymbol{w}$ is the axial vector of $\boldsymbol{W}$. As the name suggests, an infinitesimally rigid deformation is related to a small rigid deformation (see Exercise 22). For an infinitesimally rigid deformation the displacement gradient is $\nabla \boldsymbol{u} = \boldsymbol{W}$ and the infinitesimal strain tensor is given by

$$\boldsymbol{E}(\boldsymbol{X}) = \operatorname{sym} \nabla \boldsymbol{u}(\boldsymbol{X}) = \tfrac{1}{2}(\boldsymbol{W} + \boldsymbol{W}^T) = \boldsymbol{O}, \quad \forall \boldsymbol{X} \in B.$$

In view of Result 4.6, we see that an infinitesimally rigid deformation produces no strain as measured by $\boldsymbol{E}$. In particular, the change in length of an infinitesimal line segment along any coordinate direction is approximately zero, and the shear between any two coordinate directions is approximately zero at each point in B. Infinitesimally rigid deformations can be completely characterized in terms of the tensor field $\boldsymbol{E}$. In particular, it can be shown that a deformation φ is infinitesimally rigid if and only if $\boldsymbol{E}(\boldsymbol{X}) = \boldsymbol{O}$ for all $\boldsymbol{X} \in B$.

4.4 Motions

The continuous deformation of a body over the course of time is called a **motion**. The motion of a body with reference configuration B is described by a continuous map $\varphi : B \times [0, \infty) \to \mathbb{E}^3$, where for each fixed $t \geq 0$ the function $\varphi(\cdot, t) = \varphi_t : B \to \mathbb{E}^3$ is a deformation of B (see Exercise 23). At any time $t \geq 0$ the deformation φ_t maps the reference configuration B onto a configuration $B_t = \varphi_t(B)$. We call B_t the current or deformed configuration at time t (see Figure 4.7).

We assume that φ_0 is the identity map in the sense that $\varphi_0(\boldsymbol{X}) = \boldsymbol{X}$ for all $\boldsymbol{X} \in B$, which implies $B_0 = B$. Thus a motion represents a continuous deformation of a body that begins from the configuration B. The assumption of continuity implies that, during a motion, a body cannot break apart into disjoint pieces, develop holes and so on. We further assume that for each $t \geq 0$ the deformation $\varphi_t : B \to B_t$ is

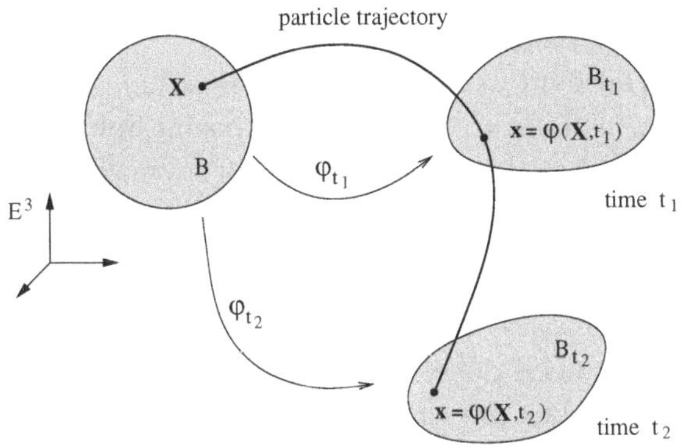

Fig. 4.7 Various configurations in the motion of a material body: B denotes the reference configuration at time $t = 0$, and B_t denotes the current configuration at time t.

admissible as described in Section 4.2. Thus we can define an inverse deformation $\boldsymbol{\psi}_t = \boldsymbol{\varphi}_t^{-1} : B_t \to B$ such that

$$\boldsymbol{X} = \boldsymbol{\psi}_t(\boldsymbol{x}) = \boldsymbol{\psi}(\boldsymbol{x}, t).$$

By properties of inverse functions, for any $t \geq 0$ we have

$$\boldsymbol{X} = \boldsymbol{\psi}_t(\boldsymbol{\varphi}_t(\boldsymbol{X})) = \boldsymbol{\psi}(\boldsymbol{\varphi}(\boldsymbol{X}, t), t) \quad \forall \boldsymbol{X} \in B,$$

and similarly

$$\boldsymbol{x} = \boldsymbol{\varphi}_t(\boldsymbol{\psi}_t(\boldsymbol{x})) = \boldsymbol{\varphi}(\boldsymbol{\psi}(\boldsymbol{x}, t), t) \quad \forall \boldsymbol{x} \in B_t$$

(see Exercise 24).

Throughout the remainder of our developments we assume that all motions $\boldsymbol{\varphi}$ and their inverses $\boldsymbol{\psi}$ are smooth in the sense that partial derivatives of all orders exist and are continuous. This is always stronger than what we need, but allows for a clean presentation.

4.4.1 Material and Spatial Fields

In our study of the motion of continuum bodies we will encounter fields defined over the current configuration B_t whose points we label by $\boldsymbol{x}$. However, since $\boldsymbol{x} = \boldsymbol{\varphi}(\boldsymbol{X}, t)$, any function of $\boldsymbol{x} \in B_t$ can also be expressed as a function of $\boldsymbol{X} \in B$. Similarly, we will encounter fields defined over the reference configuration B whose points we label by $\boldsymbol{X}$. However, since $\boldsymbol{X} = \boldsymbol{\psi}(\boldsymbol{x}, t)$, any function of $\boldsymbol{X} \in B$ can also be expressed as a function of $\boldsymbol{x} \in B_t$. To help keep track of where a field was originally defined, and how it is currently being expressed, we introduce the following definitions.

By a **material field** we mean a field expressed in terms of the points $\boldsymbol{X}$ of B; for example, $\Omega = \Omega(\boldsymbol{X}, t)$. By a **spatial field** we mean a field expressed in terms of the points $\boldsymbol{x}$ of B_t; for example, $\Gamma = \Gamma(\boldsymbol{x}, t)$. To any material field $\Omega(\boldsymbol{X}, t)$ we can associate a spatial field $\Omega_s(\boldsymbol{x}, t)$ by the relation

$$\Omega_s(\boldsymbol{x}, t) = \Omega(\boldsymbol{\psi}(\boldsymbol{x}, t), t).$$

We call Ω_s the **spatial description** of the material field Ω. Similarly, to any spatial field $\Gamma(\boldsymbol{x}, t)$ we can associate a material field $\Gamma_m(\boldsymbol{X}, t)$ by the relation

$$\Gamma_m(\boldsymbol{X}, t) = \Gamma(\boldsymbol{\varphi}(\boldsymbol{X}, t), t).$$

We call Γ_m the **material description** of the spatial field Γ (see Exercise 25).

4.4.2 Coordinate Derivatives

Throughout the remainder of our studies we will need to be careful about distinguishing between material coordinates $\boldsymbol{X} = (X_1, X_2, X_3)^T$, which label points in B, and spatial coordinates $\boldsymbol{x} = (x_1, x_2, x_3)^T$, which label points in the current configuration B_t. To distinguish between derivatives with respect to these two sets of coordinates we introduce the following notation.

We use the symbol ∇^X to denote the gradient, divergence and curl of material fields with respect to the material coordinates X_i for any fixed time $t \geq 0$. Similarly, we use the symbol ∇^x to denote the gradient, divergence and curl of spatial fields with respect to the spatial coordinates x_i for any fixed time $t \geq 0$. By extension we define the Laplacian operators $\Delta^X = \nabla^X \cdot (\nabla^X)$ and $\Delta^x = \nabla^x \cdot (\nabla^x)$.

4.4.3 Time Derivatives

In our studies we will frequently need to compute the rate of change of a given field with respect to time. By the **total time derivative** of a field we mean the rate of change of the field as measured by an observer who is tracking the motion of each particle in the body. It is important to keep in mind that, because the reference configuration B is fixed, the material coordinates $\boldsymbol{X}$ of each particle are fixed. In contrast, because the current configuration B_t changes with time, the spatial coordinates $\boldsymbol{x}$ of each particle change with time. The precise manner in which the spatial coordinates change with time is given by the motion, namely

$$\boldsymbol{x} = \boldsymbol{\varphi}(\boldsymbol{X}, t).$$

Using an overdot to denote the total time derivative, we note from the above remarks that the total time derivative of a material field $\Omega(\boldsymbol{X}, t)$ is simply

$$\dot{\Omega}(\boldsymbol{X}, t) = \frac{\partial}{\partial t}\Omega(\boldsymbol{X}, t).$$

On the other hand, because $\boldsymbol{x} = \boldsymbol{\varphi}(\boldsymbol{X}, t)$, the total time derivative of a spatial field $\Gamma(\boldsymbol{x}, t)$ is given by

$$\dot{\Gamma}(\boldsymbol{x}, t) = \left[\frac{\partial}{\partial t}\Gamma(\boldsymbol{\varphi}(\boldsymbol{X}, t), t)\right]\Big|_{\boldsymbol{X}=\boldsymbol{\psi}(\boldsymbol{x},t)} = \left[\dot{\Gamma}_m(\boldsymbol{X}, t)\right]\Big|_{\boldsymbol{X}=\boldsymbol{\psi}(\boldsymbol{x},t)}, \quad (4.16)$$

that is

$$\dot{\Gamma} = \left[\dot{\Gamma}_m\right]_s.$$

Remarks:

(1) The total time derivative of a field is often called the **material** or **substantial**, or sometimes **convective**, time derivative. It is the rate of change of the field that we would measure if we were to follow each material particle as it moves in space. Because each particle is identified with fixed material coordinates $\boldsymbol{X}$, it is the time derivative computed with $\boldsymbol{X}$ fixed.

(2) In general, notice that $\dot{\Gamma}(\boldsymbol{x}, t) \neq \frac{\partial}{\partial t}\Gamma(\boldsymbol{x}, t)$. That is, the total time derivative of a spatial field is different from the partial time derivative computed with $\boldsymbol{x}$ fixed. In this case the partial time derivative fails to account for the fact that the spatial coordinates $\boldsymbol{x}$ of a body particle change with time, namely $\boldsymbol{x} = \boldsymbol{\varphi}(\boldsymbol{X}, t)$ (see Exercise 26). □

4.4.4 Velocity and Acceleration Fields

Let $\boldsymbol{\varphi} : B \times [0, \infty) \rightarrow \boldsymbol{E}^3$ be a motion of a continuum body and consider a material particle labeled by $\boldsymbol{X}$ in the reference configuration B. At any time $t \geq 0$, this particle is labeled by $\boldsymbol{x} = \boldsymbol{\varphi}(\boldsymbol{X}, t)$ in the current configuration B_t (see Figure 4.7). We denote by $\boldsymbol{V}(\boldsymbol{X}, t)$ the **velocity** at time t of the material particle labeled by $\boldsymbol{X}$ in B. By definition of the motion we have

$$\boldsymbol{V}(\boldsymbol{X},t) = \frac{\partial}{\partial t}\boldsymbol{\varphi}(\boldsymbol{X},t).$$

Similarly, we denote by $\boldsymbol{A}(\boldsymbol{X},t)$ the **acceleration** at time t of the material particle labeled by $\boldsymbol{X}$ in B. By definition of the motion we have

$$\boldsymbol{A}(\boldsymbol{X},t) = \frac{\partial^2}{\partial t^2}\boldsymbol{\varphi}(\boldsymbol{X},t).$$

From the above definitions we see that the velocity and acceleration of material particles are naturally material fields. Frequently, however, we will need the spatial descriptions of these fields. We denote by $\boldsymbol{v}(\boldsymbol{x},t)$ the spatial description of the material velocity field $\boldsymbol{V}(\boldsymbol{X},t)$, that is

$$\boldsymbol{v}(\boldsymbol{x},t) = \boldsymbol{V}_s(\boldsymbol{x},t) = \left[\frac{\partial}{\partial t}\boldsymbol{\varphi}(\boldsymbol{X},t)\right]\bigg|_{\boldsymbol{X}=\boldsymbol{\psi}(\boldsymbol{x},t)}, \tag{4.17}$$

and we denote by $\boldsymbol{a}(\boldsymbol{x},t)$ the spatial description of the material acceleration field $\boldsymbol{A}(\boldsymbol{X},t)$, so that

$$\boldsymbol{a}(\boldsymbol{x},t) = \boldsymbol{A}_s(\boldsymbol{x},t) = \left[\frac{\partial^2}{\partial t^2}\boldsymbol{\varphi}(\boldsymbol{X},t)\right]\bigg|_{\boldsymbol{X}=\boldsymbol{\psi}(\boldsymbol{x},t)}$$

(see Exercises 27 and 28). Notice that $\boldsymbol{v}(\boldsymbol{x},t)$ and $\boldsymbol{a}(\boldsymbol{x},t)$ correspond to the velocity and acceleration of the material particle whose current coordinates are $\boldsymbol{x}$ at time t. The following result shows that the spatial velocity field can be used to derive a convenient formula for the total time derivative of an arbitrary spatial field.

Result 4.7 ***Total Time Derivative.*** *Let $\boldsymbol{\varphi} : B \times [0,\infty) \to \boldsymbol{E}^3$ be a motion of a continuum body with associated spatial velocity field $\boldsymbol{v}$, and consider an arbitrary spatial scalar field $\phi = \phi(\boldsymbol{x},t)$ and an arbitrary spatial vector field $\boldsymbol{w} = \boldsymbol{w}(\boldsymbol{x},t)$. Then the total time derivatives of ϕ and $\boldsymbol{w}$ are given by*

$$\dot{\phi} = \frac{\partial}{\partial t}\phi + \nabla^x\phi\cdot\boldsymbol{v} \quad \textit{and} \quad \dot{\boldsymbol{w}} = \frac{\partial}{\partial t}\boldsymbol{w} + (\nabla^x\boldsymbol{w})\boldsymbol{v}.$$

□

Proof By definition of the gradients of ϕ and $\boldsymbol{w}$ we have

$$\nabla^x\phi = \frac{\partial\phi}{\partial x_i}\boldsymbol{e}_i \quad \text{and} \quad \nabla^x\boldsymbol{w} = \frac{\partial w_i}{\partial x_j}\boldsymbol{e}_i\otimes\boldsymbol{e}_j,$$

where $\{e_k\}$ is any fixed frame, and by definition of $\boldsymbol{v}$ in (4.17) we have

$$\boldsymbol{v}(\boldsymbol{x},t)\Big|_{\boldsymbol{x}=\boldsymbol{\varphi}(\boldsymbol{X},t)} = \frac{\partial}{\partial t}\boldsymbol{\varphi}(\boldsymbol{X},t).$$

Using the definition of the total time derivative in (4.16) we obtain

$$\begin{aligned}
&\left[\dot{\phi}(\boldsymbol{x},t)\right]\Big|_{\boldsymbol{x}=\boldsymbol{\varphi}(\boldsymbol{X},t)} \\
&= \frac{\partial}{\partial t}\phi(\boldsymbol{\varphi}(\boldsymbol{X},t),t) \\
&= \left[\frac{\partial}{\partial t}\phi(\boldsymbol{x},t)\right]\Big|_{\boldsymbol{x}=\boldsymbol{\varphi}(\boldsymbol{X},t)} + \left[\frac{\partial}{\partial x_i}\phi(\boldsymbol{x},t)\right]\Big|_{\boldsymbol{x}=\boldsymbol{\varphi}(\boldsymbol{X},t)} \frac{\partial}{\partial t}\varphi_i(\boldsymbol{X},t) \\
&= \left[\frac{\partial}{\partial t}\phi(\boldsymbol{x},t)\right]\Big|_{\boldsymbol{x}=\boldsymbol{\varphi}(\boldsymbol{X},t)} + \left[\frac{\partial}{\partial x_i}\phi(\boldsymbol{x},t)\right]\Big|_{\boldsymbol{x}=\boldsymbol{\varphi}(\boldsymbol{X},t)} v_i(\boldsymbol{x},t)\Big|_{\boldsymbol{x}=\boldsymbol{\varphi}(\boldsymbol{X},t)}.
\end{aligned}$$

Expressing the above result in terms of spatial coordinates $\boldsymbol{x}$ we have

$$\dot{\phi}(\boldsymbol{x},t) = \frac{\partial \phi}{\partial t}(\boldsymbol{x},t) + \frac{\partial \phi}{\partial x_i}(\boldsymbol{x},t)v_i(\boldsymbol{x},t),$$

which establishes the result for ϕ. Given $\boldsymbol{w} = w_i \boldsymbol{e}_i$ we apply the above result to each scalar field $w_i(\boldsymbol{x},t)$ $(i = 1,2,3)$ to obtain

$$\dot{w}_i(\boldsymbol{x},t) = \frac{\partial w_i}{\partial t}(\boldsymbol{x},t) + \frac{\partial w_i}{\partial x_j}(\boldsymbol{x},t)v_j(\boldsymbol{x},t),$$

which implies the result for $\boldsymbol{w}$. □

Remarks:

(1) The above result shows that, if the spatial velocity $\boldsymbol{v}$ is known, then the total time derivative of a spatial field ϕ can be computed without explicit knowledge of the motion $\boldsymbol{\varphi}$ or its inverse $\boldsymbol{\psi}$ (see Exercise 29).

(2) By definition, the spatial acceleration field $\boldsymbol{a}$ of a motion satisfies $\boldsymbol{a} = \dot{\boldsymbol{v}}$, where $\boldsymbol{v}$ is the spatial velocity field of the motion. From Result 4.7 we deduce

$$\boldsymbol{a} = \frac{\partial \boldsymbol{v}}{\partial t} + (\nabla^x \boldsymbol{v})\boldsymbol{v}.$$

Thus the spatial acceleration is actually a nonlinear function of the spatial velocity and its derivatives.

(3) Many texts employ the notation $\boldsymbol{v} \cdot \nabla^x \boldsymbol{w}$ in place of $(\nabla^x \boldsymbol{w})\boldsymbol{v}$. We, however, will always use $(\nabla^x \boldsymbol{w})\boldsymbol{v}$, which denotes the usual product between the second-order tensor $\nabla^x \boldsymbol{w}$ and the vector $\boldsymbol{v}$.

□

4.5 Rate of Strain, Spin

Let $\boldsymbol{\varphi} : B \times [0, \infty) \to \mathbb{E}^3$ be a motion of a continuum body, and let B_t and $B_{t'}$ denote the configurations at times $t \geq 0$ and $t' \geq t$. Moreover, let Ω be a small ball of radius $\alpha > 0$ centered at a point $\boldsymbol{x}$ in B_t, and let Ω' be the corresponding region in $B_{t'}$ as depicted in Figure 4.8. Any quantitative measure of the rate of change of shape between Ω' and Ω in the limits $\alpha \to 0$ and $t' \to t$ is generally called **rate of strain** at $\boldsymbol{x}$ and t. Similarly, any quantitative measure of the rate of change of orientation of Ω' with respect to Ω is called **rate of rotation** or **spin**. Notice that, in contrast to the concept of strain, the concepts of rate of strain and spin are independent of the reference configuration B. For this reason rate of strain and spin play a central role in the study of fluids. In the following sections we introduce measures of rate of strain and spin and discuss some of their properties.

4.5.1 Rate of Strain Tensor L and Spin Tensor W

Let $\boldsymbol{\varphi} : B \times [0, \infty) \to \mathbb{E}^3$ be a motion of a continuum body with spatial velocity field $\boldsymbol{v}$. Then a measure of rate of strain is provided by the tensor

$$\boldsymbol{L} = \text{sym}(\nabla^x \boldsymbol{v}) = \tfrac{1}{2}(\nabla^x \boldsymbol{v} + \nabla^x \boldsymbol{v}^T).$$

We call $\boldsymbol{L}_t = \boldsymbol{L}(\cdot, t) : B_t \to \mathcal{V}^2$ the **rate of strain tensor** field associated with $\boldsymbol{v}$. Notice that, by definition, $\boldsymbol{L}$ is a spatial field and is symmetric for each point $\boldsymbol{x}$ in B_t and time t.

To any spatial velocity field $\boldsymbol{v}$ we associate another tensor field defined by

$$\boldsymbol{W} = \text{skew}(\nabla^x \boldsymbol{v}) = \tfrac{1}{2}(\nabla^x \boldsymbol{v} - \nabla^x \boldsymbol{v}^T).$$

We call $\boldsymbol{W}_t = \boldsymbol{W}(\cdot, t) : B_t \to \mathcal{V}^2$ the **spin tensor** field associated with $\boldsymbol{v}$. By definition, $\boldsymbol{W}$ is a spatial field and is skew-symmetric for each point $\boldsymbol{x}$ in B_t and time t.

Whereas $\boldsymbol{L}$ is a measure of rate of strain or distortion in a material, $\boldsymbol{W}$ is a measure of rate of rotation or spin (see Exercise 30). These

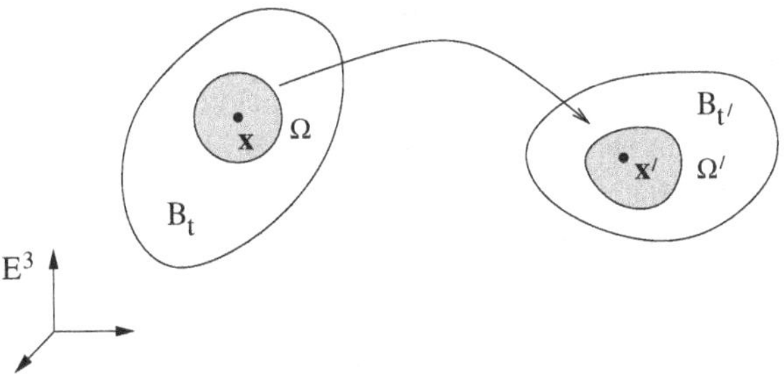

Fig. 4.8 Two configurations B_t and $B_{t'}$ in the motion of a body. The rate of change of shape of a small region Ω in B_t leads to the concept of rate of strain.

tensor fields will arise in Chapters 6 and 8 when we consider theories describing the flow of various types of fluids.

4.5.2 Interpretation of L and W

By arguments analogous to those used for the infinitesimal strain tensor, we find that the diagonal components of $\boldsymbol{L}$ quantify the *instantaneous rate of stretching* of infinitesimal line segments located at $\boldsymbol{x}$ in B_t which are aligned with the coordinate axes. Similarly, the off-diagonal components of $\boldsymbol{L}$ quantify the *instantaneous rate of shearing*. In particular, if we let $\widehat{\boldsymbol{E}}$ be the infinitesimal strain tensor associated with the small deformation $\widehat{\varphi} : B_t \to B_{t+s}$ $(0 \leq s \ll 1)$, then $\boldsymbol{L}$ is the rate of change of $\widehat{\boldsymbol{E}}$ with respect to s at $s = 0$. Equivalently, if we let $\widehat{\boldsymbol{U}}$ be the right stretch tensor associated with $\widehat{\varphi}$, then $\boldsymbol{L}$ is the rate of change of $\widehat{\boldsymbol{U}}$ with respect to s at $s = 0$.

The skew-symmetric tensor $\boldsymbol{W}$ quantifies the *instantaneous rate of rigid rotation* of an infinitesimal volume of material at $\boldsymbol{x}$ in B_t. In particular, if we let $\widehat{\boldsymbol{R}}$ be the rotation tensor in the polar decomposition of the deformation gradient $\nabla^x \widehat{\varphi}$, then $\boldsymbol{W}$ is the rate of change of $\widehat{\boldsymbol{R}}$ with respect to s at $s = 0$ (see Exercise 31).

4.5.3 Vorticity

Let $\varphi : B \times [0, \infty) \to \boldsymbol{E}^3$ be a motion of a continuum body with spatial velocity field $\boldsymbol{v}$. By the **vorticity** of a motion we mean the spatial vector field

$$\boldsymbol{w} = \nabla^x \times \boldsymbol{v}.$$

In view of Definition 2.12 and Result 1.3 we see that $\boldsymbol{w}$ is the axial vector associated with the skew-symmetric tensor $2\boldsymbol{W}$. Thus the vorticity at a point is a measure of the rate of rotation or spin at that point. In particular, from the discussion following Stokes' Theorem (Result 2.18), the vorticity at a point in a material body can be interpreted as twice the local angular velocity at that point.

4.5.4 Rigid Motions

A motion φ is called **rigid** if

$$\varphi(\boldsymbol{X},t) = \boldsymbol{c}(t) + \boldsymbol{R}(t)\boldsymbol{X}$$

for some time-dependent vector $\boldsymbol{c}(t)$ and rotation tensor $\boldsymbol{R}(t)$. For such motions we can show that the spatial velocity field can be written in the form

$$\boldsymbol{v}(\boldsymbol{x},t) = \boldsymbol{\omega}(t) \times (\boldsymbol{x} - \boldsymbol{c}(t)) + \dot{\boldsymbol{c}}(t),$$

where $\boldsymbol{\omega}(t)$ is a time-dependent vector called the **spatial angular velocity** of the motion (see Exercise 33). Equivalently, using Result 1.3 we may write

$$\boldsymbol{v}(\boldsymbol{x},t) = \boldsymbol{\Omega}(t)(\boldsymbol{x} - \boldsymbol{c}(t)) + \dot{\boldsymbol{c}}(t), \tag{4.18}$$

where $\boldsymbol{\Omega}(t)$ is the skew-symmetric second-order tensor with axial vector $\boldsymbol{\omega}(t)$. From (4.18) we find $\nabla^x \boldsymbol{v}(\boldsymbol{x},t) = \boldsymbol{\Omega}(t)$, from which we deduce $\boldsymbol{L}(\boldsymbol{x},t) = \boldsymbol{O}$, $\boldsymbol{W}(\boldsymbol{x},t) = \boldsymbol{\Omega}(t)$ and $\boldsymbol{w}(\boldsymbol{x},t) = 2\boldsymbol{\omega}(t)$ for all $t \geq 0$ and $\boldsymbol{x} \in B_t$. Thus rigid motions produce no rate of strain as measured by $\boldsymbol{L}$, but produce spin as measured by $\boldsymbol{W}$ or $\boldsymbol{w}$.

4.6 Change of Variables

Let $\varphi : B \times [0,\infty) \to \mathbb{E}^3$ be a motion of a continuum body, and for any time $t \geq 0$ consider the deformation φ_t which maps the reference configuration B onto the current configuration B_t. Throughout our developments we will frequently need to transform integrals defined over regions in B_t to integrals over corresponding regions in B. Similarly, we will frequently need to transform integrals defined over surfaces in B_t to integrals defined over corresponding surfaces in B.

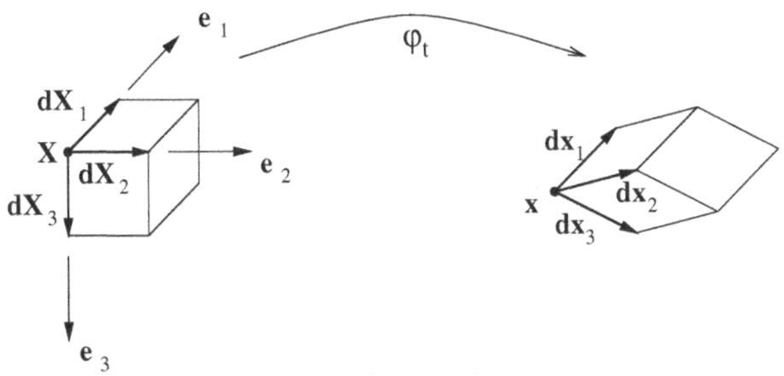

Fig. 4.9 An infinitesimal volume element in B is mapped by the deformation to an infinitesimal volume element in B_t.

4.6.1 Transformation of Volume Integrals

Here we derive a relation between an infinitesimal volume element $dV_{\boldsymbol{x}}$ at the point $\boldsymbol{x}$ in B_t, and the corresponding volume element $dV_{\boldsymbol{X}}$ at the point $\boldsymbol{X}$ in B (see Figure 4.9). We then use this relation to develop a change of variable formula for volume integrals over B_t and B.

To begin, notice that an infinitesimal volume element $dV_{\boldsymbol{X}}$ at an arbitrary point $\boldsymbol{X}$ in B can be represented in terms of a triple scalar product. In particular, let $d\boldsymbol{X}_1$, $d\boldsymbol{X}_2$ and $d\boldsymbol{X}_3$ be three infinitesimal vectors based at the point $\boldsymbol{X}$. Then the infinitesimal volume element defined by these vectors is

$$dV_{\boldsymbol{X}} = (d\boldsymbol{X}_1 \times d\boldsymbol{X}_2) \cdot d\boldsymbol{X}_3.$$

As illustrated in Figure 4.9, the three infinitesimal vectors $d\boldsymbol{X}_k$ at $\boldsymbol{X}$ are mapped by the deformation φ_t to three infinitesimal vectors $d\boldsymbol{x}_k$ at $\boldsymbol{x} = \varphi_t(\boldsymbol{X})$ in B_t. In particular, we have

$$d\boldsymbol{x}_k = \nabla^X \varphi(\boldsymbol{X}, t) d\boldsymbol{X}_k = \boldsymbol{F}(\boldsymbol{X}, t) d\boldsymbol{X}_k. \tag{4.19}$$

The infinitesimal vectors $d\boldsymbol{x}_k$ can be used to define a volume element $dV_{\boldsymbol{x}}$ via the triple scalar product, namely

$$dV_{\boldsymbol{x}} = (d\boldsymbol{x}_1 \times d\boldsymbol{x}_2) \cdot d\boldsymbol{x}_3. \tag{4.20}$$

Notice that, by properties of the determinant and the triple scalar product, we have

$$(\boldsymbol{F}\boldsymbol{u} \times \boldsymbol{F}\boldsymbol{v}) \cdot \boldsymbol{F}\boldsymbol{w} = (\det \boldsymbol{F})\, \boldsymbol{u} \times \boldsymbol{v} \cdot \boldsymbol{w}, \tag{4.21}$$

for any second-order tensor $\boldsymbol{F}$ and vectors $\boldsymbol{u}$, $\boldsymbol{v}$ and $\boldsymbol{w}$. Substituting (4.19) into (4.20) and using (4.21) we obtain

$$\begin{aligned} dV_{\boldsymbol{x}} &= (d\boldsymbol{x}_1 \times d\boldsymbol{x}_2) \cdot d\boldsymbol{x}_3 \\ &= (\boldsymbol{F}(\boldsymbol{X},t)d\boldsymbol{X}_1 \times \boldsymbol{F}(\boldsymbol{X},t)d\boldsymbol{X}_2) \cdot \boldsymbol{F}(\boldsymbol{X},t)d\boldsymbol{X}_3 \\ &= \det \boldsymbol{F}(\boldsymbol{X},t)\,(d\boldsymbol{X}_1 \times d\boldsymbol{X}_2) \cdot d\boldsymbol{X}_3 \\ &= \det \boldsymbol{F}(\boldsymbol{X},t)\, dV_{\boldsymbol{X}}. \end{aligned}$$

This expression leads to the following change of variable formula for the transformation of volume integrals.

Result 4.8 ***Transformation of Volume Integrals.*** *Let $\phi(\boldsymbol{x},t)$ be an arbitrary spatial scalar field on B_t, let Ω_t be an arbitrary subset of B_t, and let Ω be the corresponding subset of B so that $\Omega_t = \varphi_t(\Omega)$. Then*

$$\int_{\Omega_t} \phi(\boldsymbol{x},t)\, dV_{\boldsymbol{x}} = \int_{\Omega} \phi_m(\boldsymbol{X},t) \det \boldsymbol{F}(\boldsymbol{X},t)\, dV_{\boldsymbol{X}}.$$

□

Remarks:

(1) The above result provides some insight into the physical meaning of the field $\det \boldsymbol{F}$. In particular, let $\Omega_{\delta,0}$ be a ball of radius $\delta > 0$ centered at a point $\boldsymbol{X}_0 \in B$, and let $\Omega_{\delta,t} = \varphi_t(\Omega_{\delta,0})$. Then, by Result 4.8 and a Taylor expansion of $\det \boldsymbol{F}(\boldsymbol{X},t)$ about $\boldsymbol{X}_0$, we have

$$\begin{aligned} \mathrm{vol}(\Omega_{\delta,t}) &= \int_{\Omega_{\delta,t}} dV_{\boldsymbol{x}} \\ &= \int_{\Omega_{\delta,0}} \det \boldsymbol{F}(\boldsymbol{X},t)\, dV_{\boldsymbol{X}} \\ &= \int_{\Omega_{\delta,0}} [\det \boldsymbol{F}(\boldsymbol{X}_0,t) + \mathcal{O}(\delta)]\, dV_{\boldsymbol{X}} \\ &= [\det \boldsymbol{F}(\boldsymbol{X}_0,t) + \mathcal{O}(\delta)]\, \mathrm{vol}(\Omega_{\delta,0}). \end{aligned}$$

Dividing by $\mathrm{vol}(\Omega_{\delta,0})$ and taking the limit $\delta \to 0$ we deduce

$$\det \boldsymbol{F}(\boldsymbol{X}_0,t) = \lim_{\delta \to 0} \frac{\mathrm{vol}(\Omega_{\delta,t})}{\mathrm{vol}(\Omega_{\delta,0})}.$$

Hence $\det \boldsymbol{F}(\boldsymbol{X}_0,t)$ represents the local ratio of deformed volume to reference volume at a point $\boldsymbol{X}_0 \in B$ under a deformation φ_t.

(2) The field defined by $J(\boldsymbol{X},t) = \det \boldsymbol{F}(\boldsymbol{X},t)$ is typically called the **Jacobian** field for the deformation φ_t. It is a measure of *volume strain* caused by a deformation φ_t at the point $\boldsymbol{X}$ at time t. The deformation compresses material in a neighborhood of $\boldsymbol{X}$ if and only if $J(\boldsymbol{X},t) < 1$ (volume is decreased). Similarly, the deformation expands material in a neighborhood of $\boldsymbol{X}$ if and only if $J(\boldsymbol{X},t) > 1$ (volume is increased). When $J(\boldsymbol{X},t) = 1$ there is no change in material volume in a neighborhood of $\boldsymbol{X}$. In this case, the material near $\boldsymbol{X}$ is distorted in such a way that its volume is preserved. For example, it can be compressed along certain directions, and expanded along others (see also Section 4.7). Notice that $J(\boldsymbol{X},t) > 0$ for any admissible deformation.

□

4.6.2 Derivatives of Time-Dependent Integrals

Since the current configuration B_t of a body depends on time, any integral over B_t is generally time-dependent. We will frequently need to compute the time derivatives of such integrals and in this respect the following result will be useful.

Result 4.9 ***Time Derivative of Jacobian.*** *Let* $\varphi : B \times [0, \infty) \to \mathbb{E}^3$ *be a motion of a continuum body with spatial velocity field* $\boldsymbol{v}(\boldsymbol{x},t)$ *and deformation gradient field* $\boldsymbol{F}(\boldsymbol{X},t)$. *Then*

$$\frac{\partial}{\partial t}(\det \boldsymbol{F}(\boldsymbol{X},t)) = \det \boldsymbol{F}(\boldsymbol{X},t)\ \ (\nabla^x \cdot \boldsymbol{v})(\boldsymbol{x},t)\Big|_{\boldsymbol{x}=\varphi(\boldsymbol{X},t)}.$$

□

Proof Using Result 2.23 we find

$$\begin{aligned}
&\frac{\partial}{\partial t}(\mathrm{dct}\, \boldsymbol{F}(\boldsymbol{X},t)) \\
&\qquad = \det \boldsymbol{F}(\boldsymbol{X},t)\ \ \boldsymbol{F}(\boldsymbol{X},t)^{-T} : \frac{\partial}{\partial t}\boldsymbol{F}(\boldsymbol{X},t) \\
&\qquad = \det \boldsymbol{F}(\boldsymbol{X},t)\ \operatorname{tr}\left(\boldsymbol{F}(\boldsymbol{X},t)^{-1}\frac{\partial}{\partial t}\boldsymbol{F}(\boldsymbol{X},t)\right) \\
&\qquad = \det \boldsymbol{F}(\boldsymbol{X},t)\ \operatorname{tr}\left(\left\{\frac{\partial}{\partial t}\boldsymbol{F}(\boldsymbol{X},t)\right\}\boldsymbol{F}(\boldsymbol{X},t)^{-1}\right).
\end{aligned} \tag{4.22}$$

By definition of the spatial velocity field $\boldsymbol{v}$ we have

$$\boldsymbol{v}(\boldsymbol{x},t)\Big|_{\boldsymbol{x}=\boldsymbol{\varphi}(\boldsymbol{X},t)} = \frac{\partial}{\partial t}\boldsymbol{\varphi}(\boldsymbol{X},t) \quad \text{or} \quad v_i(\boldsymbol{x},t)\Big|_{\boldsymbol{x}=\boldsymbol{\varphi}(\boldsymbol{X},t)} = \frac{\partial}{\partial t}\varphi_i(\boldsymbol{X},t).$$

We next proceed to express the term $\mathrm{tr}(\{\frac{\partial}{\partial t}\boldsymbol{F}\}\boldsymbol{F}^{-1})$ in terms of $\boldsymbol{v}$. To this end, notice that

$$F_{ij}(\boldsymbol{X},t) = \frac{\partial}{\partial X_j}\varphi_i(\boldsymbol{X},t)$$

and

$$\begin{aligned}
\frac{\partial}{\partial t}\frac{\partial}{\partial X_j}\varphi_i(\boldsymbol{X},t) &= \frac{\partial}{\partial X_j}\frac{\partial}{\partial t}\varphi_i(\boldsymbol{X},t) \\
&= \frac{\partial}{\partial X_j}\left(v_i(\boldsymbol{x},t)\Big|_{\boldsymbol{x}=\boldsymbol{\varphi}(\boldsymbol{X},t)}\right) \\
&= \frac{\partial}{\partial x_k}v_i(\boldsymbol{x},t)\Big|_{\boldsymbol{x}=\boldsymbol{\varphi}(\boldsymbol{X},t)} \frac{\partial}{\partial X_j}\varphi_k(\boldsymbol{X},t).
\end{aligned}$$

This yields

$$\frac{\partial}{\partial t}F_{ij}(\boldsymbol{X},t) = \frac{\partial}{\partial x_k}v_i(\boldsymbol{x},t)\Big|_{\boldsymbol{x}=\boldsymbol{\varphi}(\boldsymbol{X},t)} F_{kj}(\boldsymbol{X},t),$$

which in tensor notation becomes

$$\frac{\partial}{\partial t}\boldsymbol{F}(\boldsymbol{X},t) = \nabla^x \boldsymbol{v}(\boldsymbol{x},t)\Big|_{\boldsymbol{x}=\boldsymbol{\varphi}(\boldsymbol{X},t)} \boldsymbol{F}(\boldsymbol{X},t).$$

From this expression we deduce

$$\left\{\frac{\partial}{\partial t}\boldsymbol{F}(\boldsymbol{X},t)\right\}\boldsymbol{F}(\boldsymbol{X},t)^{-1} = \nabla^x \boldsymbol{v}(\boldsymbol{x},t)\Big|_{\boldsymbol{x}=\boldsymbol{\varphi}(\boldsymbol{X},t)},$$

and by taking the trace we obtain

$$\begin{aligned}
\mathrm{tr}\left(\left\{\frac{\partial}{\partial t}\boldsymbol{F}(\boldsymbol{X},t)\right\}\boldsymbol{F}(\boldsymbol{X},t)^{-1}\right) &= \mathrm{tr}(\nabla^x \boldsymbol{v}(\boldsymbol{x},t))\Big|_{\boldsymbol{x}=\boldsymbol{\varphi}(\boldsymbol{X},t)} \\
&= (\nabla^x \cdot \boldsymbol{v})(\boldsymbol{x},t)\Big|_{\boldsymbol{x}=\boldsymbol{\varphi}(\boldsymbol{X},t)}.
\end{aligned}$$

Substitution of this expression into (4.22) leads to the desired result. □

The above result shows that the time derivative of the Jacobian field depends only on the Jacobian field itself and the divergence of the spatial velocity field. This observation is exploited in the following result, which provides a convenient formula for the time derivative of volume integrals over arbitrary subsets of B_t.

Result 4.10 ***Reynolds Transport Theorem.*** *Let* $\varphi : B \times [0,\infty) \to \boldsymbol{E}^3$ *be a motion of a continuum body with associated spatial velocity field* $\boldsymbol{v}(\boldsymbol{x},t)$*, let* Ω_t *be an arbitrary volume in* B_t *with surface denoted by* $\partial\Omega_t$ *and let* $\boldsymbol{n}$ *be the outward unit normal field on* $\partial\Omega_t$*. Then for any spatial scalar field* $\Phi(\boldsymbol{x},t)$ *we have*

$$\frac{d}{dt}\int_{\Omega_t} \Phi \, dV_{\boldsymbol{x}} = \int_{\Omega_t} \dot{\Phi} + \Phi(\nabla^x \cdot \boldsymbol{v}) \, dV_{\boldsymbol{x}},$$

or equivalently

$$\frac{d}{dt}\int_{\Omega_t} \Phi \, dV_{\boldsymbol{x}} = \int_{\Omega_t} \frac{\partial}{\partial t}\Phi \, dV_{\boldsymbol{x}} + \int_{\partial\Omega_t} \Phi \boldsymbol{v}\cdot\boldsymbol{n} \, dA_{\boldsymbol{x}}. \tag{4.23}$$

□

Proof See Exercise 34. □

Remarks:

(1) The above result show that, if the spatial velocity $\boldsymbol{v}$ is known, then the time derivative of a volume integral over an arbitrary subset Ω_t of B_t can be computed without explicit knowledge of the motion φ or its inverse ψ.

(2) The term *transport theorem* is motivated by the expression in (4.23). In particular, the rate of change of the integral of Φ over Ω_t is equal to the rate computed as if Ω_t were fixed in its current position, plus the rate at which Φ is *transported* across the boundary $\partial\Omega_t$ by the normal component of velocity $\boldsymbol{v}\cdot\boldsymbol{n}$. □

4.6.3 Transformation of Surface Integrals

Let Γ be a surface in B oriented by a unit normal field $\boldsymbol{N} : \Gamma \to \mathcal{V}$ and let $\gamma_t = \varphi_t(\Gamma)$ be the corresponding surface in B_t with corresponding orientation given by a unit normal field $\boldsymbol{n} : \gamma_t \to \mathcal{V}$. Given a point $\boldsymbol{X} \in \Gamma$ and an infinitesimal area element $dA_{\boldsymbol{X}}$ at $\boldsymbol{X}$, we are interested in how this area element transforms under the mapping φ_t.

For our purposes, we assume that the surface Γ is regular in the sense that, given any point $\boldsymbol{Z} \in \Gamma$, we can establish a system of **surface**

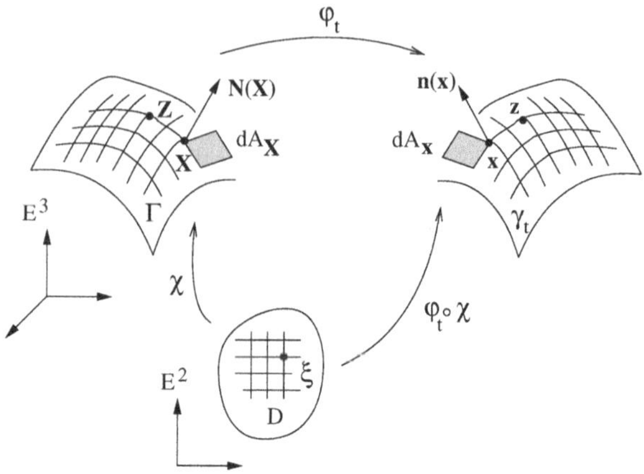

Fig. 4.10 An infinitesimal area element on the surface Γ in B is mapped by the deformation to an infinitesimal area element on the surface γ_t in B_t.

coordinates on Γ near $\boldsymbol{Z}$. Under this assumption we can find a region $D \subset \mathbb{E}^2$ and a map $\boldsymbol{\chi} : D \to \mathbb{E}^3$ such that any point $\boldsymbol{X}$ on Γ near $\boldsymbol{Z}$ can be written as

$$\boldsymbol{X} = \boldsymbol{\chi}(\boldsymbol{\xi}), \qquad \boldsymbol{\xi} = (\xi_1, \xi_2)^T \in D,$$

as depicted in Figure 4.10. In terms of the surface coordinates $\boldsymbol{\xi}$, an infinitesimal oriented area element in Γ at $\boldsymbol{X}$ is defined as

$$\boldsymbol{N}(\boldsymbol{X})\, dA_{\boldsymbol{X}} = \frac{\partial}{\partial \xi_1}\boldsymbol{\chi}(\boldsymbol{\xi}) \times \frac{\partial}{\partial \xi_2}\boldsymbol{\chi}(\boldsymbol{\xi})\, d\xi_1 d\xi_2. \tag{4.24}$$

Under the deformation $\boldsymbol{\varphi}_t$, surface coordinates on Γ near $\boldsymbol{Z}$ become surface coordinates on γ_t near $\boldsymbol{z} = \boldsymbol{\varphi}_t(\boldsymbol{Z})$. That is, any point $\boldsymbol{x}$ on γ_t near $\boldsymbol{z}$ may be written as

$$\boldsymbol{x} = \boldsymbol{\varphi}_t(\boldsymbol{\chi}(\boldsymbol{\xi})), \qquad \boldsymbol{\xi} = (\xi_1, \xi_2)^T \in D.$$

In terms of the surface coordinates $\boldsymbol{\xi}$, an infinitesimal oriented area element in γ_t at $\boldsymbol{x}$ is defined as

$$\boldsymbol{n}(\boldsymbol{x})\, dA_{\boldsymbol{x}} = \frac{\partial}{\partial \xi_1}\boldsymbol{\varphi}_t(\boldsymbol{\chi}(\boldsymbol{\xi})) \times \frac{\partial}{\partial \xi_2}\boldsymbol{\varphi}_t(\boldsymbol{\chi}(\boldsymbol{\xi}))\, d\xi_1 d\xi_2. \tag{4.25}$$

Moreover, we have

$$\begin{aligned}\frac{\partial}{\partial \xi_1}\boldsymbol{\varphi}_t(\boldsymbol{\chi}(\boldsymbol{\xi})) &= \frac{\partial}{\partial \xi_1}\varphi_t^i(\boldsymbol{\chi}(\boldsymbol{\xi}))\boldsymbol{e}_i \\ &= \left[\frac{\partial}{\partial X_j}\varphi_t^i(\boldsymbol{X})\right]\Big|_{\boldsymbol{X}=\boldsymbol{\chi}(\boldsymbol{\xi})}\frac{\partial}{\partial \xi_1}\chi_j(\boldsymbol{\xi})\boldsymbol{e}_i \\ &= \boldsymbol{F}_t(\boldsymbol{\chi}(\boldsymbol{\xi}))\frac{\partial}{\partial \xi_1}\boldsymbol{\chi}(\boldsymbol{\xi}),\end{aligned}$$

with a similar expression for $\frac{\partial}{\partial \xi_2}\boldsymbol{\varphi}_t(\boldsymbol{\chi}(\boldsymbol{\xi}))$. Using the identity

$$\boldsymbol{F}\boldsymbol{u}\times\boldsymbol{F}\boldsymbol{v} = (\det \boldsymbol{F})\boldsymbol{F}^{-T}(\boldsymbol{u}\times\boldsymbol{v}),$$

which holds for any invertible second-order tensor $\boldsymbol{F}$ and vectors $\boldsymbol{u}$ and $\boldsymbol{v}$ (see Chapter 1, Exercise 17), we can write (4.25) as

$$\begin{aligned}\boldsymbol{n}(\boldsymbol{x})\, dA_{\boldsymbol{x}} &= \left[\boldsymbol{F}_t(\boldsymbol{\chi}(\boldsymbol{\xi}))\frac{\partial}{\partial \xi_1}\boldsymbol{\chi}(\boldsymbol{\xi})\right]\times\left[\boldsymbol{F}_t(\boldsymbol{\chi}(\boldsymbol{\xi}))\frac{\partial}{\partial \xi_2}\boldsymbol{\chi}(\boldsymbol{\xi})\right]\, d\xi_1 d\xi_2 \\ &= [\det \boldsymbol{F}_t(\boldsymbol{\chi}(\boldsymbol{\xi}))]\boldsymbol{F}_t(\boldsymbol{\chi}(\boldsymbol{\xi}))^{-T}\left[\frac{\partial}{\partial \xi_1}\boldsymbol{\chi}(\boldsymbol{\xi})\times\frac{\partial}{\partial \xi_2}\boldsymbol{\chi}(\boldsymbol{\xi})\right]\, d\xi_1 d\xi_2.\end{aligned} \tag{4.26}$$

Comparing (4.26) with (4.24) we find

$$\boldsymbol{n}(\boldsymbol{x})\, dA_{\boldsymbol{x}} = (\det \boldsymbol{F}_t(\boldsymbol{X}))\boldsymbol{F}_t(\boldsymbol{X})^{-T}\boldsymbol{N}(\boldsymbol{X})\, dA_{\boldsymbol{X}},$$

where $\boldsymbol{x} = \boldsymbol{\varphi}_t(\boldsymbol{X})$. This relationship leads to the following result which relates surface integrals between the reference and deformed configurations.

Result 4.11 **Transformation of Surface Integrals.** *Let $\phi(\boldsymbol{x},t)$ be an arbitrary spatial scalar field, $\boldsymbol{w}(\boldsymbol{x},t)$ an arbitrary spatial vector field, and $\boldsymbol{T}(\boldsymbol{x},t)$ an arbitrary spatial second-order tensor field. Consider any region Ω in B which maps to a region Ω_t in B_t under a deformation $\boldsymbol{\varphi}_t$. Let $\partial\Omega$ denote the surface of Ω with outward unit normal field $\boldsymbol{N}(\boldsymbol{X})$, and let $\partial\Omega_t$ denote the surface of Ω_t with outward unit normal field $\boldsymbol{n}(\boldsymbol{x})$. Then*

$$\begin{aligned}\int_{\partial\Omega_t}\phi(\boldsymbol{x},t)\boldsymbol{n}(\boldsymbol{x})\, dA_{\boldsymbol{x}} &= \int_{\partial\Omega}\phi_m(\boldsymbol{X},t)\boldsymbol{G}(\boldsymbol{X},t)\boldsymbol{N}(\boldsymbol{X})\, dA_{\boldsymbol{X}}, \\ \int_{\partial\Omega_t}\boldsymbol{w}(\boldsymbol{x},t)\cdot\boldsymbol{n}(\boldsymbol{x})\, dA_{\boldsymbol{x}} &= \int_{\partial\Omega}\boldsymbol{w}_m(\boldsymbol{X},t)\cdot\boldsymbol{G}(\boldsymbol{X},t)\boldsymbol{N}(\boldsymbol{X})\, dA_{\boldsymbol{X}}, \\ \int_{\partial\Omega_t}\boldsymbol{T}(\boldsymbol{x},t)\boldsymbol{n}(\boldsymbol{x})\, dA_{\boldsymbol{x}} &= \int_{\partial\Omega}\boldsymbol{T}_m(\boldsymbol{X},t)\boldsymbol{G}(\boldsymbol{X},t)\boldsymbol{N}(\boldsymbol{X})\, dA_{\boldsymbol{X}},\end{aligned}$$

where $\boldsymbol{G}$ *is a material second-order tensor field defined as*

$$\boldsymbol{G}(\boldsymbol{X},t) = (\det \boldsymbol{F}(\boldsymbol{X},t))\boldsymbol{F}(\boldsymbol{X},t)^{-T}.$$

□

Notice that other types of surface integrals can also be considered and transformed in a similar way (see Exercise 35).

4.7 Volume-Preserving Motions

A motion $\varphi : B \times [0,\infty) \to \boldsymbol{E}^3$ of a continuum body is called **volume-preserving** or **isochoric** if $\text{vol}[\varphi_t(\Omega)] = \text{vol}[\Omega]$ for all $t \geq 0$ and all subsets $\Omega \subseteq B$. In particular, a motion is volume-preserving if the volume of the body, and each of its parts, remains constant throughout a motion.

Result 4.12 ***Conditions for Volume Preservation.*** *Let* $\varphi : B \times [0,\infty) \to \boldsymbol{E}^3$ *be a motion of a continuum body with spatial velocity field* $\boldsymbol{v}(\boldsymbol{x},t)$ *and deformation gradient field* $\boldsymbol{F}(\boldsymbol{X},t)$*. Then* φ *is volume-preserving if and only if* $\det \boldsymbol{F}(\boldsymbol{X},t) = 1$ *for all* $\boldsymbol{X} \in B$ *and* $t \geq 0$*, or equivalently* $(\nabla^x \cdot \boldsymbol{v})(\boldsymbol{x},t) = 0$ *for all* $\boldsymbol{x} \in B_t$ *and* $t \geq 0$*.* □

Proof See Exercise 36. □

Simple motions such as translations and rotations of body are volume-preserving because they produce no distortion. However, motions which do distort a body can also be volume-preserving (see Exercise 37).

Bibliographic Notes

Much of the material presented here on deformations, motions, strain and rate of strain can be found in the text by Gurtin (1981). A detailed analysis of these concepts in general coordinate systems is given in Ogden (1984), and a similar detailed account from a more engineering point of view can be found in Malvern (1969). The treatise by Truesdell and Toupin (1960) contains a wealth of information on these concepts and their history. For a mathematical analysis of deformations and motions within the context of Riemannian geometry and manifold theory see Marsden and Hughes (1983).

Constrained motions such as rigid and volume-preserving motions are of special interest in continuum mechanics. For example, rigid motions are of interest in the study of rigid bodies, and provide a standard against which other motions can be compared. Volume-preserving motions are of interest in the study of incompressible fluids (see Chapter 6) and solids. A complete mathematical characterization of rigid deformations and motions can be found in Gurtin (1981). For a more rigorous treatment of rigid deformations see Ciarlet (1988, Volume 1), and for a discussion of constrained motions other than rigid and volume-preserving see Ogden (1984) and Antman (1995).

The change of variable formulae for volume and surface integrals play a fundamental role in continuum mechanics and will be used repeatedly in later chapters. Discussions of these formulae can be found in standard texts on advanced multivariable calculus, for example Bourne and Kendall (1992) and Marsden and Tromba (1988). Rigorous proofs of the change of variable formulae can be found in the text by Munkres (1991).

The Reynolds Transport Theorem is of special importance in the study of fluids. In particular, it forms the basis for the so-called *control volume* method for studying fluid flow. For applications of the transport theorem in this context see White (2006) and Gurtin (1981).

Exercises

4.1 Let $B = \{\boldsymbol{X} \in \mathbb{E}^3 \mid |X_i| < 1\}$ be the reference configuration of a body as shown below, and consider the deformation $\boldsymbol{x} = \boldsymbol{\varphi}(\boldsymbol{X})$ defined in components by $x_1 = X_1$, $x_2 = X_3 + 4$, $x_3 = -X_2$.

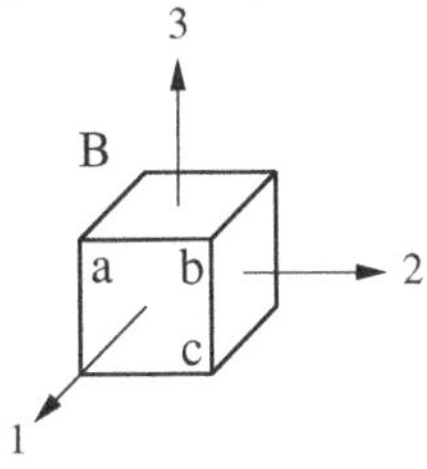

(a) Sketch the deformed configuration $B' = \boldsymbol{\varphi}(B)$ relative to the same coordinate frame and indicate the points in B' which correspond to the vertices a, b, c in B.

(b) Find the components of the displacement field $\boldsymbol{u}$.

4.2 Let $B = \{\boldsymbol{X} \in \mathbb{E}^3 \mid |X_i| < 1\}$ and $B' = \{\boldsymbol{x} \in \mathbb{E}^3 \mid |x_1| < 1,\ |x_3| < 1,\ |x_2 - 6| < 3\}$ be the reference and deformed configurations of a body as shown below, and consider a deformation $\boldsymbol{x} = \boldsymbol{\varphi}(\boldsymbol{X})$ of the form $x_1 = pX_2 + q$, $x_2 = rX_1 + s$, $x_3 = X_3$.

By comparing the faces of B' and B find constants p, q, r, s such that $B' = \varphi(B)$.

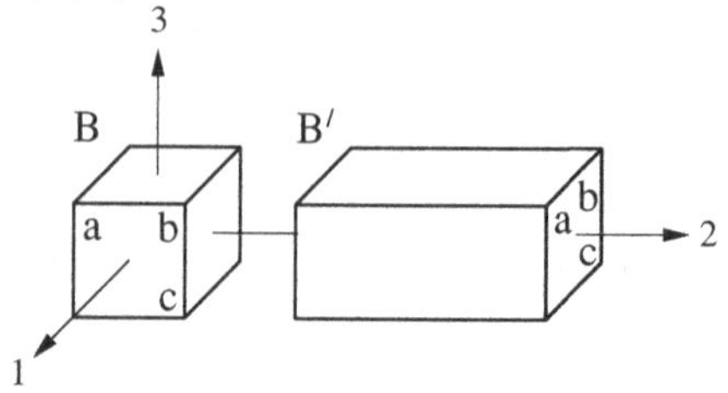

4.3 Let $B = \{\boldsymbol{X} \in \boldsymbol{E}^3 \mid |X_i| < 1\}$ be the reference configuration of a body. For each deformation $\boldsymbol{x} = \varphi(\boldsymbol{X})$ given below (i) sketch the deformed configuration B', (ii) find the components of $\nabla\varphi$, (iii) determine if φ is one-to-one, and (iv) determine if φ is admissible:

(a) $x_1 = 1 + X_1, \quad x_2 = \frac{1}{2}X_2 + 5, \quad x_3 = 2X_3,$

(b) $x_1 = X_1, \quad x_2 = 3, \quad x_3 = X_3,$

(c) $x_1 = X_1, \quad x_2 = X_2, \quad x_3 = 4 - X_3.$

4.4 Consider a deformation $\boldsymbol{x} = \varphi(\boldsymbol{X})$ defined in components by

$$x_1 = X_1^2, \quad x_2 = X_3^2, \quad x_3 = X_2 X_3.$$

(a) Find the components of $\nabla\varphi$ and determine $\det \nabla\varphi$.

(b) Is φ admissible for arbitrary B?

4.5 For each deformation $\boldsymbol{x} = \varphi(\boldsymbol{X})$ given below find the components of the deformation gradient $\boldsymbol{F}$ and determine if φ is homogeneous or non-homogeneous:

(a) $x_1 = X_1, \quad x_2 = X_2 X_3, \quad x_3 = X_3 - 1,$

(b) $x_1 = 2X_2 - 1, \quad x_2 = X_3, \quad x_3 = 3 + 5X_1,$

(c) $x_1 = \exp(X_1), \quad x_2 = -X_3, \quad x_3 = X_2.$

4.6 Let φ be a homogeneous deformation with deformation gradient $\boldsymbol{F}$. For any two points $\boldsymbol{X}$ and $\boldsymbol{Y}$ show that

$$\varphi(\boldsymbol{X}) = \varphi(\boldsymbol{Y}) + \boldsymbol{F}(\boldsymbol{X} - \boldsymbol{Y}).$$

4.7 Let $\varphi : B \to B'$ be a homogeneous deformation with deformation gradient $\boldsymbol{F}$, and let $\boldsymbol{X}(\sigma) = \boldsymbol{X}_0 + \sigma\boldsymbol{v}$ be a line segment through the point $\boldsymbol{X}_0$ in B with direction $\boldsymbol{v}$. Show that $\varphi(\boldsymbol{X}(\sigma))$ is a line segment through the point $\varphi(\boldsymbol{X}_0)$ in B' with direction $\boldsymbol{F}\boldsymbol{v}$.

4.8 One way to visualize a deformation is to show how curves etched or marked on the surface of a body appear before and after the deformation. Suppose a reference configuration B is marked with lines as follows:

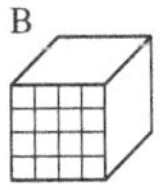

Which of the deformations visualized below appear to be homogeneous? Which appear to be non-homogeneous?

(a) 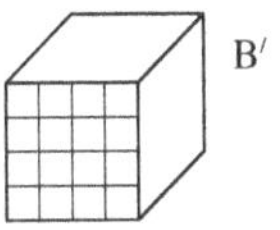

(b) 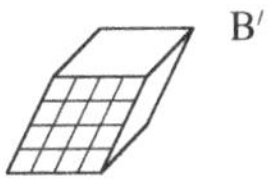

(c) 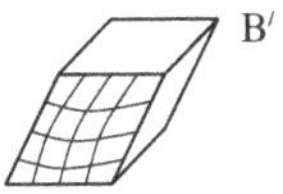

4.9 Prove Result 4.1.

4.10 Prove Result 4.2.

4.11 Prove Result 4.3.

4.12 Let $\boldsymbol{F} = \boldsymbol{R}\boldsymbol{U} = \boldsymbol{V}\boldsymbol{R}$ be the right and left polar decompositions of a deformation gradient $\boldsymbol{F}$. Show that:

(a) $\boldsymbol{U}$ and $\boldsymbol{V}$ have the same eigenvalues,

(b) if $\{\boldsymbol{u}_i\}$ is a frame of eigenvectors of $\boldsymbol{U}$, then $\{\boldsymbol{R}\boldsymbol{u}_i\}$ is a frame of eigenvectors of $\boldsymbol{V}$. Thus, in general, $\boldsymbol{U}$ and $\boldsymbol{V}$ have different eigenvectors.

4.13 Consider a deformation φ defined in components by

$$[\varphi(\boldsymbol{X})] = \begin{Bmatrix} pX_1 + a \\ qX_2 + b \\ rX_3 + c \end{Bmatrix},$$

where p, q, r, a, b, c are constants.

(a) Find conditions on p, q, r for φ to be an admissible deformation of $B = \mathbb{E}^3$.

(b) Given an arbitrary point $\boldsymbol{Y}$ find the components of $\boldsymbol{d}$, $\boldsymbol{s}$ and

$\boldsymbol{r}$ such that $\varphi = \boldsymbol{r} \circ \boldsymbol{s} \circ \boldsymbol{d}$, where $\boldsymbol{d}$ is a translation, $\boldsymbol{s}$ is a stretch from $\boldsymbol{Y}$ and $\boldsymbol{r}$ is rotation about $\boldsymbol{Y}$.

4.14 Let $\boldsymbol{Y}$ be an arbitrary point and let $\boldsymbol{Q}$ be an arbitrary rotation tensor and consider the deformation

$$\varphi(\boldsymbol{X}) = \boldsymbol{Y} + \boldsymbol{Q}(\boldsymbol{X} - \boldsymbol{Y}).$$

In particular, φ is a rotation about $\boldsymbol{Y}$. Find the deformation gradient $\boldsymbol{F}$ and the Cauchy–Green strain tensor $\boldsymbol{C}$. Does $\boldsymbol{F}$ depend on $\boldsymbol{Q}$? What about $\boldsymbol{C}$?

4.15 Suppose the deformation gradient at a point $\boldsymbol{X}_0$ in a body has components

$$[\boldsymbol{F}] = \begin{pmatrix} 1 & 0 & 0 \\ 0 & 2 & 1 \\ 0 & 1 & 2 \end{pmatrix}.$$

Find the components of the Cauchy–Green strain tensor $\boldsymbol{C}$ and the right stretch tensor $\boldsymbol{U}$.

4.16 Let $B = \{\boldsymbol{X} \in \boldsymbol{E}^3 \mid 0 < X_i < 1\}$ and consider a deformation $\boldsymbol{x} = \varphi(\boldsymbol{X})$ of the form $x_1 = pX_1$, $x_2 = qX_2$, $x_3 = rX_3$, where p, q, r are constants. Notice that, because φ is homogeneous, the Cauchy–Green strain tensor $\boldsymbol{C}$ will be constant.

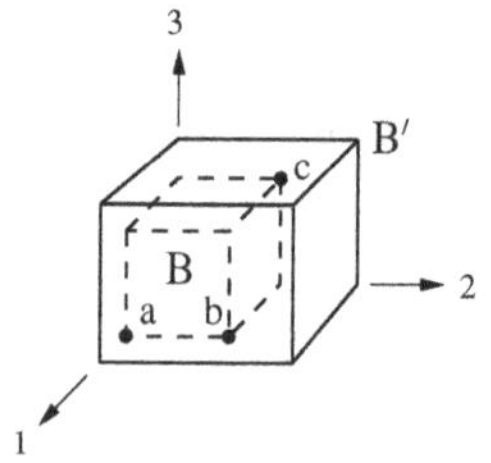

(a) Find the components of $\boldsymbol{C}$.

(b) Find the stretch λ in (i) the direction parallel to the edge $\overline{ab}$ and (ii) the direction parallel to the diagonal $\overline{ac}$.

4.17 Let $B = \{\boldsymbol{X} \in \boldsymbol{E}^3 \mid 0 < X_i < 1\}$ and consider a deformation $\boldsymbol{x} = \varphi(\boldsymbol{X})$ of the form $x_1 = X_1 + \alpha X_2$, $x_2 = X_2$, $x_3 = X_3$, where $\alpha > 0$ is a constant. Such a deformation is called a **simple shear** in the $\boldsymbol{e}_1, \boldsymbol{e}_2$-plane.

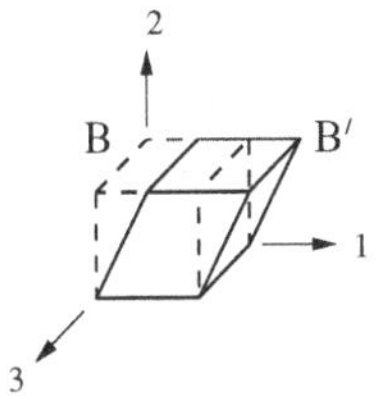

(a) Find the components of the Cauchy–Green strain tensor $\boldsymbol{C}$.

(b) Find the shear γ for the direction pair $(\boldsymbol{e}_1, \boldsymbol{e}_2)$ and for the pair $(\boldsymbol{e}_1, \boldsymbol{e}_3)$. What happens to these shears in the limits $\alpha \to 0$ and $\alpha \to \infty$?

(c) Find the extreme values of the stretch λ and the directions in which these occur. Do the directions of extreme stretch correspond to any of the diagonal directions $\pm\frac{1}{\sqrt{2}}(\boldsymbol{e}_1 + \boldsymbol{e}_2)$ or $\pm\frac{1}{\sqrt{2}}(\boldsymbol{e}_1 - \boldsymbol{e}_2)$ in the $\boldsymbol{e}_1, \boldsymbol{e}_2$-plane?

4.18 Show that the shear $\gamma(\boldsymbol{e}, \boldsymbol{d})$ between any two right principal directions $\boldsymbol{e}$ and $\boldsymbol{d}$ is zero.

4.19 Prove Result 4.5.

4.20 Let $\boldsymbol{c}$ be an arbitrary vector, $\boldsymbol{A}$ an arbitrary tensor and ϵ an arbitrary scalar. Supposing the components of $\boldsymbol{c}$ and $\boldsymbol{A}$ are of order unity and $0 \leq \epsilon \ll 1$, determine which of the following deformations are small:

(a) $\varphi(\boldsymbol{X}) = \boldsymbol{X} + \boldsymbol{c}$,

(b) $\varphi(\boldsymbol{X}) = \boldsymbol{A}\boldsymbol{X} + \boldsymbol{c}$,

(c) $\varphi(\boldsymbol{X}) = \epsilon\boldsymbol{A}\boldsymbol{X} + \boldsymbol{c}$,

(d) $\varphi(\boldsymbol{X}) = (\boldsymbol{I} + \epsilon\boldsymbol{A})\boldsymbol{X} + \boldsymbol{c}$.

4.21 Let $B = \{\boldsymbol{X} \in \mathbb{E}^3 \mid 0 < X_i < L\}$ and consider the simple shear deformation $\boldsymbol{x} = \varphi(\boldsymbol{X})$ given by $x_1 = X_1 + (\alpha/L)X_2$, $x_2 = X_2$, $x_3 = X_3$, where $\alpha > 0$ is a constant.

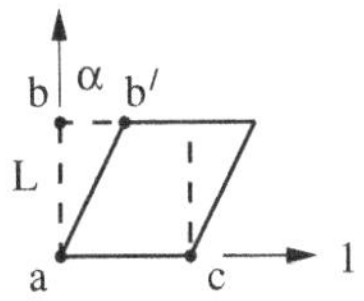

(a) Show that φ is small when $\alpha \ll L$ and find the components of the infinitesimal strain tensor $\boldsymbol{E}$.

(b) Use the series expansion $\sqrt{1+\epsilon^2} = 1 + \frac{\epsilon^2}{2} - \frac{\epsilon^4}{8} + \cdots$ to show that

$$\frac{\overline{ab'} - \overline{ab}}{\overline{ab}} = \frac{(\alpha/L)^2}{2} - \frac{(\alpha/L)^4}{8} + \cdots.$$

(Here $\overline{ab}$ denotes length.) Is this result consistent with the interpretation of E_{22} and its value found in part (a)?

(c) Use the series expansion $\arctan(\epsilon) = \epsilon - \frac{\epsilon^3}{3} + \frac{\epsilon^5}{5} - \cdots$ to show that

$$\angle bac - \angle b'ac = (\alpha/L) - \frac{(\alpha/L)^3}{3} + \frac{(\alpha/L)^5}{5} - \cdots.$$

(Here $\angle bac$ denotes angle.) Is this result consistent with the interpretation of E_{12} and its value found in part (a)?

4.22 Consider a rigid deformation of the form

$$\varphi(\boldsymbol{X}) = \boldsymbol{R}\boldsymbol{X} + \boldsymbol{c},$$

where $\boldsymbol{R}$ is a small rotation in the sense that $\boldsymbol{R} = \boldsymbol{I} + \mathcal{O}(\epsilon)$ for some $0 \le \epsilon \ll 1$. In particular, suppose

$$\boldsymbol{R} = \exp(\epsilon \boldsymbol{W}) = \boldsymbol{I} + \epsilon \boldsymbol{W} + \frac{\epsilon^2}{2}\boldsymbol{W}^2 + \cdots,$$

where $\boldsymbol{W}$ is an arbitrary skew-symmetric tensor with components of order unity (see Result 1.7).

(a) Find the displacement field $\boldsymbol{u}$ and show that φ is a small deformation for any reference configuration B.

(b) Show that, if terms of order $\mathcal{O}(\epsilon^2)$ are neglected, then φ has the form of an infinitesimally rigid deformation.

4.23 Let $B = \{\boldsymbol{X} \in \mathbb{E}^3 \mid 0 < X_i < 1\}$ and consider the motion φ defined in components by

$$[\varphi(\boldsymbol{X},t)] = \begin{Bmatrix} \exp(tX_1) + X_1 - 1 \\ \exp(tX_2) + X_2 - 1 \\ \exp(tX_3) + X_3 - 1 \end{Bmatrix}.$$

Determine the position of the boundary points $[\boldsymbol{X}_0] = (0,0,0)^T$ and $[\boldsymbol{X}_1] = (1,1,1)^T$ as a function of t.

4.24 Let $B = \mathbb{E}^3$ and consider the motion $\boldsymbol{x} = \varphi(\boldsymbol{X}, t)$ defined by

$$x_1 = e^t X_1 + X_3, \quad x_2 = X_2, \quad x_3 = X_3 - tX_1.$$

(a) Show that the inverse motion $\boldsymbol{X} = \psi(\boldsymbol{x}, t)$ is given by

$$X_1 = \frac{x_1 - x_3}{t + e^t}, \quad X_2 = x_2, \quad X_3 = \frac{tx_1 + e^t x_3}{t + e^t}.$$

(b) Verify that $\varphi(\psi(\boldsymbol{x}, t), t) = \boldsymbol{x}$ and $\psi(\varphi(\boldsymbol{X}, t), t) = \boldsymbol{X}$.

4.25 Consider the motion in Exercise 24 and let $\Omega(\boldsymbol{X}, t)$ and $\Gamma(\boldsymbol{x}, t)$ be material and spatial fields defined by

$$\Omega(\boldsymbol{X}, t) = X_1 + t, \qquad \Gamma(\boldsymbol{x}, t) = x_1 + t.$$

(a) Find the spatial description Ω_s of Ω.

(b) Find the material description Γ_m of Γ.

4.26 Consider the motion from Exercise 24 and consider the spatial field $\Gamma(\boldsymbol{x}, t) = x_1 + t$.

(a) Find the material time derivative $\dot{\Gamma}(\boldsymbol{x}, t)$.

(b) Find the partial time derivative $\frac{\partial}{\partial t}\Gamma(\boldsymbol{x}, t)$ and verify that $\dot{\Gamma}(\boldsymbol{x}, t) \neq \frac{\partial}{\partial t}\Gamma(\boldsymbol{x}, t)$.

4.27 Let $\boldsymbol{x} = \varphi(\boldsymbol{X}, t)$ be the motion considered in Exercise 24.

(a) Find the components of the material velocity $\boldsymbol{V}(\boldsymbol{X}, t)$ and the spatial velocity $\boldsymbol{v}(\boldsymbol{x}, t)$.

(b) Find the components of the material acceleration $\boldsymbol{A}(\boldsymbol{X}, t)$ and the spatial acceleration $\boldsymbol{a}(\boldsymbol{x}, t)$.

(c) Verify that $\boldsymbol{a}(\boldsymbol{x}, t) \neq \frac{\partial}{\partial t}\boldsymbol{v}(\boldsymbol{x}, t)$.

4.28 Let $B = \{\boldsymbol{X} \in \mathbb{E}^3 \mid 0 < X_i < 1\}$ and consider the motion $\boldsymbol{x} = \varphi(\boldsymbol{X}, t)$ defined by

$$x_1 = tX_2^2 + X_1, \quad x_2 = tX_1 + X_2, \quad x_3 = tX_3 + X_3.$$

(a) Find the components of the deformation gradient and find the largest number $t_c > 0$ for which $\det \boldsymbol{F}(\boldsymbol{X}, t) > 0$ for all $\boldsymbol{X} \in B$ and $t \in [0, t_c]$.

(b) Find the components of the inverse motion $\boldsymbol{X} = \psi(\boldsymbol{x}, t)$ for $t \in [0, t_c]$.

(c) Find the components of the material velocity field $\boldsymbol{V}(\boldsymbol{X},t)$ and the spatial velocity field $\boldsymbol{v}(\boldsymbol{x},t)$ for $t \in [0,t_c]$.

4.29 Let $B = \mathbb{E}^3$ and consider the motion $\boldsymbol{x} = \boldsymbol{\varphi}(\boldsymbol{X},t)$ defined by

$$x_1 = (1+t)X_1, \quad x_2 = X_2 + tX_3, \quad x_3 = X_3 - tX_2.$$

Moreover, consider the spatial field $\phi(\boldsymbol{x},t) = tx_1 + x_2$.

(a) Show that $\det \boldsymbol{F}(\boldsymbol{X},t) > 0$ for all $t \geq 0$ and find the components of the inverse motion $\boldsymbol{X} = \boldsymbol{\psi}(\boldsymbol{x},t)$ for all $t \geq 0$.

(b) Find the components of the spatial velocity field $\boldsymbol{v}(\boldsymbol{x},t)$.

(c) Find the material time derivative of ϕ using the definition $\dot{\phi} = [\dot{\phi}_m]_s$.

(d) Find the material time derivative of ϕ using Result 4.7. Do you obtain the same result as in part (c)?

4.30 Consider the deformation $\boldsymbol{x} = \boldsymbol{\varphi}(\boldsymbol{X},t)$ given by

$$\begin{aligned} x_1 &= \cos(\omega t)X_1 + \sin(\omega t)X_2, \\ x_2 &= -\sin(\omega t)X_1 + \cos(\omega t)X_2, \\ x_3 &= (1+\alpha t)X_3. \end{aligned}$$

Notice that this deformation corresponds to rotation (with rate ω) in the $\boldsymbol{e}_1, \boldsymbol{e}_2$-plane together with extension (with rate α) along the $\boldsymbol{e}_3$-axis.

(a) Find the components of the inverse motion $\boldsymbol{X} = \boldsymbol{\psi}(\boldsymbol{x},t)$.

(b) Find the components of the spatial velocity field $\boldsymbol{v}(\boldsymbol{x},t)$.

(c) Find the components of the rate of strain and spin tensors $\boldsymbol{L}(\boldsymbol{x},t)$ and $\boldsymbol{W}(\boldsymbol{x},t)$. Verify that $\boldsymbol{L}$ is determined by α, whereas $\boldsymbol{W}$ is determined by ω.

4.31 Consider a motion $\boldsymbol{\varphi} : B \times [0,\infty) \to \mathbb{E}^3$. For any fixed $t \geq 0$ let $\boldsymbol{v}$ be the spatial velocity field in the current configuration B_t and let $\boldsymbol{\psi}$ be the inverse motion. For any $s > 0$ let $\widehat{\boldsymbol{\varphi}}_s : B_t \to B_{t+s}$ be the motion which coincides with $\boldsymbol{\varphi}_{t+s} : B \to B_{t+s}$ in the sense that

$$\widehat{\boldsymbol{\varphi}}(\boldsymbol{x},s) = \boldsymbol{\varphi}(\boldsymbol{X},t+s)|_{\boldsymbol{X}=\boldsymbol{\psi}(\boldsymbol{x},t)}, \quad \forall \boldsymbol{x} \in B_t.$$

(a) Show that $\widehat{\boldsymbol{\varphi}}(\boldsymbol{x},0) = \boldsymbol{x}$ for all $\boldsymbol{x} \in B_t$.

(b) Show that $\frac{\partial}{\partial s}\widehat{\varphi}(\boldsymbol{x},0) = \boldsymbol{v}(\boldsymbol{x},t)$ for all $\boldsymbol{x} \in B_t$.

(c) Let $\widehat{\boldsymbol{F}}(\boldsymbol{x},s) = \nabla^x \widehat{\varphi}(\boldsymbol{x},s)$ be the deformation gradient associated with $\widehat{\varphi}$. Show that $\widehat{\boldsymbol{F}}(\boldsymbol{x},0) = \boldsymbol{I}$ and $\frac{\partial}{\partial s}\widehat{\boldsymbol{F}}(\boldsymbol{x},0) = \nabla^x \boldsymbol{v}(\boldsymbol{x},t)$.

(d) Let $\widehat{\boldsymbol{E}} = \mathrm{sym}(\nabla^x \widehat{\boldsymbol{u}}) = \mathrm{sym}(\widehat{\boldsymbol{F}} - \boldsymbol{I})$ be the infinitesimal strain tensor associated with $\widehat{\varphi}$. Show that

$$\boldsymbol{L}(\boldsymbol{x},t) = \frac{\partial}{\partial s}\widehat{\boldsymbol{E}}(\boldsymbol{x},0).$$

(e) Consider the right polar decomposition $\widehat{\boldsymbol{F}} = \widehat{\boldsymbol{R}}\widehat{\boldsymbol{U}}$, where $\widehat{\boldsymbol{U}}^2 = \widehat{\boldsymbol{F}}^T\widehat{\boldsymbol{F}}$. Show that

$$\boldsymbol{L}(\boldsymbol{x},t) = \frac{\partial}{\partial s}\widehat{\boldsymbol{U}}(\boldsymbol{x},0), \quad \boldsymbol{W}(\boldsymbol{x},t) = \frac{\partial}{\partial s}\widehat{\boldsymbol{R}}(\boldsymbol{x},0).$$

4.32 Consider a motion $\varphi : B \times [0,\infty) \to \mathbb{E}^3$ with spatial velocity field $\boldsymbol{v}$, spatial acceleration field $\boldsymbol{a}$, spatial vorticity field $\boldsymbol{w}$ and spatial spin field $\boldsymbol{W}$.

(a) Show that $2\boldsymbol{W}\boldsymbol{c} = \boldsymbol{w} \times \boldsymbol{c}$ for any arbitrary vector $\boldsymbol{c}$. (Thus $\boldsymbol{w}$ is the axial vector of $2\boldsymbol{W}$.)

(b) Show that the acceleration field satisfies

$$\boldsymbol{a} = \frac{\partial \boldsymbol{v}}{\partial t} + \boldsymbol{w} \times \boldsymbol{v} + \nabla^x\left(\tfrac{1}{2}|\boldsymbol{v}|^2\right).$$

4.33 Consider an arbitrary rigid motion $\varphi : B \times [0,\infty) \to \mathbb{E}^3$ of the form

$$\varphi(\boldsymbol{X},t) = \boldsymbol{R}(t)\boldsymbol{X} + \boldsymbol{c}(t),$$

where $\boldsymbol{R}(t)$ is a rotation tensor and $\boldsymbol{c}(t)$ is a vector.

(a) Find the inverse motion $\psi(\boldsymbol{x},t)$.

(b) Let $\boldsymbol{\Omega}(t) = \dot{\boldsymbol{R}}(t)\boldsymbol{R}(t)^T$. Show that $\boldsymbol{\Omega}(t)$ is skew-symmetric.

(c) Show that the spatial velocity field can be written in the form

$$\boldsymbol{v}(\boldsymbol{x},t) = \boldsymbol{\Omega}(t)(\boldsymbol{x} - \boldsymbol{c}(t)) + \dot{\boldsymbol{c}}(t).$$

4.34 Prove Result 4.10.

4.35 Show that

$$\int_{\partial\Omega_t} \boldsymbol{w}(\boldsymbol{x},t)\otimes\boldsymbol{n}(\boldsymbol{x})\, dA_{\boldsymbol{x}} = \int_{\partial\Omega} \boldsymbol{w}_m(\boldsymbol{X},t)\otimes\boldsymbol{G}(\boldsymbol{X},t)\boldsymbol{N}(\boldsymbol{X})\, dA_{\boldsymbol{X}},$$

where the tensor $\boldsymbol{G}$ and unit normals $\boldsymbol{n}$ and $\boldsymbol{N}$ are as defined in Result 4.11.

4.36 Prove Result 4.12.

4.37 Let $B = \{\boldsymbol{X} \in \mathbb{E}^3 \mid 0 < X_i < 1\}$ and consider the simple shear motion $\boldsymbol{x} = \boldsymbol{\varphi}(\boldsymbol{X},t)$ defined by

$$x_1 = X_1 + \alpha t X_2, \quad x_2 = X_2, \quad x_3 = X_3,$$

where $\alpha > 0$ is a constant. Show that $\boldsymbol{\varphi}(\boldsymbol{X},t)$ is volume-preserving.

Answers to Selected Exercises

4.1 (a) Consider an arbitrary line segment in B in the direction of $\boldsymbol{e}_3$, namely $\boldsymbol{X}(\sigma) = \alpha\boldsymbol{e}_1 + \beta\boldsymbol{e}_2 + \sigma\boldsymbol{e}_3$, where α and β are fixed and $\sigma \in (-1,1)$ is the coordinate along the segment. Under deformation, this segment becomes $\boldsymbol{x}(\sigma) = \boldsymbol{\varphi}(\boldsymbol{X}(\sigma)) = \alpha\boldsymbol{e}_1 + (\sigma+4)\boldsymbol{e}_2 - \beta\boldsymbol{e}_3$, which corresponds to a line segment in B' in the direction of $\boldsymbol{e}_2$. Notice that the line segment in B is centered at the point with coordinates $(\alpha,\beta,0)$, and the corresponding line segment in B', which is of the same length, is centered at the point with coordinates $(\alpha,4,-\beta)$. By considering different values for α and β, for example values corresponding to the vertical edges of B, we obtain the following sketch.

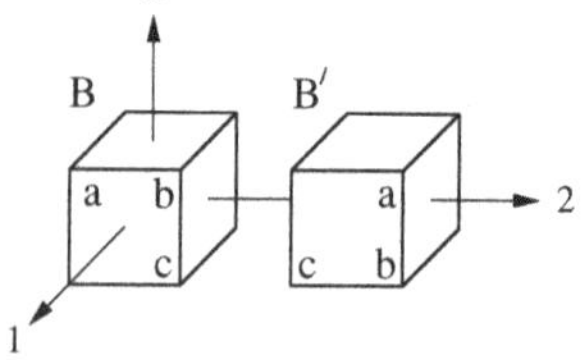

(b) $u_1 = 0$, $u_2 = X_3 + 4 - X_2$, $u_3 = -X_2 - X_3$.

4.3 In general, a deformation $\boldsymbol{\varphi} : B \to B'$ is one-to-one if and only if the equation $\boldsymbol{\varphi}(\boldsymbol{X}) = \boldsymbol{x}$ has a unique solution $\boldsymbol{X}$ for each $\boldsymbol{x}$ in the range of $\boldsymbol{\varphi}$, which is equal to B' since $\boldsymbol{\varphi}$ is onto. For homogeneous deformations of the form $\boldsymbol{\varphi}(\boldsymbol{X}) = \boldsymbol{F}\boldsymbol{X} + \boldsymbol{v}$, the equation $\boldsymbol{\varphi}(\boldsymbol{X}) = \boldsymbol{x}$ is linear and hence uniquely solvable if and only if $\boldsymbol{F} = \nabla\boldsymbol{\varphi}$ has non-zero determinant. This result is used in parts (a)–(c) below.

(a) $[\nabla\boldsymbol{\varphi}] = \text{diag}(1, \frac{1}{2}, 2)$. $\det[\nabla\boldsymbol{\varphi}] = 1$. One-to-one, admissible. (Here diag denotes a diagonal matrix.)

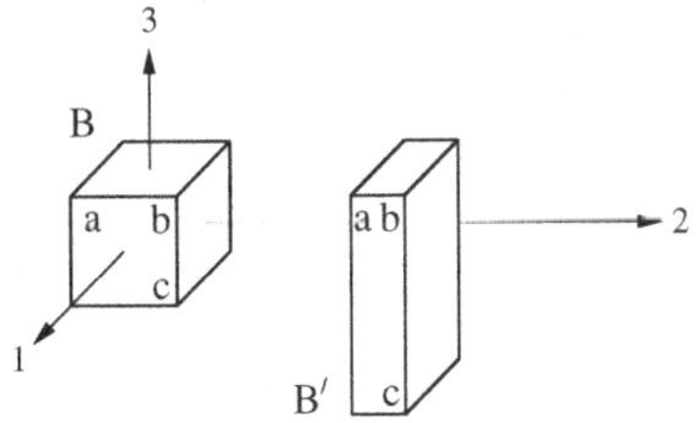

(b) $[\nabla\boldsymbol{\varphi}] = \text{diag}(1, 0, 1)$. $\det[\nabla\boldsymbol{\varphi}] = 0$. Not one-to-one thus not admissible. (B' has zero volume.)

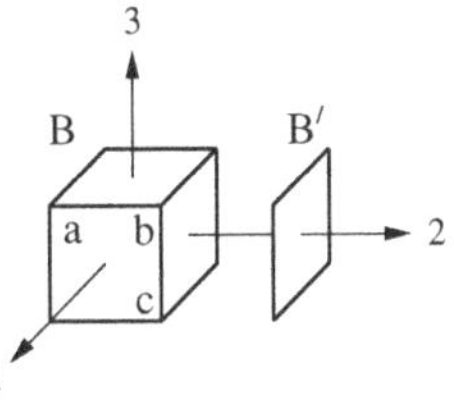

(c) $[\nabla\boldsymbol{\varphi}] = \text{diag}(1, 1, -1)$. $\det[\nabla\boldsymbol{\varphi}] = -1$. One-to-one, but not admissible. (B' is a "mirror image".)

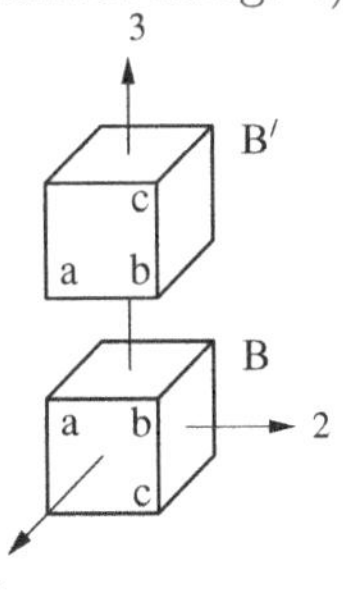

4.5 (a) $[\boldsymbol{F}] = \begin{pmatrix} 1 & 0 & 0 \\ 0 & X_3 & X_2 \\ 0 & 0 & 1 \end{pmatrix}$, $\boldsymbol{\varphi}$ is non-homogeneous.

(b) $[\boldsymbol{F}] = \begin{pmatrix} 0 & 2 & 0 \\ 0 & 0 & 1 \\ 5 & 0 & 0 \end{pmatrix}$, $\boldsymbol{\varphi}$ is homogeneous.

(c) $[F] = \begin{pmatrix} e^{X_1} & 0 & 0 \\ 0 & 0 & -1 \\ 0 & 1 & 0 \end{pmatrix}$, φ is non-homogeneous.

4.7 Since the deformation is homogeneous we have $\varphi(\boldsymbol{X}) = \boldsymbol{c} + \boldsymbol{F}\boldsymbol{X}$. Thus

$$\varphi(\boldsymbol{X}(\sigma)) = \boldsymbol{c} + \boldsymbol{F}\boldsymbol{X}_0 + \sigma\boldsymbol{F}\boldsymbol{v} = \varphi(\boldsymbol{X}_0) + \sigma\boldsymbol{F}\boldsymbol{v},$$

which is a line through the point $\varphi(\boldsymbol{X}_0)$ in B' with direction $\boldsymbol{F}\boldsymbol{v}$.

4.9 Let $\boldsymbol{Y}$ be given. Then by Exercise 4.6 we have

$$\varphi(\boldsymbol{X}) = \varphi(\boldsymbol{Y}) + \boldsymbol{F}(\boldsymbol{X} - \boldsymbol{Y}).$$

Let $\boldsymbol{d}_i(\boldsymbol{X}) = \boldsymbol{a}_i + \boldsymbol{X}$, where $\boldsymbol{a}_i$ $(i = 1, 2)$ are vectors to be determined, and let $\boldsymbol{g}(\boldsymbol{X}) = \boldsymbol{Y} + \boldsymbol{F}(\boldsymbol{X} - \boldsymbol{Y})$, where $\boldsymbol{F}$ is the deformation gradient associated with φ. Then

$$\begin{aligned}(\boldsymbol{d}_1 \circ \boldsymbol{g})(\boldsymbol{X}) &= \boldsymbol{d}_1(\boldsymbol{g}(\boldsymbol{X})) \\ &= \boldsymbol{g}(\boldsymbol{X}) + \boldsymbol{a}_1 \\ &= \boldsymbol{Y} + \boldsymbol{F}(\boldsymbol{X} - \boldsymbol{Y}) + \boldsymbol{a}_1.\end{aligned}$$

Choosing $\boldsymbol{a}_1 = \varphi(\boldsymbol{Y}) - \boldsymbol{Y}$ we obtain

$$(\boldsymbol{d}_1 \circ \boldsymbol{g})(\boldsymbol{X}) = \varphi(\boldsymbol{Y}) + \boldsymbol{F}(\boldsymbol{X} - \boldsymbol{Y}) = \varphi(\boldsymbol{X}),$$

which establishes the first decomposition. To establish the second, notice that

$$(\boldsymbol{g} \circ \boldsymbol{d}_2)(\boldsymbol{X}) = \boldsymbol{Y} + \boldsymbol{F}(\boldsymbol{X} - \boldsymbol{Y} + \boldsymbol{a}_2).$$

In this case, the choice $\boldsymbol{a}_2 = \boldsymbol{F}^{-1}\boldsymbol{a}_1$ gives the desired result. Notice that $\boldsymbol{F}^{-1}$ exists since all deformations are by assumption admissible.

4.11 Consider the stretch $\boldsymbol{s}_2 = \boldsymbol{Y} + \boldsymbol{V}(\boldsymbol{X} - \boldsymbol{Y})$. By the Spectral Decomposition Theorem the eigenvectors $\{\boldsymbol{v}_i\}$ of $\boldsymbol{V}$ can be chosen orthonormal and $\boldsymbol{V}$ can be written in the form (dropping the summation convention)

$$\boldsymbol{V} = \sum_{i=1}^{3} \lambda_i \boldsymbol{v}_i \otimes \boldsymbol{v}_i,$$

where $\lambda_i > 0$ are the associated eigenvalues. Let $\boldsymbol{h}_i$ $(i = 1, 2, 3)$ be extensions defined by

$$\boldsymbol{h}_i(\boldsymbol{X}) = \boldsymbol{Y} + \boldsymbol{F}_i(\boldsymbol{X} - \boldsymbol{Y}),$$

where (dropping the summation convention)

$$\boldsymbol{F}_i = \boldsymbol{I} + (\lambda_i - 1)\boldsymbol{v}_i \otimes \boldsymbol{v}_i. \tag{4.27}$$

Then

$$\begin{aligned}(\boldsymbol{h}_2 \circ \boldsymbol{h}_3)(\boldsymbol{X}) &= \boldsymbol{h}_2(\boldsymbol{h}_3(\boldsymbol{X}))\\ &= \boldsymbol{Y} + \boldsymbol{F}_2(\boldsymbol{h}_3(\boldsymbol{X}) - \boldsymbol{Y})\\ &= \boldsymbol{Y} + \boldsymbol{F}_2(\boldsymbol{Y} + \boldsymbol{F}_3(\boldsymbol{X} - \boldsymbol{Y}) - \boldsymbol{Y})\\ &= \boldsymbol{Y} + \boldsymbol{F}_2\boldsymbol{F}_3(\boldsymbol{X} - \boldsymbol{Y}),\end{aligned}$$

and

$$\begin{aligned}(\boldsymbol{h}_1 \circ \boldsymbol{h}_2 \circ \boldsymbol{h}_3)(\boldsymbol{X}) &= \boldsymbol{h}_1(\,(\boldsymbol{h}_2 \circ \boldsymbol{h}_3)(\boldsymbol{X})\,)\\ &= \boldsymbol{Y} + \boldsymbol{F}_1((\boldsymbol{h}_2 \circ \boldsymbol{h}_3)(\boldsymbol{X}) - \boldsymbol{Y})\\ &= \boldsymbol{Y} + \boldsymbol{F}_1(\boldsymbol{Y} + \boldsymbol{F}_2\boldsymbol{F}_3(\boldsymbol{X} - \boldsymbol{Y}) - \boldsymbol{Y})\\ &= \boldsymbol{Y} + \boldsymbol{F}_1\boldsymbol{F}_2\boldsymbol{F}_3(\boldsymbol{X} - \boldsymbol{Y}).\end{aligned}$$

Using (4.27) we find

$$\begin{aligned}\boldsymbol{F}_2\boldsymbol{F}_3 &= (\boldsymbol{I} + (\lambda_2 - 1)\boldsymbol{v}_2 \otimes \boldsymbol{v}_2)(\boldsymbol{I} + (\lambda_3 - 1)\boldsymbol{v}_3 \otimes \boldsymbol{v}_3)\\ &= \boldsymbol{I} + (\lambda_2 - 1)\boldsymbol{v}_2 \otimes \boldsymbol{v}_2 + (\lambda_3 - 1)\boldsymbol{v}_3 \otimes \boldsymbol{v}_3\\ &\qquad + (\lambda_2 - 1)(\lambda_3 - 1)(\boldsymbol{v}_2 \otimes \boldsymbol{v}_2)(\boldsymbol{v}_3 \otimes \boldsymbol{v}_3)\\ &= \boldsymbol{I} + (\lambda_2 - 1)\boldsymbol{v}_2 \otimes \boldsymbol{v}_2 + (\lambda_3 - 1)\boldsymbol{v}_3 \otimes \boldsymbol{v}_3\\ &\qquad + (\lambda_2 - 1)(\lambda_3 - 1)(\boldsymbol{v}_2 \cdot \boldsymbol{v}_3)\boldsymbol{v}_2 \otimes \boldsymbol{v}_3\\ &= \boldsymbol{I} + (\lambda_2 - 1)\boldsymbol{v}_2 \otimes \boldsymbol{v}_2 + (\lambda_3 - 1)\boldsymbol{v}_3 \otimes \boldsymbol{v}_3,\end{aligned}$$

and, by similar manipulations, we obtain

$$\begin{aligned}\boldsymbol{F}_1\boldsymbol{F}_2\boldsymbol{F}_3 = \boldsymbol{I} &+ (\lambda_1 - 1)\boldsymbol{v}_1 \otimes \boldsymbol{v}_1\\ &+ (\lambda_2 - 1)\boldsymbol{v}_2 \otimes \boldsymbol{v}_2 + (\lambda_3 - 1)\boldsymbol{v}_3 \otimes \boldsymbol{v}_3.\end{aligned} \tag{4.28}$$

Since the eigenvectors $\{\boldsymbol{v}_i\}$ form an orthonormal basis we can represent $\boldsymbol{I}$ in this basis as

$$\boldsymbol{I} = I_{ij}\boldsymbol{v}_i \otimes \boldsymbol{v}_j \quad \text{where} \quad I_{ij} = \boldsymbol{v}_i \cdot \boldsymbol{I}\boldsymbol{v}_j = \boldsymbol{v}_i \cdot \boldsymbol{v}_j = \delta_{ij}.$$

In particular, we have

$$\boldsymbol{I} = \delta_{ij}\boldsymbol{v}_i \otimes \boldsymbol{v}_j = \boldsymbol{v}_1 \otimes \boldsymbol{v}_1 + \boldsymbol{v}_2 \otimes \boldsymbol{v}_2 + \boldsymbol{v}_3 \otimes \boldsymbol{v}_3.$$

Using this result in (4.28) we find

$$\boldsymbol{F}_1\boldsymbol{F}_2\boldsymbol{F}_3 = \lambda_1\boldsymbol{v}_1 \otimes \boldsymbol{v}_1 + \lambda_2\boldsymbol{v}_2 \otimes \boldsymbol{v}_2 + \lambda_3\boldsymbol{v}_3 \otimes \boldsymbol{v}_3 = \boldsymbol{V},$$

which implies

$$\begin{aligned}(\boldsymbol{h}_1 \circ \boldsymbol{h}_2 \circ \boldsymbol{h}_3)(\boldsymbol{X}) &= \boldsymbol{Y} + \boldsymbol{F}_1\boldsymbol{F}_2\boldsymbol{F}_3(\boldsymbol{X} - \boldsymbol{Y}) \\ &= \boldsymbol{Y} + \boldsymbol{V}(\boldsymbol{X} - \boldsymbol{Y}) \\ &= \boldsymbol{s}_2(\boldsymbol{X}).\end{aligned}$$

This establishes the result for the stretch $\boldsymbol{s}_2$. The result for $\boldsymbol{s}_1$ follows in a similar manner.

4.13 (a) We have $\boldsymbol{\varphi}(\boldsymbol{X}) = \boldsymbol{F}\boldsymbol{X} + \boldsymbol{c}$, where $[\boldsymbol{F}] = \mathrm{diag}(p, q, r)$ and $[\boldsymbol{c}] = (a, b, c)^T$. (Here diag denotes a diagonal matrix.) $\boldsymbol{\varphi}$ is one-to-one if and only if the equation $\boldsymbol{\varphi}(\boldsymbol{X}) = \boldsymbol{x}$ has a unique solution $\boldsymbol{X} \in B$ for each $\boldsymbol{x} \in B'$. By a result from linear algebra, this requires $\det[\boldsymbol{F}] \neq 0$. In order for $\boldsymbol{\varphi}$ to be admissible we require furthermore that $\det[\boldsymbol{F}] > 0$. Since $\det[\boldsymbol{F}] = pqr$ we obtain the condition $pqr > 0$.

(b) Comparing the desired decomposition $\boldsymbol{\varphi} = \boldsymbol{r} \circ \boldsymbol{s} \circ \boldsymbol{d}$ with the results outlined in the Translation-Fixed Point and Stretch-Rotation Decompositions, we deduce that $\boldsymbol{r}$ and $\boldsymbol{s}$ must be the rotation and stretch defined by

$$\boldsymbol{r}(\boldsymbol{X}) = \boldsymbol{Y} + \boldsymbol{R}(\boldsymbol{X} - \boldsymbol{Y}), \qquad \boldsymbol{s}(\boldsymbol{X}) = \boldsymbol{Y} + \boldsymbol{U}(\boldsymbol{X} - \boldsymbol{Y}),$$

where $\boldsymbol{R}$ and $\boldsymbol{U}$ are the tensors in the right polar decomposition $\boldsymbol{F} = \boldsymbol{R}\boldsymbol{U}$. Since

$$[\boldsymbol{F}]^T[\boldsymbol{F}] = \begin{pmatrix} p^2 & 0 & 0 \\ 0 & q^2 & 0 \\ 0 & 0 & r^2 \end{pmatrix}$$

is in diagonal (spectral) form we obtain

$$[\boldsymbol{U}] = \sqrt{[\boldsymbol{F}]^T[\boldsymbol{F}]} = \begin{pmatrix} |p| & 0 & 0 \\ 0 & |q| & 0 \\ 0 & 0 & |r| \end{pmatrix}.$$

Here we have used the fact that $\sqrt{p^2} = |p|$ and so on. For any number $\eta \neq 0$ let $\mathrm{sign}(\eta) = \eta/|\eta|$. Then

$$[\boldsymbol{R}] = [\boldsymbol{F}][\boldsymbol{U}]^{-1} = \begin{pmatrix} \mathrm{sign}(p) & 0 & 0 \\ 0 & \mathrm{sign}(q) & 0 \\ 0 & 0 & \mathrm{sign}(r) \end{pmatrix},$$

and we obtain

$$[\boldsymbol{r}(\boldsymbol{X})] = [\boldsymbol{Y}] + [\boldsymbol{R}]([\boldsymbol{X}] - [\boldsymbol{Y}]) = \left\{ \begin{array}{c} Y_1 + \text{sign}(p)(X_1 - Y_1) \\ Y_2 + \text{sign}(q)(X_2 - Y_2) \\ Y_3 + \text{sign}(r)(X_3 - Y_3) \end{array} \right\}$$

and

$$[\boldsymbol{s}(\boldsymbol{X})] = [\boldsymbol{Y}] + [\boldsymbol{U}]([\boldsymbol{X}] - [\boldsymbol{Y}]) = \left\{ \begin{array}{c} Y_1 + |p|(X_1 - Y_1) \\ Y_2 + |q|(X_2 - Y_2) \\ Y_3 + |r|(X_3 - Y_3) \end{array} \right\}.$$

It remains to determine the translation $\boldsymbol{d}$. To this end, let $\boldsymbol{d}(\boldsymbol{X}) = \boldsymbol{X} + \boldsymbol{a}$, where $\boldsymbol{a}$ is a vector to be determined. A straightforward computation shows that

$$(\boldsymbol{r} \circ \boldsymbol{s} \circ \boldsymbol{d})(\boldsymbol{X}) = \boldsymbol{Y} + \boldsymbol{F}(\boldsymbol{a} - \boldsymbol{Y}) + \boldsymbol{F}\boldsymbol{X}.$$

Setting $(\boldsymbol{r} \circ \boldsymbol{s} \circ \boldsymbol{d})(\boldsymbol{X}) = \boldsymbol{\varphi}(\boldsymbol{X})$, where $\boldsymbol{\varphi}(\boldsymbol{X}) = \boldsymbol{F}\boldsymbol{X} + \boldsymbol{c}$, we obtain $\boldsymbol{Y} + \boldsymbol{F}(\boldsymbol{a} - \boldsymbol{Y}) = \boldsymbol{c}$, which implies

$$\boldsymbol{a} = \boldsymbol{Y} + \boldsymbol{F}^{-1}(\boldsymbol{c} - \boldsymbol{Y}).$$

Thus we get

$$[\boldsymbol{d}(\boldsymbol{X})] = [\boldsymbol{X}] + [\boldsymbol{a}] = \left\{ \begin{array}{c} X_1 + Y_1 + p^{-1}(a - Y_1) \\ X_2 + Y_2 + q^{-1}(b - Y_2) \\ X_3 + Y_3 + r^{-1}(c - Y_3) \end{array} \right\},$$

which completes the desired decomposition.

4.15 By direct calculation

$$[\boldsymbol{C}] = [\boldsymbol{F}]^T[\boldsymbol{F}] = \begin{pmatrix} 1 & 0 & 0 \\ 0 & 5 & 4 \\ 0 & 4 & 5 \end{pmatrix}.$$

To determine $[\boldsymbol{U}]$ we need the eigenvalues $\mu_i > 0$ and corresponding (orthonormal) eigenvectors $[\boldsymbol{c}_i]$ of $[\boldsymbol{C}]$. By direct calculation we find $\{\mu_1, \mu_2, \mu_3\} = \{1, 1, 9\}$ and

$$[\boldsymbol{c}_1] = (1, 0, 0)^T, \quad [\boldsymbol{c}_2] = \frac{1}{\sqrt{2}}(0, -1, 1)^T, \quad [\boldsymbol{c}_3] = \frac{1}{\sqrt{2}}(0, 1, 1)^T.$$

Since $[\boldsymbol{C}] = \sum_{i=1}^{3} \mu_i [\boldsymbol{c}_i][\boldsymbol{c}_i]^T$ (no summation convention), the Tensor Square Root Theorem gives

$$[\boldsymbol{U}] = \sqrt{[\boldsymbol{C}]} = \sum_{i=1}^{3} \sqrt{\mu_i}\, [\boldsymbol{c}_i][\boldsymbol{c}_i]^T = \begin{pmatrix} 1 & 0 & 0 \\ 0 & 2 & 1 \\ 0 & 1 & 2 \end{pmatrix}.$$

For this problem, because $[\boldsymbol{F}]$ itself is symmetric positive-definite, we find that $[\boldsymbol{U}]$ is equal to $[\boldsymbol{F}]$. In general, however, $[\boldsymbol{U}]$ and $[\boldsymbol{F}]$ will be different.

4.17 (a) For the given deformation we find

$$[\boldsymbol{F}] = \begin{pmatrix} 1 & \alpha & 0 \\ 0 & 1 & 0 \\ 0 & 0 & 1 \end{pmatrix},$$

which gives

$$[\boldsymbol{C}] = [\boldsymbol{F}]^T[\boldsymbol{F}] = \begin{pmatrix} 1 & \alpha & 0 \\ \alpha & 1+\alpha^2 & 0 \\ 0 & 0 & 1 \end{pmatrix}.$$

(b) By definition, $\gamma(\boldsymbol{e}_1, \boldsymbol{e}_2) = \Theta(\boldsymbol{e}_1, \boldsymbol{e}_2) - \theta(\boldsymbol{e}_1, \boldsymbol{e}_2)$, where $\Theta = \frac{\pi}{2}$ is the angle between $\boldsymbol{e}_1$ and $\boldsymbol{e}_2$, and θ is defined by

$$\cos\theta(\boldsymbol{e}_1, \boldsymbol{e}_2) = \frac{[\boldsymbol{e}_1]^T[\boldsymbol{C}][\boldsymbol{e}_2]}{\sqrt{[\boldsymbol{e}_1]^T[\boldsymbol{C}][\boldsymbol{e}_1]}\,\sqrt{[\boldsymbol{e}_2]^T[\boldsymbol{C}][\boldsymbol{e}_2]}} = \frac{\alpha}{\sqrt{1+\alpha^2}}.$$

This gives $\gamma(\boldsymbol{e}_1, \boldsymbol{e}_2) = \frac{\pi}{2} - \arccos(\frac{\alpha}{\sqrt{1+\alpha^2}})$. A similar calculation shows that $\gamma(\boldsymbol{e}_1, \boldsymbol{e}_3) = 0$. Since $\arccos(0) = \frac{\pi}{2}$ and $\arccos(1) = 0$ we deduce that $\gamma(\boldsymbol{e}_1, \boldsymbol{e}_2) \to 0$ as $\alpha \to 0$ and $\gamma(\boldsymbol{e}_1, \boldsymbol{e}_2) \to \frac{\pi}{2}$ as $\alpha \to \infty$. There is no shear between $\boldsymbol{e}_1$ and $\boldsymbol{e}_3$ for any α.

(c) A direct calculation shows the eigenvalues λ_i^2 of $\boldsymbol{C}$ (squares of the principal stretches) are

$$\begin{aligned} \lambda_1^2 &= 1 + \tfrac{1}{2}\alpha^2 + \alpha\sqrt{1+\tfrac{1}{4}\alpha^2}, \\ \lambda_2^2 &= 1, \\ \lambda_3^2 &= 1 + \tfrac{1}{2}\alpha^2 - \alpha\sqrt{1+\tfrac{1}{4}\alpha^2}, \end{aligned}$$

with corresponding non-normalized eigenvectors

$$\begin{aligned} [\boldsymbol{v}_1] &= \left(\sqrt{1+\tfrac{1}{4}\alpha^2} - \tfrac{1}{2}\alpha, 1, 0\right)^T, \\ [\boldsymbol{v}_2] &= (0,0,1)^T, \\ [\boldsymbol{v}_3] &= \left(\sqrt{1+\tfrac{1}{4}\alpha^2} + \tfrac{1}{2}\alpha, -1, 0\right)^T. \end{aligned}$$

Since $\lambda_1 > 1$ and $\lambda_3 < 1$ for all $\alpha > 0$ we deduce that the maximum stretch is λ_1 which occurs in a direction parallel to $\boldsymbol{v}_1$, and the minimum stretch is λ_3 which occurs in a direction parallel to $\boldsymbol{v}_3$. Notice that $\boldsymbol{v}_1$ and $\boldsymbol{v}_3$ are generally not parallel to any of the diagonal directions $\pm\frac{1}{\sqrt{2}}(1,1,0)^T$ and $\pm\frac{1}{\sqrt{2}}(1,-1,0)^T$ in the $\boldsymbol{e}_1, \boldsymbol{e}_2$-plane.

4.19 The result for the diagonal components follows from the relations $\lambda(\boldsymbol{e}) = \sqrt{\boldsymbol{e} \cdot \boldsymbol{C}\boldsymbol{e}}$ and $C_{ii} = \boldsymbol{e}_i \cdot \boldsymbol{C}\boldsymbol{e}_i$ (no sum). Putting these together gives $C_{ii} = \lambda^2(\boldsymbol{e}_i)$ (no sum). To establish the result for the off-diagonal components C_{ij} $(i \neq j)$ we begin with the relation

$$\cos\theta(\boldsymbol{e}_i, \boldsymbol{e}_j) = \frac{\boldsymbol{e}_i \cdot \boldsymbol{C}\boldsymbol{e}_j}{\sqrt{\boldsymbol{e}_i \cdot \boldsymbol{C}\boldsymbol{e}_i}\,\sqrt{\boldsymbol{e}_j \cdot \boldsymbol{C}\boldsymbol{e}_j}},$$

which implies

$$C_{ij} = \lambda(\boldsymbol{e}_i)\lambda(\boldsymbol{e}_j)\cos\theta(\boldsymbol{e}_i, \boldsymbol{e}_j) \quad \text{(no sum, } i \neq j).$$

Using the fact $\gamma(\boldsymbol{e}_i, \boldsymbol{e}_j) = \Theta(\boldsymbol{e}_i, \boldsymbol{e}_j) - \theta(\boldsymbol{e}_i, \boldsymbol{e}_j)$, where $\Theta(\boldsymbol{e}_i, \boldsymbol{e}_j) = \frac{\pi}{2}$ is the angle between $\boldsymbol{e}_i$ and $\boldsymbol{e}_j$ $(i \neq j)$, we obtain

$$\begin{aligned} C_{ij} &= \lambda(\boldsymbol{e}_i)\lambda(\boldsymbol{e}_j)\cos\theta(\boldsymbol{e}_i, \boldsymbol{e}_j) \\ &= \lambda(\boldsymbol{e}_i)\lambda(\boldsymbol{e}_j)\cos\left(\frac{\pi}{2} - \gamma(\boldsymbol{e}_1, \boldsymbol{e}_2)\right) \\ &= \lambda(\boldsymbol{e}_i)\lambda(\boldsymbol{e}_j)\sin\gamma(\boldsymbol{e}_i, \boldsymbol{e}_j), \end{aligned}$$

which establishes the result.

4.21 (a) In components, the displacement field is $u_1 = (\alpha/L)X_2$, $u_2 = 0$, $u_3 = 0$ and the displacement gradient field is

$$[\nabla \boldsymbol{u}] = \begin{pmatrix} 0 & \alpha/L & 0 \\ 0 & 0 & 0 \\ 0 & 0 & 0 \end{pmatrix}.$$

All components $\partial u_i/\partial X_j$ will be small provided $\epsilon = \alpha/L \ll 1$. By definition

$$[\boldsymbol{E}] = \tfrac{1}{2}([\nabla \boldsymbol{u}] + [\nabla \boldsymbol{u}]^T) = \begin{pmatrix} 0 & \alpha/2L & 0 \\ \alpha/2L & 0 & 0 \\ 0 & 0 & 0 \end{pmatrix}.$$

Notice that $\nabla \boldsymbol{u}$ and $\boldsymbol{E}$ are constant because the deformation is homogeneous.

(b) Since $\overline{ab'} = \sqrt{\alpha^2 + L^2}$ we get

$$\frac{\overline{ab'} - \overline{ab}}{\overline{ab}} = \sqrt{1 + \epsilon^2} - 1 = \frac{\epsilon^2}{2} - \frac{\epsilon^4}{8} + \cdots,$$

where $\epsilon = \alpha/L$. Neglecting terms of order $\mathcal{O}(\epsilon^2)$ we find

$$\frac{\overline{ab'} - \overline{ab}}{\overline{ab}} \approx 0,$$

which is consistent with the value of E_{22} found in part (a). In particular, E_{22} is equal to the change in length of the line segment $\overline{ab}$ up to order $\mathcal{O}(\epsilon^2)$.

(c) Let $\gamma = \angle bac - \angle b'ac$. Then since $\tan\gamma = \alpha/L$ we find

$$\gamma = \arctan(\epsilon) = \epsilon - \frac{\epsilon^3}{3} + \frac{\epsilon^5}{5} - \cdots,$$

where $\epsilon = \alpha/L$. Neglecting terms of order $\mathcal{O}(\epsilon^2)$ we find

$$\gamma \approx \alpha/L,$$

which is consistent with the value E_{12} found in part (a). In particular, E_{12} is equal to half the change in angle between the line segments $\overline{ac}$ and $\overline{ab}$ up to order $\mathcal{O}(\epsilon^2)$.

4.23 For $[\boldsymbol{X}_0] = (0,0,0)^T$ we get

$$[\boldsymbol{\varphi}(\boldsymbol{X}_0, t)] = (0,0,0)^T, \quad \forall t \geq 0,$$

which implies that the material point initially at $\boldsymbol{X}_0$ remains fixed for all time. For $[\boldsymbol{X}_1] = (1,1,1)^T$ we get

$$[\boldsymbol{\varphi}(\boldsymbol{X}_1, t)] = (e^t, e^t, e^t)^T, \quad \forall t \geq 0,$$

which implies that the material point initially at $\boldsymbol{X}_1$ moves along the straight line through the point $\boldsymbol{X}_1$ with direction $(1,1,1)^T$.

4.25 (a) By definition, $\Omega_s(\boldsymbol{x}, t) = \Omega(\boldsymbol{X}, t)|_{\boldsymbol{X}=\boldsymbol{\psi}(\boldsymbol{x},t)}$. Thus

$$\Omega_s(\boldsymbol{x}, t) = (X_1 + t)|_{\boldsymbol{X}=\boldsymbol{\psi}(\boldsymbol{x},t)} = \frac{x_1 - x_3}{t + e^t} + t.$$

(b) By definition, $\Gamma_m(\boldsymbol{X}, t) = \Gamma(\boldsymbol{x}, t)|_{\boldsymbol{x}=\boldsymbol{\varphi}(\boldsymbol{X},t)}$. Thus

$$\Gamma_m(\boldsymbol{X}, t) = (x_1 + t)|_{\boldsymbol{x}=\boldsymbol{\varphi}(\boldsymbol{X},t)} = e^t X_1 + X_3 + t.$$

4.27 (a) By definition, $\boldsymbol{V} = \dot{\boldsymbol{\varphi}}$ and $\boldsymbol{v} = \boldsymbol{V}_s$. Thus

$$[\boldsymbol{V}(\boldsymbol{X}, t)] = \left[\frac{\partial}{\partial t}\boldsymbol{\varphi}(\boldsymbol{X}, t)\right] = (e^t X_1, 0, -X_1)^T,$$

and

$$[\boldsymbol{v}(\boldsymbol{x},t)] = [\boldsymbol{V}(\boldsymbol{X},t)]|_{\boldsymbol{X}=\boldsymbol{\psi}(\boldsymbol{x},t)} = \left(\frac{e^t(x_1-x_3)}{t+e^t}, 0, \frac{(x_3-x_1)}{t+e^t}\right)^T.$$

(b) By definition, $\boldsymbol{A} = \dot{\boldsymbol{V}}$ and $\boldsymbol{a} = \boldsymbol{A}_s$. Thus

$$[\boldsymbol{A}(\boldsymbol{X},t)] = \left[\frac{\partial}{\partial t}\boldsymbol{V}(\boldsymbol{X},t)\right] = (e^t X_1, 0, 0)^T,$$

and

$$[\boldsymbol{a}(\boldsymbol{x},t)] = [\boldsymbol{A}(\boldsymbol{X},t)]|_{\boldsymbol{X}=\boldsymbol{\psi}(\boldsymbol{x},t)} = \left(\frac{e^t(x_1-x_3)}{t+e^t}, 0, 0\right)^T.$$

(c) From parts (a) and (b) we notice by inspection that $\boldsymbol{a}(\boldsymbol{x},t) \neq \frac{\partial}{\partial t}\boldsymbol{v}(\boldsymbol{x},t)$.

4.29 (a) The given motion is homogeneous of the form $\boldsymbol{x} = \boldsymbol{\varphi}(\boldsymbol{X},t) = \boldsymbol{F}(t)\boldsymbol{X}$ with

$$[\boldsymbol{F}(t)] = \begin{pmatrix} 1+t & 0 & 0 \\ 0 & 1 & t \\ 0 & -t & 1 \end{pmatrix}, \qquad \det[\boldsymbol{F}(t)] = (1+t)(1+t^2).$$

In particular, the deformation gradient field is independent of $\boldsymbol{X}$ and $\det[\boldsymbol{F}(t)] > 0$ for all $t \geq 0$. Due to the homogeneous form of the motion, the inverse motion is given by $\boldsymbol{X} = \boldsymbol{\psi}(\boldsymbol{x},t) = \boldsymbol{F}(t)^{-1}\boldsymbol{x}$, which in components is

$$X_1 = \frac{x_1}{1+t}, \quad X_2 = \frac{x_2 - tx_3}{1+t^2}, \quad X_3 = \frac{x_3 + tx_2}{1+t^2}.$$

(b) In components, the material velocity field is

$$[\boldsymbol{V}(\boldsymbol{X},t)] = (X_1, X_3, -X_2)^T,$$

and the spatial velocity field is

$$[\boldsymbol{v}(\boldsymbol{x},t)] = \left(\frac{x_1}{1+t}, \frac{x_3+tx_2}{1+t^2}, \frac{tx_3-x_2}{1+t^2}\right)^T.$$

(c) By definition of the material time derivative of a spatial field we have $\dot{\phi} = [\dot{\phi}_m]_s$. Since

$$\phi_m(\boldsymbol{X},t) = [tx_1 + x_2]|_{\boldsymbol{x}=\boldsymbol{\varphi}(\boldsymbol{X},t)} = (t+t^2)X_1 + X_2 + tX_3$$

and

$$\dot{\phi}_m(\boldsymbol{X}, t) = (1+2t)X_1 + X_3,$$

we get

$$\dot{\phi}(\boldsymbol{x}, t) = [(1+2t)X_1 + X_3]|_{\boldsymbol{X}=\boldsymbol{\psi}(\boldsymbol{x},t)} = \frac{(1+2t)x_1}{1+t} + \frac{x_3 + tx_2}{1+t^2}.$$

(d) The material time derivative of ϕ can also be computed from the formula $\dot{\phi} = \frac{\partial}{\partial t}\phi + \nabla^x \phi \cdot \boldsymbol{v}$, where $\boldsymbol{v}$ is the spatial velocity field. Since

$$\frac{\partial}{\partial t}\phi(\boldsymbol{x}, t) = x_1 \quad \text{and} \quad [\nabla^x \phi(\boldsymbol{x}, t)] = (t, 1, 0)^T,$$

we get

$$\dot{\phi}(\boldsymbol{x}, t) = x_1 + \frac{tx_1}{1+t} + \frac{x_3 + tx_2}{1+t^2},$$

which agrees with the result in part(c).

4.31 (a) Since $\boldsymbol{\psi}(\boldsymbol{x}, t)$ is the inverse motion to $\boldsymbol{\varphi}(\boldsymbol{X}, t)$ we have

$$\widehat{\boldsymbol{\varphi}}(\boldsymbol{x}, 0) = \boldsymbol{\varphi}(\boldsymbol{X}, t)|_{\boldsymbol{X}=\boldsymbol{\psi}(\boldsymbol{x},t)} = \boldsymbol{\varphi}(\boldsymbol{\psi}(\boldsymbol{x}, t), t) = \boldsymbol{x}.$$

(b) Taking the derivative with respect to s we get

$$\frac{\partial}{\partial s}\widehat{\boldsymbol{\varphi}}(\boldsymbol{x}, s) = \frac{\partial}{\partial t}\boldsymbol{\varphi}(\boldsymbol{X}, t+s)|_{\boldsymbol{X}=\boldsymbol{\psi}(\boldsymbol{x},t)} = \boldsymbol{V}(\boldsymbol{X}, t+s)|_{\boldsymbol{X}=\boldsymbol{\psi}(\boldsymbol{x},t)},$$

and setting $s = 0$ gives

$$\frac{\partial}{\partial s}\widehat{\boldsymbol{\varphi}}(\boldsymbol{x}, 0) = \boldsymbol{V}(\boldsymbol{X}, t)|_{\boldsymbol{X}=\boldsymbol{\psi}(\boldsymbol{x},t)} = \boldsymbol{v}(\boldsymbol{x}, t).$$

(c) Since $\widehat{\boldsymbol{\varphi}}(\boldsymbol{x}, 0) = \boldsymbol{x}$ we have

$$\widehat{\boldsymbol{F}}(\boldsymbol{x}, 0) = \nabla^x \widehat{\boldsymbol{\varphi}}(\boldsymbol{x}, 0) = \boldsymbol{I}.$$

Moreover, since $\frac{\partial}{\partial s}\widehat{\boldsymbol{\varphi}}(\boldsymbol{x}, 0) = \boldsymbol{v}(\boldsymbol{x}, t)$ we have

$$\begin{aligned}\frac{\partial}{\partial s}\widehat{\boldsymbol{F}}(\boldsymbol{x}, 0) &= \frac{\partial}{\partial s}(\nabla^x \widehat{\boldsymbol{\varphi}}(\boldsymbol{x}, s))|_{s=0} \\ &= \nabla^x \left(\frac{\partial}{\partial s}\widehat{\boldsymbol{\varphi}}(\boldsymbol{x}, s)|_{s=0}\right) = \nabla^x \boldsymbol{v}(\boldsymbol{x}, t).\end{aligned}$$

(d) Using the results from part (c) we have

$$\frac{\partial}{\partial s}\widehat{\boldsymbol{E}}(\boldsymbol{x}, 0) = \text{sym}\left(\frac{\partial}{\partial s}\widehat{\boldsymbol{F}}(\boldsymbol{x}, 0)\right) = \text{sym}(\nabla^x \boldsymbol{v}(\boldsymbol{x}, t)) = \boldsymbol{L}(\boldsymbol{x}, t).$$

(e) Using the relation $\widehat{\boldsymbol{U}}^2 = \widehat{\boldsymbol{F}}^T\widehat{\boldsymbol{F}}$ we have

$$\Big(\frac{\partial}{\partial s}\widehat{\boldsymbol{U}}\Big)\widehat{\boldsymbol{U}} + \widehat{\boldsymbol{U}}\Big(\frac{\partial}{\partial s}\widehat{\boldsymbol{U}}\Big) = \Big(\frac{\partial}{\partial s}\widehat{\boldsymbol{F}}\Big)^T\widehat{\boldsymbol{F}} + \widehat{\boldsymbol{F}}^T\Big(\frac{\partial}{\partial s}\widehat{\boldsymbol{F}}\Big).$$

Setting $s = 0$ and using the results from part (c) gives

$$\frac{\partial}{\partial s}\widehat{\boldsymbol{U}}(\boldsymbol{x},0) = \tfrac{1}{2}(\nabla^x\boldsymbol{v}(\boldsymbol{x},t)^T + \nabla^x\boldsymbol{v}(\boldsymbol{x},t)) = \boldsymbol{L}(\boldsymbol{x},t).$$

Similarly, using the relation $\widehat{\boldsymbol{R}}\widehat{\boldsymbol{U}} = \widehat{\boldsymbol{F}}$ we have

$$\Big(\frac{\partial}{\partial s}\widehat{\boldsymbol{R}}\Big)\widehat{\boldsymbol{U}} + \widehat{\boldsymbol{R}}\Big(\frac{\partial}{\partial s}\widehat{\boldsymbol{U}}\Big) = \frac{\partial}{\partial s}\widehat{\boldsymbol{F}}.$$

Setting $s = 0$ and using the results from part (c) gives

$$\frac{\partial}{\partial s}\widehat{\boldsymbol{R}}(\boldsymbol{x},0) = \frac{\partial}{\partial s}\widehat{\boldsymbol{F}}(\boldsymbol{x},0) - \frac{\partial}{\partial s}\widehat{\boldsymbol{U}}(\boldsymbol{x},0) = \text{skew}(\nabla^x\boldsymbol{v}(\boldsymbol{x},t)),$$

which is the desired result.

4.33 (a) Due to the homogeneous form of the motion $\boldsymbol{x} = \boldsymbol{\varphi}(\boldsymbol{X},t) = \boldsymbol{R}(t)\boldsymbol{X} + \boldsymbol{c}(t)$, the inverse motion is given by

$$\boldsymbol{X} = \boldsymbol{\psi}(\boldsymbol{x},t) = \boldsymbol{R}(t)^T(\boldsymbol{x} - \boldsymbol{c}(t)),$$

where $\boldsymbol{R}(t)^T = \boldsymbol{R}(t)^{-1}$ since $\boldsymbol{R}(t)$ is a rotation.

(b) Using the fact that $\boldsymbol{I} = \boldsymbol{R}(t)\boldsymbol{R}(t)^T$ for all $t \geq 0$ we have

$$\boldsymbol{O} = \dot{\boldsymbol{R}}(t)\boldsymbol{R}(t)^T + \boldsymbol{R}(t)\dot{\boldsymbol{R}}(t)^T = \boldsymbol{\Omega}(t) + \boldsymbol{\Omega}(t)^T,$$

which shows that $\boldsymbol{\Omega}(t)$ is skew-symmetric for each $t \geq 0$.

(c) By definition, the material velocity field is $\boldsymbol{V}(\boldsymbol{X},t) = \dot{\boldsymbol{R}}(t)\boldsymbol{X} + \dot{\boldsymbol{c}}(t)$ and the spatial velocity field is

$$\begin{aligned}\boldsymbol{v}(\boldsymbol{x},t) &= \boldsymbol{V}(\boldsymbol{X},t)|_{\boldsymbol{X}=\boldsymbol{\psi}(\boldsymbol{x},t)}\\ &= \dot{\boldsymbol{R}}(t)\boldsymbol{R}(t)^T(\boldsymbol{x} - \boldsymbol{c}(t)) + \dot{\boldsymbol{c}}(t)\\ &= \boldsymbol{\Omega}(t)(\boldsymbol{x} - \boldsymbol{c}(t)) + \dot{\boldsymbol{c}}(t),\end{aligned}$$

which is the desired result.

4.35 Let $\boldsymbol{c}$ be an arbitrary vector and consider the spatial tensor field $\boldsymbol{w}\otimes\boldsymbol{c}$. Using the change of variable formula for tensor fields we have

$$\int_{\partial\Omega_t}(\boldsymbol{w}\otimes\boldsymbol{c})\boldsymbol{n}\,dA_{\boldsymbol{x}} = \int_{\partial\Omega}(\boldsymbol{w}_m\otimes\boldsymbol{c})\boldsymbol{G}\boldsymbol{N}\,dA_{\boldsymbol{X}}.$$

Noting that $[\boldsymbol{b}\otimes\boldsymbol{c}]\boldsymbol{a} = [\boldsymbol{b}\otimes\boldsymbol{a}]\boldsymbol{c}$ for any vectors $\boldsymbol{a}$, $\boldsymbol{b}$ and $\boldsymbol{c}$ we get

$$\int_{\partial\Omega_t} (\boldsymbol{w}\otimes\boldsymbol{n})\boldsymbol{c}\, dA_{\boldsymbol{x}} = \int_{\partial\Omega} (\boldsymbol{w}_m \otimes \boldsymbol{G}\boldsymbol{N})\boldsymbol{c}\, dA_{\boldsymbol{X}}.$$

The result now follows by the arbitrariness of $\boldsymbol{c}$.

4.37 For the given deformation we find

$$[\boldsymbol{F}] = \begin{pmatrix} 1 & \alpha t & 0 \\ 0 & 1 & 0 \\ 0 & 0 & 1 \end{pmatrix},$$

which gives $\det[\boldsymbol{F}] = 1$ for all $\boldsymbol{X} \in \mathcal{B}$ and $t \geq 0$. Thus the motion is volume-preserving.

5
Balance Laws

In this chapter we state various axioms which form the basis for a thermo-mechanical theory of continuum bodies. These axioms provide a set of balance laws which describe how the mass, momentum, energy and entropy of a body change in time under prescribed external influences. We first state these laws in global or integral form, then derive various corresponding local statements, primarily in the form of partial differential equations. The balance laws stated here apply to all bodies regardless of their constitution. In Chapters 6–9 these laws are specialized to various classes of bodies with specific material properties, via constitutive models.

The important ideas in this chapter are: (i) the balance laws of mass, momentum, energy and entropy for continuum bodies; (ii) the difference between the integral form of a law and its local Eulerian and Lagrangian forms; (iii) the axiom of material frame-indifference and its role in constitutive modeling; (iv) the idea of a material constraint and its implications for the stress field in a body; (v) the balance laws relevant to the isothermal modeling of continuum bodies.

5.1 Motivation

In order to motivate the contents of this chapter it is useful to recall some basic ideas from the mechanics of particle systems. To this end, we consider a system of N particles with masses m_i and positions $\boldsymbol{x}_i$ as illustrated in Figure 5.1. It will be helpful to think of these particles as the atoms making up a continuum body. We suppose that every distinct pair of particles ($j \neq i$) interacts with an energy U_{ij}, which is assumed to be symmetric in the sense that $U_{ij} = U_{ji}$, and we denote the interaction force of particle j on i by $\boldsymbol{f}_{ij}^{\mathrm{int}} = -\nabla^{\boldsymbol{x}_i} U_{ij}$. We further suppose that

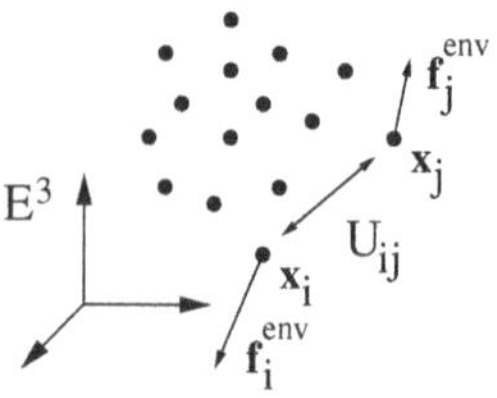

Fig. 5.1 Illustration of a particle system. $\boldsymbol{x}_i$ denotes the position of particle i, $\boldsymbol{f}_i^{\mathrm{env}}$ denotes an environmental force on particle i, and U_{ij} denotes a pairwise interaction energy between particles i and j.

each particle is subject to an environmental force $\boldsymbol{f}_i^{\mathrm{env}}$. For simplicity, we assume that no other forces act on or within the system.

A complete description of the motion of the particle system is given by the equations (no summation convention)

$$\dot{m}_i = 0, \qquad m_i \ddot{\boldsymbol{x}}_i = \boldsymbol{f}_i^{\mathrm{env}} + \sum_{\substack{j=1 \\ j \neq i}}^{N} \boldsymbol{f}_{ij}^{\mathrm{int}}, \qquad \forall i = 1, \ldots, N. \tag{5.1}$$

The first equation states that the mass of each particle is constant. The second equation states that each particle obeys **Newton's Second Law**. That is, the mass of each particle times its acceleration is equal to the resultant force on it.

Under suitable assumptions on the interaction energies U_{ij}, the above equations can be used to derive balance laws for mass, linear momentum, angular momentum and energy for any subset of the particle system. To this end, consider an arbitrary subset I of $\{1, \ldots, N\}$ and let Ω_t denote the current configuration of the particles with labels $i \in I$. To the configuration Ω_t at any time $t \geq 0$ we associate a total mass $M[\Omega_t]$, linear momentum $\boldsymbol{l}[\Omega_t]$, angular momentum $\boldsymbol{j}[\Omega_t]$, internal energy $U[\Omega_t]$ and kinetic energy $K[\Omega_t]$ as follows

$$\begin{gathered} M[\Omega_t] = \sum_{i \in I} m_i, \quad \boldsymbol{l}[\Omega_t] = \sum_{i \in I} m_i \dot{\boldsymbol{x}}_i, \quad \boldsymbol{j}[\Omega_t] = \sum_{i \in I} \boldsymbol{x}_i \times m_i \dot{\boldsymbol{x}}_i, \\ U[\Omega_t] = \sum_{\substack{i,j \in I \\ i<j}} U_{ij}, \quad K[\Omega_t] = \sum_{i \in I} \tfrac{1}{2} m_i |\dot{\boldsymbol{x}}_i|^2. \end{gathered} \tag{5.2}$$

Assuming U_{ij} depends on $\boldsymbol{x}_i$ and $\boldsymbol{x}_j$ only through the pairwise distance $|\boldsymbol{x}_i - \boldsymbol{x}_j|$ and interactions are symmetric as described before, we deduce the following equations from (5.1)

$$\frac{d}{dt} M[\Omega_t] = 0, \tag{5.3}$$

$$\frac{d}{dt}\boldsymbol{l}[\Omega_t] = \sum_{i \in I} \left[\boldsymbol{f}_i^{\text{env}} + \sum_{j \notin I} \boldsymbol{f}_{ij}^{\text{int}} \right], \tag{5.4}$$

$$\frac{d}{dt}\boldsymbol{j}[\Omega_t] = \sum_{i \in I} \boldsymbol{x}_i \times \left[\boldsymbol{f}_i^{\text{env}} + \sum_{j \notin I} \boldsymbol{f}_{ij}^{\text{int}} \right], \tag{5.5}$$

$$\frac{d}{dt}(U[\Omega_t] + K[\Omega_t]) = \sum_{i \in I} \dot{\boldsymbol{x}}_i \cdot \left[\boldsymbol{f}_i^{\text{env}} + \sum_{j \notin I} \boldsymbol{f}_{ij}^{\text{int}} \right]. \tag{5.6}$$

These equations state that the mass of Ω_t does not change with time, the rate of change of linear momentum is equal to the resultant external force on Ω_t, the rate of change of angular momentum is equal to the resultant external torque on Ω_t and the rate of change of the sum of internal and kinetic energy is equal to the power of external forces on Ω_t (see Exercises 1 and 2).

Equations (5.3)–(5.6) are typically referred to as **balance laws**. They describe how the mass, linear momentum, angular momentum and energy of Ω_t may be conserved, increased or decreased depending on external influences. For example, consider the case when Ω_t corresponds to the entire system of particles, equivalently $I = \{1, \ldots, N\}$, and suppose all environmental forces are zero. In this case, equations (5.3)–(5.6) imply that the mass, linear momentum, angular momentum and energy of Ω_t would all be conserved throughout any motion. That is, these quantities would remain constant.

The four balance laws (5.3)–(5.6) provide a fundamental starting point in the modeling of continuum bodies. The balance laws for mass, linear momentum and angular momentum can be generalized to continuum bodies in a straightforward manner: we essentially replace the summation by an integral. The balance law for energy can also be generalized; however, the situation is more complicated. Continuum modeling cannot represent the detail that occurs at length scales comparable to, or shorter than, inter-atomic spacings. In particular, the velocity at a point in a continuum body is to be interpreted as the average velocity of individual atoms in the vicinity of that point.

Fluctuations in the velocities of the individual atoms about the mean are not represented in the continuum velocity field. They are associated with time and length scales on which the model no longer applies. To account for situations in which the energy of these fluctuations is significant, it is necessary to introduce new concepts at the continuum level: **temperature**, to measure the size of the velocity fluctuations, and **heat**, to measure the energy of these fluctuations. Temperature and heat will play an important role in the generalization of (5.6) to continuum bodies.

When suitably generalized, the four balance laws (5.3)–(5.6) form the basis of a thermo-mechanical theory of continuum bodies. In this case, the four laws are essential postulates. That is, while (5.3)–(5.6) are all implied by (5.1) for systems of particles under special assumptions, there is no such implication in general when the assumptions are removed. For this reason, we simply postulate these laws in the continuum case and note that their validity has been verified by experiment. A fifth, purely thermal, balance law will also be postulated in the continuum case. This law, which is motivated by the thermodynamics of homogeneous processes, involves the concept of entropy and will take the form of an inequality. The four balance laws corresponding to (5.3)–(5.6), together with the entropy inequality, will provide a complete thermo-mechanical theory of continuum bodies.

5.2 Balance Laws in Integral Form

In this section we state the balance laws for mass, momentum, energy and entropy as they apply to any open subset of a continuum body. We begin with a statement of the conservation of mass, and then proceed to the laws of inertia which connect the rates of change of the total linear and angular momentum of a body to external forces applied to it. Introducing the notion of internal energy and heating of a continuum body, and the net working of external forces, we state the laws of thermodynamics which describe how various forms of energy, such as thermal and mechanical, are interconverted.

5.2.1 Conservation of Mass and Laws of Inertia

Consider a continuum body with reference configuration B undergoing a motion $\boldsymbol{\varphi} : B \times [0, \infty) \to \mathbb{E}^3$. Let B_t denote the current configuration, and let $\rho(\boldsymbol{x}, t)$ denote the mass density and $\boldsymbol{v}(\boldsymbol{x}, t)$ the velocity at a point $\boldsymbol{x} \in B_t$ at time $t \geq 0$. Moreover, let Ω_t denote an arbitrary open subset of B_t. Motivated by (5.2) we define the **mass** of Ω_t by

$$\text{mass}[\Omega_t] = \int_{\Omega_t} \rho(\boldsymbol{x}, t)\, dV_{\boldsymbol{x}},$$

the **linear momentum** by

$$\boldsymbol{l}[\Omega_t] = \int_{\Omega_t} \rho(\boldsymbol{x},t)\boldsymbol{v}(\boldsymbol{x},t)\, dV_{\boldsymbol{x}},$$

and the **angular momentum** with respect to a point $\boldsymbol{z}$ by

$$\boldsymbol{j}[\Omega_t]_{\boldsymbol{z}} = \int_{\Omega_t} (\boldsymbol{x}-\boldsymbol{z}) \times \rho(\boldsymbol{x},t)\boldsymbol{v}(\boldsymbol{x},t)\, dV_{\boldsymbol{x}}.$$

In the absence of chemical reactions and relativistic effects, the balance law stated in (5.3) is generalized to continuum bodies as follows.

Axiom 5.1 Conservation of Mass. *The mass of any open subset of a continuum body does not change as the body changes place and shape. That is*

$$\frac{d}{dt}\,\text{mass}[\Omega_t] = 0, \qquad \forall \Omega_t \subseteq B_t.$$

□

Suppose Ω_t is subject to an external traction field $\boldsymbol{t}(\boldsymbol{x},t)$ and an external body force field per unit mass $\boldsymbol{b}(\boldsymbol{x},t)$ (see Chapter 3). Then the resultant force on Ω_t is

$$\boldsymbol{r}[\Omega_t] = \int_{\Omega_t} \rho(\boldsymbol{x},t)\boldsymbol{b}(\boldsymbol{x},t)\, dV_{\boldsymbol{x}} + \int_{\partial\Omega_t} \boldsymbol{t}(\boldsymbol{x},t)\, dA_{\boldsymbol{x}},$$

and the resultant torque on Ω_t with respect to $\boldsymbol{z}$ is

$$\boldsymbol{\tau}[\Omega_t]_{\boldsymbol{z}} = \int_{\Omega_t} (\boldsymbol{x}-\boldsymbol{z}) \times \rho(\boldsymbol{x},t)\boldsymbol{b}(\boldsymbol{x},t)\, dV_{\boldsymbol{x}} + \int_{\partial\Omega_t} (\boldsymbol{x}-\boldsymbol{z}) \times \boldsymbol{t}(\boldsymbol{x},t)\, dA_{\boldsymbol{x}}.$$

The balance laws stated in (5.4) and (5.5) are generalized to continuum bodies as follows.

Axiom 5.2 Laws of Inertia. *With respect to a fixed frame of reference, the rate of change of linear momentum of any open subset of a continuum body equals the resultant force applied to it, and the rate of change of angular momentum about the origin equals the resultant torque about the origin. That is*

$$\frac{d}{dt}\boldsymbol{l}[\Omega_t] = \boldsymbol{r}[\Omega_t] \quad \textit{and} \quad \frac{d}{dt}\boldsymbol{j}[\Omega_t]_{\boldsymbol{o}} = \boldsymbol{\tau}[\Omega_t]_{\boldsymbol{o}}, \qquad \forall \Omega_t \subseteq B_t. \tag{5.7}$$

□

The above laws are postulated to hold for every motion and every subset of a continuum body. Beginning from these laws it is also possible to derive alternative forms of the angular momentum equation. In particular, from (5.7) we deduce that the angular momentum of Ω_t also satisfies the equations

$$\frac{d}{dt}\boldsymbol{j}[\Omega_t]_{\boldsymbol{z}} = \boldsymbol{\tau}[\Omega_t]_{\boldsymbol{z}} \quad \text{and} \quad \frac{d}{dt}\boldsymbol{j}[\Omega_t]_{\boldsymbol{x}_{\mathrm{com}}} = \boldsymbol{\tau}[\Omega_t]_{\boldsymbol{x}_{\mathrm{com}}}, \qquad (5.8)$$

where $\boldsymbol{z}$ is any fixed point and $\boldsymbol{x}_{\mathrm{com}}$ is the generally time-dependent mass center of Ω_t (see Exercises 3 and 5). This is analogous to the equilibrium situation outlined in Chapter 3. The forms of the angular momentum equation in (5.8) are typically used in the study of rigid bodies. The linear momentum equation in (5.7) can also be phrased in terms of the mass center of Ω_t (see Exercise 4).

5.2.2 First and Second Laws of Thermodynamics

5.2.2.1 Temperature and Heat

In the most basic sense, **temperature** is a physical property of matter that is based on our natural perception of hot and cold. In the modeling of continuum bodies we assume that a temperature is defined at each point in a body at each time. In particular, we assume the existence of an absolute **temperature field** $\theta(\boldsymbol{x}, t) > 0$ defined for $\boldsymbol{x} \in B_t$ and $t \geq 0$. The temperature at a point in a continuum body is to be interpreted as a measure of the velocity fluctuations of individual atoms in the vicinity of that point. Thus, while continuum velocity is a measure of the mean, continuum temperature is a measure of the variation of atomistic velocities near a point.

By the **thermal energy** or **heat content** of a body we mean an energy associated with the velocity fluctuations of individual atoms in the body. In particular, heat is an energy associated with temperature. Physical experience tells us that material bodies are capable of interconverting heat and mechanical work. Deforming a body can affect its temperature and thereby its heat content. Similarly, adding heat to a body by raising its temperature may set the body in motion, for example it may expand. Because of this interconversion, heat may be expressed in the same units as mechanical work.

The heat content of a body can be affected in two basic ways. First, heat can be produced or consumed throughout the volume of a body by mechanical, chemical or electromagnetic mechanisms. For example, an electrical current can produce heat as it flows through the volume of an isolated body. Second, heat can be transferred from one body to another through physical contact or through thermal radiation at a surface. For example, when a warm body is put in physical contact with a cool one, experience tells us that the warm body will get cooler (lose heat) and the cool body will get warmer (gain heat). In this case there is a transfer of heat per unit area of the contact surface. Transfer through contact can be further classified as conduction or convection depending on the circumstances and the nature of the bodies involved.

Given an arbitrary open subset Ω_t of B_t we define the **net heating** $Q[\Omega_t]$ as the rate per unit time at which heat is being added to Ω_t. In analogy with our treatment of forces (see Chapter 3), we assume that the net heating can be decomposed into a body heating $Q_b[\Omega_t]$ and a surface heating $Q_s[\Omega_t]$ such that

$$Q[\Omega_t] = Q_b[\Omega_t] + Q_s[\Omega_t]. \tag{5.9}$$

We assume the body heating $Q_b[\Omega_t]$ is a continuous function of volume. In particular, we assume there exists a **heat supply field** per unit volume $\hat{r}(\boldsymbol{x}, t) \in I\!R$ such that

$$Q_b[\Omega_t] = \int_{\Omega_t} \hat{r}(\boldsymbol{x}, t)\, dV_{\boldsymbol{x}}.$$

For convenience, we define a heat supply field per unit mass $r(\boldsymbol{x}, t) \in I\!R$ by

$$r(\boldsymbol{x}, t) = \rho(\boldsymbol{x}, t)^{-1} \hat{r}(\boldsymbol{x}, t),$$

so that

$$Q_b[\Omega_t] = \int_{\Omega_t} \rho(\boldsymbol{x}, t) r(\boldsymbol{x}, t)\, dV_{\boldsymbol{x}}. \tag{5.10}$$

We assume the surface heating $Q_s[\Omega_t]$ is a continuous function of surface area. In particular, we assume there exists a **heat transfer field** per unit area $h(\boldsymbol{x}, t) \in I\!R$ such that

$$Q_s[\Omega_t] = \int_{\partial\Omega_t} h(\boldsymbol{x}, t)\, dA_{\boldsymbol{x}}. \tag{5.11}$$

Moreover, we assume the heat transfer field is of the form

$$h(\boldsymbol{x}, t) = -\boldsymbol{q}(\boldsymbol{x}, t) \cdot \boldsymbol{n}(\boldsymbol{x}), \tag{5.12}$$

where $\boldsymbol{n}$ is the outward unit normal field for the surface $\partial\Omega_t$ and $\boldsymbol{q}$ is the Fourier–Stokes **heat flux vector field** in B_t.

The direction and intensity of heat flow across any surface in a body is determined by the vector field $\boldsymbol{q}$. The negative sign in (5.12) is due to the fact that $\boldsymbol{n}$ is chosen to be the outward as opposed to inward unit normal. In particular, from (5.12) we deduce that h is positive when $\boldsymbol{q}$ points into Ω_t. The relation between the surface heat transfer and the heat flux vector is analogous to the relation between the surface traction and the Cauchy stress tensor (see Result 3.3). In particular, it can be shown that, under mild assumptions, both traction and heat transfer must necessarily be linear in the surface normal.

Substituting (5.10), (5.11) and (5.12) into (5.9) we obtain

$$Q[\Omega_t] = \int_{\Omega_t} \rho(\boldsymbol{x},t) r(\boldsymbol{x},t)\, dV_{\boldsymbol{x}} - \int_{\partial\Omega_t} \boldsymbol{q}(\boldsymbol{x},t)\cdot\boldsymbol{n}(\boldsymbol{x})\, dA_{\boldsymbol{x}}. \tag{5.13}$$

Notice that heat is absorbed or released from Ω_t depending on whether $Q[\Omega_t]$ is positive or negative.

5.2.2.2 *Kinetic Energy, Power of External Forces and Net Working*

Consider an arbitrary open subset Ω_t of B_t with external traction field $\boldsymbol{t}(\boldsymbol{x},t)$ and external body force field per unit mass $\boldsymbol{b}(\boldsymbol{x},t)$. We define the **kinetic energy** of Ω_t by

$$K[\Omega_t] = \int_{\Omega_t} \tfrac{1}{2}\rho(\boldsymbol{x},t)|\boldsymbol{v}(\boldsymbol{x},t)|^2\, dV_{\boldsymbol{x}},$$

and the **power of external forces** on Ω_t by

$$\mathcal{P}[\Omega_t] = \int_{\Omega_t} \rho(\boldsymbol{x},t)\boldsymbol{b}(\boldsymbol{x},t)\cdot\boldsymbol{v}(\boldsymbol{x},t)\, dV_{\boldsymbol{x}} + \int_{\partial\Omega_t} \boldsymbol{t}(\boldsymbol{x},t)\cdot\boldsymbol{v}(\boldsymbol{x},t)\, dA_{\boldsymbol{x}}.$$

By the **net working** $\mathcal{W}[\Omega_t]$ of external forces on Ω_t, we mean the mechanical power delivered by these forces that is not used up in producing motion, that is

$$\mathcal{W}[\Omega_t] = \mathcal{P}[\Omega_t] - \frac{d}{dt}K[\Omega_t]. \tag{5.14}$$

If $\mathcal{W}[\Omega_t] = 0$ throughout a time interval, then all the mechanical energy delivered to the body by external forces is used to produce motion. In particular, it is converted to kinetic energy. If $\mathcal{W}[\Omega_t] > 0$ throughout a time interval, then some of the mechanical energy delivered to the body is stored or converted to some form other than kinetic energy. Conversely, if $\mathcal{W}[\Omega_t] < 0$, then energy stored in the body is released as work against the external forces or as motion.

5.2.2.3 Internal Energy and The First Law

The energy content of a continuum body not associated with kinetic energy is called **internal energy**. The internal energy of a body represents a store of energy which may be increased in various ways, for example by adding heat to the body or performing mechanical work on it. Once stored, internal energy can be released in any form, for example heat, mechanical work or motion.

For our purposes, we assume the internal energy of a body consists only of thermal (heat) and mechanical (elastic) energy. In particular, we neglect such things as chemical and electromagnetic energy, which in general would also contribute to the internal energy. Neglecting these other contributions is tantamount to assuming that they are constants independent of any net heating and working. In the analysis of some problems it is often possible and convenient to introduce a potential energy associated with external forces. Such an energy should not be confused with internal energy. A brief discussion of potential energy is given below.

Given an arbitrary open subset Ω_t of B_t we denote its internal energy by $U[\Omega_t]$. We assume $U[\Omega_t]$ is a continuous function of volume. In particular, we assume there exists an **internal energy density field** per unit volume $\hat{\phi}(\boldsymbol{x},t) \in I\!R$ such that

$$U[\Omega_t] = \int_{\Omega_t} \hat{\phi}(\boldsymbol{x},t)\, dV_{\boldsymbol{x}}.$$

For convenience, we define an internal energy density field per unit mass $\phi(\boldsymbol{x},t) \in I\!R$ by

$$\phi(\boldsymbol{x},t) = \rho^{-1}(\boldsymbol{x},t)\hat{\phi}(\boldsymbol{x},t),$$

so that

$$U[\Omega_t] = \int_{\Omega_t} \rho(\boldsymbol{x},t)\phi(\boldsymbol{x},t)\, dV_{\boldsymbol{x}}.$$

Considering only thermo-mechanical energy, the balance law stated in (5.6) is generalized to continuum bodies as follows.

Axiom 5.3 ***First Law of Thermodynamics.*** *The rate of change of internal energy of any open subset of a continuum body equals the sum of the net heating and net working applied to it. That is*

$$\frac{d}{dt}U[\Omega_t] = Q[\Omega_t] + \mathcal{W}[\Omega_t], \qquad \forall \Omega_t \subseteq B_t.$$

In view of (5.14), *we have*

$$\frac{d}{dt}(U[\Omega_t] + K[\Omega_t]) = Q[\Omega_t] + \mathcal{P}[\Omega_t], \qquad \forall \Omega_t \subseteq B_t. \tag{5.15}$$

□

There are various fundamental differences between the continuum energy balance law (5.15) and the corresponding particle law (5.6). For example, the kinetic energy in a continuum model is an energy associated with a local mean velocity of individual particles (atoms). In contrast, the kinetic energy in a particle model is an energy associated with individual velocities. Furthermore, the internal energy in a continuum model has both mechanical (elastic) and thermal (heat) contributions. In contrast, the internal energy in a particle model has only a mechanical contribution. In general, the energy balance law (5.6) for particles does not explicitly contain any quantities related to heat. Similar fundamental differences can also be identified between the continuum and particle balance laws pertaining to momentum.

In some cases, the power of external forces can be written in the form

$$\mathcal{P}[\Omega_t] = -\frac{d}{dt}G[\Omega_t],$$

where $G[\Omega_t]$ is called a **potential energy for external forces**. In this special case the energy balance law (5.15) becomes

$$\frac{d}{dt}(U[\Omega_t] + K[\Omega_t] + G[\Omega_t]) = Q[\Omega_t].$$

This says that the rate of change of the total energy $U + K + G$ is equal to the net heating Q. When the net heating is zero, this total energy is conserved throughout a motion.

5.2.2.4 Entropy and the Second Law

The energy balance law provides a relation between the rate of change of internal energy and the net heating and working of external influences on a body. While this relation implies that the rates must always be balanced, it places no fundamental restriction on the rates themselves. For example, in any time interval in which the net working is zero, the energy balance law does not limit the rate at which a body may absorb heat and store it as internal energy. Similarly, it does not limit the rate at which a body may release internal energy in the form of heat. In its simplest form, the Second Law of Thermodynamics expresses the fact that each body has a limit on the rate at which heat can be absorbed, but has no limit on the rate at which heat can be released.

The Second Law has a long and complicated history. It has its origins in the study of homogeneous systems undergoing reversible processes, and its extension to the general framework of continuum mechanics is not entirely settled. Here we adopt a form of the Second Law called the Clausius–Duhem inequality which will suffice for our purposes. In particular, we introduce this inequality primarily to illustrate how thermodynamic considerations place restrictions on the constitutive relations of a body, and how various statements about dissipation and irreversibility may be deduced from it.

To motivate the form of the law given here, we first consider a homogeneous body Ω_t with uniform temperature $\Theta[\Omega_t]$ for each $t \geq 0$. For such a body the Second Law postulates the existence of an a-priori least upper bound $\Xi[\Omega_t]$ for the net heating $Q[\Omega_t]$ so that

$$Q[\Omega_t] \leq \Xi[\Omega_t]. \tag{5.16}$$

The quantity $\Xi[\Omega_t]$ in general depends on properties of the body Ω_t. When the net working is zero, we deduce from the First Law that $\frac{d}{dt}U[\Omega_t] = Q[\Omega_t]$. Thus the quantity $\Xi[\Omega_t]$ can also be interpreted as the maximum rate at which a body can store internal energy in the absence of net working.

By the **entropy** of a homogeneous body Ω_t we mean a quantity $\mathcal{H}[\Omega_t]$ defined, up to an additive constant, by

$$\frac{d}{dt}\mathcal{H}[\Omega_t] = \frac{\Xi[\Omega_t]}{\Theta[\Omega_t]}.$$

Thus the entropy of a body is a quantity whose rate of change at any instant of time is equal to the upper heating bound per unit temperature. In particular, the rate of change of entropy is a measure of a body's ability to absorb heat. At the atomic level, entropy can be interpreted as a measure of disorder, that is, the multitude of atomic configurations compatible with prescribed values of macroscopic variables.

In terms of entropy, the bound in (5.16) becomes

$$\frac{d}{dt}\mathcal{H}[\Omega_t] \geq \frac{Q[\Omega_t]}{\Theta[\Omega_t]}. \tag{5.17}$$

This is a classic form of the Second Law for homogeneous systems called the **Clausius–Planck inequality**. The irreversibility of natural processes is reflected in the fact that

$$\frac{d}{dt}\mathcal{H}[\Omega_t] \geq 0 \quad \text{when} \quad Q[\Omega_t] = 0.$$

In particular, the entropy of a body favors increase, even when the net heating vanishes as in the case of an isolated body.

We next drop the assumption of homogeneity and consider an extension of the Clausius–Planck inequality to a general continuum body. Just as with internal energy, we assume the entropy $\mathcal{H}[\Omega_t]$ of a body is a continuous function of volume. In particular, we assume there exists an **entropy density field** per unit volume $\hat{\eta}(\boldsymbol{x}, t) \in \mathbb{R}$ such that

$$\mathcal{H}[\Omega_t] = \int_{\Omega_t} \hat{\eta}(\boldsymbol{x}, t)\, dV_{\boldsymbol{x}}.$$

For convenience, we introduce an entropy density per unit mass by

$$\eta(\boldsymbol{x}, t) = \rho^{-1}(\boldsymbol{x}, t)\hat{\eta}(\boldsymbol{x}, t),$$

so that

$$\mathcal{H}[\Omega_t] = \int_{\Omega_t} \rho(\boldsymbol{x}, t)\eta(\boldsymbol{x}, t)\, dV_{\boldsymbol{x}}.$$

A natural generalization of (5.17) to spatially inhomogeneous systems may be stated as follows.

Axiom 5.4 ***Second Law of Thermodynamics.*** *The rate of entropy production in any open subset of a continuum body is bounded below by the heating per unit temperature. That is*

$$\frac{d}{dt}\mathcal{H}[\Omega_t] \geq \int_{\Omega_t} \frac{\rho(\boldsymbol{x}, t) r(\boldsymbol{x}, t)}{\theta(\boldsymbol{x}, t)}\, dV_{\boldsymbol{x}} - \int_{\partial\Omega_t} \frac{\boldsymbol{q}(\boldsymbol{x}, t) \cdot \boldsymbol{n}(\boldsymbol{x})}{\theta(\boldsymbol{x}, t)}\, dA_{\boldsymbol{x}}, \quad \forall \Omega_t \subseteq B_t.$$

□

The above formulation of the Second Law is called the **Clausius–Duhem inequality**. We assume that it holds for every motion and every subset of a continuum body and study the consequences. As we shall see, the above inequality places certain restrictions on the constitutive relations of a body, and will lead to various statements about the dissipation of energy and flow of heat. While the precise form of the Second Law in continuum mechanics is not settled, we expect similar consequences from any other inequality which might replace Axiom 5.4.

5.2.3 Integral Versus Local Balance Laws

In the above developments, we have summarized balance laws for mass, momentum, energy and entropy in terms of arbitrary open subsets Ω_t of the current configuration B_t of a continuum body undergoing a motion $\varphi : B \times [0,\infty) \to \mathbb{E}^3$. We next exploit the Localization Theorem (see Result 2.19) to obtain local forms of these balance laws in terms of partial differential equations. When the local form of a balance law is formulated in terms of the current coordinates $\boldsymbol{x} \in B_t$ and time t, we say that it is in **spatial** or **Eulerian** form. When a balance law is formulated in terms of the reference coordinates $\boldsymbol{X} \in B$ and time t, we say that it is in **material** or **Lagrangian** form. In the next two sections we develop local statements of the balance laws in both Eulerian and Lagrangian forms.

5.3 Localized Eulerian Form of Balance Laws

Consider a continuum body with reference configuration B undergoing a motion $\varphi : B \times [0,\infty) \to \mathbb{E}^3$. As before, we denote the current configuration by $B_t = \varphi_t(B)$, and we assume $B_0 = B$ so that φ_0 is the identity (see Figure 4.7). For each $t \geq 0$ we assume the deformation $\varphi_t : B \to B_t$ is admissible, and we further assume $\varphi(\boldsymbol{X}, t)$ is smooth in the sense that partial derivatives of all orders exist and are continuous. This is much stronger than what we need to derive local statements, but it simplifies the exposition.

5.3.1 Conservation of Mass

Consider an arbitrary open subset Ω_t of B_t and let Ω be the corresponding subset in B so that $\Omega_t = \varphi_t(\Omega)$. Then Axiom 5.1 implies

$$\text{mass}[\Omega_t] = \text{mass}[\Omega_0],$$

where $\Omega_0 = \Omega$. In view of Result 4.8 we have

$$\text{mass}[\Omega_t] = \int_{\Omega_t} \rho(\boldsymbol{x}, t)\, dV_{\boldsymbol{x}} = \int_{\Omega} \rho_m(\boldsymbol{X}, t) \det \boldsymbol{F}(\boldsymbol{X}, t)\, dV_{\boldsymbol{X}},$$

where ρ_m is the material description of the spatial field ρ, that is, $\rho_m(\boldsymbol{X}, t) = \rho(\varphi(\boldsymbol{X}, t), t)$. Also, since $\varphi(\boldsymbol{X}, 0) = \boldsymbol{X}$, we have

$$\text{mass}[\Omega_0] = \int_{\Omega} \rho(\boldsymbol{X}, 0)\, dV_{\boldsymbol{X}} = \int_{\Omega} \rho_0(\boldsymbol{X})\, dV_{\boldsymbol{X}},$$

where $\rho_0(\boldsymbol{X}) = \rho(\boldsymbol{X}, 0)$. Thus conservation of mass requires

$$\int_\Omega [\rho_m(\boldsymbol{X}, t) \det \boldsymbol{F}(\boldsymbol{X}, t) - \rho_0(\boldsymbol{X})] \, dV_{\boldsymbol{X}} = 0,$$

for any time $t \geq 0$ and open set Ω in B. By the Localization Theorem we deduce that

$$\rho_m(\boldsymbol{X}, t) \det \boldsymbol{F}(\boldsymbol{X}, t) = \rho_0(\boldsymbol{X}), \quad \forall \boldsymbol{X} \in B, \ t \geq 0. \tag{5.18}$$

As stated above, the law of conservation of mass is in the material or Lagrangian form. We next convert this to the spatial or Eulerian form. Taking the time derivative of both sides of (5.18) gives

$$\left\{ \frac{\partial}{\partial t} \rho(\boldsymbol{\varphi}(\boldsymbol{X}, t), t) \right\} \det \boldsymbol{F}(\boldsymbol{X}, t) + \rho(\boldsymbol{\varphi}(\boldsymbol{X}, t), t) \left\{ \frac{\partial}{\partial t} \det \boldsymbol{F}(\boldsymbol{X}, t) \right\} = 0.$$

Using Result 4.9 for the time-derivative of the Jacobian field $\det \boldsymbol{F}$ we obtain

$$\left\{ \frac{\partial}{\partial t} \rho(\boldsymbol{\varphi}(\boldsymbol{X}, t), t) \right\} \det \boldsymbol{F}(\boldsymbol{X}, t) + \rho(\boldsymbol{\varphi}(\boldsymbol{X}, t), t) \det \boldsymbol{F}(\boldsymbol{X}, t) (\nabla^x \cdot \boldsymbol{v})(\boldsymbol{x}, t) \Big|_{\boldsymbol{x} = \boldsymbol{\varphi}(\boldsymbol{X}, t)} = 0.$$

Dividing through by $\det \boldsymbol{F}$, which is positive for any admissible motion, and using Result 4.7 for the total time derivative of a spatial field we deduce

$$\frac{\partial}{\partial t} \rho(\boldsymbol{x}, t) + \nabla^x \rho(\boldsymbol{x}, t) \cdot \boldsymbol{v}(\boldsymbol{x}, t) + \rho(\boldsymbol{x}, t) (\nabla^x \cdot \boldsymbol{v})(\boldsymbol{x}, t) = 0.$$

Finally, with the aid of the identity $\nabla^x \cdot (\phi \boldsymbol{w}) = \nabla^x \phi \cdot \boldsymbol{w} + \phi (\nabla^x \cdot \boldsymbol{w})$, which holds for an arbitrary scalar field ϕ and vector field $\boldsymbol{w}$, we obtain the following result.

Result 5.5 ***Conservation of Mass in Eulerian Form.*** *Let $\boldsymbol{\varphi} : B \times [0, \infty) \to \boldsymbol{E}^3$ be a motion of a continuum body with associated spatial velocity field $\boldsymbol{v}(\boldsymbol{x}, t)$ and spatial mass density field $\rho(\boldsymbol{x}, t)$. Then conservation of mass requires*

$$\frac{\partial}{\partial t} \rho + \nabla^x \cdot (\rho \boldsymbol{v}) = 0, \quad \forall \boldsymbol{x} \in B_t, \ t \geq 0.$$

Equivalently, by definition of the total time derivative, we have

$$\dot{\rho} + \rho \nabla^x \cdot \boldsymbol{v} = 0, \quad \forall \boldsymbol{x} \in B_t, \ t \geq 0.$$

□

For a related result derived under different assumptions see Exercise 6. Because the current configuration B_t of a continuum body depends on time, any integral over B_t will also depend on time. We next exploit Result 5.5 to derive a simple expression for the time derivative of integrals over B_t expressed relative to the mass density ρ.

Result 5.6 *Time Derivative of Integrals Relative to Mass.* *Let $\boldsymbol{\varphi} : B \times [0, \infty) \to \mathbb{E}^3$ be a motion of a continuum body with associated spatial velocity field $\boldsymbol{v}(\boldsymbol{x}, t)$ and spatial mass density field $\rho(\boldsymbol{x}, t)$. Let $\Phi(\boldsymbol{x}, t)$ be any spatial scalar, vector or second-order tensor field, and let Ω_t be an arbitrary open subset of B_t. Then*

$$\frac{d}{dt} \int_{\Omega_t} \Phi(\boldsymbol{x}, t) \rho(\boldsymbol{x}, t) \, dV_{\boldsymbol{x}} = \int_{\Omega_t} \dot{\Phi}(\boldsymbol{x}, t) \rho(\boldsymbol{x}, t) \, dV_{\boldsymbol{x}}.$$

□

Proof A slight generalization of Result 4.8 from scalar fields to vector and second-order tensor fields gives

$$\int_{\Omega_t} \Phi(\boldsymbol{x}, t) \rho(\boldsymbol{x}, t) \, dV_{\boldsymbol{x}} = \int_{\Omega} \Phi(\boldsymbol{\varphi}(\boldsymbol{X}, t), t) \rho(\boldsymbol{\varphi}(\boldsymbol{X}, t), t) \det \boldsymbol{F}(\boldsymbol{X}, t) \, dV_{\boldsymbol{X}},$$

where Ω is the corresponding subset of B, that is, $\Omega_t = \boldsymbol{\varphi}_t(\Omega)$. Using conservation of mass in the form (5.18), we obtain

$$\int_{\Omega_t} \Phi(\boldsymbol{x}, t) \rho(\boldsymbol{x}, t) \, dV_{\boldsymbol{x}} = \int_{\Omega} \Phi(\boldsymbol{\varphi}(\boldsymbol{X}, t), t) \rho_0(\boldsymbol{X}) \, dV_{\boldsymbol{X}}.$$

Noting that the region Ω in B does not depend on time, we take time derivatives and obtain

$$\begin{aligned} \frac{d}{dt} \int_{\Omega_t} \Phi(\boldsymbol{x}, t) \rho(\boldsymbol{x}, t) \, dV_{\boldsymbol{x}} &= \frac{d}{dt} \int_{\Omega} \Phi(\boldsymbol{\varphi}(\boldsymbol{X}, t), t) \rho_0(\boldsymbol{X}) \, dV_{\boldsymbol{X}} \\ &= \int_{\Omega} \frac{\partial}{\partial t} \Phi(\boldsymbol{\varphi}(\boldsymbol{X}, t), t) \rho_0(\boldsymbol{X}) \, dV_{\boldsymbol{X}} \\ &= \int_{\Omega} \left\{ \frac{\partial}{\partial t} \Phi(\boldsymbol{\varphi}(\boldsymbol{X}, t), t) \right\} \rho(\boldsymbol{\varphi}(\boldsymbol{X}, t), t) \det \boldsymbol{F}(\boldsymbol{X}, t) \, dV_{\boldsymbol{X}}, \end{aligned}$$

where the last line follows from (5.18). Using the relationship

$$\dot{\Phi}(\boldsymbol{x}, t) \Big|_{\boldsymbol{x} = \boldsymbol{\varphi}(\boldsymbol{X}, t)} = \frac{\partial}{\partial t} \Phi(\boldsymbol{\varphi}(\boldsymbol{X}, t), t),$$

together with Result 4.8, we get

$$\begin{aligned}
\frac{d}{dt}\int_{\Omega_t} &\Phi(\boldsymbol{x},t)\rho(\boldsymbol{x},t)\,dV_{\boldsymbol{x}} \\
&= \int_{\Omega} \dot{\Phi}(\boldsymbol{x},t)\Big|_{\boldsymbol{x}=\boldsymbol{\varphi}(\boldsymbol{X},t)} \rho(\boldsymbol{\varphi}(\boldsymbol{X},t),t)\det \boldsymbol{F}(\boldsymbol{X},t)\,dV_{\boldsymbol{X}} \\
&= \int_{\Omega_t} \dot{\Phi}(\boldsymbol{x},t)\rho(\boldsymbol{x},t)\,dV_{\boldsymbol{x}},
\end{aligned}$$

which is the desired result. □

5.3.2 *Balance of Linear Momentum*

As stated in Axiom 5.2, the balance law for linear momentum for an arbitrary open subset Ω_t of B_t is

$$\begin{aligned}
\frac{d}{dt}\int_{\Omega_t} &\rho(\boldsymbol{x},t)\boldsymbol{v}(\boldsymbol{x},t)\,dV_{\boldsymbol{x}} \\
&= \int_{\partial\Omega_t} \boldsymbol{t}(\boldsymbol{x},t)\,dA_{\boldsymbol{x}} + \int_{\Omega_t}\rho(\boldsymbol{x},t)\boldsymbol{b}(\boldsymbol{x},t)\,dV_{\boldsymbol{x}},
\end{aligned} \tag{5.19}$$

where $\rho(\boldsymbol{x},t)$ is the spatial mass density field, $\boldsymbol{v}(\boldsymbol{x},t)$ is the spatial velocity field, $\boldsymbol{b}(\boldsymbol{x},t)$ is the spatial body force field per unit mass and $\boldsymbol{t}(\boldsymbol{x},t)$ is the traction field on the surface $\partial\Omega_t$ in B_t.

From Chapter 3 we recall that the traction field on any material surface in the current configuration B_t is determined by the Cauchy stress field $\boldsymbol{S}(\boldsymbol{x},t)$ as

$$\boldsymbol{t}(\boldsymbol{x},t) = \boldsymbol{S}(\boldsymbol{x},t)\boldsymbol{n}(\boldsymbol{x}),$$

where $\boldsymbol{n}(\boldsymbol{x})$ is the outward unit normal field on the surface. Thus we can rewrite (5.19) as (omitting the arguments $\boldsymbol{x}$ and t for clarity)

$$\frac{d}{dt}\int_{\Omega_t}\rho\boldsymbol{v}\,dV_{\boldsymbol{x}} = \int_{\partial\Omega_t}\boldsymbol{S}\boldsymbol{n}\,dA_{\boldsymbol{x}} + \int_{\Omega_t}\rho\boldsymbol{b}\,dV_{\boldsymbol{x}}.$$

Using Result 5.6 and the Divergence Theorem for second-order tensor fields we obtain

$$\int_{\Omega_t}\rho\dot{\boldsymbol{v}}\,dV_{\boldsymbol{x}} = \int_{\Omega_t}\left[\nabla^x\cdot\boldsymbol{S} + \rho\boldsymbol{b}\right]\,dV_{\boldsymbol{x}}.$$

The fact that this must hold for an arbitrary open subset Ω_t of B_t, together with the Localization Theorem, leads to the following result.

Result 5.7 Law of Linear Momentum in Eulerian Form. *Let $\varphi : B \times [0,\infty) \to \mathbb{E}^3$ be a motion of a continuum body with associated spatial velocity field $\boldsymbol{v}(\boldsymbol{x},t)$ and spatial mass density field $\rho(\boldsymbol{x},t)$. Then balance of linear momentum requires*

$$\rho\dot{\boldsymbol{v}} = \nabla^x \cdot \boldsymbol{S} + \rho\boldsymbol{b}, \quad \forall \boldsymbol{x} \in B_t,\ t \geq 0,$$

where $\boldsymbol{S}(\boldsymbol{x},t)$ is the Cauchy stress field and $\boldsymbol{b}(\boldsymbol{x},t)$ is the spatial body force per unit mass. □

The above equations may be viewed as a generalization to the dynamic case of the equilibrium (balance of forces) condition described in Result 3.5. In particular, the condition of equilibrium corresponds to the special case when $\dot{\boldsymbol{v}} = \boldsymbol{0}$ for all $\boldsymbol{x} \in B_t$ and $t \geq 0$.

Remarks:

(1) By Result 4.7 we have

$$\dot{\boldsymbol{v}} = \frac{\partial}{\partial t}\boldsymbol{v} + (\nabla^x \boldsymbol{v})\boldsymbol{v}.$$

Thus the balance of linear momentum equation may be written in the equivalent form

$$\rho\left[\frac{\partial}{\partial t}\boldsymbol{v} + (\nabla^x \boldsymbol{v})\boldsymbol{v}\right] = \nabla^x \cdot \boldsymbol{S} + \rho\boldsymbol{b}.$$

(2) As mentioned in Chapter 4, many texts use the notation $\boldsymbol{v} \cdot \nabla^x \boldsymbol{v}$ in place of $(\nabla^x \boldsymbol{v})\boldsymbol{v}$. We, however, will always use $(\nabla^x \boldsymbol{v})\boldsymbol{v}$, which denotes the application of the second-order tensor $\nabla^x \boldsymbol{v}$ to the vector $\boldsymbol{v}$.

□

5.3.3 Balance of Angular Momentum

As stated in Axiom 5.2, the balance law for angular momentum (about the origin) for an arbitrary open subset Ω_t of B_t is

$$\begin{aligned}\frac{d}{dt}\int_{\Omega_t} \boldsymbol{x} \times \rho(\boldsymbol{x},t)\boldsymbol{v}(\boldsymbol{x},t)\, dV_{\boldsymbol{x}} \\ = \int_{\partial\Omega_t} \boldsymbol{x} \times \boldsymbol{t}(\boldsymbol{x},t)\, dA_{\boldsymbol{x}} + \int_{\Omega_t} \boldsymbol{x} \times \rho(\boldsymbol{x},t)\boldsymbol{b}(\boldsymbol{x},t)\, dV_{\boldsymbol{x}},\end{aligned} \tag{5.20}$$

where $\rho(\boldsymbol{x},t)$ is the spatial mass density field, $\boldsymbol{v}(\boldsymbol{x},t)$ is the spatial velocity field, $\boldsymbol{b}(\boldsymbol{x},t)$ is the spatial body force field per unit mass and $\boldsymbol{t}(\boldsymbol{x},t)$ is the traction field on the surface $\partial\Omega_t$ in B_t.

To simplify the left-hand side of (5.20) we notice that the relation $\boldsymbol{x} = \boldsymbol{\varphi}(\boldsymbol{X},t)$ implies $\dot{\boldsymbol{x}} = \boldsymbol{v}(\boldsymbol{x},t)$. Thus

$$\frac{d}{dt}[\boldsymbol{x} \times \boldsymbol{v}(\boldsymbol{x},t)] = \dot{\boldsymbol{x}} \times \boldsymbol{v}(\boldsymbol{x},t) + \boldsymbol{x} \times \dot{\boldsymbol{v}}(\boldsymbol{x},t) = \boldsymbol{x} \times \dot{\boldsymbol{v}}(\boldsymbol{x},t),$$

and by Result 5.6 we get

$$\begin{aligned}\frac{d}{dt}\int_{\Omega_t} \rho(\boldsymbol{x},t)[\boldsymbol{x} \times \boldsymbol{v}(\boldsymbol{x},t)]\, dV_{\boldsymbol{x}} &= \int_{\Omega_t} \rho(\boldsymbol{x},t)\frac{d}{dt}[\boldsymbol{x} \times \boldsymbol{v}(\boldsymbol{x},t)]\, dV_{\boldsymbol{x}} \\ &= \int_{\Omega_t} \rho(\boldsymbol{x},t)[\boldsymbol{x} \times \dot{\boldsymbol{v}}(\boldsymbol{x},t)]\, dV_{\boldsymbol{x}}.\end{aligned}$$

Substituting the above result into (5.20) and using the definition of the Cauchy stress field $\boldsymbol{S}(\boldsymbol{x},t)$ we obtain

$$\begin{aligned}&\int_{\Omega_t} \rho(\boldsymbol{x},t)[\boldsymbol{x} \times \dot{\boldsymbol{v}}(\boldsymbol{x},t)]\, dV_{\boldsymbol{x}} \\ &\quad = \int_{\partial\Omega_t} \boldsymbol{x} \times \boldsymbol{S}(\boldsymbol{x},t)\boldsymbol{n}\, dA_{\boldsymbol{x}} + \int_{\Omega_t} \rho(\boldsymbol{x},t)[\boldsymbol{x} \times \boldsymbol{b}(\boldsymbol{x},t)]\, dV_{\boldsymbol{x}}.\end{aligned}$$

By employing Result 5.7 the above expression may be reduced to

$$\int_{\partial\Omega_t} \boldsymbol{x} \times \boldsymbol{S}(\boldsymbol{x},t)\boldsymbol{n}\, dA_{\boldsymbol{x}} - \int_{\Omega_t} \boldsymbol{x} \times (\nabla^{x} \cdot \boldsymbol{S})(\boldsymbol{x},t)\, dV_{\boldsymbol{x}} = \boldsymbol{0},$$

which must hold for an arbitrary open subset Ω_t of B_t. Proceeding as in the proof of Result 3.5 we arrive at the following result.

Result 5.8 ***Law of Angular Momentum in Eulerian Form.*** *Let $\boldsymbol{\varphi} : B \times [0,\infty) \to \boldsymbol{E}^3$ be a motion of a continuum body and let $\boldsymbol{S}(\boldsymbol{x},t)$ denote the Cauchy stress field in B_t. Then balance of angular momentum requires*

$$\boldsymbol{S}^T = \boldsymbol{S}, \quad \forall \boldsymbol{x} \in B_t,\ t \geq 0.$$

□

Thus balance of angular momentum requires that the Cauchy stress field be symmetric at each point and time in the motion of a body. Notice that this result is identical to the one obtained in Result 3.5 of Chapter 3, which was based only on equilibrium arguments. Thus the Cauchy stress field must be symmetric whether a body is in motion or in equilibrium. For a related result derived under different assumptions see Exercise 7.

5.3.4 Characterization of Net Working

Before proceeding to study localized versions of the First and Second Laws of Thermodynamics, we use Results 5.7 and 5.8 to derive a relation between the rate of change of kinetic energy and the power of external and internal forces. As we will see, this relation will lead to an explicit expression for the net working on a body.

To begin, consider Result 5.7 and take the dot product with the spatial velocity field $\boldsymbol{v}(\boldsymbol{x},t)$ to obtain

$$\rho\boldsymbol{v}\cdot\dot{\boldsymbol{v}} = (\nabla^x\cdot\boldsymbol{S})\cdot\boldsymbol{v} + \rho\boldsymbol{b}\cdot\boldsymbol{v}, \tag{5.21}$$

for all $\boldsymbol{x}\in B_t$ and $t\geq 0$. Integration of (5.21) over an arbitrary open subset Ω_t of B_t yields

$$\int_{\Omega_t}\rho\boldsymbol{v}\cdot\dot{\boldsymbol{v}}\,dV_{\boldsymbol{x}} = \int_{\Omega_t}\left[(\nabla^x\cdot\boldsymbol{S})\cdot\boldsymbol{v} + \rho\boldsymbol{b}\cdot\boldsymbol{v}\right]\,dV_{\boldsymbol{x}}.$$

Using the identity in Result 2.11, namely

$$\nabla^x\cdot(\boldsymbol{S}^T\boldsymbol{v}) = (\nabla^x\cdot\boldsymbol{S})\cdot\boldsymbol{v} + \boldsymbol{S}:\nabla^x\boldsymbol{v},$$

together with the Divergence Theorem, we obtain

$$\int_{\Omega_t}\rho\boldsymbol{v}\cdot\dot{\boldsymbol{v}}\,dV_{\boldsymbol{x}}$$
$$= \int_{\Omega_t}-\boldsymbol{S}:\nabla^x\boldsymbol{v}\,dV_{\boldsymbol{x}} + \int_{\partial\Omega_t}\boldsymbol{S}^T\boldsymbol{v}\cdot\boldsymbol{n}\,dA_{\boldsymbol{x}} + \int_{\Omega_t}\rho\boldsymbol{b}\cdot\boldsymbol{v}\,dV_{\boldsymbol{x}}.$$

From Results 5.8 and 1.13 we deduce $\boldsymbol{S}:\nabla^x\boldsymbol{v} = \boldsymbol{S}:\boldsymbol{L}$, where $\boldsymbol{L} = \mathrm{sym}(\nabla^x\boldsymbol{v})$ is the rate of strain field. Substitution of this result into the above expression yields

$$\int_{\Omega_t}\rho\boldsymbol{v}\cdot\dot{\boldsymbol{v}}\,dV_{\boldsymbol{x}} + \int_{\Omega_t}\boldsymbol{S}:\boldsymbol{L}\,dV_{\boldsymbol{x}} = \int_{\partial\Omega_t}\boldsymbol{v}\cdot\boldsymbol{S}\boldsymbol{n}\,dA_{\boldsymbol{x}} + \int_{\Omega_t}\rho\boldsymbol{b}\cdot\boldsymbol{v}\,dV_{\boldsymbol{x}},$$

which leads to the following result.

Result 5.9 ***Net Working in Eulerian Form.*** *Let $\varphi : B\times[0,\infty)\to\boldsymbol{E}^3$ be a motion of a continuum body with associated Cauchy stress field $\boldsymbol{S}(\boldsymbol{x},t)$ and rate of strain field $\boldsymbol{L}(\boldsymbol{x},t)$, and let Ω_t be an arbitrary open* *subset of B_t. Then*

$$\frac{d}{dt}K[\Omega_t] + \int_{\Omega_t} \boldsymbol{S} : \boldsymbol{L} \, dV_{\boldsymbol{x}} = \mathcal{P}[\Omega_t], \quad \forall t \geq 0,$$

where $K[\Omega_t]$ is the total kinetic energy and $\mathcal{P}[\Omega_t]$ is the power of external forces on Ω_t. Thus, in view of (5.14)*, the net working $\mathcal{W}[\Omega_t]$ is given by*

$$\mathcal{W}[\Omega_t] = \int_{\Omega_t} \boldsymbol{S} : \boldsymbol{L} \, dV_{\boldsymbol{x}}, \quad \forall t \geq 0.$$

□

The quantity $\boldsymbol{S} : \boldsymbol{L}$ is typically called the **stress power** associated with a motion. It corresponds to the rate of work done by internal forces (stresses) at each point in a continuum body. The above result shows that the net working on any open subset of a body is equal to the integral of the stress power over that subset.

5.3.5 First Law of Thermodynamics

As stated in Axiom 5.3, the balance law of energy for an arbitrary open subset Ω_t of B_t is

$$\frac{d}{dt}\int_{\Omega_t} \rho\phi \, dV_{\boldsymbol{x}} = \int_{\Omega_t} \boldsymbol{S} : \boldsymbol{L} \, dV_{\boldsymbol{x}} - \int_{\partial\Omega_t} \boldsymbol{q} \cdot \boldsymbol{n} \, dA_{\boldsymbol{x}} + \int_{\Omega_t} \rho r \, dV_{\boldsymbol{x}},$$

where the expressions in Result 5.9 and equation (5.13) for the net working and heating, respectively, have been used. Here $\rho(\boldsymbol{x}, t)$ is the mass density field, $\phi(\boldsymbol{x}, t)$ is the internal energy field per unit mass, $\boldsymbol{q}(\boldsymbol{x}, t)$ is the Fourier–Stokes heat flux vector field, $r(\boldsymbol{x}, t)$ is the heat supply field per unit mass, $\boldsymbol{L}(\boldsymbol{x}, t)$ is the rate of strain field and $\boldsymbol{S}(\boldsymbol{x}, t)$ is the Cauchy stress field in B_t.

Using Result 5.6 together with the Divergence Theorem we obtain

$$\int_{\Omega_t} \rho\dot{\phi} \, dV_{\boldsymbol{x}} = \int_{\Omega_t} \boldsymbol{S} : \boldsymbol{L} \, dV_{\boldsymbol{x}} - \int_{\Omega_t} \nabla^x \cdot \boldsymbol{q} \, dV_{\boldsymbol{x}} + \int_{\Omega_t} \rho r \, dV_{\boldsymbol{x}}.$$

The fact that the above statement must hold for an arbitrary open subset Ω_t of B_t, together with the Localization Theorem, leads to the following result.

Result 5.10 ***Law of Energy in Eulerian Form.*** *Let $\varphi : B \times [0,\infty) \to \boldsymbol{E}^3$ be a motion of a continuum body. Then*

$$\rho\dot{\phi} = \boldsymbol{S} : \boldsymbol{L} - \nabla^x \cdot \boldsymbol{q} + \rho r, \quad \forall \boldsymbol{x} \in B_t,\ t \geq 0.$$

□

5.3.6 Second Law of Thermodynamics

As stated in Axiom 5.4, the Clausius–Duhem form of the Second Law of Thermodynamics for an arbitrary open subset Ω_t of B_t is

$$\frac{d}{dt}\int_{\Omega_t} \rho\eta \, dV_{\boldsymbol{x}} \geq \int_{\Omega_t} \frac{\rho r}{\theta}\, dV_{\boldsymbol{x}} - \int_{\partial\Omega_t} \frac{\boldsymbol{q}\cdot\boldsymbol{n}}{\theta}\, dA_{\boldsymbol{x}},$$

where $\rho(\boldsymbol{x},t)$ is the mass density field, $\eta(\boldsymbol{x},t)$ is the entropy field per unit mass, $\boldsymbol{q}(\boldsymbol{x},t)$ is the Fourier–Stokes heat flux vector field, $r(\boldsymbol{x},t)$ is the heat supply field per unit mass and $\theta(\boldsymbol{x},t)$ is the (absolute) temperature field in B_t. An application of the Divergence Theorem followed by the Localization Theorem leads to the following local statement.

Result 5.11 ***Clausius–Duhem Inequality in Eulerian Form.*** *Let $\varphi : B \times [0,\infty) \to \boldsymbol{E}^3$ be a motion of a continuum body. Then*

$$\rho\dot{\eta} \geq \theta^{-1}\rho r - \nabla^x \cdot (\theta^{-1}\boldsymbol{q}), \quad \forall \boldsymbol{x} \in B_t,\ t \geq 0. \tag{5.22}$$

□

By expanding the divergence term in (5.22) and multiplying through by θ we obtain

$$\theta\rho\dot{\eta} \geq \rho r - \nabla^x \cdot \boldsymbol{q} + \theta^{-1}\boldsymbol{q}\cdot\nabla^x\theta. \tag{5.23}$$

The above expression can be written in the equivalent form

$$\delta - \theta^{-1}\boldsymbol{q}\cdot\nabla^x\theta \geq 0, \tag{5.24}$$

where $\delta = \theta\rho\dot{\eta} - (\rho r - \nabla^x \cdot \boldsymbol{q})$ is called the **internal dissipation density field** per unit volume. It represents the difference between the local rate of entropy increase and the local heating. In other words, δ is a measure of the local rate of entropy increase due to mechanisms other than the local supply and transfer of heat.

Various statements about the internal dissipation δ and the heat flux $\boldsymbol{q}$ can be made on the basis of (5.24). For example, at any point at which the temperature satisfies $\nabla^x\theta = \boldsymbol{0}$, the internal dissipation must satisfy $\delta \geq 0$. Thus, the internal dissipation is everywhere non-negative in bodies with spatially homogeneous temperature. From (5.24) we also deduce that, at any point at which $\delta = 0$, the heat flux and temperature must satisfy $\boldsymbol{q} \cdot \nabla^x \theta \leq 0$. Thus, the heat flux vector must everywhere make an obtuse angle with the temperature gradient in bodies with no internal dissipation. In particular, heat must flow from "hot" to "cold" in such bodies.

To study the consequences of the Clausius–Duhem inequality for various constitutive models introduced later, it is convenient to introduce the quantity

$$\psi(\boldsymbol{x}, t) = \phi(\boldsymbol{x}, t) - \theta(\boldsymbol{x}, t)\eta(\boldsymbol{x}, t), \tag{5.25}$$

which is called the **free energy density field** per unit mass. According to the classic theory of homogeneous systems undergoing reversible processes, free energy is that portion of the internal energy available for performing work at constant temperature. Result 5.11 can be restated in terms free energy as follows.

Result 5.12 ***Reduced Clausius–Duhem Inequality in Eulerian Form.*** *Let $\varphi : B \times [0, \infty) \to \boldsymbol{E}^3$ be a motion of a continuum body with spatial free energy field $\psi(\boldsymbol{x}, t)$. Then*

$$\rho\dot{\psi} \leq \boldsymbol{S} : \boldsymbol{L} - \rho\eta\dot{\theta} - \theta^{-1}\boldsymbol{q} \cdot \nabla^x \theta, \quad \forall \boldsymbol{x} \in B_t,\ t \geq 0. \tag{5.26}$$

□

Proof From (5.25) we obtain $\theta\dot{\eta} = \dot{\phi} - \dot{\theta}\eta - \dot{\psi}$ (see Exercise 8). Multiplying this expression by ρ and using Result 5.10 to eliminate $\rho\dot{\phi}$ we get

$$\rho\theta\dot{\eta} = \boldsymbol{S} : \boldsymbol{L} - \nabla^x \cdot \boldsymbol{q} + \rho r - \rho\dot{\theta}\eta - \rho\dot{\psi}.$$

Substituting this expression into (5.23) leads to the desired result. □

Various statements about the free energy can be deduced from Result 5.12. For example, in bodies with constant, spatially homogeneous temperature we must have $\rho\dot{\psi} \leq \boldsymbol{S} : \boldsymbol{L}$. If we further assume that a body can only undergo reversible processes in the sense that equality is always achieved in the Clausius–Duhem inequality, we obtain $\rho\dot{\psi} = \boldsymbol{S} : \boldsymbol{L}$. Thus for such bodies the rate of change of free energy is equal to the stress power. The term "reduced" in Result 5.12 reflects the fact that, when considered within the classic case of homogeneous bodies, the inequality in (5.26) is independent of the heat supply r and heat flux $\boldsymbol{q}$ in contrast to (5.22).

5.3.7 Summary

In the Eulerian description of the motion of a general continuum body there are 22 basic unknown fields:

$\varphi_i(\boldsymbol{X}, t)$	3 components of motion
$v_i(\boldsymbol{x}, t)$	3 components of velocity
$\rho(\boldsymbol{x}, t)$	1 mass density
$S_{ij}(\boldsymbol{x}, t)$	9 components of stress
$\theta(\boldsymbol{x}, t)$	1 temperature
$q_i(\boldsymbol{x}, t)$	3 components of heat flux
$\phi(\boldsymbol{x}, t)$	1 internal energy per unit mass
$\eta(\boldsymbol{x}, t)$	1 entropy per unit mass.

To determine these unknown fields we have the following 11 equations:

$[v_i]_m = \frac{\partial}{\partial t}\varphi_i$	3 kinematical
$\frac{\partial}{\partial t}\rho + (\rho v_i)_{,i} = 0$	1 mass
$\rho \dot{v}_i = S_{ij,j} + \rho b_i$	3 linear momentum
$S_{ij} = S_{ji}$	3 independent angular momentum
$\rho \dot{\phi} = S_{ij} v_{i,j} - q_{i,i} + \rho r$	1 energy.

Remarks:

(1) The motion map φ is often not needed within the Eulerian framework. In particular, when the current placement $\mathcal{B}_t$ is known or specified for all $t \geq 0$, the balance laws can typically be applied without explicit knowledge of φ. The number of unknowns is then reduced to 19 (φ disappears), and the number of equations is reduced to 8 (kinematical equations disappear).

(2) Since the number of unknowns is greater than the number of equations by 11, extra equations are required to close the system. As we will see later, system closure can be obtained by the introduction of **constitutive equations** which relate $(\boldsymbol{S}, \boldsymbol{q}, \phi, \eta)$ to

$(\rho, \boldsymbol{v}, \theta)$. Such relations reflect the specific material properties of a body (see Exercise 9).

(3) The Second Law of Thermodynamics (Clausius–Duhem inequality) does not provide an equation with which to determine unknown fields. Instead, we interpret the Second Law as providing a restriction on constitutive relations. In particular, the constitutive relations of a body must guarantee that the Second Law is satisfied in every possible motion of that body.

(4) When thermal effects are neglected, the number of unknowns is further reduced from 19 to 13 ($\boldsymbol{q}$, ϕ, η and θ disappear), and the number of equations is reduced from 8 to 7 (balance of energy disappears). In this case, system closure can be obtained by 6 constitutive equations which relate $\boldsymbol{S}$ to $(\rho, \boldsymbol{v})$. □

5.4 Localized Lagrangian Form of Balance Laws

Again we consider a continuum body with reference configuration B undergoing a motion $\boldsymbol{\varphi} : B \times [0, \infty) \to \mathbb{E}^3$. We denote the current configuration by $B_t = \boldsymbol{\varphi}_t(B)$, and we assume $B_0 = B$ so that $\boldsymbol{\varphi}_0$ is the identity. Since the deformation $\boldsymbol{\varphi}_t : B \to B_t$ is assumed to be admissible for each $t \geq 0$, we have a bijection between B and B_t for each $t \geq 0$. Thus, by a change of variable, any balance law stated in terms of $\boldsymbol{x} \in B_t$ can be expressed in terms of $\boldsymbol{X} \in B$. This leads to the material or Lagrangian form of the balance laws.

5.4.1 Conservation of Mass

As before, consider an arbitrary open subset Ω_t of B_t, and let Ω be the corresponding subset in B so that $\Omega_t = \boldsymbol{\varphi}_t(\Omega)$. Then Axiom 5.1 implies

$$\mathrm{mass}[\Omega_t] = \mathrm{mass}[\Omega_0],$$

where $\Omega_0 = \Omega$. In view of the arguments leading to (5.18) we have the following result.

Result 5.13 ***Conservation of Mass in Lagrangian Form.*** *Let $\boldsymbol{\varphi} : B \times [0, \infty) \to \mathbb{E}^3$ be a motion of a continuum body with associated deformation gradient field $\boldsymbol{F}(\boldsymbol{X}, t)$ and spatial mass density field $\rho(\boldsymbol{x}, t)$.*

Let $\rho_0(\boldsymbol{X})$ be the mass density field in the reference configuration B. Then conservation of mass requires

$$\rho_m(\boldsymbol{X},t)\det \boldsymbol{F}(\boldsymbol{X},t) = \rho_0(\boldsymbol{X}), \quad \forall \boldsymbol{X} \in B,\ t \geq 0,$$

where $\rho_m(\boldsymbol{X},t)$ is the material description of the spatial field $\rho(\boldsymbol{x},t)$. □

5.4.2 Balance of Linear Momentum

There are various ways to develop the Lagrangian form of the balance law for linear momentum. We may proceed directly from the integral form of the law in Axiom 5.2. Alternatively, we may integrate the local Eulerian form in Result 5.7 over an arbitrary open subset Ω_t of B_t, change variables to obtain an integral over an open subset Ω of B, and then localize. Here we follow the first approach.

As stated in Axiom 5.2, the balance law for linear momentum for an arbitrary open subset Ω_t of B_t is

$$\frac{d}{dt}\boldsymbol{l}[\Omega_t] = \boldsymbol{r}[\Omega_t], \tag{5.27}$$

where $\boldsymbol{l}[\Omega_t]$ is the linear momentum of Ω_t defined by

$$\boldsymbol{l}[\Omega_t] = \int_{\Omega_t} \rho(\boldsymbol{x},t)\boldsymbol{v}(\boldsymbol{x},t)\, dV_{\boldsymbol{x}},$$

and $\boldsymbol{r}[\Omega_t]$ is the resultant external force on Ω_t defined by

$$\boldsymbol{r}[\Omega_t] = \int_{\Omega_t} \rho(\boldsymbol{x},t)\boldsymbol{b}(\boldsymbol{x},t)\, dV_{\boldsymbol{x}} + \int_{\partial\Omega_t} \boldsymbol{S}(\boldsymbol{x},t)\boldsymbol{n}(\boldsymbol{x})\, dA_{\boldsymbol{x}}.$$

Performing a change of variable in the integral for $\boldsymbol{l}[\Omega_t]$ we obtain

$$\begin{aligned} \boldsymbol{l}[\Omega_t] &= \int_{\Omega_t} \rho(\boldsymbol{x},t)\boldsymbol{v}(\boldsymbol{x},t)\, dV_{\boldsymbol{x}} \\ &= \int_{\Omega} \rho(\boldsymbol{\varphi}(\boldsymbol{X},t),t)\boldsymbol{v}(\boldsymbol{\varphi}(\boldsymbol{X},t),t)\det \boldsymbol{F}(\boldsymbol{X},t)\, dV_{\boldsymbol{X}} \\ &= \int_{\Omega} \rho_0(\boldsymbol{X})\dot{\boldsymbol{\varphi}}(\boldsymbol{X},t)\, dV_{\boldsymbol{X}}, \end{aligned} \tag{5.28}$$

where we have used Result 5.13. Performing a similar change of variable in the integrals for $\boldsymbol{r}[\Omega_t]$ we obtain

$$\begin{aligned}\boldsymbol{r}[\Omega_t] &= \int_{\partial\Omega_t} \boldsymbol{S}(\boldsymbol{x},t)\boldsymbol{n}(\boldsymbol{x})\, dA_{\boldsymbol{x}} + \int_{\Omega_t} \rho(\boldsymbol{x},t)\boldsymbol{b}(\boldsymbol{x},t)\, dV_{\boldsymbol{x}} \\ &= \int_{\partial\Omega} (\det \boldsymbol{F}(\boldsymbol{X},t))\, \boldsymbol{S}(\boldsymbol{\varphi}(\boldsymbol{X},t),t)\boldsymbol{F}(\boldsymbol{X},t)^{-T}\boldsymbol{N}(\boldsymbol{X})\, dA_{\boldsymbol{X}} \\ &\qquad + \int_{\Omega} \det \boldsymbol{F}(\boldsymbol{X},t)\rho(\boldsymbol{\varphi}(\boldsymbol{X},t),t)\boldsymbol{b}(\boldsymbol{\varphi}(\boldsymbol{X},t),t)\, dV_{\boldsymbol{X}},\end{aligned}$$

where we have used Result 4.11 for the transformation of the surface integral. In particular, $\boldsymbol{n}(\boldsymbol{x})$ is the unit outward normal field on $\partial\Omega_t$ and $\boldsymbol{N}(\boldsymbol{X})$ is the unit outward normal field on $\partial\Omega$. The following definition will help simplify our developments.

Definition 5.14 *Let $\boldsymbol{S}(\boldsymbol{x},t)$ be the Cauchy stress field in B_t associated with a motion $\boldsymbol{\varphi} : B \times [0,\infty) \to \boldsymbol{E}^3$. Then by the* **nominal** *or* **first Piola–Kirchhoff stress** *field associated with $\boldsymbol{\varphi}$ we mean*

$$\boldsymbol{P}(\boldsymbol{X},t) = (\det \boldsymbol{F}(\boldsymbol{X},t))\, \boldsymbol{S}_m(\boldsymbol{X},t)\boldsymbol{F}(\boldsymbol{X},t)^{-T}, \quad \forall \boldsymbol{X} \in B,\ t \geq 0,$$

and by the **second Piola–Kirchhoff stress** *field associated with $\boldsymbol{\varphi}$ we mean*

$$\boldsymbol{\Sigma}(\boldsymbol{X},t) = \boldsymbol{F}(\boldsymbol{X},t)^{-1}\boldsymbol{P}(\boldsymbol{X},t), \quad \forall \boldsymbol{X} \in B,\ t \geq 0.$$

□

To proceed, let $\boldsymbol{b}_m$ be the material description of the spatial body force field $\boldsymbol{b}$. Then, using Result 5.13 and Definition 5.14, we find

$$\boldsymbol{r}[\Omega_t] = \int_{\partial\Omega} \boldsymbol{P}(\boldsymbol{X},t)\boldsymbol{N}(\boldsymbol{X})\, dA_{\boldsymbol{X}} + \int_{\Omega} \rho_0(\boldsymbol{X})\boldsymbol{b}_m(\boldsymbol{X},t)\, dV_{\boldsymbol{X}}. \qquad (5.29)$$

Substituting (5.28) and (5.29) into (5.27) we obtain

$$\begin{aligned}&\frac{d}{dt}\int_{\Omega} \rho_0(\boldsymbol{X})\dot{\boldsymbol{\varphi}}(\boldsymbol{X},t)\, dV_{\boldsymbol{X}} \\ &\qquad = \int_{\partial\Omega} \boldsymbol{P}(\boldsymbol{X},t)\boldsymbol{N}(\boldsymbol{X})\, dA_{\boldsymbol{X}} + \int_{\Omega} \rho_0(\boldsymbol{X})\boldsymbol{b}_m(\boldsymbol{X},t)\, dV_{\boldsymbol{X}}.\end{aligned}$$

By the Divergence Theorem and the fact that Ω is independent of time we get

$$\begin{aligned}&\int_{\Omega} \rho_0(\boldsymbol{X})\ddot{\boldsymbol{\varphi}}(\boldsymbol{X},t)\, dV_{\boldsymbol{X}} \\ &\qquad = \int_{\Omega} [(\nabla^X \cdot \boldsymbol{P})(\boldsymbol{X},t) + \rho_0(\boldsymbol{X})\boldsymbol{b}_m(\boldsymbol{X},t)]\, dV_{\boldsymbol{X}}.\end{aligned}$$

Because we have a bijection between B and B_t, any balance law which holds for an arbitrary subset Ω_t of B_t must also hold for an arbitrary subset Ω of B. This fact leads to the following result.

Result 5.15 Law of Linear Momentum in Lagrangian Form. *Let $\varphi : B \times [0,\infty) \to \mathbb{E}^3$ be a motion of a continuum body and let $\rho_0(\boldsymbol{X})$ denote the mass density field in the reference configuration B. Then balance of linear momentum requires*

$$\rho_0 \ddot{\varphi} = \nabla^X \cdot \boldsymbol{P} + \rho_0 \boldsymbol{b}_m, \quad \forall \boldsymbol{X} \in B,\ t \geq 0,$$

where $\boldsymbol{P}(\boldsymbol{X},t)$ is the first Piola–Kirchhoff stress field and $\boldsymbol{b}_m(\boldsymbol{X},t)$ is the material description of the spatial body force field $\boldsymbol{b}(\boldsymbol{x},t)$. □

5.4.3 Balance of Angular Momentum

The following is a straightforward consequence of Result 5.8 and Definition 5.14.

Result 5.16 Law of Angular Momentum in Lagrangian Form. *Let $\varphi : B \times [0,\infty) \to \mathbb{E}^3$ be a motion of a continuum body with associated deformation gradient field $\boldsymbol{F}(\boldsymbol{X},t)$. Then balance of angular momentum requires*

$$\boldsymbol{P}\boldsymbol{F}^T = \boldsymbol{F}\boldsymbol{P}^T \quad \text{or equivalently} \quad \boldsymbol{\Sigma}^T = \boldsymbol{\Sigma}, \quad \forall \boldsymbol{X} \in B,\ t \geq 0,$$

where $\boldsymbol{P}(\boldsymbol{X},t)$ is the first and $\boldsymbol{\Sigma}(\boldsymbol{X},t)$ is the second Piola–Kirchhoff stress field associated with φ. □

5.4.4 Characterization of Net Working

Before proceeding to study localized Lagrangian versions of the First and Second Laws of Thermodynamics, we use Result 5.15 to derive a relation between the rate of change of kinetic energy and the power of external and internal forces. As before, this will lead to an explicit expression for the net working on a body.

To begin, consider an arbitrary open subset Ω_t of B_t, and let Ω be the corresponding subset in B. Then the kinetic energy of Ω_t is defined as

$$K[\Omega_t] = \int_{\Omega_t} \tfrac{1}{2}\rho(\boldsymbol{x},t)|\boldsymbol{v}(\boldsymbol{x},t)|^2 \, dV_{\boldsymbol{x}},$$

and the power of external forces on Ω_t is defined as

$$\mathcal{P}[\Omega_t] = \int_{\Omega_t} \rho(\boldsymbol{x},t)\boldsymbol{b}(\boldsymbol{x},t)\cdot\boldsymbol{v}(\boldsymbol{x},t)\,dV_{\boldsymbol{x}} + \int_{\partial\Omega_t} \boldsymbol{S}(\boldsymbol{x},t)^T\boldsymbol{v}(\boldsymbol{x},t)\cdot\boldsymbol{n}(\boldsymbol{x})\,dA_{\boldsymbol{x}}.$$

Notice that in the expression for $\mathcal{P}[\Omega_t]$ we have made use of the identity $\boldsymbol{S}^T\boldsymbol{v}\cdot\boldsymbol{n} = \boldsymbol{v}\cdot\boldsymbol{S}\boldsymbol{n}$.

Performing a change of variable in the integral for $K[\Omega_t]$ and using Result 5.13 we obtain

$$K[\Omega_t] = \int_{\Omega} \tfrac{1}{2}\rho_0(\boldsymbol{X})|\dot{\boldsymbol{\varphi}}(\boldsymbol{X},t)|^2\,dV_{\boldsymbol{X}}. \tag{5.30}$$

Similarly, performing a change of variable in the integrals for $\mathcal{P}[\Omega_t]$ we obtain (omitting the arguments $\boldsymbol{X}$ and t for brevity)

$$\begin{aligned}\mathcal{P}[\Omega_t] &= \int_{\Omega} \rho_0\boldsymbol{b}_m\cdot\boldsymbol{v}_m\,dV_{\boldsymbol{X}} + \int_{\partial\Omega} (\det\boldsymbol{F})\,\boldsymbol{S}_m^T\boldsymbol{v}_m\cdot\boldsymbol{F}^{-T}\boldsymbol{N}\,dA_{\boldsymbol{X}}\\ &= \int_{\Omega} \rho_0\boldsymbol{b}_m\cdot\boldsymbol{v}_m\,dV_{\boldsymbol{X}} + \int_{\partial\Omega} (\det\boldsymbol{F})\,\boldsymbol{v}_m\cdot\boldsymbol{S}_m\boldsymbol{F}^{-T}\boldsymbol{N}\,dA_{\boldsymbol{X}},\end{aligned}$$

which, by definition of the material velocity field $\dot{\boldsymbol{\varphi}}$ and the first Piola–Kirchhoff stress field $\boldsymbol{P}$, yields

$$\begin{aligned}\mathcal{P}[\Omega_t] &= \int_{\Omega} \rho_0\boldsymbol{b}_m\cdot\dot{\boldsymbol{\varphi}}\,dV_{\boldsymbol{X}} + \int_{\partial\Omega} \dot{\boldsymbol{\varphi}}\cdot\boldsymbol{P}\boldsymbol{N}\,dA_{\boldsymbol{X}}\\ &= \int_{\Omega} \rho_0\boldsymbol{b}_m\cdot\dot{\boldsymbol{\varphi}}\,dV_{\boldsymbol{X}} + \int_{\partial\Omega} \boldsymbol{P}^T\dot{\boldsymbol{\varphi}}\cdot\boldsymbol{N}\,dA_{\boldsymbol{X}}.\end{aligned}$$

Furthermore, by use of the Divergence Theorem together with Result 2.11, we obtain

$$\begin{aligned}\mathcal{P}[\Omega_t] &= \int_{\Omega} \left[\rho_0\boldsymbol{b}_m\cdot\dot{\boldsymbol{\varphi}} + \nabla^X\cdot(\boldsymbol{P}^T\dot{\boldsymbol{\varphi}})\right]\,dV_{\boldsymbol{X}}\\ &= \int_{\Omega} \left[(\nabla^X\cdot\boldsymbol{P} + \rho_0\boldsymbol{b}_m)\cdot\dot{\boldsymbol{\varphi}} + \boldsymbol{P}:\nabla^X\dot{\boldsymbol{\varphi}}\right]\,dV_{\boldsymbol{X}}.\end{aligned} \tag{5.31}$$

Taking the time derivative of (5.30) and using Result 5.15 to substitute for $\rho_0\ddot{\boldsymbol{\varphi}}$ gives

$$\frac{d}{dt}K[\Omega_t] = \int_{\Omega} \rho_0\dot{\boldsymbol{\varphi}}\cdot\ddot{\boldsymbol{\varphi}}\,dV_{\boldsymbol{X}} = \int_{\Omega} (\nabla^X\cdot\boldsymbol{P} + \rho_0\boldsymbol{b}_m)\cdot\dot{\boldsymbol{\varphi}}\,dV_{\boldsymbol{X}}.$$

Substituting (5.31) into the right-hand side of this expression, and using the fact that

$$\nabla^X\dot{\boldsymbol{\varphi}}(\boldsymbol{X},t) = \nabla^X\frac{\partial}{\partial t}\boldsymbol{\varphi}(\boldsymbol{X},t) = \frac{\partial}{\partial t}\nabla^X\boldsymbol{\varphi}(\boldsymbol{X},t) = \dot{\boldsymbol{F}}(\boldsymbol{X},t),$$

we arrive at the following result.

Result 5.17 Net Working in Lagrangian Form. *Let $\varphi : B \times [0,\infty) \to \mathbb{E}^3$ be a motion of a continuum body with deformation gradient $\boldsymbol{F}(\boldsymbol{X},t)$ and first Piola–Kirchhoff stress $\boldsymbol{P}(\boldsymbol{X},t)$, and let Ω_t be an arbitrary open subset of B_t with corresponding subset Ω in B. Then*

$$\frac{d}{dt}K[\Omega_t] + \int_\Omega \boldsymbol{P} : \dot{\boldsymbol{F}}\, dV_{\boldsymbol{X}} = \mathcal{P}[\Omega_t], \quad \forall t \geq 0,$$

where $K[\Omega_t]$ is the total kinetic energy and $\mathcal{P}[\Omega_t]$ is the power of external forces on Ω_t. Thus, in view of (5.14), the net working $\mathcal{W}[\Omega_t]$ is given by

$$\mathcal{W}[\Omega_t] = \int_\Omega \boldsymbol{P} : \dot{\boldsymbol{F}}\, dV_{\boldsymbol{X}}, \quad \forall t \geq 0.$$

□

A comparison of Results 5.17 and 5.9 shows that $\boldsymbol{P} : \dot{\boldsymbol{F}}$ and $\boldsymbol{S} : \boldsymbol{L}$ each provide a measure of the rate of work (power) done by internal forces (stresses) in a continuum body. While the quantity $\boldsymbol{S} : \boldsymbol{L}$ measures the stress power per unit volume of the current configuration B_t, the quantity $\boldsymbol{P} : \dot{\boldsymbol{F}}$ measures the stress power per unit volume of the reference configuration B (see Exercise 10).

5.4.5 First Law of Thermodynamics

As stated in Axiom 5.3, the balance law of energy for an arbitrary open subset Ω_t of B_t is

$$\frac{d}{dt}U[\Omega_t] = Q[\Omega_t] + \mathcal{W}[\Omega_t], \tag{5.32}$$

where $U[\Omega_t]$ is the internal energy of Ω_t defined as

$$U[\Omega_t] = \int_{\Omega_t} \rho(\boldsymbol{x},t)\phi(\boldsymbol{x},t)\, dV_{\boldsymbol{x}},$$

$Q[\Omega_t]$ is the net heating of Ω_t defined as

$$Q[\Omega_t] = \int_{\Omega_t} \rho(\boldsymbol{x},t)r(\boldsymbol{x},t)\, dV_{\boldsymbol{x}} - \int_{\partial\Omega_t} \boldsymbol{q}(\boldsymbol{x},t)\cdot\boldsymbol{n}(\boldsymbol{x})\, dA_{\boldsymbol{x}},$$

and $\mathcal{W}[\Omega_t]$ is the net working of external forces on Ω_t.

Performing a change of variable in the integral for $U[\Omega_t]$ we obtain

$$U[\Omega_t] = \int_\Omega \rho_0(\boldsymbol{X})\Phi(\boldsymbol{X},t)\, dV_{\boldsymbol{X}}, \tag{5.33}$$

where $\Phi(\boldsymbol{X},t) = \phi_m(\boldsymbol{X},t)$ is the material description of the spatial internal energy field $\phi(\boldsymbol{x},t)$ per unit mass. Similarly, performing a change of variable in the integrals for $Q[\Omega_t]$ we obtain (omitting the arguments $\boldsymbol{X}$ and t for brevity)

$$Q[\Omega_t] = \int_{\Omega} \rho_0 r_m \, dV_{\boldsymbol{X}} - \int_{\partial\Omega} (\det \boldsymbol{F})\boldsymbol{q}_m \cdot \boldsymbol{F}^{-T}\boldsymbol{N} \, dA_{\boldsymbol{X}}.$$

If we introduce a material heat flux vector field by

$$\boldsymbol{Q}(\boldsymbol{X},t) = (\det \boldsymbol{F}(\boldsymbol{X},t))\boldsymbol{F}(\boldsymbol{X},t)^{-1}\boldsymbol{q}_m(\boldsymbol{X},t), \tag{5.34}$$

and a material heat supply field by

$$R(\boldsymbol{X},t) = r_m(\boldsymbol{X},t),$$

then the net heating can be written in the convenient form

$$Q[\Omega_t] = \int_{\Omega} \rho_0 R \, dV_{\boldsymbol{X}} - \int_{\partial\Omega} \boldsymbol{Q} \cdot \boldsymbol{N} \, dA_{\boldsymbol{X}}. \tag{5.35}$$

Substituting (5.35) and (5.33) into (5.32), and using Result 5.17 together with the Divergence Theorem, we obtain the following result.

Result 5.18 ***Law of Energy in Lagrangian Form.*** *Let $\varphi : B \times [0,\infty) \to \boldsymbol{E}^3$ be a motion of a continuum body with associated material internal energy field $\Phi(\boldsymbol{X},t)$, material heat flux vector field $\boldsymbol{Q}(\boldsymbol{X},t)$ and material heat supply field $R(\boldsymbol{X},t)$. Then*

$$\rho_0 \dot{\Phi} = \boldsymbol{P} : \dot{\boldsymbol{F}} - \nabla^X \cdot \boldsymbol{Q} + \rho_0 R, \quad \forall \boldsymbol{X} \in B, \; t \geq 0.$$

□

5.4.6 Second Law of Thermodynamics

As with the balance law for linear momentum, there are various ways to develop the Lagrangian form of the Second Law of Thermodynamics (Clausius–Duhem inequality). We may proceed directly from the integral form in Axiom 5.4. Alternatively, we may integrate the local Eulerian form in Result 5.11 over an arbitrary open subset Ω_t of B_t, change variables to obtain an integral over the corresponding subset Ω of B, and then localize. Here we follow the second approach.

Integration of the expression in Result 5.11 over an arbitrary open subset Ω_t of B_t yields

$$\int_{\Omega_t} \rho\dot{\eta} \, dV_{\boldsymbol{x}} \geq \int_{\Omega_t} \frac{\rho r}{\theta} \, dV_{\boldsymbol{x}} - \int_{\partial\Omega_t} \frac{\boldsymbol{q}\cdot\boldsymbol{n}}{\theta} \, dA_{\boldsymbol{x}},$$

where $\boldsymbol{n}$ is the unit outward normal on $\partial\Omega_t$. By performing a change of variable in the above integrals we obtain

$$\int_{\Omega} \rho_0 \dot{\eta}_m \, dV_{\boldsymbol{X}} \geq \int_{\Omega} \frac{\rho_0 R}{\Theta} \, dV_{\boldsymbol{X}} - \int_{\partial\Omega} \frac{\boldsymbol{Q} \cdot \boldsymbol{N}}{\Theta} \, dA_{\boldsymbol{X}},$$

where $\Theta = \theta_m$ is the material description of the spatial temperature field θ, η_m is the material description of the spatial entropy field η, and $\boldsymbol{Q}$ is the material heat flux vector field defined in (5.34). The Divergence Theorem together with the Localization Theorem then lead to the following result.

Result 5.19 ***Clausius–Duhem Inequality in Lagrangian Form.*** *Let* $\varphi : B \times [0, \infty) \to \boldsymbol{E}^3$ *be a motion of a continuum body. Then*

$$\rho_0 \dot{\eta}_m \geq \Theta^{-1} \rho_0 R - \nabla^X \cdot (\Theta^{-1} \boldsymbol{Q}), \quad \forall \boldsymbol{X} \in B, \ t \geq 0.$$

□

As in the Eulerian case, it is useful to introduce a free energy density per unit mass

$$\Psi(\boldsymbol{X}, t) = \Phi(\boldsymbol{X}, t) - \Theta(\boldsymbol{X}, t) \eta_m(\boldsymbol{X}, t).$$

Using the free energy, and arguments similar to those leading to Result 5.12, we obtain the following result (see Exercise 11).

Result 5.20 ***Reduced Clausius–Duhem Inequality in Lagrangian Form.*** *Let* $\varphi : B \times [0, \infty) \to \boldsymbol{E}^3$ *be a motion of a continuum body. Then*

$$\rho_0 \dot{\Psi} \leq \boldsymbol{P} : \dot{\boldsymbol{F}} - \rho_0 \eta_m \dot{\Theta} - \Theta^{-1} \boldsymbol{Q} \cdot \nabla^X \Theta, \quad \forall \boldsymbol{X} \in B, \ t \geq 0.$$

□

All the remarks about the Eulerian formulation of the Clausius–Duhem inequality regarding internal dissipation, flow of heat and free energy carry over to this formulation.

5.4.7 Summary

In the Lagrangian description of the motion of a general continuum body there are 21 basic unknown fields:

$\varphi_i(\boldsymbol{X}, t)$	3 components of motion

$V_i(\boldsymbol{X}, t)$	3 components of velocity
$P_{ij}(\boldsymbol{X}, t)$	9 components of stress
$\Theta(\boldsymbol{X}, t)$	1 material temperature
$Q_i(\boldsymbol{X}, t)$	3 components of heat flux
$\Phi(\boldsymbol{X}, t)$	1 internal energy per unit mass
$\eta_m(\boldsymbol{X}, t)$	1 entropy per unit mass.

To determine these unknown fields we have the following 10 equations:

$V_i = \frac{\partial}{\partial t}\varphi_i$	3 kinematical
$\rho_0 \dot{V}_i = P_{ij,j} + \rho_0 [b_i]_m$	3 linear momentum
$P_{ik}F_{jk} = F_{ik}P_{jk}$	3 independent angular momentum
$\rho_0 \dot{\Phi} = P_{ij}\dot{F}_{ij} - Q_{i,i} + \rho_0 R$	1 energy.

Remarks:

(1) In contrast with the Eulerian formulation, the mass density field in the Lagrangian formulation is a known quantity. It is just the prescribed density ρ_0 of the body in the reference configuration.

(2) In many situations the material velocity field $\boldsymbol{V}$ is not needed explicitly. In this case, the number of unknowns is reduced to 18 ($\boldsymbol{V}$ disappears), and the number of equations is reduced to 7 (kinematical equations disappear).

(3) Since the number of unknowns is greater than the number of equations by 11, extra equations are required to close the system. As we will see later, system closure can be achieved by the introduction of constitutive equations which relate $(\boldsymbol{\Sigma}, \boldsymbol{Q}, \Phi, \eta_m)$ to (φ, Θ), where $\boldsymbol{\Sigma} = \boldsymbol{F}^{-1}\boldsymbol{P}$ is the (symmetric) second Piola–Kirchhoff stress field.

(4) As mentioned earlier, the Second Law (Clausius–Duhem inequality) does not provide an equation with which to determine unknown fields. Instead, we interpret the Second Law as providing a restriction on constitutive relations. In particular, the constitutive relations of a body must guarantee that the Second Law is satisfied in every possible motion of that body (see Exercise 12).

(5) When thermal effects are neglected, the number of unknowns

is further reduced from 18 to 12 ($\boldsymbol{Q}$, Φ, η_m and Θ disappear), and the number of equations is reduced from 7 to 6 (balance of energy disappears). In this case, system closure can be obtained by 6 constitutive equations relating $\boldsymbol{\Sigma}$ to φ. □

5.5 Frame-Indifference

In this section we introduce the idea of a superposed rigid motion and use it to define the concept of frame-indifference. We then state an axiom which asserts that certain fields in continuum mechanics are frame-indifferent, or equivalently, independent of observer. As we will see, this axiom will place strict limitations on the constitutive equations of a material body.

5.5.1 Superposed Rigid Motions

If two observers in different reference frames witness the motion of a body, then the two motions so witnessed, when referred to a common frame, must be related by a superposed rigid motion as defined next.

Definition 5.21 *Two motions $\varphi, \varphi^* : B \times [0,\infty) \to \mathbb{E}^3$ of a body are related by a* **superposed rigid motion** *if*

$$\varphi^*(\boldsymbol{X},t) = \boldsymbol{Q}(t)\varphi(\boldsymbol{X},t) + \boldsymbol{c}(t), \quad \forall \boldsymbol{X} \in B,\ t \geq 0, \tag{5.36}$$

for some rotation $\boldsymbol{Q}(t)$ and vector $\boldsymbol{c}(t)$. □

In the above definition, the functions $\boldsymbol{Q}(t)$ and $\boldsymbol{c}(t)$ describe the movement of one observer relative to the other. In particular, $\boldsymbol{Q}(t)$ describes the relative rotation, whereas $\boldsymbol{c}(t)$ describes the relative translation. Thus different pairs of observers are described by different functions $\boldsymbol{Q}(t)$ and $\boldsymbol{c}(t)$. Throughout our developments we implicitly assume that any two observers measure time relative to the same clock. A more general relation between φ and φ^* would be necessary to represent observers with different clocks.

The relation between various measures of strain and rate of strain as seen by two different observers can be derived from (5.36). In particular, let $\boldsymbol{v}(\boldsymbol{x},t)$ and $\boldsymbol{F}(\boldsymbol{X},t)$ be the spatial velocity and deformation gradient fields associated with φ, and let $\boldsymbol{v}^*(\boldsymbol{x}^*,t)$ and $\boldsymbol{F}^*(\boldsymbol{X},t)$ be the corresponding quantities associated with φ^*. Moreover, let $\boldsymbol{C}$, $\boldsymbol{VR}$ and $\boldsymbol{RU}$ be the Cauchy–Green strain tensor and the left and right polar decompositions associated with $\boldsymbol{F}$, and let $\boldsymbol{C}^*$, $\boldsymbol{V}^*\boldsymbol{R}^*$ and $\boldsymbol{R}^*\boldsymbol{U}^*$ be the corresponding quantities associated with $\boldsymbol{F}^*$. Finally, let $\boldsymbol{L}$ be the rate of strain field associated with $\boldsymbol{v}$, and let $\boldsymbol{L}^*$ be the rate of strain associated with $\boldsymbol{v}^*$. Then from (5.36) we deduce the following result (see Exercise 13).

Result 5.22 ***Effect of Superposed Rigid Motion.*** *The relation between the material fields* $(\boldsymbol{F}, \boldsymbol{R}, \boldsymbol{U}, \boldsymbol{V}, \boldsymbol{C})$ *and* $(\boldsymbol{F}^*, \boldsymbol{R}^*, \boldsymbol{U}^*, \boldsymbol{V}^*, \boldsymbol{C}^*)$ *is given by*

$$\boldsymbol{F}^* = \boldsymbol{Q}\boldsymbol{F}, \quad \boldsymbol{R}^* = \boldsymbol{Q}\boldsymbol{R},$$

$$\boldsymbol{U}^* = \boldsymbol{U}, \quad \boldsymbol{V}^* = \boldsymbol{Q}\boldsymbol{V}\boldsymbol{Q}^T, \quad \boldsymbol{C}^* = \boldsymbol{C},$$

for all $\boldsymbol{X} \in B$ *and* $t \geq 0$. *The relation between the spatial fields* $(\nabla^x \boldsymbol{v}, \boldsymbol{L})$ *and* $(\nabla^{x^*} \boldsymbol{v}^*, \boldsymbol{L}^*)$ *is given by*

$$\nabla^{x^*} \boldsymbol{v}^*(\boldsymbol{x}^*, t)\Big|_{\boldsymbol{x}^* = \boldsymbol{g}(\boldsymbol{x},t)} = \boldsymbol{Q}(t)\nabla^x \boldsymbol{v}(\boldsymbol{x}, t)\boldsymbol{Q}(t)^T + \dot{\boldsymbol{Q}}(t)\boldsymbol{Q}(t)^T,$$

$$\boldsymbol{L}^*(\boldsymbol{x}^*, t)\Big|_{\boldsymbol{x}^* = \boldsymbol{g}(\boldsymbol{x},t)} = \boldsymbol{Q}(t)\boldsymbol{L}(\boldsymbol{x}, t)\boldsymbol{Q}(t)^T,$$

for all $\boldsymbol{x} \in B_t$ *and* $t \geq 0$. *Here* $\dot{\boldsymbol{Q}}(t)$ *denotes the derivative of* $\boldsymbol{Q}(t)$. □

5.5.2 *Axiom of Frame-Indifference*

At any time $t \geq 0$, let B_t denote the current placement of B under the motion $\boldsymbol{\varphi}$, let B_t^* denote the current placement under $\boldsymbol{\varphi}^*$, and let $\boldsymbol{g}_t : B_t \to B_t^*$ denote the rigid motion defined by (5.36), that is

$$\boldsymbol{g}_t(\boldsymbol{x}) = \boldsymbol{g}(\boldsymbol{x}, t) = \boldsymbol{Q}(t)\boldsymbol{x} + \boldsymbol{c}(t).$$

Let $\phi(\boldsymbol{x}, t)$ be a scalar field, $\boldsymbol{w}(\boldsymbol{x}, t)$ a vector field and $\boldsymbol{S}(\boldsymbol{x}, t)$ a second-order tensor field associated with the body in its placement B_t whose points we denote by $\boldsymbol{x}$. Moreover, let $\phi^*(\boldsymbol{x}^*, t)$, $\boldsymbol{w}^*(\boldsymbol{x}^*, t)$ and $\boldsymbol{S}^*(\boldsymbol{x}^*, t)$ be the corresponding fields associated with the body in its placement B_t^* whose points we denote by $\boldsymbol{x}^*$.

Definition 5.23 *The fields* ϕ, $\boldsymbol{w}$ *and* $\boldsymbol{S}$ *are called* **frame-indifferent** *if for all superposed rigid motions* $\boldsymbol{g}_t : B_t \to B_t^*$ *we have*

$$\begin{aligned}
\phi^*(\boldsymbol{x}^*, t) &= \phi(\boldsymbol{x}, t), \\
\boldsymbol{w}^*(\boldsymbol{x}^*, t) &= \boldsymbol{Q}(t)\boldsymbol{w}(\boldsymbol{x}, t), \\
\boldsymbol{S}^*(\boldsymbol{x}^*, t) &= \boldsymbol{Q}(t)\boldsymbol{S}(\boldsymbol{x}, t)\boldsymbol{Q}(t)^T,
\end{aligned}$$

for all $\boldsymbol{x} \in B_t$ *and* $t \geq 0$. *Here* $\boldsymbol{x}^* = \boldsymbol{g}(\boldsymbol{x}, t)$, *that is*

$$\boldsymbol{x}^* = \boldsymbol{Q}(t)\boldsymbol{x} + \boldsymbol{c}(t).$$

□

The above notion of frame-indifference captures the idea that some physical quantities associated with a body are inherent to the body and independent of observer. For example, any two observers in different reference frames should agree on the mass density and temperature of a continuum body. They should also agree, after an appropriate change of basis, on the components of the heat flux vector and stress tensor fields. These ideas are made precise by saying that the mass density, temperature, heat flux and Cauchy stress fields are all frame-indifferent in the sense of Definition 5.23. Not all fields, however, are frame-indifferent. For example, two observers in relative motion will generally disagree on the velocity and acceleration fields in a body. Common experience and intuition suggest the following assumption.

Axiom 5.24 ***Material Frame-Indifference.*** *The spatial mass density* $\rho(\boldsymbol{x}, t)$, *temperature* $\theta(\boldsymbol{x}, t)$, *entropy per unit mass* $\eta(\boldsymbol{x}, t)$, *internal energy per unit mass* $\phi(\boldsymbol{x}, t)$, *Cauchy stress* $\boldsymbol{S}(\boldsymbol{x}, t)$ *and Fourier heat flux* $\boldsymbol{q}(\boldsymbol{x}, t)$ *are all frame-indifferent fields.* □

The above axiom places strict limitations on the constitutive equations used to describe the material properties of a given body. For example, suppose the internal energy, heat flux and Cauchy stress are modeled by constitutive equations of the form

$$\begin{aligned}
\phi(\boldsymbol{x}, t) &= \widehat{\phi}(\rho(\boldsymbol{x}, t), \theta(\boldsymbol{x}, t), \boldsymbol{L}(\boldsymbol{x}, t)), \\
\boldsymbol{q}(\boldsymbol{x}, t) &= \widehat{\boldsymbol{q}}(\rho(\boldsymbol{x}, t), \theta(\boldsymbol{x}, t), \boldsymbol{L}(\boldsymbol{x}, t)), \\
\boldsymbol{S}(\boldsymbol{x}, t) &= \widehat{\boldsymbol{S}}(\rho(\boldsymbol{x}, t), \theta(\boldsymbol{x}, t), \boldsymbol{L}(\boldsymbol{x}, t)),
\end{aligned}$$

where $\widehat{\phi}$, $\widehat{\boldsymbol{q}}$ and $\widehat{\boldsymbol{S}}$ are given functions. Then in order to comply with Axiom 5.24 these functions must have the property (omitting the arguments $\boldsymbol{x}$, $\boldsymbol{x}^*$ and t for brevity)

$$\begin{aligned}
\widehat{\phi}(\rho^*, \theta^*, \boldsymbol{L}^*) &= \widehat{\phi}(\rho, \theta, \boldsymbol{L}), \\
\widehat{\boldsymbol{q}}(\rho^*, \theta^*, \boldsymbol{L}^*) &= \boldsymbol{Q}\widehat{\boldsymbol{q}}(\rho, \theta, \boldsymbol{L}), \\
\widehat{\boldsymbol{S}}(\rho^*, \theta^*, \boldsymbol{L}^*) &= \boldsymbol{Q}\widehat{\boldsymbol{S}}(\rho, \theta, \boldsymbol{L})\boldsymbol{Q}^T.
\end{aligned}$$

Since $\rho^* = \rho$ and $\theta^* = \theta$ by Axiom 5.24, and $\boldsymbol{L}^* = \boldsymbol{Q}\boldsymbol{L}\boldsymbol{Q}^T$ by Result 5.22, we deduce that the functions $\widehat{\phi}$, $\widehat{\boldsymbol{q}}$ and $\widehat{\boldsymbol{S}}$ must satisfy

$$\begin{aligned}\widehat{\phi}(\rho,\theta,\boldsymbol{Q}\boldsymbol{L}\boldsymbol{Q}^T) &= \widehat{\phi}(\rho,\theta,\boldsymbol{L}),\\ \widehat{\boldsymbol{q}}(\rho,\theta,\boldsymbol{Q}\boldsymbol{L}\boldsymbol{Q}^T) &= \boldsymbol{Q}\widehat{\boldsymbol{q}}(\rho,\theta,\boldsymbol{L}),\\ \widehat{\boldsymbol{S}}(\rho,\theta,\boldsymbol{Q}\boldsymbol{L}\boldsymbol{Q}^T) &= \boldsymbol{Q}\widehat{\boldsymbol{S}}(\rho,\theta,\boldsymbol{L})\boldsymbol{Q}^T,\end{aligned}$$

for all rotations $\boldsymbol{Q}$ and all admissible values of ρ, θ and $\boldsymbol{L}$. Thus the axiom of material frame-indifference imposes severe restrictions on the possible forms of $\widehat{\phi}$, $\widehat{\boldsymbol{q}}$ and $\widehat{\boldsymbol{S}}$. Functions violating these restrictions would deliver different results when employed by different observers (see Exercise 14). For this reason, frame-indifference is imposed on constitutive equations either exactly or, for linearized models, approximately.

5.6 Material Constraints

Physical experience tells us that certain materials greatly resist certain types of deformations. For example, liquids such as water greatly resist changes in volume, but are otherwise very deformable. We usually model such materials by placing a-priori restrictions or constraints on their motions.

Definition 5.25 *A continuum body is said to be subject to a simple* **material** *or* **internal constraint** *if every motion $\boldsymbol{\varphi} : B \times [0,\infty) \to \boldsymbol{E}^3$ must satisfy an equation of the form*

$$\gamma(\boldsymbol{F}(\boldsymbol{X},t)) = 0, \quad \forall \boldsymbol{X} \in B,\ t \geq 0, \tag{5.37}$$

where $\boldsymbol{F}(\boldsymbol{X},t)$ is the deformation gradient and $\gamma : \mathcal{V}^2 \to \boldsymbol{R}$ is a specified function.

□

As alluded to above, resistance to volume change is an important example of a material constraint. Materials which admit no volume change are called **incompressible**. Such materials can experience only volume-preserving motions. In view of Result 4.12, an appropriate constraint function for an incompressible material is

$$\gamma(\boldsymbol{F}) = \det \boldsymbol{F} - 1. \tag{5.38}$$

When this function is substituted into (5.37) we obtain $\det \boldsymbol{F}(\boldsymbol{X},t) = 1$ for all $\boldsymbol{X} \in B$ and $t \geq 0$, which implies that the local volume of material in a neighborhood of each point is constant in time. Result 4.12 also shows that an equivalent formulation of this constraint is $\nabla^x \cdot \boldsymbol{v}(\boldsymbol{x},t) = 0$

for all $\boldsymbol{x} \in B_t$ and $t \geq 0$. In particular, this divergence condition implies $\det \boldsymbol{F}(\boldsymbol{X},t) = 1$ for all $t > 0$ provided it holds at $t = 0$.

Material constraints must be enforced or maintained by appropriate stresses. While many different systems of stresses could presumably enforce a given constraint, we make the following assumption.

Axiom 5.26 ***Stress Fields in Constrained Materials.*** *The first Piola–Kirchhoff stress field $\boldsymbol{P}(\boldsymbol{X},t)$ in a continuum body subject to a material constraint $\gamma(\boldsymbol{F}(\boldsymbol{X},t)) = 0$ can be decomposed into two parts*

$$\boldsymbol{P}(\boldsymbol{X},t) = \boldsymbol{P}^{(r)}(\boldsymbol{X},t) + \boldsymbol{P}^{(a)}(\boldsymbol{X},t).$$

Here $\boldsymbol{P}^{(a)}(\boldsymbol{X},t)$ is an active stress field determined by a constitutive equation, and $\boldsymbol{P}^{(r)}(\boldsymbol{X},t)$ is a reactive stress field whose stress power is zero in all possible motions, that is

$$\boldsymbol{P}^{(r)}(\boldsymbol{X},t) : \dot{\boldsymbol{F}}(\boldsymbol{X},t) = 0, \quad \forall \boldsymbol{X} \in B,\ t \geq 0. \tag{5.39}$$

□

From (5.39) we can deduce the most general form of the reactive stress $\boldsymbol{P}^{(r)}$ associated with a given constraint. The condition that $\gamma(\boldsymbol{F}(\boldsymbol{X},t))$ be constant in time for all $\boldsymbol{X} \in B$ is equivalent to the condition

$$0 = \frac{\partial}{\partial t}\gamma(\boldsymbol{F}(\boldsymbol{X},t)) = D\gamma(\boldsymbol{F}(\boldsymbol{X},t)) : \dot{\boldsymbol{F}}(\boldsymbol{X},t), \quad \forall \boldsymbol{X} \in B,\ t \geq 0.$$

Thus all motions satisfying the constraint have the property that $\dot{\boldsymbol{F}}$ is orthogonal to $D\gamma(\boldsymbol{F})$ with respect to the standard inner product in the space of second-order tensors. In particular, $\dot{\boldsymbol{F}}$ must lie in the hyperplane orthogonal to $D\gamma(\boldsymbol{F})$ for each $\boldsymbol{X} \in B$ and $t \geq 0$. Here $D\gamma(\boldsymbol{F})$ denotes the derivative of γ at $\boldsymbol{F}$ (see Definition 2.20).

The fact that (5.39) must hold for all possible motions satisfying the constraint implies that $\boldsymbol{P}^{(r)}$ must be orthogonal to the hyperplane defined by $D\gamma(\boldsymbol{F})$. In particular, $\boldsymbol{P}^{(r)}$ must be parallel to $D\gamma(\boldsymbol{F})$ for each $\boldsymbol{X} \in B$ and $t \geq 0$. Thus the most general form for the reactive stress is

$$\boldsymbol{P}^{(r)}(\boldsymbol{X},t) = q(\boldsymbol{X},t)D\gamma(\boldsymbol{F}(\boldsymbol{X},t)),$$

where $q(\boldsymbol{X},t)$ is an unknown scalar field referred to as the **multiplier**. It is the unknown part of the reactive stress that enforces the constraint $\gamma(\boldsymbol{F}(\boldsymbol{X},t)) = 0$. For example, for incompressible materials defined by (5.38), we deduce (see Result 2.22)

$$\boldsymbol{P}^{(r)}(\boldsymbol{X},t) = q(\boldsymbol{X},t)\det(\boldsymbol{F}(\boldsymbol{X},t))\boldsymbol{F}(\boldsymbol{X},t)^{-T}. \tag{5.40}$$

Using Definition 5.14 we can translate the content of Axiom 5.26 from the first Piola–Kirchhoff to the Cauchy stress field. In particular, the Cauchy stress in a body subject to a material constraint can be decomposed as

$$\boldsymbol{S}(\boldsymbol{x},t) = \boldsymbol{S}^{(r)}(\boldsymbol{x},t) + \boldsymbol{S}^{(a)}(\boldsymbol{x},t),$$

where $\boldsymbol{S}^{(r)}$ is a reactive Cauchy stress corresponding to $\boldsymbol{P}^{(r)}$, and $\boldsymbol{S}^{(a)}$ is an active Cauchy stress corresponding to $\boldsymbol{P}^{(a)}$. For example, for incompressible materials, we deduce from (5.40) that the reactive Cauchy stress takes the form

$$\boldsymbol{S}^{(r)}(\boldsymbol{x},t) = -p(\boldsymbol{x},t)\boldsymbol{I},$$

where $-p(\boldsymbol{x},t) = q_s(\boldsymbol{x},t)$ is the spatial description of the material field $q(\boldsymbol{X},t)$. Thus the reactive Cauchy stress is spherical (see Chapter 3) and the multiplier field can be interpreted as a pressure.

Axiom 5.24 asserts that the Cauchy stress field in a body is frame-indifferent. In the earlier case of a body with no constraint, we showed that this axiom places a restriction on the constitutive equation for $\boldsymbol{S}(\boldsymbol{x},t)$. In the present case of a body with a constraint, we interpret this axiom as placing a restriction on the possible forms of the constraint function γ and the constitutive equation for the active stress $\boldsymbol{S}^{(a)}$. In particular, we will always assume that $\gamma(\boldsymbol{F}(\boldsymbol{X},t))$ and $\boldsymbol{S}^{(a)}(\boldsymbol{x},t)$ are both frame-indifferent. These assumptions, together with the frame-indifference of $\boldsymbol{S}(\boldsymbol{x},t)$, imply that the multiplier field is frame-indifferent.

5.7 Isothermal Considerations

Many classic models of fluids and solids are purely mechanical. Here we introduce the notion of an isothermal process as a means of describing such models within the framework of this chapter, and discuss the balance laws relevant to the isothermal modeling of continuum bodies.

Definition 5.27 *By a* **thermo-mechanical process** *for a continuum body we mean a pair of functions* (φ, Θ), *where* $\varphi(\boldsymbol{X},t)$ *is an admissible motion and* $\Theta(\boldsymbol{X},t)$ *is a material temperature field. A process is called* **isothermal** *if* $\Theta(\boldsymbol{X},t)$ *is constant independent of* $\boldsymbol{X}$ *and* t, *equivalently, if the spatial temperature* $\theta(\boldsymbol{x},t)$ *is constant independent of* $\boldsymbol{x}$ *and* t. □

Our basic assumption in the isothermal modeling of a continuum body is that the body can experience only isothermal processes. In particular, for every motion φ we assume the material temperature Θ is constant for all $\boldsymbol{X} \in B$ and $t \geq 0$. This assumption is physically reasonable for bodies with negligible internal and external sources of heat. Internal sources of heat are expected to be negligible when the body is subject to mild rates of strain. External sources of heat are expected to be negligible when the temperature of the body environment is nearly uniform, constant and equal to that of the body. When these conditions are violated the isothermal assumption is no longer expected to be valid.

The balance laws of mass, linear momentum and angular momentum and the axiom of material frame-indifference are fundamental in the isothermal modeling of continuum bodies. The balance law of energy is not directly relevant because it contains quantities, namely the internal energy, heat flux vector and heat supply, which are typically left unspecified in isothermal models. If any two of these quantities were to be known, then the energy balance law could be used to gain information about the third. For simplicity, in our analysis of isothermal models in Chapters 6 and 7 we leave all thermal quantities including entropy unspecified and make no reference to the energy balance law. At a first glance it appears that the Clausius–Duhem inequality (Second Law of Thermodynamics) is also not directly relevant. However, we show below that it leads to a purely mechanical energy inequality for an important class of bodies.

Definition 5.28 *A thermo-mechanical process* (φ, Θ) *for a body is said to be* **closed** *in an interval* $[t_0, t_1]$ *if* $\varphi(\boldsymbol{X}, t_1) = \varphi(\boldsymbol{X}, t_0)$, $\dot{\varphi}(\boldsymbol{X}, t_1) = \dot{\varphi}(\boldsymbol{X}, t_0)$ *and* $\Theta(\boldsymbol{X}, t_1) = \Theta(\boldsymbol{X}, t_0)$ *for all* $\boldsymbol{X} \in B$. *A body is called* **energetically passive** *if for every closed process its material free energy density satisfies* $\Psi(\boldsymbol{X}, t_1) - \Psi(\boldsymbol{X}, t_0) \geq 0$ *for all* $\boldsymbol{X} \in B$. □

Thus a process is closed in an interval $[t_0, t_1]$ if all particles in a body have the same position, velocity and temperature at time t_1 as they had at time t_0. Moreover, a body is energetically passive if the net change in its free energy density is non-negative in any closed process. For example, when Ψ depends only on the current value of the fields φ, $\dot{\varphi}$ and Θ, and not on their past history, the net change in Ψ is zero. The next result is a consequence of the reduced Clausius–Duhem inequality in either Lagrangian or Eulerian form (Results 5.20 and 5.12).

Result 5.29 ***Mechanical Energy Inequality.*** *Consider an energetically passive body with reference configuration* B. *Then for any closed* *isothermal process in an interval* $[t_0, t_1]$ *we must have*

$$\int_{t_0}^{t_1} \boldsymbol{P}(\boldsymbol{X},t) : \dot{\boldsymbol{F}}(\boldsymbol{X},t)\, dt \geq 0, \qquad \forall \boldsymbol{X} \in B, \tag{5.41}$$

where $\boldsymbol{P}(\boldsymbol{X},t)$ *is the first Piola–Kirchhoff stress field and* $\boldsymbol{F}(\boldsymbol{X},t)$ *is the deformation gradient. Equivalently*

$$\int_{t_0}^{t_1} (\det \boldsymbol{F}(\boldsymbol{X},t))\, \boldsymbol{S}_m(\boldsymbol{X},t) : \boldsymbol{L}_m(\boldsymbol{X},t)\, dt \geq 0, \qquad \forall \boldsymbol{X} \in B, \tag{5.42}$$

where $\boldsymbol{S}(\boldsymbol{x},t)$ *is the Cauchy stress field and* $\boldsymbol{L}(\boldsymbol{x},t)$ *is the rate of strain field.* □

Proof For any thermo-mechanical process the reduced Clausius–Duhem inequality in Lagrangian form (Result 5.20) states

$$\rho_0 \dot{\Psi} \leq \boldsymbol{P} : \dot{\boldsymbol{F}} - \rho_0 \eta_m \dot{\Theta} - \Theta^{-1} \boldsymbol{Q} \cdot \nabla^X \Theta, \quad \forall \boldsymbol{X} \in B,\ t \geq 0.$$

If the process is isothermal, we get $\dot{\Theta} = 0$ and $\nabla^X \Theta = \boldsymbol{0}$, and integrating over a time interval $[t_0, t_1]$ gives

$$\rho_0(\boldsymbol{X})[\Psi(\boldsymbol{X},t_1) - \Psi(\boldsymbol{X},t_0)] \leq \int_{t_0}^{t_1} \boldsymbol{P}(\boldsymbol{X},t) : \dot{\boldsymbol{F}}(\boldsymbol{X},t)\, dt.$$

Additionally, if the process is closed and the body is energetically passive, the left-hand side is non-negative, which yields the result in (5.41). Similarly, for any process the reduced Clausius–Duhem inequality in Eulerian form (Result 5.12) states

$$\rho \dot{\psi} \leq \boldsymbol{S} : \boldsymbol{L} - \rho \eta \dot{\theta} - \theta^{-1} \boldsymbol{q} \cdot \nabla^x \theta, \quad \forall \boldsymbol{x} \in B_t,\ t \geq 0.$$

If the process is isothermal, we get $\dot{\theta} = 0$ and $\nabla^x \theta = \boldsymbol{0}$, which implies

$$\rho \dot{\psi} \leq \boldsymbol{S} : \boldsymbol{L}, \quad \forall \boldsymbol{x} \in B_t,\ t \geq 0.$$

In material form this reads

$$\rho_m \dot{\Psi} \leq \boldsymbol{S}_m : \boldsymbol{L}_m, \quad \forall \boldsymbol{X} \in B,\ t \geq 0,$$

where $\Psi = \psi_m$. By conservation of mass (Result 5.13) we have $\rho_m = \rho_0 / \det \boldsymbol{F}$. Substituting this into the above inequality and integrating as before leads to the result in (5.42). □

Result 5.29 implies that, in any closed isothermal process, an energetically passive body may consume mechanical energy but cannot create it. In particular, for any open subset Ω_t of B_t we find

$$\int_{t_0}^{t_1} \mathcal{P}[\Omega_t]\, dt \geq 0, \tag{5.43}$$

where $\mathcal{P}[\Omega_t]$ is the power of external forces on Ω_t. This inequality may be deduced by integrating (5.41) over an arbitrary subset Ω of B and using Result 5.17 together with the definition of a closed process. Equivalently, we may integrate (5.42) and use Result 5.9. The integral in (5.43) corresponds to the work of external forces on Ω_t. The fact that this integral is non-negative implies that Ω_t can only consume energy.

As with the Clausius–Duhem inequality, we view Result 5.29 as a constitutive restriction. In particular, under the assumption that a body is energetically passive, the constitutive equation for the stress field must guarantee that the mechanical energy inequality is satisfied in every closed isothermal process. Thus in the isothermal modeling of continuum bodies we have three basic balance laws pertaining to mass, linear momentum and angular momentum. Moreover, we have two basic constitutive restrictions pertaining to material frame-indifference and the mechanical energy inequality. Notice that the latter applies only to a particular class of bodies and processes.

Bibliographic Notes

Much of the material presented here on the mechanical balance laws can be found in Truesdell (1991) and Gurtin (1981), and much of the material on the thermal balance laws can be found in Truesdell (1984). A comprehensive account of both the mechanical and thermal balance laws can be found in the treatise by Truesdell and Toupin (1960). More concise accounts can be found in Šilhavý (1997), Antman (1995) and Marsden and Hughes (1983) from a mathematical point of view, and in Spencer (1992), Mase (1970) and Malvern (1969) from an engineering point of view.

The definition of entropy and the form of the Second Law presented here is adapted from Truesdell (1984). That text contains a historical overview of these concepts beginning from the classic theory of homogeneous, reversible systems to the modern theory of continuum mechanics. In particular, various issues surrounding the generalization of the Second Law from the classic theory to the present, and its role as a restriction on constitutive relations, are highlighted there. The role of statistical theories in the development of modern thermodynamics is also discussed.

The axiom on material frame-indifference presented here asserts that various quantities associated with a continuum body are invariant under changes of observer as represented by superposed rigid motions. We interpret this axiom as placing restrictions on possible constitutive relations for such quantities. In many texts, this axiom is stated directly as a restriction on the constitutive relations. These two views are mathematically equivalent. The one adopted here is convenient because it does not require a discussion of constitutive theory as a prerequisite. Some authors choose to state the axiom of frame-indifference using superposed motions defined by general orthogonal as opposed to proper orthogonal (rotation) tensors. For a discussion see Truesdell (1991).

Throughout our developments we have assumed that the underlying fields are smooth. A discussion of the balance laws of linear and angular momentum in the non-smooth case can be found in Antman (1995) and references therein. A comprehensive treatment of jump conditions across surfaces of discontinuity in continuum mechanics is given in Šilhavý (1997) and Truesdell and Toupin (1960).

Our presentation on material constraints is adapted from Truesdell (1991). A detailed account of this theory can be found in Antman (1995). The result on the mechanical energy inequality given here is based on the discussion of internal dissipation given in Truesdell (1984).

Exercises

5.1 Let $W_{12}(\boldsymbol{x}, \boldsymbol{y})$ and $W_{21}(\boldsymbol{y}, \boldsymbol{x})$ be scalar-valued functions of two vector variables $\boldsymbol{x}$ and $\boldsymbol{y}$, and let $\boldsymbol{r} = \boldsymbol{x} - \boldsymbol{y}$.

(a) Show $\nabla^{\boldsymbol{x}}|\boldsymbol{r}| = \dfrac{\boldsymbol{r}}{|\boldsymbol{r}|}$ and $\nabla^{\boldsymbol{y}}|\boldsymbol{r}| = -\dfrac{\boldsymbol{r}}{|\boldsymbol{r}|}$.

(b) Assuming $W_{12}(\boldsymbol{x}, \boldsymbol{y}) = \widehat{W}(|\boldsymbol{r}|) = W_{21}(\boldsymbol{y}, \boldsymbol{x})$ show

$$\nabla^{\boldsymbol{x}} W_{12} = \frac{\widehat{W}'(|\boldsymbol{r}|)}{|\boldsymbol{r}|}\boldsymbol{r}, \qquad \nabla^{\boldsymbol{y}} W_{21} = -\frac{\widehat{W}'(|\boldsymbol{r}|)}{|\boldsymbol{r}|}\boldsymbol{r}.$$

Here $\widehat{W}'(s)$ denotes the derivative of $\widehat{W}(s)$.

5.2 Use the results of Exercise 1 to show that (5.1) implies the particle balance laws in (5.3)–(5.6) under the stated assumptions on the interaction energies, namely

$$U_{ij}(\boldsymbol{x}_i, \boldsymbol{x}_j) = \widehat{U}_{ij}(|\boldsymbol{x}_i - \boldsymbol{x}_j|) = U_{ji}(\boldsymbol{x}_j, \boldsymbol{x}_i), \qquad \text{(no sum)}.$$

5.3 Use Axiom 5.2 and the definitions of $\boldsymbol{j}[\Omega_t]_{\boldsymbol{z}}$, $\boldsymbol{l}[\Omega_t]$, $\boldsymbol{\tau}[\Omega_t]_{\boldsymbol{z}}$ and $\boldsymbol{r}[\Omega_t]$ to show that the angular momentum of a body Ω_t satisfies

$$\frac{d}{dt}\boldsymbol{j}[\Omega_t]_{\boldsymbol{z}} = \boldsymbol{\tau}[\Omega_t]_{\boldsymbol{z}},$$

where $\boldsymbol{z}$ is any fixed point.

5.4 Use Axioms 5.1 and 5.2 to show that

$$M\dot{\boldsymbol{x}}_{\text{com}} = \boldsymbol{l}[\Omega_t], \qquad M\ddot{\boldsymbol{x}}_{\text{com}} = \boldsymbol{r}[\Omega_t],$$

where M is the mass and $\boldsymbol{x}_{\text{com}}$ is the center of mass of Ω_t.

Remark: The second result is known as the **Center of Mass Theorem**. It states that the motion of the center of mass of an arbitrary body Ω_t is the same as that of a particle having the same mass as Ω_t, located at the center of mass of Ω_t and subject to the same resultant force as Ω_t.

5.5 Use the results of Exercise 4 to extend the result of Exercise 3. In particular, show that the angular momentum of a body Ω_t also satisfies

$$\frac{d}{dt}\boldsymbol{j}[\Omega_t]_{\boldsymbol{x}_{\text{com}}} = \boldsymbol{\tau}[\Omega_t]_{\boldsymbol{x}_{\text{com}}},$$

where $\boldsymbol{x}_{\text{com}}$ is the center of mass of Ω_t.

5.6 Rather than being constant as asserted in Axiom 5.1, suppose the mass of an arbitrary open subset Ω_t of B_t satisfies the growth law

$$\frac{d}{dt}\,\text{mass}[\Omega_t] = g[\Omega_t], \tag{5.44}$$

where $g[\Omega_t]$ is a net growth rate of the form

$$g[\Omega_t] = \int_{\Omega_t} \gamma(\boldsymbol{x},t)\rho(\boldsymbol{x},t)\,dV_{\boldsymbol{x}}.$$

Here $\gamma(\boldsymbol{x},t)$ is a prescribed spatial growth rate per unit mass. Show that the local Eulerian form of (5.44) is

$$\frac{\partial}{\partial t}\rho + \nabla^x \cdot (\rho\boldsymbol{v}) = \gamma\rho \quad \forall \boldsymbol{x} \in B_t,\ t \geq 0.$$

5.7 Here we generalize our basic definitions of angular momentum and torque and study the resulting form of the balance of angular momentum equation. In particular, suppose the angular momentum (about origin) of an arbitrary open subset Ω_t is given by

$$\boldsymbol{j}[\Omega_t]_{\boldsymbol{o}} = \int_{\Omega_t} \rho\boldsymbol{\eta} + \boldsymbol{x} \times \rho\boldsymbol{v}\, dV_{\boldsymbol{x}},$$

where $\boldsymbol{\eta}(\boldsymbol{x},t)$ is an intrinsic angular momentum field per unit mass, and suppose the resultant torque (about origin) of external influences on Ω_t is given by

$$\boldsymbol{\tau}[\Omega_t]_{\boldsymbol{o}} = \int_{\Omega_t} \rho\boldsymbol{h} + \boldsymbol{x} \times \rho\boldsymbol{b}\, dV_{\boldsymbol{x}} + \int_{\partial\Omega_t} \boldsymbol{m} + \boldsymbol{x} \times \boldsymbol{t}\, dA_{\boldsymbol{x}},$$

where $\boldsymbol{h}(\boldsymbol{x},t)$ is an intrinsic body torque field per unit mass and $\boldsymbol{m}(\boldsymbol{x},t)$ is an intrinsic surface torque field per unit area. Notice that the usual definitions of angular momentum and torque are recovered in the case when $\boldsymbol{\eta}$, $\boldsymbol{h}$ and $\boldsymbol{m}$ all vanish. In direct analogy to the relation $\boldsymbol{t} = \boldsymbol{S}\boldsymbol{n}$ for surface traction, we assume $\boldsymbol{m} = \boldsymbol{M}\boldsymbol{n}$ where $\boldsymbol{M}(\boldsymbol{x},t)$ is a second-order tensor field analogous to the Cauchy stress $\boldsymbol{S}(\boldsymbol{x},t)$.

(a) Show that the local Eulerian form of the law of angular momentum in Axiom 5.2 is

$$\rho\dot{\boldsymbol{\eta}} = \nabla^x \cdot \boldsymbol{M} + \boldsymbol{\xi} + \rho\boldsymbol{h}, \quad \forall \boldsymbol{x} \in B_t,\ t \geq 0,$$

where $\boldsymbol{\xi}(\boldsymbol{x},t) = \epsilon_{ijk} S_{kj}(\boldsymbol{x},t)\boldsymbol{e}_i$. (Hint: The linear momentum equation is unchanged.)

(b) Assuming $\boldsymbol{M}$ and $\boldsymbol{h}$ are identically zero, show that $\dot{\boldsymbol{\eta}}(\boldsymbol{x},t) = \boldsymbol{0}$ if and only if $\boldsymbol{S}(\boldsymbol{x},t)$ is symmetric.

5.8 Let $\boldsymbol{\varphi} : B \times [0,\infty) \to \mathbb{E}^3$ be a motion of a continuum body with spatial mass density field $\rho(\boldsymbol{x},t)$ and spatial velocity field $\boldsymbol{v}(\boldsymbol{x},t)$. Let $\psi(\boldsymbol{x},t)$ and $\phi(\boldsymbol{x},t)$ be arbitrary scalar fields and let $\boldsymbol{w}(\boldsymbol{x},t)$ be an arbitrary vector field defined on B_t.

(a) Show $(\ln\psi)^{\bullet} = \psi^{-1}\dot{\psi}$ and $(\psi\phi)^{\bullet} = \dot{\psi}\phi + \psi\dot{\phi}$, where a dot denotes the total or material time derivative.

(b) Use part (a) and Result 5.5 to show

$$(\rho^{-1}\phi)^{\bullet} = \rho^{-1}\left[\frac{\partial}{\partial t}\phi + \nabla^x \cdot (\phi\boldsymbol{v})\right].$$

(c) Use the result in (b) to show

$$(\rho^{-1}\boldsymbol{w})^{\bullet} = \rho^{-1}\left[\frac{\partial}{\partial t}\boldsymbol{w} + \nabla^x \cdot (\boldsymbol{w} \otimes \boldsymbol{v})\right].$$

5.9 Consider a material model where the Cauchy stress tensor is given by

$$\boldsymbol{S}(\boldsymbol{x},t) = -p(\boldsymbol{x},t)\boldsymbol{I}.$$

Here $p(\boldsymbol{x},t)$ is a scalar field called pressure. Show that Results 5.5, 5.7 and 5.10 imply

$$\begin{aligned}
&\frac{\partial \rho}{\partial t} + \nabla^x \cdot (\rho \boldsymbol{v}) = 0,\\
&\frac{\partial}{\partial t}(\rho \boldsymbol{v}) + \nabla^x \cdot (\rho \boldsymbol{v} \otimes \boldsymbol{v} + p\boldsymbol{I}) = \rho \boldsymbol{b},\\
&\frac{\partial}{\partial t}(\rho E) + \nabla^x \cdot (\rho E \boldsymbol{v} + p\boldsymbol{v}) = \rho r - \nabla^x \cdot \boldsymbol{q} + \rho \boldsymbol{b} \cdot \boldsymbol{v},
\end{aligned}$$

where $E = \phi + \frac{1}{2}|\boldsymbol{v}|^2$ is the total energy density field. Notice that, once constitutive equations relating $(p, \boldsymbol{q}, \phi)$ to $(\rho, \boldsymbol{v}, \theta)$ are specified, we obtain a closed system of equations for $(\rho, \boldsymbol{v}, \theta)$.

5.10 Show that the stress power $\boldsymbol{P} : \dot{\boldsymbol{F}}$ per unit reference volume can be written as

$$\boldsymbol{P} : \dot{\boldsymbol{F}} = \tfrac{1}{2}\boldsymbol{\Sigma} : \dot{\boldsymbol{C}},$$

where $\boldsymbol{\Sigma}$ is the second Piola–Kirchhoff stress tensor and $\boldsymbol{C}$ is the right Cauchy strain tensor (see Chapter 4).

5.11 Derive Result 5.20.

5.12 Here we study the energy balance equation in Result 5.18 and the Clausius–Duhem inequality in Result 5.20 for a body undergoing an arbitrary rigid motion

$$\boldsymbol{\varphi}(\boldsymbol{X},t) = \boldsymbol{\Lambda}(t)\boldsymbol{X} + \boldsymbol{c}(t), \quad \forall \boldsymbol{X} \in B,\ t \geq 0,$$

where $\boldsymbol{\Lambda}(t)$ is a rotation and $\boldsymbol{c}(t)$ is a vector. Notice that $\boldsymbol{\Lambda}(t) = \boldsymbol{I}$ and $\boldsymbol{c}(t) = \boldsymbol{0}$ for all $t \geq 0$ corresponds to the case of a body at rest in its reference configuration B.

(a) Use the result of Exercise 10 to show that the stress power $\boldsymbol{P} : \dot{\boldsymbol{F}}$ per unit reference volume vanishes identically for an arbitrary rigid motion.

(b) Suppose the internal energy density and heat flux fields are given by constitutive relations of the form $\Phi = \alpha\Theta$ and $\boldsymbol{Q} =$

$-\kappa\nabla^{X}\Theta$, where α and κ are scalar constants. Show that the energy balance equation reduces to

$$\rho_0\alpha\frac{\partial\Theta}{\partial t} = \kappa\Delta^{X}\Theta + \rho_0 R, \quad \forall \boldsymbol{X} \in B,\ t \geq 0.$$

This is a linear partial differential equation for $\Theta(\boldsymbol{X}, t)$ known as the **Heat Equation**.

(c) Suppose the constitutive relation for Φ is dropped and replaced by constitutive relations for the free energy Ψ and entropy η_m of the form

$$\Psi = \widehat{\Psi}(\Theta), \qquad \eta_m = -\frac{d\widehat{\Psi}}{d\Theta}(\Theta),$$

where $\widehat{\Psi}$ is a given function. In this case show that the energy balance equation reduces to

$$-\rho_0\Theta\frac{d^2\widehat{\Psi}}{d\Theta^2}(\Theta)\frac{\partial\Theta}{\partial t} = \kappa\Delta^{X}\Theta + \rho_0 R, \quad \forall \boldsymbol{X} \in B,\ t \geq 0.$$

This is a nonlinear partial differential equation for $\Theta(\boldsymbol{X}, t)$.

(d) Assuming constitutive relations as in part (c), show that the Clausius–Duhem inequality in Result 5.20 is satisfied for arbitrary rigid motions and temperature fields only if $\kappa \geq 0$. Thus a model with $\kappa < 0$ would violate the Second Law of Thermodynamics.

5.13 Derive Result 5.22.

5.14 With the aid of Result 5.22 determine which of the following constitutive equations for the Cauchy stress field $\boldsymbol{S}$ are frame-indifferent. Here μ, λ and γ are arbitrary constants.

(a) $\boldsymbol{S} = 2\mu\nabla^{x}\boldsymbol{v}$. (b) $\boldsymbol{S} = \mu\boldsymbol{L}$. (c) $\boldsymbol{S} = \lambda\rho^{\gamma}\boldsymbol{I}$.

5.15 Consider a material body with spatial velocity field $\boldsymbol{v}$ and associated spin tensor field $\boldsymbol{W} = \frac{1}{2}(\nabla^{x}\boldsymbol{v} - \nabla^{x}\boldsymbol{v}^T)$. Let $\boldsymbol{T}$ be an arbitrary frame-indifferent spatial tensor field on the body, so $\boldsymbol{T}^* = \boldsymbol{Q}\boldsymbol{T}\boldsymbol{Q}^T$ for any superposed rigid motion defined by $\boldsymbol{Q}(t)$. Let $\boldsymbol{J}_{\text{ma}}(\boldsymbol{T})$ denote the material or total time derivative of $\boldsymbol{T}$ defined by

$$\boldsymbol{J}_{\text{ma}}(\boldsymbol{T}) = \dot{\boldsymbol{T}},$$

and let $\boldsymbol{J}_{\text{co}}(\boldsymbol{T}, \boldsymbol{W})$ denote the co-rotational or Jaumann time derivative of $\boldsymbol{T}$ with respect to $\boldsymbol{W}$ defined by (see Chapter 2, Exercise 14)

$$\boldsymbol{J}_{\text{co}}(\boldsymbol{T}, \boldsymbol{W}) = \dot{\boldsymbol{T}} + \boldsymbol{W}^T \boldsymbol{T} + \boldsymbol{T}\boldsymbol{W}.$$

(a) For any superposed rigid motion defined by a rotation tensor $\boldsymbol{Q}(t)$ show that

$$\boldsymbol{W}^* = \boldsymbol{Q}\boldsymbol{W}\boldsymbol{Q}^T + \dot{\boldsymbol{Q}}\boldsymbol{Q}^T.$$

(b) Show that $\boldsymbol{J}_{\text{co}}(\boldsymbol{T}, \boldsymbol{W})$ is frame-indifferent in the sense that

$$\boldsymbol{J}_{\text{co}}(\boldsymbol{T}^*, \boldsymbol{W}^*) = \boldsymbol{Q}\boldsymbol{J}_{\text{co}}(\boldsymbol{T}, \boldsymbol{W})\boldsymbol{Q}^T,$$

whereas $\boldsymbol{J}_{\text{ma}}(\boldsymbol{T})$ is not frame-indifferent in the sense that

$$\boldsymbol{J}_{\text{ma}}(\boldsymbol{T}^*) \neq \boldsymbol{Q}\boldsymbol{J}_{\text{ma}}(\boldsymbol{T})\boldsymbol{Q}^T.$$

Answers to Selected Exercises

5.1 (a) In any coordinate frame $\{\boldsymbol{e}_k\}$ we have $|\boldsymbol{r}|^2 = \sum_{q=1}^{3}(x_q - y_q)^2$, which gives

$$\frac{\partial |\boldsymbol{r}|^2}{\partial x_k} = 2(x_k - y_k).$$

Moreover, by the chain rule, the left-hand side is $2|\boldsymbol{r}|\partial|\boldsymbol{r}|/\partial x_k$. Thus

$$\frac{\partial |\boldsymbol{r}|}{\partial x_k} = \frac{(x_k - y_k)}{|\boldsymbol{r}|} \quad \text{and} \quad \nabla^{\boldsymbol{x}}|\boldsymbol{r}| = \sum_{k=1}^{3} \frac{\partial |\boldsymbol{r}|}{\partial x_k}\boldsymbol{e}_k = \frac{\boldsymbol{r}}{|\boldsymbol{r}|}.$$

The second result follows in a similar way, namely

$$\frac{\partial |\boldsymbol{r}|}{\partial y_k} = -\frac{(x_k - y_k)}{|\boldsymbol{r}|} \quad \text{and} \quad \nabla^{\boldsymbol{y}}|\boldsymbol{r}| = \sum_{k=1}^{3} \frac{\partial |\boldsymbol{r}|}{\partial y_k}\boldsymbol{e}_k = -\frac{\boldsymbol{r}}{|\boldsymbol{r}|}.$$

(b) From the relation $W_{12}(\boldsymbol{x}, \boldsymbol{y}) = \widehat{W}(|\boldsymbol{r}|)$ and the chain rule we get

$$\nabla^{\boldsymbol{x}} W_{12} = \sum_{k=1}^{3} \frac{\partial W_{12}}{\partial x_k}\boldsymbol{e}_k = \sum_{k=1}^{3} \widehat{W}' \frac{\partial |\boldsymbol{r}|}{\partial x_k}\boldsymbol{e}_k = \frac{\widehat{W}'}{|\boldsymbol{r}|}\boldsymbol{r}.$$

Similarly, since $W_{21}(\boldsymbol{y}, \boldsymbol{x}) = \widehat{W}(|\boldsymbol{r}|)$ we get

$$\nabla^{\boldsymbol{y}} W_{21} = \sum_{k=1}^{3} \frac{\partial W_{21}}{\partial y_k}\boldsymbol{e}_k = \sum_{k=1}^{3} \widehat{W}' \frac{\partial |\boldsymbol{r}|}{\partial y_k}\boldsymbol{e}_k = -\frac{\widehat{W}'}{|\boldsymbol{r}|}\boldsymbol{r}.$$

5.3 From the definitions of $\boldsymbol{j}[\Omega_t]_{\boldsymbol{z}}$, $\boldsymbol{l}[\Omega_t]$, $\boldsymbol{\tau}[\Omega_t]_{\boldsymbol{z}}$ and $\boldsymbol{r}[\Omega_t]$ we deduce

$$\boldsymbol{j}[\Omega_t]_{\boldsymbol{z}} = \boldsymbol{j}[\Omega_t]_{\boldsymbol{o}} - \boldsymbol{z} \times \boldsymbol{l}[\Omega_t], \qquad \boldsymbol{\tau}[\Omega_t]_{\boldsymbol{z}} = \boldsymbol{\tau}[\Omega_t]_{\boldsymbol{o}} - \boldsymbol{z} \times \boldsymbol{r}[\Omega_t].$$

Differentiating the first equation and subtracting the second gives

$$\begin{aligned}&\frac{d}{dt}\boldsymbol{j}[\Omega_t]_{\boldsymbol{z}} - \boldsymbol{\tau}[\Omega_t]_{\boldsymbol{z}} \\ &\qquad = \frac{d}{dt}\boldsymbol{j}[\Omega_t]_{\boldsymbol{o}} - \boldsymbol{\tau}[\Omega_t]_{\boldsymbol{o}} - \frac{d}{dt}\Big(\boldsymbol{z} \times \boldsymbol{l}[\Omega_t]\Big) + \boldsymbol{z} \times \boldsymbol{r}[\Omega_t].\end{aligned}$$

The desired result follows from the laws of inertia and the fact that $\boldsymbol{z}$ is constant.

5.5 Proceeding just as in Exercise 3 we obtain

$$\begin{aligned}&\frac{d}{dt}\boldsymbol{j}[\Omega_t]_{\boldsymbol{x}_{\mathrm{com}}} - \boldsymbol{\tau}[\Omega_t]_{\boldsymbol{x}_{\mathrm{com}}} \\ &\qquad = \frac{d}{dt}\boldsymbol{j}[\Omega_t]_{\boldsymbol{o}} - \boldsymbol{\tau}[\Omega_t]_{\boldsymbol{o}} - \frac{d}{dt}\Big(\boldsymbol{x}_{\mathrm{com}} \times \boldsymbol{l}[\Omega_t]\Big) + \boldsymbol{x}_{\mathrm{com}} \times \boldsymbol{r}[\Omega_t].\end{aligned}$$

Expanding the derivative of the cross product and using the laws of inertia to eliminate terms we get

$$\frac{d}{dt}\boldsymbol{j}[\Omega_t]_{\boldsymbol{x}_{\mathrm{com}}} - \boldsymbol{\tau}[\Omega_t]_{\boldsymbol{x}_{\mathrm{com}}} = -\dot{\boldsymbol{x}}_{\mathrm{com}} \times \boldsymbol{l}[\Omega_t].$$

The result follows from the fact that $\dot{\boldsymbol{x}}_{\mathrm{com}} \times \boldsymbol{l}[\Omega_t] = \boldsymbol{0}$. In particular, $\dot{\boldsymbol{x}}_{\mathrm{com}}$ is parallel to $\boldsymbol{l}[\Omega_t]$ by the results of Exercise 4.

5.7 (a) For any open subset Ω_t of B_t the axiom on angular momentum states

$$\frac{d}{dt}\boldsymbol{j}[\Omega_t]_{\boldsymbol{o}} = \boldsymbol{\tau}[\Omega_t]_{\boldsymbol{o}}.$$

From the given expression for $\boldsymbol{\tau}[\Omega_t]_{\boldsymbol{o}}$, together with the relations $\boldsymbol{t} = \boldsymbol{S}\boldsymbol{n}$ and $\boldsymbol{m} = \boldsymbol{M}\boldsymbol{n}$, we have (omitting arguments $\boldsymbol{x}$ and t for brevity)

$$\boldsymbol{\tau}[\Omega_t]_{\boldsymbol{o}} = \int_{\Omega_t} \rho\boldsymbol{h} + \boldsymbol{x} \times \rho\boldsymbol{b}\, dV_{\boldsymbol{x}} + \int_{\partial\Omega_t} \boldsymbol{M}\boldsymbol{n} + \boldsymbol{x} \times \boldsymbol{S}\boldsymbol{n}\, dA_{\boldsymbol{x}}.$$

By the Divergence Theorem we note that

$$\begin{aligned}\int_{\partial\Omega_t} \boldsymbol{x} \times \boldsymbol{S}\boldsymbol{n}\, dA_{\boldsymbol{x}} &= \int_{\partial\Omega_t} \epsilon_{ijk} x_j S_{kl} n_l \boldsymbol{e}_i \, dA_{\boldsymbol{x}} \\ &= \int_{\Omega_t} (\epsilon_{ijk} x_j S_{kl})_{,l} \boldsymbol{e}_i \, dV_{\boldsymbol{x}} \\ &= \int_{\Omega_t} (\epsilon_{ijk} S_{kj} + \epsilon_{ijk} x_j S_{kl,l}) \boldsymbol{e}_i \, dV_{\boldsymbol{x}} \\ &= \int_{\Omega_t} \boldsymbol{\xi} + \boldsymbol{x} \times \nabla^x \cdot \boldsymbol{S} \, dV_{\boldsymbol{x}},\end{aligned}$$

where we introduce $\boldsymbol{\xi} = \epsilon_{ijk} S_{kj} \boldsymbol{e}_i$ for convenience. From the definition of $\boldsymbol{j}[\Omega_t]_{\boldsymbol{o}}$ and the result on time derivatives of integrals relative to mass density we get

$$\frac{d}{dt}\boldsymbol{j}[\Omega_t]_{\boldsymbol{o}} = \int_{\Omega_t} \rho\dot{\boldsymbol{\eta}} + \rho(\dot{\boldsymbol{x}} \times \boldsymbol{v} + \boldsymbol{x} \times \dot{\boldsymbol{v}}) \, dV_{\boldsymbol{x}}.$$

Substituting the above result into the axiom of angular momentum and using the fact that $\dot{\boldsymbol{x}} \times \boldsymbol{v} = \boldsymbol{0}$ we get

$$\begin{aligned}&\int_{\Omega_t} \rho\dot{\boldsymbol{\eta}} + \boldsymbol{x} \times \rho\dot{\boldsymbol{v}} \, dV_{\boldsymbol{x}} \\ &\qquad = \int_{\Omega_t} \rho\boldsymbol{h} + \boldsymbol{x} \times \rho\boldsymbol{b} \, dV_{\boldsymbol{x}} + \int_{\partial\Omega_t} \boldsymbol{M}\boldsymbol{n} + \boldsymbol{x} \times \boldsymbol{S}\boldsymbol{n} \, dA_{\boldsymbol{x}}.\end{aligned}$$

Using the Divergence Theorem on the surface integrals and collecting terms we arrive at

$$\int_{\Omega_t} \rho\dot{\boldsymbol{\eta}} - \nabla^x \cdot \boldsymbol{M} - \boldsymbol{\xi} - \rho\boldsymbol{h} + \boldsymbol{x} \times [\rho\dot{\boldsymbol{v}} - \nabla^x \cdot \boldsymbol{S} - \rho\boldsymbol{b}] \, dV_{\boldsymbol{x}} = \boldsymbol{0}.$$

Because our definitions of linear momentum and resultant force have not changed, the axiom of linear momentum still implies $\rho\dot{\boldsymbol{v}} = \nabla^x \cdot \boldsymbol{S} + \rho\boldsymbol{b}$. Using this fact, together with the Localization Theorem, we conclude

$$\rho\dot{\boldsymbol{\eta}} = \nabla^x \cdot \boldsymbol{M} + \boldsymbol{\xi} + \rho\boldsymbol{h}, \quad \forall \boldsymbol{x} \in B_t,\ t \geq 0,$$

which is the desired result.

(b) In the case when $\boldsymbol{M}$ and $\boldsymbol{h}$ are identically zero we get $\rho\dot{\boldsymbol{\eta}} = \boldsymbol{\xi}$. Thus, for any $\boldsymbol{x} \in B_t$ and $t \geq 0$ we deduce that $\dot{\boldsymbol{\eta}}(\boldsymbol{x}, t) = \boldsymbol{0}$ if and only if $\boldsymbol{\xi}(\boldsymbol{x}, t) = \boldsymbol{0}$. Just as in the analysis of the equilibrium equations in Chapter 3, we find that $\boldsymbol{\xi}(\boldsymbol{x}, t) = \boldsymbol{0}$ if and only if $\boldsymbol{S}(\boldsymbol{x}, t)$ is symmetric.

5.9 The first equation is simply the mass conservation equation. To establish the second we notice that

$$\begin{aligned}
&\frac{\partial}{\partial t}(\rho \boldsymbol{v}) + \nabla^x \cdot (\rho \boldsymbol{v} \otimes \boldsymbol{v} + p\boldsymbol{I}) \\
&= \frac{\partial}{\partial t}(\rho \boldsymbol{v}) + \nabla^x \cdot (\boldsymbol{v} \otimes (\rho \boldsymbol{v}) + p\boldsymbol{I}) \\
&= \frac{\partial \rho}{\partial t}\boldsymbol{v} + \rho \frac{\partial \boldsymbol{v}}{\partial t} + (\nabla^x \boldsymbol{v})\rho \boldsymbol{v} + \nabla^x \cdot (\rho \boldsymbol{v})\boldsymbol{v} + \nabla^x p \\
&= \left(\frac{\partial \rho}{\partial t} + \nabla^x \cdot (\rho \boldsymbol{v})\right)\boldsymbol{v} + \nabla^x p + \rho\left(\frac{\partial \boldsymbol{v}}{\partial t} + (\nabla^x \boldsymbol{v})\boldsymbol{v}\right) \\
&= \nabla^x p + \rho \dot{\boldsymbol{v}},
\end{aligned}$$

where the last line follows from mass conservation. Using $\boldsymbol{S} = -p\boldsymbol{I}$ and the balance of linear momentum equation we obtain

$$\frac{\partial}{\partial t}(\rho \boldsymbol{v}) + \nabla^x \cdot (\rho \boldsymbol{v} \otimes \boldsymbol{v} + p\boldsymbol{I}) = -\nabla^x \cdot \boldsymbol{S} + \rho \dot{\boldsymbol{v}} = \rho \boldsymbol{b},$$

which is the desired result. To establish the third equation we notice that

$$\begin{aligned}
&\frac{\partial}{\partial t}(\rho E) + \nabla^x \cdot (\rho E \boldsymbol{v} + p\boldsymbol{v}) \\
&\qquad = \frac{\partial}{\partial t}(\rho E) + \nabla^x(\rho E) \cdot \boldsymbol{v} + (\rho E)\nabla^x \cdot \boldsymbol{v} + \nabla^x \cdot (p\boldsymbol{v}) \\
&\qquad = (\rho E)^{\bullet} + (\rho E)\nabla^x \cdot \boldsymbol{v} + \nabla^x \cdot (p\boldsymbol{v}) \\
&\qquad = (\dot{\rho} + \rho \nabla^x \cdot \boldsymbol{v})E + \rho \dot{E} - \nabla^x \cdot (\boldsymbol{S}\boldsymbol{v}),
\end{aligned}$$

where the last line follows from $\boldsymbol{S} = -p\boldsymbol{I}$. By mass conservation we have $\dot{\rho} + \rho \nabla^x \cdot \boldsymbol{v} = 0$. Using this result, together with the definition $E = \phi + \frac{1}{2}|\boldsymbol{v}|^2$ and the symmetry of $\boldsymbol{S}$, we obtain

$$\begin{aligned}
&\frac{\partial}{\partial t}(\rho E) + \nabla^x \cdot (\rho E \boldsymbol{v} + p\boldsymbol{v}) \\
&\qquad = \rho(\dot{\phi} + \dot{\boldsymbol{v}} \cdot \boldsymbol{v}) - (\nabla^x \cdot \boldsymbol{S}) \cdot \boldsymbol{v} - \boldsymbol{S} : \nabla^x \boldsymbol{v} \\
&\qquad = (\rho \dot{\phi} - \boldsymbol{S} : \boldsymbol{L}) + (\rho \dot{\boldsymbol{v}} - \nabla^x \cdot \boldsymbol{S}) \cdot \boldsymbol{v}.
\end{aligned}$$

Finally, substituting for $\rho\dot{\phi}$ and $\rho\dot{\boldsymbol{v}}$ from the balance of energy and linear momentum equations we get

$$\frac{\partial}{\partial t}(\rho E) + \nabla^x \cdot (\rho E \boldsymbol{v} + p\boldsymbol{v}) = -\nabla^x \cdot \boldsymbol{q} + \rho r + \rho \boldsymbol{b} \cdot \boldsymbol{v},$$

which is the desired result.

5.11 From the Clausius–Duhem inequality in Lagrangian form we have

$$\rho_0 \dot{\eta}_m \geq \Theta^{-1} \rho_0 R - \nabla^X \cdot (\Theta^{-1} \boldsymbol{Q}).$$

Expanding the divergence term and multiplying through by Θ we obtain

$$\Theta \rho_0 \dot{\eta}_m \geq \rho_0 R - \nabla^X \cdot \boldsymbol{Q} + \Theta^{-1} \boldsymbol{Q} \cdot \nabla^X \Theta. \tag{5.45}$$

From the definition of the free energy we have $\Theta \dot{\eta}_m = \dot{\Phi} - \dot{\Theta} \eta_m - \dot{\Psi}$. Multiplying this expression by ρ_0 and using the balance of energy equation to eliminate $\rho_0 \dot{\Phi}$ we get

$$\rho_0 \Theta \dot{\eta}_m = \boldsymbol{P} : \dot{\boldsymbol{F}} - \nabla^X \cdot \boldsymbol{Q} + \rho_0 R - \rho_0 \dot{\Theta} \eta_m - \rho_0 \dot{\Psi}.$$

Substituting this expression into (5.45) leads to

$$\rho_0 \dot{\Psi} \leq \boldsymbol{P} : \dot{\boldsymbol{F}} - \rho_0 \eta_m \dot{\Theta} - \Theta^{-1} \boldsymbol{Q} \cdot \nabla^X \Theta,$$

which is the desired result.

5.13 Since $\varphi^* = \boldsymbol{Q}(t) \varphi + \boldsymbol{c}(t)$ we have

$$\boldsymbol{F}^*(\boldsymbol{X}, t) = \boldsymbol{Q}(t) \boldsymbol{F}(\boldsymbol{X}, t).$$

By definition of the Cauchy strain tensor we have

$$\boldsymbol{C}^* = (\boldsymbol{F}^*)^T \boldsymbol{F}^* = (\boldsymbol{Q}\boldsymbol{F})^T (\boldsymbol{Q}\boldsymbol{F}) = \boldsymbol{F}^T \boldsymbol{Q}^T \boldsymbol{Q} \boldsymbol{F} = \boldsymbol{F}^T \boldsymbol{F} = \boldsymbol{C},$$

and from this result we obtain

$$\boldsymbol{U}^* = \sqrt{\boldsymbol{C}^*} = \sqrt{\boldsymbol{C}} = \boldsymbol{U}.$$

In view of the relations $\boldsymbol{F}^* = \boldsymbol{R}^* \boldsymbol{U}^*$ and $\boldsymbol{F} = \boldsymbol{R}\boldsymbol{U}$, and the results $\boldsymbol{F}^* = \boldsymbol{Q}\boldsymbol{F}$ and $\boldsymbol{U}^* = \boldsymbol{U}$, we obtain $\boldsymbol{R}^* \boldsymbol{U} = \boldsymbol{Q}\boldsymbol{R}\boldsymbol{U}$. Hence, since $\boldsymbol{U}$ is invertible, we have

$$\boldsymbol{R}^* = \boldsymbol{Q}\boldsymbol{R}.$$

Similarly, given the relations $\boldsymbol{F}^* = \boldsymbol{V}^* \boldsymbol{R}^*$ and $\boldsymbol{F} = \boldsymbol{V}\boldsymbol{R}$, and the results $\boldsymbol{F}^* = \boldsymbol{Q}\boldsymbol{F}$ and $\boldsymbol{R}^* = \boldsymbol{Q}\boldsymbol{R}$, we obtain $\boldsymbol{V}^* \boldsymbol{Q}\boldsymbol{R} = \boldsymbol{Q}\boldsymbol{V}\boldsymbol{R}$. Hence, since $\boldsymbol{R}$ is invertible, we have

$$\boldsymbol{V}^* = \boldsymbol{Q}\boldsymbol{V}\boldsymbol{Q}^T.$$

From Chapter 4 we recall that the spatial velocity and material deformation gradients for a motion φ are related as

$$(\nabla^x \boldsymbol{v})(\boldsymbol{x}, t)\Big|_{\boldsymbol{x} = \varphi(\boldsymbol{X}, t)} = \dot{\boldsymbol{F}}(\boldsymbol{X}, t) \boldsymbol{F}(\boldsymbol{X}, t)^{-1}.$$

Similarly, for the motion φ^* we have

$$(\nabla^{x^*} \boldsymbol{v}^*)(\boldsymbol{x}^*, t)\Big|_{\boldsymbol{x}^* = \varphi^*(\boldsymbol{X}, t)} = \dot{\boldsymbol{F}}^*(\boldsymbol{X}, t)\boldsymbol{F}^*(\boldsymbol{X}, t)^{-1}.$$

Since

$$\dot{\boldsymbol{F}}^*(\boldsymbol{X}, t) = \dot{\boldsymbol{Q}}(t)\boldsymbol{F}(\boldsymbol{X}, t) + \boldsymbol{Q}(t)\dot{\boldsymbol{F}}(\boldsymbol{X}, t),$$

and

$$\begin{aligned}\boldsymbol{F}^*(\boldsymbol{X}, t)^{-1} &= (\boldsymbol{Q}(t)\boldsymbol{F}(\boldsymbol{X}, t))^{-1} \\ &= \boldsymbol{F}(\boldsymbol{X}, t)^{-1}\boldsymbol{Q}(t)^{-1} = \boldsymbol{F}(\boldsymbol{X}, t)^{-1}\boldsymbol{Q}(t)^T,\end{aligned}$$

we have

$$\begin{aligned}&(\nabla^{x^*} \boldsymbol{v}^*)(\boldsymbol{x}^*, t)\Big|_{\boldsymbol{x}^* = \varphi^*(\boldsymbol{X}, t)} \\ &\quad = \left[\dot{\boldsymbol{Q}}(t)\boldsymbol{F}(\boldsymbol{X}, t) + \boldsymbol{Q}(t)\dot{\boldsymbol{F}}(\boldsymbol{X}, t)\right]\boldsymbol{F}(\boldsymbol{X}, t)^{-1}\boldsymbol{Q}(t)^T \\ &\quad = \dot{\boldsymbol{Q}}(t)\boldsymbol{Q}(t)^T + \boldsymbol{Q}(t)\dot{\boldsymbol{F}}(\boldsymbol{X}, t)\boldsymbol{F}(\boldsymbol{X}, t)^{-1}\boldsymbol{Q}(t)^T \\ &\quad = \dot{\boldsymbol{Q}}(t)\boldsymbol{Q}(t)^T + \boldsymbol{Q}(t)(\nabla^x \boldsymbol{v})(\boldsymbol{x}, t)\Big|_{\boldsymbol{x} = \varphi(\boldsymbol{X}, t)}\boldsymbol{Q}(t)^T.\end{aligned}$$

Using the fact that $\boldsymbol{x}^* = \varphi^*(\boldsymbol{X}, t) = \boldsymbol{g}(\boldsymbol{x}, t)\Big|_{\boldsymbol{x} = \varphi(\boldsymbol{X}, t)}$ we obtain

$$(\nabla^{x^*} \boldsymbol{v}^*)(\boldsymbol{x}^*, t)\Big|_{\boldsymbol{x}^* = \boldsymbol{g}(\boldsymbol{x}, t)} = \boldsymbol{Q}(t)(\nabla^x \boldsymbol{v})(\boldsymbol{x}, t)\boldsymbol{Q}(t)^T + \dot{\boldsymbol{Q}}(t)\boldsymbol{Q}(t)^T,$$

which is the desired result. From the relation $\boldsymbol{Q}(t)\boldsymbol{Q}(t)^T = \boldsymbol{I}$ we deduce $\boldsymbol{O} = \frac{d}{dt}\left\{\boldsymbol{Q}(t)\boldsymbol{Q}(t)^T\right\} = \dot{\boldsymbol{Q}}(t)\boldsymbol{Q}(t)^T + \boldsymbol{Q}(t)\dot{\boldsymbol{Q}}(t)^T$, which implies

$$\dot{\boldsymbol{Q}}(t)\boldsymbol{Q}(t)^T = -\boldsymbol{Q}(t)\dot{\boldsymbol{Q}}(t)^T.$$

From this result and the fact $\dot{\boldsymbol{Q}}(t)^T = \frac{d}{dt}\left\{\boldsymbol{Q}(t)^T\right\} = \left\{\frac{d}{dt}\boldsymbol{Q}(t)\right\}^T$ we obtain

$$\dot{\boldsymbol{Q}}(t)\boldsymbol{Q}(t)^T = -\left\{\dot{\boldsymbol{Q}}(t)\boldsymbol{Q}(t)^T\right\}^T,$$

which implies $\dot{\boldsymbol{Q}}(t)\boldsymbol{Q}(t)^T$ is skew. Finally, from the definition

$L^* = \mathrm{sym}(\nabla^{x^*} v^*)$ we get

$$\begin{aligned}
L^*(x^*,t)\Big|_{x^*=g(x,t)} \\
&= \tfrac{1}{2}(\nabla^{x^*} v^*)(x^*,t)\Big|_{x^*=g(x,t)} + \tfrac{1}{2}((\nabla^{x^*} v^*)(x^*,t))^T\Big|_{x^*=g(x,t)} \\
&= \tfrac{1}{2}Q(t)(\nabla^x v)(x,t)Q(t)^T + \tfrac{1}{2}\dot{Q}(t)Q(t)^T \\
&\quad + \tfrac{1}{2}Q(t)((\nabla^x v)(x,t))^T Q(t)^T + \tfrac{1}{2}\left\{\dot{Q}(t)Q(t)^T\right\}^T \\
&= \tfrac{1}{2}Q(t)[\,(\nabla^x v)(x,t) + ((\nabla^x v)(x,t))^T\,]Q(t)^T \\
&= Q(t)L(x,t)Q(t)^T,
\end{aligned}$$

which is the desired result.

5.15 (a) From the result on the effects of superposed rigid motion we have $\nabla^{x^*} v^* = Q\nabla^x v Q^T + \dot{Q}Q^T$, and taking the transpose gives $(\nabla^{x^*} v^*)^T = Q\nabla^x v^T Q^T + Q\dot{Q}^T$. From these expressions we get

$$W^* = QWQ^T + \frac{\dot{Q}Q^T - Q\dot{Q}^T}{2}.$$

From $QQ^T = I$ we deduce $\dot{Q}Q^T + Q\dot{Q}^T = O$. Substituting this result into the expression above gives

$$W^* = QWQ^T + \dot{Q}Q^T, \tag{5.46}$$

which is the desired result.

(b) Since T is frame indifferent we have $T^* = QTQ^T$, and since $J_{\mathrm{ma}}(T) = \dot{T}$ we have

$$J_{\mathrm{ma}}(T^*) = (T^*)^{\bullet} = Q\dot{T}Q^T + \dot{Q}TQ^T + QT\dot{Q}^T.$$

By (5.46), and the fact that W and W^* are skew, we get $W^*Q = QW + \dot{Q}$ and $-Q^T W^* = -WQ^T + \dot{Q}^T$. Thus

$$\begin{aligned}
J_{\mathrm{ma}}(T^*) &= QJ_{\mathrm{ma}}(T)Q^T \\
&\quad + [W^*Q - QW]TQ^T + QT[WQ^T - Q^T W^*]
\end{aligned}$$

This shows that $J_{\mathrm{ma}}(T^*) \neq QJ_{\mathrm{ma}}(T)Q^T$ for general superposed rigid motions as claimed. To investigate the co-rotated derivative $J_{\mathrm{co}}(T, W)$ we note from above that

$$\begin{aligned}
(T^*)^{\bullet} &= Q[\dot{T} - WT + TW]Q^T + W^*QTQ^T - QTQ^T W^* \\
&= Q[\dot{T} - WT + TW]Q^T + W^*T^* - T^*W^*,
\end{aligned}$$

which implies

$$(\boldsymbol{T}^*)^\bullet - \boldsymbol{W}^*\boldsymbol{T}^* + \boldsymbol{T}^*\boldsymbol{W}^* = \boldsymbol{Q}[\dot{\boldsymbol{T}} - \boldsymbol{W}\boldsymbol{T} + \boldsymbol{T}\boldsymbol{W}]\boldsymbol{Q}^T.$$

Thus we obtain $\boldsymbol{J}_{\rm co}(\boldsymbol{T}^*, \boldsymbol{W}^*) = \boldsymbol{Q}\boldsymbol{J}_{\rm co}(\boldsymbol{T}, \boldsymbol{W})\boldsymbol{Q}^T$ as claimed.

6

Isothermal Fluid Mechanics

In this chapter we consider several applications of the Eulerian balance laws derived in Chapter 5. Both because it is physically reasonable in many circumstances, and because it simplifies the exposition, we neglect thermal effects here; they are studied in some detail in Chapter 8. Considering only the isothermal case there are 16 basic unknown fields in the Eulerian description of a continuum body:

$\varphi_i(\boldsymbol{X},t)$	3 components of motion
$v_i(\boldsymbol{x},t)$	3 components of velocity
$\rho(\boldsymbol{x},t)$	1 mass density
$S_{ij}(\boldsymbol{x},t)$	9 components of stress.

To determine these unknown fields we have the following 10 equations:

$[v_i]_m = \frac{\partial}{\partial t}\varphi_i$	3 kinematical
$\frac{\partial}{\partial t}\rho + (\rho v_i)_{,i} = 0$	1 mass
$\rho\dot{v}_i = S_{ij,j} + \rho b_i$	3 linear momentum
$S_{ij} = S_{ji}$	3 independent angular momentum.

Thus 6 additional equations are required to balance the number of equations and unknowns. These are provided by constitutive equations that relate the 6 independent components of the Cauchy stress field $\boldsymbol{S}$ to the variables ρ, $\boldsymbol{v}$ and $\boldsymbol{\varphi}$. In the case of a material subject to an internal constraint, there is an additional equation which defines the constraint, together with an additional unknown multiplier field which enforces it. Both constitutive and constraint equations, if any, must be consistent with the axiom of material frame-indifference. Moreover, assuming a material is energetically passive, the constitutive equation must also satisfy the mechanical energy inequality as discussed in Chapter 5.

In this chapter we study constitutive models that relate the Cauchy stress $\boldsymbol{S}$ to the spatial mass density ρ and the spatial velocity gradient $\nabla^x \boldsymbol{v}$. Such models are typically used to describe the behavior of various types of fluids. Because each such constitutive model is independent of $\boldsymbol{\varphi}$, a closed system of equations for ρ, $\boldsymbol{v}$ and $\boldsymbol{S}$ is provided by the balance equations for mass, linear momentum and angular momentum. As we will see, these equations provide a complete description of motion for bodies occupying regions of space with prescribed boundaries. When boundaries are not known a-priori, then $\boldsymbol{\varphi}$ and the kinematical equation must in general also be considered.

The specific constitutive models considered in this chapter are those for: (i) ideal incompressible fluids, (ii) elastic fluids and (iii) Newtonian incompressible fluids. Models (i) and (ii) describe inviscid fluids, whereas (iii) describes a viscous fluid. Models (i) and (iii) describe incompressible fluids, whereas (ii) describes a compressible fluid. For each model we summarize the governing equations, give an example of a typical initial-boundary value problem, and study various qualitative properties. In the context of model (ii) we also introduce the concept of linearization, a technique that will be used repeatedly in later chapters.

6.1 Ideal Fluids

In this section we study the constitutive model for an ideal fluid. We show that the model is frame-indifferent and satisfies the mechanical energy inequality, and then discuss a standard initial-boundary value problem. We end with some results concerning irrotational motions.

6.1.1 Definition

A continuum body with reference configuration B is said to be an **ideal fluid** if:

(1) The reference mass density field $\rho_0(\boldsymbol{X})$ is uniform in the sense that $\rho_0(\boldsymbol{X}) = \rho_0 > 0$ (constant).

(2) The material is incompressible (see Section 5.6), which means that the spatial velocity field must satisfy

$$\nabla^x \cdot \boldsymbol{v}(\boldsymbol{x}, t) = 0.$$

(3) The Cauchy stress field is spherical or Eulerian (see Section 3.5), which means that there is a scalar field $p(\boldsymbol{x},t)$ called the pressure such that

$$\boldsymbol{S}(\boldsymbol{x},t) = -p(\boldsymbol{x},t)\boldsymbol{I}.$$

In view of the discussion in Section 5.6, the pressure field in Property (3) can be identified as the multiplier for the constraint in Property (2). In particular, the stress field in an ideal fluid is purely reactive and is determined entirely by the incompressibility constraint. Property (3) implies that the stress field is also necessarily symmetric. Thus the balance equation for angular momentum (Result 5.8) is automatically satisfied and will not be considered further.

Properties (1) and (2) and the conservation of mass equation (Result 5.5) imply that the spatial mass density is uniform in space and constant in time, in particular

$$\rho(\boldsymbol{x},t) = \rho_0 > 0, \quad \forall \boldsymbol{x} \in B_t, \ t \geq 0.$$

This conclusion may be reached by substituting the incompressibility condition $\nabla^x \cdot \boldsymbol{v} = 0$ into the conservation of mass equation, which yields

$$\dot{\rho}(\boldsymbol{x},t) = 0, \quad \forall \boldsymbol{x} \in B_t, \ t \geq 0.$$

By definition of the total or material time derivative we have

$$\frac{\partial}{\partial t}\rho(\boldsymbol{\varphi}(\boldsymbol{X},t),t) = 0, \quad \forall \boldsymbol{X} \in B, \ t \geq 0.$$

From this result, together with the convention $\boldsymbol{\varphi}(\boldsymbol{X},0) = \boldsymbol{X}$ and the definition of $\rho_0(\boldsymbol{X})$, we get

$$\rho(\boldsymbol{\varphi}(\boldsymbol{X},t),t) = \rho(\boldsymbol{X},0) = \rho_0(\boldsymbol{X}), \quad \forall \boldsymbol{X} \in B, \ t \geq 0.$$

The stated result then follows from the condition on $\rho_0(\boldsymbol{X})$ and the relation $\boldsymbol{x} = \boldsymbol{\varphi}(\boldsymbol{X},t)$.

6.1.2 Euler Equations

Setting $\rho(\boldsymbol{x},t) = \rho_0$ and $\boldsymbol{S}(\boldsymbol{x},t) = -p(\boldsymbol{x},t)\boldsymbol{I}$ in the balance of linear momentum equation (Result 5.7) and the conservation of mass equation (Result 5.5), we obtain a closed system of equations for the spatial velocity and pressure fields in a body of ideal fluid. In particular, we obtain

$$\rho_0\dot{\boldsymbol{v}} = \nabla^x \cdot (-p\boldsymbol{I}) + \rho_0\boldsymbol{b} \quad \text{and} \quad \nabla^x \cdot \boldsymbol{v} = 0,$$

where $\boldsymbol{b}(\boldsymbol{x},t)$ is a prescribed spatial body force field per unit mass. Notice that the conservation of mass equation reduces to the incompressibility constraint on the spatial velocity field.

In view of Result 4.7 we have

$$\dot{\boldsymbol{v}} = \frac{\partial}{\partial t}\boldsymbol{v} + (\nabla^x \boldsymbol{v})\boldsymbol{v}.$$

Also, it is straightforward to verify that

$$\nabla^x \cdot (-p\boldsymbol{I}) = -\nabla^x p.$$

Thus the spatial velocity and pressure fields in a body of ideal fluid with reference configuration B must satisfy the following equations for all $\boldsymbol{x} \in B_t$ and $t \geq 0$

$$\boxed{\begin{aligned} &\rho_0 \left[\frac{\partial}{\partial t}\boldsymbol{v} + (\nabla^x \boldsymbol{v})\boldsymbol{v}\right] = -\nabla^x p + \rho_0 \boldsymbol{b}, \\ &\nabla^x \cdot \boldsymbol{v} = 0. \end{aligned}} \tag{6.1}$$

These equations are known as the **Euler Equations** for an ideal fluid.

Remarks:

(1) In an ideal fluid there is no explicit equation that relates pressure to velocity or density. Instead, the pressure field is a fundamental unknown that must be determined simultaneously with the velocity field. In particular, the pressure field can be identified as the multiplier associated with the constraint of incompressibility (see Section 5.6).

(2) The equations in (6.1) can determine pressure only to within an additive function of time. That is, if the pair $\boldsymbol{v}(\boldsymbol{x},t)$, $p(\boldsymbol{x},t)$ satisfies (6.1), then so does $\boldsymbol{v}(\boldsymbol{x},t)$, $p(\boldsymbol{x},t)+f(t)$ for any arbitrary scalar-valued function $f(t)$. This follows from the fact that $\nabla^x(p+f) = \nabla^x p$.

(3) In an ideal fluid there are no shear stresses. In particular, because the Cauchy stress field is spherical, the traction $\boldsymbol{t}$ on a surface with normal $\boldsymbol{n}$ is itself normal to the surface since $\boldsymbol{t} = \boldsymbol{S}\boldsymbol{n} = -p\boldsymbol{n}$ (see Section 3.5). □

6.1.3 Frame-Indifference Considerations

As discussed in Section 5.6, the Cauchy stress field in a body subject to a material constraint can be decomposed as

$$\boldsymbol{S}(\boldsymbol{x},t) = \boldsymbol{S}^{(r)}(\boldsymbol{x},t) + \boldsymbol{S}^{(a)}(\boldsymbol{x},t),$$

where $\boldsymbol{S}^{(r)}$ is a reactive stress determined by the constraint, and $\boldsymbol{S}^{(a)}$ is an active stress determined by a given constitutive equation. For an ideal fluid we have

$$\boldsymbol{S}^{(r)}(\boldsymbol{x},t) = -p(\boldsymbol{x},t)\boldsymbol{I}, \qquad \boldsymbol{S}^{(a)}(\boldsymbol{x},t) = \boldsymbol{O},$$

where p is the multiplier associated with the incompressibility constraint $\nabla^x \cdot \boldsymbol{v} = 0$, or equivalently, $\det \boldsymbol{F} = 1$. This constraint can be written in the standard material form

$$\gamma(\boldsymbol{F}(\boldsymbol{X},t)) = 0,$$

where $\gamma(\boldsymbol{F}) = \det \boldsymbol{F} - 1$.

For a constrained model to be consistent with the axiom of material frame-indifference (Axiom 5.24) it is sufficient that the constraint field $\gamma(\boldsymbol{F}(\boldsymbol{X},t))$ and active stress field $\boldsymbol{S}^{(a)}(\boldsymbol{x},t)$ be frame-indifferent in the sense of Definition 5.23. To determine if this is the case, let B_t and B_t^* denote two configurations of a body of ideal fluid that are related by a superposed rigid motion

$$\boldsymbol{x}^* = \boldsymbol{g}(\boldsymbol{x},t) = \boldsymbol{Q}(t)\boldsymbol{x} + \boldsymbol{c}(t),$$

where $\boldsymbol{Q}(t)$ is an arbitrary rotation tensor and $\boldsymbol{c}(t)$ is an arbitrary vector. Let $\gamma(\boldsymbol{F}(\boldsymbol{X},t))$ and $\boldsymbol{S}^{(a)}(\boldsymbol{x},t)$ denote the constraint and active stress fields associated with B_t, and let $\gamma(\boldsymbol{F}^*(\boldsymbol{X},t))$ and $\boldsymbol{S}^{(a)*}(\boldsymbol{x}^*,t)$ denote the corresponding fields associated with B_t^*. Recall that, by Result 5.22, we have $\boldsymbol{F}^*(\boldsymbol{X},t) = \boldsymbol{Q}(t)\boldsymbol{F}(\boldsymbol{X},t)$. Using the definitions of γ and $\boldsymbol{S}^{(a)}$ (the active stress is identically zero), together with the fact that the determinant of a rotation is unity, we deduce (omitting the arguments $\boldsymbol{x}^*$, $\boldsymbol{x}$, $\boldsymbol{X}$ and t for brevity)

$$\begin{aligned} \gamma(\boldsymbol{F}^*) &= \det \boldsymbol{F}^* - 1 = \det \boldsymbol{F} - 1 = \gamma(\boldsymbol{F}), \\ \boldsymbol{S}^{(a)^*} &= \boldsymbol{O} = \boldsymbol{Q}[\boldsymbol{O}]\boldsymbol{Q}^T = \boldsymbol{Q}\boldsymbol{S}^{(a)}\boldsymbol{Q}^T. \end{aligned}$$

Thus the material model for an ideal fluid is frame-indifferent.

6.1.4 Mechanical Energy Considerations

Here we assume the isothermal model of an ideal fluid is energetically passive as discussed in Section 5.7 and show that it satisfies the mechanical energy inequality (Result 5.29).

Since $\boldsymbol{S} = -p\boldsymbol{I}$ and $\boldsymbol{L} = \text{sym}(\nabla^x \boldsymbol{v})$, for any motion of an ideal fluid we have

$$\boldsymbol{S} : \boldsymbol{L} = -p\boldsymbol{I} : \text{sym}(\nabla^x \boldsymbol{v}).$$

Using the facts that $\boldsymbol{I} : \boldsymbol{A} = \text{tr}\, \boldsymbol{A}$ and $\text{tr}(\text{sym}(\boldsymbol{A})) = \text{tr}\, \boldsymbol{A}$ for any second-order tensor $\boldsymbol{A}$, we get

$$\boldsymbol{S} : \boldsymbol{L} = -p\, \text{tr}(\nabla^x \boldsymbol{v}) = -p\nabla^x \cdot \boldsymbol{v} = 0, \quad \forall \boldsymbol{x} \in B_t,\ t \geq 0,$$

where the last equality follows from the incompressibility condition. For any time interval $[t_0, t_1]$ the above result implies

$$\int_{t_0}^{t_1} (\det \boldsymbol{F}(\boldsymbol{X}, t))\, \boldsymbol{S}_m(\boldsymbol{X}, t) : \boldsymbol{L}_m(\boldsymbol{X}, t)\, dt = 0, \qquad \forall \boldsymbol{X} \in B,$$

where $\boldsymbol{F}$ is the deformation gradient. This shows that the model of an ideal fluid is consistent with the mechanical energy inequality. In particular, the inequality is satisfied for any isothermal process, closed or not.

6.1.5 Initial-Boundary Value Problems

An **initial-boundary value problem** for a body of ideal fluid is a set of equations that describe the motion of the body subject to specified initial conditions in B at time $t = 0$, and boundary conditions on ∂B_t for $t \geq 0$. The Eulerian form of the balance laws for an ideal fluid are particularly well-suited for those problems in which the body occupies a fixed region D of space. We typically assume D is a bounded open set as shown in Figure 6.1. However, it is also useful in applications to consider unbounded open sets such as the exterior of the region shown in the figure, or the whole of Euclidean space.

A standard initial-boundary value problem for a body of ideal fluid occupying a fixed region D can be stated as follows: Find $\boldsymbol{v} : D \times [0, T] \to$

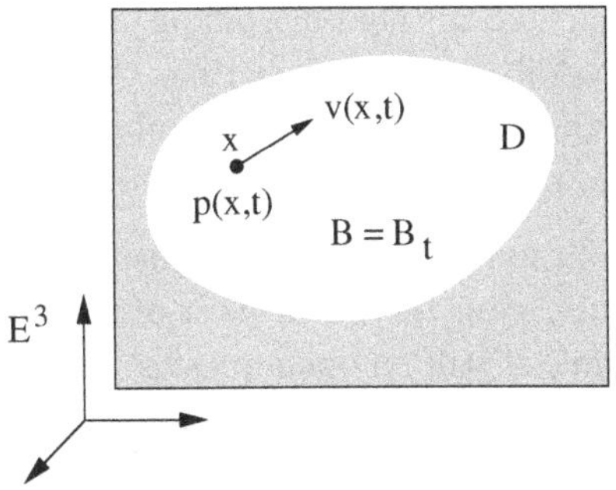

Fig. 6.1 Ideal fluid body occupying a fixed region D (cavity).

$\mathcal{V}$ and $p : D \times [0,T] \to I\!R$ such that

$$\boxed{\begin{array}{ll} \rho_0 \left[\frac{\partial}{\partial t}\boldsymbol{v} + (\nabla^x \boldsymbol{v})\boldsymbol{v}\right] = -\nabla^x p + \rho_0 \boldsymbol{b} & \text{in } \ D \times [0,T] \\ \nabla^x \cdot \boldsymbol{v} = 0 & \text{in } \ D \times [0,T] \\ \boldsymbol{v} \cdot \boldsymbol{n} = 0 & \text{in } \partial D \times [0,T] \\ \boldsymbol{v}(\cdot, 0) = \boldsymbol{v}_0(\cdot) & \text{in } \ D. \end{array}} \tag{6.2}$$

In the above system, ρ_0 is a prescribed constant mass density and $\boldsymbol{b}$ is a prescribed spatial body force per unit mass. Equation $(6.2)_1$ is the balance of linear momentum equation and $(6.2)_2$ is the material incompressibility constraint, which in the current case is equivalent to the conservation of mass equation. Equation $(6.2)_3$ is a **boundary condition** which expresses the fact that fluid cannot cross the boundary of D and $(6.2)_4$ is an **initial condition**. Here $\boldsymbol{v}_0$ is a given field which corresponds to the fluid velocity field at time $t = 0$. The specified field $\boldsymbol{v}_0$ should be compatible with the incompressibility and boundary conditions in the sense that $\nabla^x \cdot \boldsymbol{v}_0 = 0$ in D and $\boldsymbol{v}_0 \cdot \boldsymbol{n} = 0$ on ∂D.

Remarks:

(1) The system in (6.2) is a nonlinear initial-boundary value problem for the fields $(\boldsymbol{v}, p)$. This system is guaranteed to have a solution on some finite time interval $[0, T]$ provided the region D is sufficiently regular, and the field $\boldsymbol{v}_0$ is sufficiently smooth and compatible with the incompressibility and boundary conditions.

(2) When D is bounded, the field $\boldsymbol{v}$ is unique, whereas p is unique only to within an additive function of time. When D is unbounded, additional boundary conditions at infinity are typically imposed which result in uniqueness for both $\boldsymbol{v}$ and p.

(3) Although a solution to (6.2) can be guaranteed on a finite time interval depending on the given data of a problem, global existence for all time is still an open question. Moreover, while some special exact solutions to (6.2) are known, numerical approximation is generally required to obtain quantitative information about solutions.

(4) For simplicity, we assume that solutions are continuously differentiable in all variables as many times as are required for the calculations we perform. □

6.1.6 Motion Map, Other Boundary Conditions

For problems in which a fluid body occupies a fixed region D, the fluid velocity and pressure fields $(\boldsymbol{v}, p)$ can be determined from (6.2) independently of the motion map $\boldsymbol{\varphi}$. In particular, $\boldsymbol{\varphi}$ does not enter the problem because the current placement of the body is known for all $t \geq 0$, namely, $B_t = D$. If desired, $\boldsymbol{\varphi}$ can be determined from $\boldsymbol{v}$ via the equations

$$\left.\begin{aligned} \frac{\partial}{\partial t}\boldsymbol{\varphi}(\boldsymbol{X},t) &= \boldsymbol{v}(\boldsymbol{\varphi}(\boldsymbol{X},t),t) \\ \boldsymbol{\varphi}(\boldsymbol{X},0) &= \boldsymbol{X} \end{aligned}\right\}, \quad \forall \boldsymbol{X} \in D,\ t \in [0,T]. \tag{6.3}$$

Equation $(6.3)_1$ is the definition of the spatial velocity in terms of the motion map (see Chapter 4) and $(6.3)_2$ is an initial condition which expresses the convention that $B_0 = B$. Thus the solution $\boldsymbol{\varphi}(\boldsymbol{X}, t)$ of this system is the current position of the material particle whose initial position was $\boldsymbol{X} \in D$.

Two different families of curves are typically used to visualize the motion of a fluid body. By a **pathline** we mean the trajectory of an individual fluid particle for $t \geq 0$. In particular, pathlines are the curves defined by $\boldsymbol{\varphi}(\boldsymbol{X}, t)$ with $\boldsymbol{X}$ fixed. Pathlines give a Lagrangian picture of fluid motion. A small, suspended particle following a pathline is typically called a **passive tracer**. By a **streamline** we mean an integral curve of the vector field $\boldsymbol{v}(\boldsymbol{x}, t)$ with t fixed. In particular, streamlines are the solution curves of the differential equation $\frac{d}{ds}\boldsymbol{y} = \boldsymbol{v}(\boldsymbol{y}(s), t)$. For each fixed $t \geq 0$, streamlines provide a global portrait of the spatial velocity field $\boldsymbol{v}(\boldsymbol{x},t)$. While the set of streamlines for a motion is generally different from the set of pathlines, the two sets coincide when the spatial velocity $\boldsymbol{v}(\boldsymbol{x},t)$ is independent of t.

Problems in which a fluid body does not occupy a fixed region of space can also be considered. For example, a fluid (liquid) in an open container exposed to the environment will generally not occupy a fixed region of space; it may develop waves on its exposed surface. In such problems, $(\boldsymbol{v},p)$ cannot generally be determined independently of φ. In particular, φ would be an additional unknown required to describe the shape of the exposed surface, and $(6.3)_1$ would be included as an additional equation. Moreover, a boundary condition on pressure (traction) would typically be imposed on the exposed surface.

The boundary condition $\boldsymbol{v}\cdot\boldsymbol{n} = 0$ is appropriate for describing the interface between an ideal fluid and a fixed, impermeable solid. If the solid were not fixed, but instead had a prescribed velocity field $\boldsymbol{\vartheta}$, then an appropriate boundary condition would be $\boldsymbol{v}\cdot\boldsymbol{n} = \boldsymbol{\vartheta}\cdot\boldsymbol{n}$. As mentioned above, other boundary conditions can be considered when the interface is not fluid-solid, or when the motion of the boundary is not known a-priori. Also, different conditions could be imposed on different parts of the boundary.

6.1.7 Irrotational Motion, Bernoulli's Theorem

Here we introduce the idea of an irrotational motion and classify the properties of such motions. We show that the spatial velocity and pressure fields in an irrotational motion of an ideal fluid satisfy a much simpler system of equations than those in (6.2).

Definition 6.1 *Let $\varphi : B \times [0,\infty) \to \boldsymbol{E}^3$ be a motion of a continuum body with spatial velocity field $\boldsymbol{v}$ and spin field $\boldsymbol{W} = \mathrm{skew}(\nabla^x \boldsymbol{v})$. Then φ is said to be* **irrotational** *if*

$$\boldsymbol{W}(\boldsymbol{x},t) = \boldsymbol{O}, \quad \forall \boldsymbol{x} \in B_t,\ t \geq 0,$$

or equivalently

$$(\nabla^x \times \boldsymbol{v})(\boldsymbol{x},t) = \boldsymbol{0}, \quad \forall \boldsymbol{x} \in B_t,\ t \geq 0, \tag{6.4}$$

where $\nabla^x \times \boldsymbol{v}$ is the vorticity. □

Thus a motion is said to be irrotational if the spatial vorticity field is identically zero. In view of Stokes' Theorem (Result 2.18), an irrotational motion is to be interpreted as one in which material particles experience no net rotation. In particular, the circulation of material around the edge of an arbitrary open surface in space (a disc for example) must vanish. Similar interpretations of irrotational motions can be drawn from the properties of the spin tensor outlined in Section 4.5.2.

The following result is a classic theorem of vector analysis. For brevity we omit the proof, but we note that the sufficiency part of the result is clear since $\nabla^x \times (\nabla^x \phi) = \mathbf{0}$ for any scalar field ϕ.

Result 6.2 ***Velocity Potential for Irrotational Motion.*** *Assume B_t is simply connected for all $t \geq 0$. Then a smooth vector field $\boldsymbol{v}$ satisfies (6.4) if and only if there is a spatial field ϕ such that*

$$\boldsymbol{v}(\boldsymbol{x}, t) = \nabla^x \phi(\boldsymbol{x}, t), \quad \forall \boldsymbol{x} \in B_t, \ t \geq 0.$$

The scalar field ϕ is called a **potential** *for $\boldsymbol{v}$.* □

Remarks:

(1) A set $\Omega \subseteq \boldsymbol{E}^3$ is called **connected** if any two points in Ω can be connected by a curve in Ω. In addition, if any closed curve in Ω can be continuously deformed to a point without leaving Ω, then it is called **simply connected**.

(2) Any vector field whose curl vanishes identically as in (6.4) is said to be **irrotational**. In this sense, the spatial velocity field for an irrotational motion is irrotational. □

The following result establishes an important property of the spatial acceleration field for an irrotational motion.

Result 6.3 ***Acceleration Field for Irrotational Motion.*** *Let $\boldsymbol{\varphi} : B \times [0, \infty) \to \boldsymbol{E}^3$ be a motion of a continuum body with spatial velocity field $\boldsymbol{v}$ and spin field $\boldsymbol{W}$. Then*

$$\begin{aligned}\dot{\boldsymbol{v}} &= \frac{\partial}{\partial t}\boldsymbol{v} + \tfrac{1}{2}\nabla^x (|\boldsymbol{v}|^2) + 2\boldsymbol{W}\boldsymbol{v} \\ &= \frac{\partial}{\partial t}\boldsymbol{v} + \tfrac{1}{2}\nabla^x (|\boldsymbol{v}|^2) + (\nabla^x \times \boldsymbol{v}) \times \boldsymbol{v}.\end{aligned} \tag{6.5}$$

Thus, for an irrotational motion, we have

$$\dot{\boldsymbol{v}} = \frac{\partial}{\partial t}\boldsymbol{v} + \tfrac{1}{2}\nabla^x (|\boldsymbol{v}|^2). \tag{6.6}$$

□

Proof From Result 4.7 the spatial acceleration field is given by

$$\dot{\boldsymbol{v}} = \frac{\partial}{\partial t}\boldsymbol{v} + (\nabla^x \boldsymbol{v})\boldsymbol{v}, \tag{6.7}$$

and by definition of $\boldsymbol{W}$ we have

$$2\boldsymbol{W}\boldsymbol{v} = [\nabla^x \boldsymbol{v} - (\nabla^x \boldsymbol{v})^T]\boldsymbol{v}.$$

Using the fact that

$$(\nabla^x \boldsymbol{v})^T \boldsymbol{v} = v_{i,j} v_i \boldsymbol{e}_j = \tfrac{1}{2}(v_i v_i)_{,j} \boldsymbol{e}_j = \tfrac{1}{2}\nabla^x(|\boldsymbol{v}|^2),$$

we find

$$2\boldsymbol{W}\boldsymbol{v} = (\nabla^x \boldsymbol{v})\boldsymbol{v} - \tfrac{1}{2}\nabla^x(|\boldsymbol{v}|^2). \tag{6.8}$$

Solving (6.8) for $(\nabla^x \boldsymbol{v})\boldsymbol{v}$ and substituting the result into (6.7) leads to the first equation in (6.5). The second equation in (6.5) follows from the definition of $\nabla^x \times \boldsymbol{v}$ (see Definition 2.12), and the result in (6.6) follows from the definition of an irrotational motion. □

Not all motions of an ideal fluid are irrotational. However, the next result shows that any motion which begins from a suitable initial velocity, for example a uniform velocity, will be irrotational for all time provided that the body force field per unit mass is conservative. A spatial force field $\boldsymbol{b}$ is called **conservative** if there exists a spatial scalar field β such that

$$\boldsymbol{b}(\boldsymbol{x}, t) = -\nabla^x \beta(\boldsymbol{x}, t).$$

Result 6.4 ***Criterion for Irrotational Motion in Ideal Fluids.*** *Let $\varphi : B \times [0, \infty) \to \mathbb{E}^3$ be a motion of a body of ideal fluid with spatial velocity field $\boldsymbol{v}$ and constant mass density ρ_0, and suppose the body is subject to a conservative body force field per unit mass $\boldsymbol{b} = -\nabla^x \beta$. If $\boldsymbol{v}$ is irrotational at $t = 0$, then it is irrotational for all $t \geq 0$.* □

Proof To establish the result let $\boldsymbol{w} = \nabla^x \times \boldsymbol{v}$. Then, taking the curl of $(6.1)_1$, and employing $(6.5)_2$ and the fact that the curl of a gradient is zero, we find that $\boldsymbol{w}$ satisfies

$$\rho_0 \left[\frac{\partial \boldsymbol{w}}{\partial t} + \nabla^x \times (\boldsymbol{w} \times \boldsymbol{v})\right](\boldsymbol{x}, t) = \boldsymbol{0}. \tag{6.9}$$

Moreover, by assumption on $\boldsymbol{v}$, we have the initial condition

$$\boldsymbol{w}(\boldsymbol{x}, 0) = \boldsymbol{0}.$$

Expanding the curl operations in (6.9) and dividing through by the constant $\rho_0 > 0$ gives

$$\frac{\partial \boldsymbol{w}}{\partial t} + (\nabla^x \boldsymbol{w})\boldsymbol{v} - (\nabla^x \boldsymbol{v})\boldsymbol{w} + \boldsymbol{w}(\nabla^x \cdot \boldsymbol{v}) - \boldsymbol{v}(\nabla^x \cdot \boldsymbol{w}) = \boldsymbol{0}.$$

Using the facts that $\nabla^x \cdot \boldsymbol{v} = 0$ (because the flow is incompressible) and $\nabla^x \cdot \boldsymbol{w} = 0$ (because the divergence of a curl is always zero) we deduce

$$\dot{\boldsymbol{w}} - (\nabla^x \boldsymbol{v})\boldsymbol{w} = \boldsymbol{0}. \tag{6.10}$$

We next show that $\boldsymbol{w}(\boldsymbol{x}, t) = \boldsymbol{0}$ is the unique solution of (6.10) with initial condition $\boldsymbol{w}(\boldsymbol{x}, 0) = \boldsymbol{0}$. Rather than appeal to the general theory of linear partial differential equations, we establish this result by direct calculation using a convenient change of variables. To this end, let $\boldsymbol{w}_m(\boldsymbol{X}, t)$ be the material description of $\boldsymbol{w}(\boldsymbol{x}, t)$, so $\boldsymbol{w}_m(\boldsymbol{X}, t) = \boldsymbol{w}(\boldsymbol{\varphi}(\boldsymbol{X}, t), t)$, and let $\boldsymbol{\xi}(\boldsymbol{X}, t)$ be the material field defined by

$$\boldsymbol{\xi} = \boldsymbol{F}^{-1}\boldsymbol{w}_m, \quad \text{equivalently} \quad \boldsymbol{w}_m = \boldsymbol{F}\boldsymbol{\xi}. \tag{6.11}$$

Then from $(6.11)_2$ we have

$$\frac{\partial}{\partial t}\boldsymbol{w}_m = \Big(\frac{\partial}{\partial t}\boldsymbol{F}\Big)\boldsymbol{\xi} + \boldsymbol{F}\Big(\frac{\partial}{\partial t}\boldsymbol{\xi}\Big),$$

and solving this equation for $\frac{\partial}{\partial t}\boldsymbol{\xi}$ we obtain, after eliminating $\boldsymbol{\xi}$

$$\frac{\partial}{\partial t}\boldsymbol{\xi} = \boldsymbol{F}^{-1}\left[\frac{\partial}{\partial t}\boldsymbol{w}_m - \Big(\frac{\partial}{\partial t}\boldsymbol{F}\Big)\boldsymbol{F}^{-1}\boldsymbol{w}_m\right]. \tag{6.12}$$

Since $\frac{\partial}{\partial t}\boldsymbol{w}_m = (\dot{\boldsymbol{w}})_m$ and $(\frac{\partial}{\partial t}\boldsymbol{F})\boldsymbol{F}^{-1} = (\nabla^x \boldsymbol{v})_m$, we find that the right-hand side of (6.12) vanishes by (6.10). Thus $\boldsymbol{\xi}$ satisfies the simple equation

$$\frac{\partial}{\partial t}\boldsymbol{\xi}(\boldsymbol{X}, t) = \boldsymbol{0}.$$

From this we deduce $\boldsymbol{\xi}(\boldsymbol{X}, t) = \boldsymbol{\xi}(\boldsymbol{X}, 0)$ for all $t \geq 0$ and $\boldsymbol{X} \in B$. Moreover, since $\boldsymbol{F}(\boldsymbol{X}, 0) = \boldsymbol{I}$ we have $\boldsymbol{\xi}(\boldsymbol{X}, 0) = \boldsymbol{w}_m(\boldsymbol{X}, 0)$. When these results are combined with $(6.11)_2$ we find

$$\begin{aligned} \boldsymbol{w}_m(\boldsymbol{X}, t) &= \boldsymbol{F}(\boldsymbol{X}, t)\boldsymbol{\xi}(\boldsymbol{X}, t) \\ &= \boldsymbol{F}(\boldsymbol{X}, t)\boldsymbol{\xi}(\boldsymbol{X}, 0) \\ &= \boldsymbol{F}(\boldsymbol{X}, t)\boldsymbol{w}_m(\boldsymbol{X}, 0). \end{aligned}$$

Thus, if a motion is initially irrotational in the sense that $\boldsymbol{w}_m(\boldsymbol{X}, 0) = \boldsymbol{0}$ for all $\boldsymbol{X} \in B$, then it is forever irrotational in the sense that $\boldsymbol{w}_m(\boldsymbol{X}, t) = \boldsymbol{0}$ for all $t \geq 0$ and $\boldsymbol{X} \in B$. □

The next result shows that the balance of linear momentum equation for an ideal fluid can be reduced to a particularly simple form when the motion is irrotational.

Result 6.5 ***Bernoulli's Theorem for Ideal Fluids.*** *Let $\varphi : B \times [0,\infty) \to \boldsymbol{E}^3$ be a motion of a body of ideal fluid with constant mass density ρ_0. For all $t \geq 0$ suppose the motion is irrotational with spatial velocity $\boldsymbol{v} = \nabla^x \phi$, and suppose the body is subject to a conservative body force per unit mass $\boldsymbol{b} = -\nabla^x \beta$. Then balance of linear momentum implies*

$$\nabla^x \left(\frac{\partial}{\partial t}\phi + \tfrac{1}{2}|\boldsymbol{v}|^2 + \rho_0^{-1} p + \beta \right) = \boldsymbol{0}, \tag{6.13}$$

or equivalently

$$\frac{\partial}{\partial t}\phi + \tfrac{1}{2}|\boldsymbol{v}|^2 + \rho_0^{-1} p + \beta = f, \tag{6.14}$$

for some function $f(t)$ depending on the motion. □

Proof By Result 6.3, and the assumptions $\boldsymbol{v} = \nabla^x \phi$ and $\boldsymbol{b} = -\nabla^x \beta$, the balance of linear momentum equation $(6.1)_1$ can be written as

$$\rho_0 \left[\frac{\partial}{\partial t}(\nabla^x \phi) + \tfrac{1}{2}\nabla^x(|\boldsymbol{v}|^2) \right] = -\nabla^x p - \rho_0 \nabla^x \beta.$$

The result in (6.13) follows from the fact that the operators $\frac{\partial}{\partial t}$ and ∇^x commute. The result in (6.14) follows from the definition of ∇^x. □

From the above result we deduce that the spatial velocity field $\boldsymbol{v}$ and pressure field p in an irrotational motion of an ideal fluid can be determined from a much simpler system of equations than those in (6.2). In particular, suppose as before that the body occupies a fixed region D of space and consider the representation $\boldsymbol{v} = \nabla^x \phi$ guaranteed by Result 6.2. Then from $(6.2)_{2,3}$ we find that ϕ must satisfy the equation

$$\Delta^x \phi = 0, \quad \forall \boldsymbol{x} \in D,\ t \geq 0, \tag{6.15}$$

together with the boundary condition

$$\nabla^x \phi \cdot \boldsymbol{n} = 0, \quad \forall \boldsymbol{x} \in \partial D,\ t \geq 0. \tag{6.16}$$

From $(6.2)_1$ we find that p must satisfy (6.14) for all $\boldsymbol{x} \in D$ and $t \geq 0$. The initial condition in $(6.2)_4$ cannot be specified arbitrarily. It must be compatible with the irrotational motion determined by (6.15) and (6.16).

Remarks:

(1) The system in (6.15) and (6.16) is a linear boundary-value problem for the scalar field ϕ. This system can determine ϕ only to within an additive function of time, which has no effect on the velocity field $\boldsymbol{v}$ since $\boldsymbol{v} = \nabla^x \phi$. Equation (6.15) is called **Laplace's equation** for ϕ.

(2) When D is bounded, the system in (6.15) and (6.16) admits only the trivial solution $\phi = 0$ up to an additive function of time, which implies $\boldsymbol{v} = \boldsymbol{0}$. Thus the only possible irrotational motion of an ideal fluid in a bounded domain with a fixed boundary is the trivial one.

(3) When D is unbounded, the system in (6.15) and (6.16) is typically augmented with boundary conditions at infinity, for example uniform flow and decay conditions, which lead to unique, non-trivial ϕ and $\boldsymbol{v}$. Notice that ϕ and $\boldsymbol{v}$ can depend on time through these boundary conditions.

(4) The pressure field p corresponding to a velocity field $\boldsymbol{v}$ can be found from (6.14) up to an unknown function $f(t)$. This function can be determined when the value of p is known at some reference point $\boldsymbol{x}_*$ for all $t \geq 0$. Thus the difference $p(\boldsymbol{x}, t) - p(\boldsymbol{x}_*, t)$ can be determined for all $\boldsymbol{x} \in D$ and $t \geq 0$.

(5) Any spatial field that is independent of t is referred to as **steady**. In the case when all fields are steady, the pressure difference $p(\boldsymbol{x}) - p(\boldsymbol{x}_*)$ is given by the simple relation

$$p - p_* = \tfrac{1}{2}\rho_0(|\boldsymbol{v}_*|^2 - |\boldsymbol{v}|^2) + \rho_0(\beta_* - \beta),$$

where p_*, $\boldsymbol{v}_*$ and β_* denote values at $\boldsymbol{x}_*$. □

6.2 Elastic Fluids

In this section we study the constitutive model for an elastic fluid, which is a simple model for an isothermal compressible fluid. We show that the model is frame-indifferent and satisfies the mechanical energy inequality, and discuss a standard initial-boundary value problem. We also describe results pertaining to irrotational motions and prove a generalized form of Bernoulli's Theorem. We end with a derivation of the linearized equations of motion for small disturbances.

6.2.1 Definition

A continuum body with reference configuration B is said to be an **elastic fluid** if:

(1) The Cauchy stress field is spherical or Eulerian (see Section 3.5), which means that there is a scalar field $p(\boldsymbol{x}, t)$ called the pressure such that

$$\boldsymbol{S}(\boldsymbol{x}, t) = -p(\boldsymbol{x}, t)\boldsymbol{I}.$$

(2) The pressure field $p(\boldsymbol{x}, t)$ is related to the spatial mass density field $\rho(\boldsymbol{x}, t)$ by the equation

$$p(\boldsymbol{x}, t) = \pi(\rho(\boldsymbol{x}, t)),$$

where $\pi : \mathbb{R}_+ \to \mathbb{R}$ is a given function satisfying $\pi'(s) > 0$ for all $s > 0$. The constitutive relation $p = \pi(\rho)$ is typically called an **equation of state**.

Properties (1) and (2) imply that the Cauchy stress field in an elastic fluid is entirely determined by the spatial mass density. Property (1) also implies that the stress field is necessarily symmetric. Thus the balance equation for angular momentum (Result 5.8) is automatically satisfied and will not be considered further. The assumption on π' in Property (2) implies that, under isothermal conditions, pressure increases with density. This assumption is physically reasonable in many circumstances. As we will see later, it leads to a well-defined notion of the "speed of sound" in an elastic fluid.

6.2.2 Elastic Fluid Equations

A closed system of equations for the spatial velocity and density fields in a body of elastic fluid is provided by the balance of linear momentum equation (Result 5.7) and the conservation of mass equation (Result 5.5). In particular, by setting $\boldsymbol{S} = -p\boldsymbol{I}$ in these equations we obtain

$$\rho\dot{\boldsymbol{v}} = \nabla^x \cdot (\quad p\boldsymbol{I}) + \rho\boldsymbol{b} \quad \text{and} \quad \frac{\partial}{\partial t}\rho + \nabla^x \cdot (\rho\boldsymbol{v}) = 0,$$

where $p = \pi(\rho)$ and $\boldsymbol{b}(\boldsymbol{x}, t)$ is a prescribed spatial body force field per unit mass.

In view of Result 4.7 we have

$$\dot{\boldsymbol{v}} = \frac{\partial}{\partial t}\boldsymbol{v} + (\nabla^x \boldsymbol{v})\boldsymbol{v}.$$

Also, it is straightforward to verify that

$$\nabla^x \cdot (-p\boldsymbol{I}) = -\nabla^x p = -\nabla^x [\pi(\rho)] = -\pi'(\rho)\nabla^x \rho.$$

Thus the spatial velocity and density fields in a body of elastic fluid with reference configuration B must satisfy the following equations for all $\boldsymbol{x} \in B_t$ and $t \geq 0$

$$\boxed{\begin{aligned} &\rho\left[\frac{\partial}{\partial t}\boldsymbol{v} + (\nabla^x \boldsymbol{v})\boldsymbol{v}\right] = -\pi'(\rho)\nabla^x \rho + \rho\boldsymbol{b}, \\ &\frac{\partial}{\partial t}\rho + \nabla^x \cdot (\rho\boldsymbol{v}) = 0. \end{aligned}} \tag{6.17}$$

These equations are known as the **Elastic Fluid Equations**.

Remarks:

(1) In contrast to an ideal fluid, an elastic fluid is compressible. It may experience volume changes and its spatial mass density is not necessarily constant. Because there is no incompressibility constraint, the pressure field p in an elastic fluid is not an additional unknown. Rather, it is completely determined by the mass density ρ.

(2) Just as for an ideal fluid, an elastic fluid can support no shear stresses. In particular, because the Cauchy stress field is spherical, the traction $\boldsymbol{t}$ on a surface with normal $\boldsymbol{n}$ is itself normal to the surface since $\boldsymbol{t} = \boldsymbol{S}\boldsymbol{n} = -p\boldsymbol{n}$ (see Section 3.5). □

6.2.3 Frame-Indifference Considerations

An elastic fluid model is defined by the constitutive equation

$$\boldsymbol{S}(\boldsymbol{x}, t) = -\pi(\rho(\boldsymbol{x}, t))\boldsymbol{I}, \tag{6.18}$$

where π is a given function and ρ is the spatial mass density. In order for this model to be consistent with the axiom of material frame-indifference (Axiom 5.24) it must deliver a stress field that is frame-indifferent in the sense of Definition 5.23. To determine if this is the case, let B_t and B_t^* denote two configurations of a body of elastic fluid that are related by a superposed rigid motion

$$\boldsymbol{x}^* = \boldsymbol{g}(\boldsymbol{x}, t) = \boldsymbol{Q}(t)\boldsymbol{x} + \boldsymbol{c}(t),$$

where $\boldsymbol{Q}(t)$ is an arbitrary rotation tensor and $\boldsymbol{c}(t)$ is an arbitrary vector. Let $\boldsymbol{S}(\boldsymbol{x},t)$ denote the stress field associated with B_t, and let $\boldsymbol{S}^*(\boldsymbol{x}^*,t)$ denote the stress field associated with B_t^*. Recall that, by Axiom 5.24, we have $\rho^*(\boldsymbol{x}^*,t) = \rho(\boldsymbol{x},t)$, where $\boldsymbol{x}^* = \boldsymbol{g}(\boldsymbol{x},t)$. Using the relation in (6.18) we deduce (omitting the arguments $\boldsymbol{x}^*$, $\boldsymbol{x}$ and t for brevity)

$$\boldsymbol{S}^* = -\pi(\rho^*)\boldsymbol{I} = \boldsymbol{Q}[-\pi(\rho)\boldsymbol{I}]\boldsymbol{Q}^T = \boldsymbol{Q}\boldsymbol{S}\boldsymbol{Q}^T.$$

Thus the constitutive model for an elastic fluid is frame-indifferent.

6.2.4 Mechanical Energy Considerations

Here we assume the isothermal model of an elastic fluid is energetically passive as discussed in Section 5.7 and show that it satisfies the mechanical energy inequality (Result 5.29) for any constitutive relation of the form $p = \pi(\rho)$.

Since $\boldsymbol{S} = -\pi(\rho)\boldsymbol{I}$ and $\boldsymbol{L} = \mathrm{sym}(\nabla^x \boldsymbol{v})$, for any motion of an elastic fluid we have

$$\boldsymbol{S} : \boldsymbol{L} = -\pi(\rho)\boldsymbol{I} : \mathrm{sym}(\nabla^x \boldsymbol{v}),$$

and by the same arguments as in the ideal case we find

$$\boldsymbol{S} : \boldsymbol{L} = -\pi(\rho)\,\mathrm{tr}(\nabla^x \boldsymbol{v}) = -\pi(\rho)\nabla^x \cdot \boldsymbol{v}, \quad \forall \boldsymbol{x} \in B_t,\ t \geq 0.$$

In material form this reads

$$\boldsymbol{S}_m : \boldsymbol{L}_m = -\pi(\rho_m)(\nabla^x \cdot \boldsymbol{v})_m, \quad \forall \boldsymbol{X} \in B,\ t \geq 0.$$

Since $(\nabla^x \cdot \boldsymbol{v})_m \det \boldsymbol{F} = \frac{\partial}{\partial t}(\det \boldsymbol{F})$ by Result 4.9, and $\rho_m = \rho_0 / \det \boldsymbol{F}$ by Result 5.13, we get

$$(\det \boldsymbol{F})\, \boldsymbol{S}_m : \boldsymbol{L}_m = -\pi\Big(\frac{\rho_0}{\det \boldsymbol{F}}\Big)\frac{\partial}{\partial t}(\det \boldsymbol{F}), \quad \forall \boldsymbol{X} \in B,\ t \geq 0. \tag{6.19}$$

Let $g(\boldsymbol{X},s)$, $s > 0$, be any function defined by $g'(\boldsymbol{X},s) = -\pi\big(\frac{\rho_0(\boldsymbol{X})}{s}\big)$, where the prime denotes a derivative with respect to the argument s. Then, by the chain rule, equation (6.19) can be written as

$$(\det \boldsymbol{F})\, \boldsymbol{S}_m : \boldsymbol{L}_m = g'(\boldsymbol{X}, \det \boldsymbol{F})\frac{\partial}{\partial t}(\det \boldsymbol{F}) = \frac{\partial}{\partial t} g(\boldsymbol{X}, \det \boldsymbol{F}).$$

For any time interval $[t_0, t_1]$ the above result implies

$$\begin{aligned}\int_{t_0}^{t_1} &(\det \boldsymbol{F}(\boldsymbol{X},t))\, \boldsymbol{S}_m(\boldsymbol{X},t) : \boldsymbol{L}_m(\boldsymbol{X},t)\, dt \\ &= g(\boldsymbol{X}, \det \boldsymbol{F}(\boldsymbol{X},t_1)) - g(\boldsymbol{X}, \det \boldsymbol{F}(\boldsymbol{X},t_0)), \quad \forall \boldsymbol{X} \in B.\end{aligned} \tag{6.20}$$

For any closed isothermal process in $[t_0, t_1]$ we have $\varphi(\boldsymbol{X}, t_1) = \varphi(\boldsymbol{X}, t_0)$ for all $\boldsymbol{X} \in B$, which implies $\det \boldsymbol{F}(\boldsymbol{X}, t_1) = \det \boldsymbol{F}(\boldsymbol{X}, t_0)$ for all $\boldsymbol{X} \in B$. Thus the right-hand side of (6.20) vanishes and the model of an elastic fluid satisfies the mechanical energy inequality for any constitutive relation $p = \pi(\rho)$.

6.2.5 Initial-Boundary Value Problems

An initial-boundary value problem for a body of elastic fluid is a set of equations that describe the motion of the body subject to specified initial conditions in B at time $t = 0$, and boundary conditions on ∂B_t for $t \geq 0$. Just as for an ideal fluid, the Eulerian form of the balance laws for an elastic fluid are particularly well-suited for those problems in which the body occupies a fixed region D of space. In this case the spatial velocity field $\boldsymbol{v}$ and the spatial mass density field ρ can be determined independently of the motion φ. As before, we typically assume D is a bounded open set as shown in Figure 6.1. However, it is also useful in applications to consider unbounded open sets such as the exterior of the region shown in the figure, or the whole of Euclidean space.

A standard initial-boundary value problem for a body of elastic fluid occupying a fixed region D can be stated as follows: Find $\boldsymbol{v} : D \times [0, T] \to \mathcal{V}$ and $\rho : D \times [0, T] \to I\!\!R$ such that

$$\boxed{\begin{array}{ll} \rho\left[\frac{\partial}{\partial t}\boldsymbol{v} + (\nabla^x \boldsymbol{v})\boldsymbol{v}\right] = -\pi'(\rho)\nabla^x \rho + \rho \boldsymbol{b} & \text{in} \quad D \times [0, T] \\ \frac{\partial}{\partial t}\rho + \nabla^x \cdot (\rho \boldsymbol{v}) = 0 & \text{in} \quad D \times [0, T] \\ \boldsymbol{v} \cdot \boldsymbol{n} = 0 & \text{in} \ \partial D \times [0, T] \\ \boldsymbol{v}(\cdot, 0) = \boldsymbol{v}_0(\cdot) & \text{in} \quad D \\ \rho(\cdot, 0) = \rho_0(\cdot) & \text{in} \quad D. \end{array}} \tag{6.21}$$

In the above system, π is a prescribed function and $\boldsymbol{b}$ is a prescribed spatial body force field per unit mass. Equation $(6.21)_1$ is the balance of linear momentum equation and $(6.21)_2$ is the conservation of mass equation. Equation $(6.21)_3$ is a boundary condition which expresses the fact that fluid cannot cross the boundary of D and $(6.21)_{4,5}$ are initial conditions for the spatial velocity and mass density fields, respectively. The specified field $\boldsymbol{v}_0$ should be compatible with the boundary condition in the sense that $\boldsymbol{v}_0 \cdot \boldsymbol{n} = 0$ on ∂D.

Remarks:

(1) The system in (6.21) is a nonlinear initial-boundary value problem for the fields $(\boldsymbol{v}, \rho)$. We expect this system to have a solution on some finite time interval $[0, T]$ under mild assumptions on D, $\boldsymbol{v}_0$ and ρ_0. However, general questions of existence and uniqueness for all time are difficult.

(2) In some applications we find that solutions of (6.21) are not smooth due to the appearance of **shock waves**. These are surfaces across which one or more components of a solution has a jump discontinuity. A proper analysis of such solutions requires the introduction of an appropriate weak form of the balance laws. Here, however, we assume that solutions are continuously differentiable as many times as are required for the calculations we perform.

(3) The boundary condition $\boldsymbol{v} \cdot \boldsymbol{n} = 0$ is appropriate for describing the interface between an elastic fluid and a fixed, impermeable solid. If the solid were not fixed, but instead had a prescribed velocity field $\boldsymbol{\vartheta}$, then an appropriate boundary condition would be $\boldsymbol{v}\cdot\boldsymbol{n} = \boldsymbol{\vartheta}\cdot\boldsymbol{n}$. Other boundary conditions can also be considered when the interface is not fluid-solid, or when the motion of the boundary is not known a-priori. Also, different conditions could be imposed on different parts of the boundary. □

6.2.6 Irrotational Motion, Generalized Bernoulli's Theorem

Here we restrict attention to irrotational motions and establish a generalized version of Bernoulli's Theorem for elastic fluids. We show that the spatial velocity and density fields in an irrotational motion of an elastic fluid satisfy a simpler system of equations than those in (6.21).

To begin, let $\gamma : \mathbb{R}_+ \to \mathbb{R}$ be the function defined by

$$\gamma(s) = \int_a^s \frac{\pi'(\xi)}{\xi}\, d\xi,$$

where $a > 0$ is an arbitrary constant that will play no role in our developments. From this definition we get

$$\gamma'(s) = \frac{\pi'(s)}{s}, \quad \forall s > 0, \tag{6.22}$$

and by the assumption $\pi'(s) > 0$ we note that $\gamma'(s) > 0$ for all $s > 0$. Thus γ has an inverse which we denote by ζ. In particular, we have

$$\zeta(\gamma(s)) = s, \quad \forall s > 0.$$

The following result shows that, just as for an ideal fluid, any motion of an elastic fluid which begins from a suitable initial velocity, for example a uniform velocity, will be irrotational for all time provided that the body force field per unit mass is conservative.

Result 6.6 ***Criterion for Irrotational Motion in Elastic Fluids.*** *Let $\varphi : B \times [0,\infty) \to \boldsymbol{E}^3$ be a motion of a body of elastic fluid with spatial velocity $\boldsymbol{v}$ and mass density ρ, and suppose the body is subject to a conservative body force field per unit mass $\boldsymbol{b} = -\nabla^x \beta$. If $\boldsymbol{v}$ is irrotational at $t = 0$, then it is irrotational for all $t \geq 0$.* □

Proof Dividing through by $\rho > 0$ in $(6.17)_1$ gives

$$\frac{\partial}{\partial t}\boldsymbol{v} + (\nabla^x \boldsymbol{v})\boldsymbol{v} = -\frac{\pi'(\rho)}{\rho}\nabla^x \rho + \boldsymbol{b},$$

and using (6.22) and the assumption $\boldsymbol{b} = -\nabla^x \beta$ we obtain

$$\frac{\partial}{\partial t}\boldsymbol{v} + (\nabla^x \boldsymbol{v})\boldsymbol{v} = -\nabla^x \gamma(\rho) - \nabla^x \beta. \tag{6.23}$$

Next, let $\boldsymbol{w} = \nabla^x \times \boldsymbol{v}$ and take the curl of (6.23). Employing Result 6.3, and the fact that the curl of a gradient is zero, we find that $\boldsymbol{w}$ satisfies the equation

$$\frac{\partial \boldsymbol{w}}{\partial t} + \nabla^x \times (\boldsymbol{w} \times \boldsymbol{v}) = \boldsymbol{0}. \tag{6.24}$$

Moreover, by assumption on $\boldsymbol{v}$, we have the initial condition

$$\boldsymbol{w}(\boldsymbol{x}, 0) = \boldsymbol{0}.$$

Expanding the curl operations in (6.24) gives

$$\frac{\partial \boldsymbol{w}}{\partial t} + (\nabla^x \boldsymbol{w})\boldsymbol{v} - (\nabla^x \boldsymbol{v})\boldsymbol{w} + \boldsymbol{w}(\nabla^x \cdot \boldsymbol{v}) - \boldsymbol{v}(\nabla^x \cdot \boldsymbol{w}) = \boldsymbol{0},$$

and using the fact that $\nabla^x \cdot \boldsymbol{w} = 0$ (the divergence of a curl is always zero) we deduce

$$\dot{\boldsymbol{w}} - (\nabla^x \boldsymbol{v})\boldsymbol{w} + (\nabla^x \cdot \boldsymbol{v})\boldsymbol{w} = \boldsymbol{0}.$$

Introducing the material field $\boldsymbol{\xi}$ as in the proof of Result 6.4, we find that $\boldsymbol{\xi}$ must satisfy the equation

$$\frac{\partial}{\partial t}\boldsymbol{\xi} = -(\nabla^x \cdot \boldsymbol{v})_m \boldsymbol{\xi}.$$

For each $\boldsymbol{X} \in B$ this is a linear, non-autonomous ordinary differential equation for $\boldsymbol{\xi}(\boldsymbol{X}, t)$. Since $\boldsymbol{\xi}(\boldsymbol{X}, t) = \boldsymbol{0}$ is the unique solution with initial condition $\boldsymbol{\xi}(\boldsymbol{X}, 0) = \boldsymbol{0}$, we deduce that $\boldsymbol{w}_m(\boldsymbol{X}, t) = \boldsymbol{0}$ for all $t \geq 0$ and $\boldsymbol{X} \in B$ provided $\boldsymbol{w}_m(\boldsymbol{X}, 0) = \boldsymbol{0}$ for all $\boldsymbol{X} \in B$, which is the desired result. □

The next result shows that, just as for an ideal fluid, the balance of linear momentum equation for an elastic fluid can be reduced to a particularly simple form when the motion is irrotational.

Result 6.7 ***Bernoulli's Theorem for Elastic Fluids.*** *Let $\varphi : B \times [0, \infty) \to \boldsymbol{E}^3$ be a motion of a body of elastic fluid with spatial mass density ρ. For all $t \geq 0$ suppose the motion is irrotational with spatial velocity $\boldsymbol{v} = \nabla^x \phi$, and suppose the body is subject to a conservative body force per unit mass $\boldsymbol{b} = -\nabla^x \beta$. Then balance of linear momentum implies*

$$\nabla^x \left(\frac{\partial}{\partial t}\phi + \tfrac{1}{2}|\boldsymbol{v}|^2 + \gamma(\rho) + \beta \right) = \boldsymbol{0}, \tag{6.25}$$

or equivalently

$$\frac{\partial}{\partial t}\phi + \tfrac{1}{2}|\boldsymbol{v}|^2 + \gamma(\rho) + \beta = f, \tag{6.26}$$

for some function $f(t)$ depending on the motion. □

Proof By Result 6.3, and the assumptions $\boldsymbol{v} = \nabla^x \phi$ and $\boldsymbol{b} = -\nabla^x \beta$, the balance of linear momentum equation $(6.17)_1$ can be written as

$$\rho \left[\frac{\partial}{\partial t}(\nabla^x \phi) + \tfrac{1}{2}\nabla^x (|\boldsymbol{v}|^2) \right] = -\pi'(\rho)\nabla^x \rho - \rho \nabla^x \beta.$$

Dividing by $\rho > 0$ and using the definition of the function γ we get

$$\frac{\partial}{\partial t}(\nabla^x \phi) + \tfrac{1}{2}\nabla^x (|\boldsymbol{v}|^2) = -\gamma'(\rho)\nabla^x \rho - \nabla^x \beta.$$

The result in (6.25) follows from the relation $\gamma'(\rho)\nabla^x \rho = \nabla^x \gamma(\rho)$ and the fact that the operators $\frac{\partial}{\partial t}$ and ∇^x commute. The result in (6.26) follows from the definition of ∇^x. □

From the above result we deduce that the spatial velocity field $\boldsymbol{v}$ and density field ρ in an irrotational motion of an elastic fluid can be determined from a simpler system of equations than those in (6.21). In particular, suppose as before that the body occupies a fixed region D of space and consider the representation $\boldsymbol{v} = \nabla^x \phi$ guaranteed by Result 6.2. Then from $(6.21)_{2,3}$ we find that ϕ and ρ must satisfy the equation

$$\frac{\partial}{\partial t}\rho + \nabla^x \cdot (\rho \nabla^x \phi) = 0, \quad \forall \boldsymbol{x} \in D,\ t \geq 0, \tag{6.27}$$

together with the boundary condition

$$\nabla^x \phi \cdot \boldsymbol{n} = 0, \quad \forall \boldsymbol{x} \in \partial D,\ t \geq 0. \tag{6.28}$$

From $(6.21)_1$ we find that ρ must satisfy (6.26) for all $\boldsymbol{x} \in D$ and $t \geq 0$. In particular, using the inverse function ζ, we may solve (6.26) for ρ to get

$$\rho = \zeta(\gamma(\rho)) = \zeta\Big(f - \frac{\partial}{\partial t}\phi - \tfrac{1}{2}|\nabla^x \phi|^2 - \beta\Big). \tag{6.29}$$

The initial conditions in $(6.21)_{4,5}$ in general cannot be specified arbitrarily when seeking this form of solution. They must be compatible with the irrotational motion determined by (6.27), (6.28) and (6.29).

Remarks:

(1) When ρ is eliminated using (6.29), the system in (6.27) and (6.28) forms a nonlinear initial-boundary value problem for the scalar field ϕ. This system is noticeably more complicated compared to the corresponding system in (6.15) and (6.16) for an ideal fluid. Questions of existence and uniqueness of solutions for such systems are in general very difficult.

(2) In the case when all fields are steady, the partial differential equation for ϕ in (6.27) reduces to

$$\nabla^x \cdot (\rho \nabla^x \phi) = 0, \quad \forall \boldsymbol{x} \in D, \tag{6.30}$$

and the equation for ρ in (6.29) reduces to

$$\rho = \zeta\Big(f - \tfrac{1}{2}|\nabla^x \phi|^2 - \beta\Big), \tag{6.31}$$

where f is a constant determined by the motion. In particular, if the values of ρ and $\boldsymbol{v}$ are known at a reference point $\boldsymbol{x}_*$, then from (6.26) we deduce

$$f = \tfrac{1}{2}|\boldsymbol{v}_*|^2 + \gamma(\rho_*) + \beta_*.$$

Thus, in the steady case, we obtain a nonlinear boundary-value problem for the field ϕ defined by (6.30), (6.31) and (6.28). □

6.2.7 Linearization

Consider a body of elastic fluid that occupies a fixed region D as shown in Figure 6.1. Suppose the fluid is initially at rest, that the mass density is initially uniform, and that there are no body forces so

$$\boldsymbol{v}_0(\boldsymbol{x}) = \boldsymbol{0}, \quad \rho_0(\boldsymbol{x}) = \rho^* > 0, \quad \boldsymbol{b}(\boldsymbol{x}, t) = \boldsymbol{0}.$$

Under these assumptions, the equations in (6.21) have a uniform solution for all $t \geq 0$, namely

$$\boldsymbol{v}(\boldsymbol{x}, t) = \boldsymbol{0}, \quad \rho(\boldsymbol{x}, t) = \rho^*.$$

Such a solution is called a **quiescent state.**

Suppose now we consider the initial-boundary value problem (6.21) with initial data close to a quiescent state in the sense that

$$|\boldsymbol{v}_0(\boldsymbol{x})| = \mathcal{O}(\epsilon), \quad |\rho_0(\boldsymbol{x}) - \rho^*| = \mathcal{O}(\epsilon),$$

where $0 \leq \epsilon \ll 1$ is a small parameter. Furthermore, suppose the boundary of D now vibrates with normal velocity $\vartheta = \mathcal{O}(\epsilon)$, so that the boundary condition in (6.21) becomes

$$\boldsymbol{v} \cdot \boldsymbol{n} = \vartheta = \mathcal{O}(\epsilon).$$

In this case, it is reasonable to expect that solutions of (6.21) will deviate only slightly from a quiescent state in the sense that

$$|\boldsymbol{v}(\boldsymbol{x}, t)| = \mathcal{O}(\epsilon), \quad |\rho(\boldsymbol{x}, t) - \rho^*| = \mathcal{O}(\epsilon).$$

Our goal is to derive a simplified set of equations describing such motions.

To begin, we note that if the initial and boundary conditions in (6.21) depend on a small parameter ϵ as discussed above, then the velocity and density fields within the body will also depend on ϵ. We denote this dependence by writing $\boldsymbol{v}^\epsilon$ and ρ^ϵ. Expanding $\boldsymbol{v}^\epsilon$ and ρ^ϵ in a power series in ϵ we have

$$\begin{aligned} \boldsymbol{v}^\epsilon &= \boldsymbol{0} + \epsilon \boldsymbol{v}^{(1)} + \epsilon^2 \boldsymbol{v}^{(2)} + \mathcal{O}(\epsilon^3), \\ \rho^\epsilon &= \rho^* + \epsilon \rho^{(1)} + \epsilon^2 \rho^{(2)} + \mathcal{O}(\epsilon^3), \end{aligned} \tag{6.32}$$

where $\boldsymbol{v}^{(k)}$ and $\rho^{(k)}$ $(k = 1, 2, \ldots)$ are unknown fields. Since $\boldsymbol{v}^\epsilon$ and ρ^ϵ must satisfy the balance equations $(6.21)_{1,2}$ we have

$$\begin{aligned} \rho^\epsilon \left[\frac{\partial}{\partial t}\boldsymbol{v}^\epsilon + (\nabla^x \boldsymbol{v}^\epsilon)\boldsymbol{v}^\epsilon\right] &= -\pi'(\rho^\epsilon)\nabla^x \rho^\epsilon, \\ \frac{\partial}{\partial t}\rho^\epsilon + \nabla^x \cdot (\rho^\epsilon \boldsymbol{v}^\epsilon) &= 0. \end{aligned} \tag{6.33}$$

Substituting (6.32) into $(6.33)_1$ gives

$$\begin{aligned} \Big(\rho^* + \epsilon\rho^{(1)} + \mathcal{O}(\epsilon^2)\Big) \Big[\frac{\partial}{\partial t}\Big(\epsilon \boldsymbol{v}^{(1)} + \mathcal{O}(\epsilon^2)\Big) \\ + \Big(\nabla^x (\epsilon \boldsymbol{v}^{(1)} + \mathcal{O}(\epsilon^2))\Big)\Big(\epsilon \boldsymbol{v}^{(1)} + \mathcal{O}(\epsilon^2)\Big)\Big] \\ = -\pi'(\rho^* + \mathcal{O}(\epsilon))\nabla^x \Big(\rho^* + \epsilon\rho^{(1)} + \mathcal{O}(\epsilon^2)\Big), \end{aligned} \tag{6.34a}$$

and substituting (6.32) into $(6.33)_2$ gives

$$\begin{aligned} \frac{\partial}{\partial t}\Big(\rho^* + \epsilon\rho^{(1)} + \mathcal{O}(\epsilon^2)\Big) \\ + \nabla^x \cdot \Big((\rho^* + \epsilon\rho^{(1)} + \mathcal{O}(\epsilon^2))(\epsilon \boldsymbol{v}^{(1)} + \mathcal{O}(\epsilon^2))\Big) = 0. \end{aligned} \tag{6.34b}$$

Notice that similar expansions can be written for the boundary and initial conditions in $(6.21)_{3,4,5}$. Assuming that these expansions and (6.34) hold for each power of ϵ, and using the fact that

$$\pi'(\rho^* + \mathcal{O}(\epsilon)) = \pi'(\rho^*) + \mathcal{O}(\epsilon),$$

we collect terms involving the first power of ϵ and arrive at the following problem: Find $\boldsymbol{v}^{(1)} : D \times [0, T] \to \mathcal{V}$ and $\rho^{(1)} : D \times [0, T] \to \mathbb{R}$ such that

$$\boxed{\begin{array}{ll} \frac{\partial}{\partial t}\boldsymbol{v}^{(1)} + \frac{1}{\rho^*}\pi'(\rho^*)\nabla^x \rho^{(1)} = \mathbf{0} & \text{in} \quad D \times [0, T] \\ \frac{\partial}{\partial t}\rho^{(1)} + \rho^* \nabla^x \cdot \boldsymbol{v}^{(1)} = 0 & \text{in} \quad D \times [0, T] \\ \boldsymbol{v}^{(1)} \cdot \boldsymbol{n} = \vartheta & \text{in } \partial D \times [0, T] \\ \boldsymbol{v}^{(1)}(\cdot, 0) = \boldsymbol{v}_0^{(1)}(\cdot) & \text{in} \quad D \\ \rho^{(1)}(\cdot, 0) = \rho_0^{(1)}(\cdot) & \text{in} \quad D. \end{array}} \tag{6.35}$$

These equations are typically called the **Acoustic Equations**. They are an approximate system of balance equations that are appropriate for describing small deviations from a quiescent state in an elastic fluid.

Remarks:

(1) The system in (6.35) is a linear initial-boundary value problem for the disturbance fields $(\boldsymbol{v}^{(1)}, \rho^{(1)})$. We expect this system to have a unique solution on any given time interval $[0, T]$ under mild assumptions on D, $\boldsymbol{v}_0$ and ρ_0.

(2) Eliminating the velocity field $\boldsymbol{v}^{(1)}$ between $(6.35)_1$ and $(6.35)_2$ we find that the density disturbance field $\rho^{(1)}$ satisfies

$$\frac{\partial^2}{\partial t^2}\rho^{(1)} = \pi'(\rho^*)\Delta^x \rho^{(1)}. \tag{6.36}$$

This equation is typically referred to as the **Wave Equation**.

(3) The constitutive assumption $\pi' > 0$ guarantees that solutions of (6.36) are wave-like for any $\rho^* > 0$. In this respect, the quantity $\sqrt{\pi'(\rho^*)}$ corresponds to the wave speed. It is usually called the **speed of sound** in the body at density ρ^*. □

6.3 Newtonian Fluids

In this section we study the constitutive model for an incompressible Newtonian (viscous) fluid. We show that the model is frame-indifferent and satisfies the mechanical energy inequality, and then discuss a standard initial-boundary value problem.

6.3.1 Definition

A continuum body with reference configuration B is said to be an incompressible **Newtonian fluid** if:

(1) The reference mass density field $\rho_0(\boldsymbol{X})$ is uniform in the sense that $\rho_0(\boldsymbol{X}) = \rho_0 > 0$ (constant).

(2) The material is incompressible (see Section 5.6), which means that the spatial velocity field must satisfy

$$\nabla^x \cdot \boldsymbol{v} = 0.$$

(3) The Cauchy stress field is **Newtonian**, which means that there is a scalar field $p(\boldsymbol{x}, t)$ called the pressure and a constant fourth-order tensor C such that

$$\boldsymbol{S} = -p\boldsymbol{I} + \mathsf{C}(\nabla^x \boldsymbol{v}),$$

where $\nabla^x \boldsymbol{v}(\boldsymbol{x},t)$ is the spatial velocity gradient. The fourth-order tensor $\mathbf{C}$ is assumed to satisfy the left minor symmetry condition

$$(\mathbf{C}(\boldsymbol{A}))^T = \mathbf{C}(\boldsymbol{A}), \quad \forall \boldsymbol{A} \in \mathcal{V}^2, \tag{6.37}$$

and the trace condition

$$\operatorname{tr} \mathbf{C}(\boldsymbol{A}) = 0, \quad \forall \boldsymbol{A} \in \mathcal{V}^2, \ \operatorname{tr} \boldsymbol{A} = 0. \tag{6.38}$$

As in the case of an ideal fluid, Properties (1) and (2) and the conservation of mass equation (Result 5.5) imply that the spatial mass density is uniform in space and constant in time, in particular, $\rho(\boldsymbol{x},t) = \rho_0 > 0$. In view of the discussion in Section 5.6, the pressure field in Property (3) can be identified as the multiplier for the constraint in Property (2). In particular, the stress field in a Newtonian fluid has both a reactive and an active part. The reactive part $-p\boldsymbol{I}$ is determined by the incompressibility constraint, whereas the active part $\mathbf{C}(\nabla^x \boldsymbol{v})$ is determined by the spatial velocity gradient.

Condition (6.37) in Property (3) implies that the stress field is necessarily symmetric. Thus the balance equation for angular momentum (Result 5.8) is automatically satisfied and will not be considered further. Condition (6.38) in Property (3) implies that $p = -\frac{1}{3} \operatorname{tr} \boldsymbol{S}$ when $\operatorname{tr} \nabla^x \boldsymbol{v} = \nabla^x \cdot \boldsymbol{v} = 0$. Thus the multiplier p associated with the incompressibility constraint can be interpreted as the physical pressure. As a result, $-p\boldsymbol{I}$ and $\mathbf{C}(\nabla^x \boldsymbol{v})$ correspond to the spherical and deviatoric stress tensors, respectively (see Section 3.5.4).

The axiom of material frame-indifference places strict limitations on the possible form of $\mathbf{C}$. In particular, in Result 6.8 below we show that $\mathbf{C}$ must be of the form

$$\mathbf{C}(\nabla^x \boldsymbol{v}) = 2\mu \operatorname{sym}(\nabla^x \boldsymbol{v}),$$

where μ is a constant called the **absolute viscosity** of the fluid. Arguments based on the mechanical energy inequality imply that $\mu \geq 0$. Notice that the incompressible Newtonian fluid model reduces to the ideal fluid model when $\mu = 0$.

6.3.2 Navier–Stokes Equations

Setting $\rho = \rho_0$ and $\boldsymbol{S} = -p\boldsymbol{I} + 2\mu \operatorname{sym}(\nabla^x \boldsymbol{v})$ in the balance of linear momentum equation (Result 5.7) and the conservation of mass equation (Result 5.5), we obtain a closed system of equations for the spatial velocity and pressure fields in a body of Newtonian fluid. In particular, we obtain

$$\rho_0 \dot{\boldsymbol{v}} = \nabla^x \cdot (-p\boldsymbol{I} + 2\mu \,\mathrm{sym}[\nabla^x \boldsymbol{v}]) + \rho_0 \boldsymbol{b} \quad \text{and} \quad \nabla^x \cdot \boldsymbol{v} = 0,$$

where $\boldsymbol{b}$ is a prescribed spatial body force field per unit mass. Notice that the conservation of mass equation reduces to the incompressibility constraint on the spatial velocity field.

In view of Result 4.7 we have

$$\dot{\boldsymbol{v}} = \frac{\partial}{\partial t}\boldsymbol{v} + (\nabla^x \boldsymbol{v})\boldsymbol{v},$$

and from the definition of $\boldsymbol{S}$ and $\mathrm{sym}[\nabla^x \boldsymbol{v}]$ we get

$$\nabla^x \cdot \boldsymbol{S} = -\nabla^x p + \mu \nabla^x \cdot (\nabla^x \boldsymbol{v}) + \mu \nabla^x \cdot (\nabla^x \boldsymbol{v})^T.$$

Using the definition of the divergence of a second-order tensor we find $\nabla^x \cdot (\nabla^x \boldsymbol{v}) = \Delta^x \boldsymbol{v}$ and $\nabla^x \cdot (\nabla^x \boldsymbol{v})^T = \nabla^x (\nabla^x \cdot \boldsymbol{v})$. These results together with the condition $\nabla^x \cdot \boldsymbol{v} = 0$ imply

$$\nabla^x \cdot \boldsymbol{S} = -\nabla^x p + \mu \Delta^x \boldsymbol{v}.$$

Thus the spatial velocity and pressure fields in a body of Newtonian fluid with reference configuration B must satisfy the following equations for all $\boldsymbol{x} \in B_t$ and $t \geq 0$

$$\boxed{\begin{aligned} &\rho_0 \left[\frac{\partial}{\partial t}\boldsymbol{v} + (\nabla^x \boldsymbol{v})\boldsymbol{v}\right] = \mu \Delta^x \boldsymbol{v} - \nabla^x p + \rho_0 \boldsymbol{b}, \\ &\nabla^x \cdot \boldsymbol{v} = 0. \end{aligned}} \tag{6.39}$$

These equations are known as the **Navier–Stokes Equations** for a Newtonian fluid.

Remarks:

(1) As for an ideal fluid, there is no explicit equation that relates pressure to velocity or density in a Newtonian fluid. Instead, the pressure field is a fundamental unknown that must be determined simultaneously with the velocity field.

(2) The equations in (6.39) can determine pressure only to within an additive function of time. That is, if the pair $\boldsymbol{v}(\boldsymbol{x}, t)$, $p(\boldsymbol{x}, t)$ satisfies (6.39), then so does $\boldsymbol{v}(\boldsymbol{x}, t)$, $p(\boldsymbol{x}, t) + f(t)$ for any arbitrary scalar-valued function $f(t)$. This follows from the fact that $\nabla^x (p + f) = \nabla^x p$.

(3) The main difference between a Newtonian fluid and an ideal fluid is that a Newtonian fluid is able to develop shear stresses. In particular, the traction $\boldsymbol{t}$ on a surface with normal $\boldsymbol{n}$ is not necessarily normal to the surface since $\boldsymbol{t} = \boldsymbol{S}\boldsymbol{n} = -p\boldsymbol{n} + 2\mu\,\mathrm{sym}(\nabla^x \boldsymbol{v})\boldsymbol{n}$.

(4) From (6.39) we see that a Newtonian fluid model reduces to the ideal fluid model in the case when $\mu = 0$. When μ is small, a Newtonian fluid can often be approximated by an ideal fluid in regions of motion far from solid boundaries. Close to such boundaries a Newtonian fluid develops **boundary layers** in which the viscous shear stresses become dominant.

(5) An important quantity arising in the study of flows described by equations (6.39) is the **Reynolds number** $Re = \rho_0 \vartheta_* \ell_* / \mu$ (a dimensionless constant), where ρ_0 is the mass density, ϑ_* is a characteristic speed and ℓ_* is a characteristic length associated with the flow. Flows with a low Reynolds number are typically smooth or **laminar**, while flows with a high Reynolds number are typically fluctuating or **turbulent**.

(6) When the spatial acceleration term is neglected, the equations in (6.39) take the form

$$\begin{aligned} &\mu \Delta^x \boldsymbol{v} - \nabla^x p + \rho_0 \boldsymbol{b} = \boldsymbol{0}, \\ &\nabla^x \cdot \boldsymbol{v} = 0. \end{aligned}$$

These linear equations are typically referred to as the **Stokes Equations**. They provide an approximate system of balance equations appropriate for motions which are nearly steady and slow, and which have small velocity gradients. □

6.3.3 Frame-Indifference Considerations

As discussed in Section 5.6, the Cauchy stress field in a body subject to a material constraint can be decomposed as

$$\boldsymbol{S}(\boldsymbol{x}, t) = \boldsymbol{S}^{(r)}(\boldsymbol{x}, t) + \boldsymbol{S}^{(a)}(\boldsymbol{x}, t),$$

where $\boldsymbol{S}^{(r)}$ is a reactive stress determined by the constraint, and $\boldsymbol{S}^{(a)}$ is an active stress determined by a given constitutive equation. For a Newtonian fluid we have

$$\boldsymbol{S}^{(r)}(\boldsymbol{x}, t) = -p(\boldsymbol{x}, t)\boldsymbol{I}, \qquad \boldsymbol{S}^{(a)}(\boldsymbol{x}, t) = \mathsf{C}(\nabla^x \boldsymbol{v}(\boldsymbol{x}, t)),$$

where p is the multiplier associated with the incompressibility constraint $\nabla^x \cdot \boldsymbol{v} = 0$, or equivalently $\det \boldsymbol{F} = 1$. This constraint can be written in the standard material form

$$\gamma(\boldsymbol{F}(\boldsymbol{X}, t)) = 0,$$

where $\gamma(\boldsymbol{F}) = \det \boldsymbol{F} - 1$.

For a constrained model to be consistent with the axiom of material frame-indifference (Axiom 5.24) it is sufficient that the constraint field $\gamma(\boldsymbol{F}(\boldsymbol{X}, t))$ and active stress field $\boldsymbol{S}^{(a)}(\boldsymbol{x}, t)$ be frame-indifferent in the sense of Definition 5.23. Previous arguments used for an ideal fluid show that $\gamma(\boldsymbol{F}(\boldsymbol{X}, t))$ is frame-indifferent. Conditions under which $\boldsymbol{S}^{(a)}(\boldsymbol{x}, t)$ is frame-indifferent are established in the following result.

Result 6.8 ***Frame-Indifference of Newtonian Active Stress.*** *The constitutive model for the active stress in a Newtonian fluid*

$$\boldsymbol{S}^{(a)} = \mathbf{C}(\nabla^x \boldsymbol{v}),$$

is frame-indifferent if and only if

$$\mathbf{C}(\nabla^x \boldsymbol{v}) = 2\mu \,\mathrm{sym}(\nabla^x \boldsymbol{v}) = 2\mu \boldsymbol{L}, \tag{6.40}$$

where $\boldsymbol{L} = \mathrm{sym}(\nabla^x \boldsymbol{v})$ *is the rate of strain tensor and* μ *is a scalar constant.* □

Proof For simplicity, we only show sufficiency of (6.40). To begin, notice that the fourth-order tensor $\mathbf{C}$ defined in (6.40) satisfies the required properties in (6.37) and (6.38). In particular, (6.37) is satisfied since

$$(\mathbf{C}(\boldsymbol{A}))^T = (2\mu \,\mathrm{sym}(\boldsymbol{A}))^T = 2\mu \,\mathrm{sym}(\boldsymbol{A}) = \mathbf{C}(\boldsymbol{A}).$$

Using the fact that $\mathrm{tr}(\mathrm{sym}(\boldsymbol{A})) = \mathrm{tr}\, \boldsymbol{A}$ for any second-order tensor $\boldsymbol{A}$, we find that the condition in (6.38) is also satisfied since

$$\mathrm{tr}\, \mathbf{C}(\boldsymbol{A}) = 2\mu \,\mathrm{tr}(\mathrm{sym}(\boldsymbol{A})) = 2\mu \,\mathrm{tr}\, \boldsymbol{A} = 0 \quad \text{when} \quad \mathrm{tr}\, \boldsymbol{A} = 0.$$

To show that the constitutive model defined by (6.40) is consistent with the axiom of material frame-indifference, let B_t and B_t^* denote two configurations of a body of Newtonian fluid that are related by a superposed rigid motion

$$\boldsymbol{x}^* = \boldsymbol{g}(\boldsymbol{x}, t) = \boldsymbol{Q}(t)\boldsymbol{x} + \boldsymbol{c}(t),$$

where $\boldsymbol{Q}(t)$ is an arbitrary rotation tensor and $\boldsymbol{c}(t)$ is an arbitrary vector. Let $\boldsymbol{S}^{(a)}(\boldsymbol{x}, t)$ denote the active stress in B_t, and let $\boldsymbol{S}^{(a)*}(\boldsymbol{x}^*, t)$

denote the active stress in B_t^*. By Result 5.22 we have $\boldsymbol{L}^*(\boldsymbol{x}^*, t) = \boldsymbol{Q}(t)\boldsymbol{L}(\boldsymbol{x}, t)\boldsymbol{Q}(t)^T$, where $\boldsymbol{x}^* = \boldsymbol{g}(\boldsymbol{x}, t)$, and from this we deduce (omitting the arguments $\boldsymbol{x}^*$, $\boldsymbol{x}$ and t for brevity)

$$\boldsymbol{S}^{(a)*} = 2\mu\boldsymbol{L}^* = 2\mu\boldsymbol{Q}\boldsymbol{L}\boldsymbol{Q}^T = \boldsymbol{Q}\boldsymbol{S}^{(a)}\boldsymbol{Q}^T.$$

Thus the constitutive relation for the active stress is frame-indifferent. □

6.3.4 Mechanical Energy Considerations

Here we assume the isothermal model of a Newtonian fluid is energetically passive as discussed in Section 5.7 and show that it satisfies the mechanical energy inequality (Result 5.29) under an appropriate restriction on μ.

Since $\boldsymbol{S} = -p\boldsymbol{I} + 2\mu\boldsymbol{L}$, for any motion of a Newtonian fluid we have

$$\boldsymbol{S} : \boldsymbol{L} = -p\boldsymbol{I} : \boldsymbol{L} + 2\mu\boldsymbol{L} : \boldsymbol{L}.$$

Using the fact that $\boldsymbol{L} = \mathrm{sym}(\nabla^x \boldsymbol{v})$ we get $\boldsymbol{I} : \boldsymbol{L} = \mathrm{tr}(\mathrm{sym}(\nabla^x \boldsymbol{v})) = \nabla^x \cdot \boldsymbol{v}$, which vanishes in view of the incompressibility constraint. Thus

$$\boldsymbol{S} : \boldsymbol{L} = 2\mu\boldsymbol{L} : \boldsymbol{L},$$

and for any time interval $[t_0, t_1]$ we obtain

$$\int_{t_0}^{t_1} (\det \boldsymbol{F})\, \boldsymbol{S}_m : \boldsymbol{L}_m \, dt = 2\mu \int_{t_0}^{t_1} (\det \boldsymbol{F})\, \boldsymbol{L}_m : \boldsymbol{L}_m \, dt.$$

The mechanical energy inequality requires that the left-hand side be non-negative for every closed isothermal process in $[t_0, t_1]$. Since the conditions $\det \boldsymbol{F} > 0$ holds for all admissible processes, and since $\boldsymbol{L}_m : \boldsymbol{L}_m \geq 0$ by properties of the tensor inner-product, we conclude that a Newtonian fluid model satisfies the mechanical energy inequality if and only if $\mu \geq 0$.

6.3.5 Initial-Boundary Value Problems

An initial-boundary value problem for a body of Newtonian fluid is a set of equations that describe the motion of the body subject to specified initial conditions in B at time $t = 0$, and boundary conditions on ∂B_t for $t \geq 0$. Just as for ideal and elastic fluids, the Eulerian form of the balance laws for a Newtonian fluid are particularly well-suited for those problems in which the body occupies a fixed region D of space. In this case the spatial velocity field $\boldsymbol{v}$ and the spatial pressure field p can be determined independently of the motion $\boldsymbol{\varphi}$. As before, we typically assume D is a bounded open set as shown in Figure 6.1. However, it is also useful in applications to consider unbounded open sets such as the exterior of the region shown in the figure, or the whole of Euclidean space.

A standard initial-boundary value problem for a body of Newtonian fluid occupying a fixed region D can be stated as follows: Find $\boldsymbol{v} : D \times [0,T] \to \mathcal{V}$ and $p : D \times [0,T] \to I\!R$ such that

$$\boxed{\begin{array}{ll} \rho_0 \left[\frac{\partial}{\partial t}\boldsymbol{v} + (\nabla^x \boldsymbol{v})\boldsymbol{v}\right] = \mu \Delta^x \boldsymbol{v} - \nabla^x p + \rho_0 \boldsymbol{b} & \text{in} \quad D \times [0,T] \\ \nabla^x \cdot \boldsymbol{v} = 0 & \text{in} \quad D \times [0,T] \\ \boldsymbol{v} = \boldsymbol{0} & \text{in } \partial D \times [0,T] \\ \boldsymbol{v}(\cdot, 0) = \boldsymbol{v}_0(\cdot) & \text{in} \quad D. \end{array}} \tag{6.41}$$

In the above system, ρ_0 is a prescribed constant mass density and $\boldsymbol{b}$ is a prescribed spatial body force per unit mass. Equation $(6.41)_1$ is the balance of linear momentum equation and $(6.41)_2$ is the material incompressibility constraint, which in the current case is equivalent to the conservation of mass equation. Equation $(6.41)_3$ is the so-called **no-slip boundary condition** that expresses the fact that the fluid cannot cross or slip along the boundary of D and $(6.41)_4$ is an initial condition. Here $\boldsymbol{v}_0$ is a given field which corresponds to the fluid velocity field at time $t = 0$. The specified field $\boldsymbol{v}_0$ should be compatible with the incompressibility and boundary conditions in the sense that $\nabla^x \cdot \boldsymbol{v}_0 = 0$ in D and $\boldsymbol{v}_0 = \boldsymbol{0}$ on ∂D.

Remarks:

(1) The system in (6.41) is a nonlinear initial-boundary value problem for the fields $(\boldsymbol{v}, p)$. This system is guaranteed to have a solution on some finite time interval $[0,T]$ provided the region D is sufficiently regular, and the field $\boldsymbol{v}_0$ is sufficiently smooth and compatible with the incompressibility and boundary conditions.

(2) When D is bounded, the field $\boldsymbol{v}$ is unique, whereas p is unique only to within an additive function of time. When D is unbounded, additional boundary conditions at infinity are typically imposed, which result in uniqueness for both $\boldsymbol{v}$ and p. For simplicity, we as-

sume that solutions are continuously differentiable in all variables as many times as are required for the calculations we perform.

(3) Although a solution to (6.41) can be guaranteed on a finite time interval depending on the given data of a problem, global existence for all time is still an open question. Moreover, just as for the ideal and elastic fluid equations, numerical approximation is generally required to obtain quantitative information about solutions.

(4) The standard boundary condition for a Newtonian fluid is different than the one for ideal and elastic fluids. In the Newtonian case all components of velocity, normal as well as tangential, must vanish at a fixed boundary. Just as for ideal and elastic fluids, the normal component vanishes because fluid cannot penetrate the boundary. In contrast to ideal and elastic fluids, the tangential components vanish because of viscous shear stresses.

(5) The boundary condition $\boldsymbol{v} = \mathbf{0}$ on ∂D is appropriate for describing the interface between a Newtonian fluid and a fixed, impermeable solid. If the solid were not fixed, but instead had a prescribed velocity field $\boldsymbol{\vartheta}$, then an appropriate boundary condition would be $\boldsymbol{v} = \boldsymbol{\vartheta}$ on ∂D. Other boundary conditions can also be considered when the interface is not fluid-solid, or when the motion of the boundary is not known a-priori. Also, different conditions could be imposed on different parts of the boundary. □

6.4 Kinetic Energy of Fluid Motion

Here we derive some results on the dissipation of kinetic energy in fluid motion. We restrict attention to motions with conservative body forces in regions with fixed boundaries, and show that an ideal fluid has no mechanism for dissipation while a Newtonian fluid does. We begin by presenting two results that will prove useful in our developments. The first result establishes an integral identity involving the gradient and Laplacian of a smooth vector field.

Result 6.9 ***Integration by Parts.*** *Let $\boldsymbol{v}$ be a smooth vector field in a regular, bounded region D in $\mathbb{E}^3$ and suppose $\boldsymbol{v} = \mathbf{0}$ on ∂D. Then*

$$\int_D (\Delta^x \boldsymbol{v}) \cdot \boldsymbol{v}\, dV_{\boldsymbol{x}} = -\int_D (\nabla^x \boldsymbol{v}) : (\nabla^x \boldsymbol{v})\, dV_{\boldsymbol{x}}.$$

□

Proof From the definition of the vector Laplacian (see Definition 2.14) we have

$$
\begin{aligned}
(\Delta^x \boldsymbol{v}) \cdot \boldsymbol{v} = v_{i,jj} v_i &= (v_{i,j} v_i)_{,j} - v_{i,j} v_{i,j} \\
&= \nabla^x \cdot ((\nabla^x \boldsymbol{v})^T \boldsymbol{v}) - (\nabla^x \boldsymbol{v}) : (\nabla^x \boldsymbol{v}).
\end{aligned}
$$

Integrating this expression over D yields

$$
\begin{aligned}
\int_D (\Delta^x \boldsymbol{v}) \cdot \boldsymbol{v} \, dV_{\boldsymbol{x}} &= \int_D \nabla^x \cdot ((\nabla^x \boldsymbol{v})^T \boldsymbol{v}) \, dV_{\boldsymbol{x}} - \int_D (\nabla^x \boldsymbol{v}) : (\nabla^x \boldsymbol{v}) \, dV_{\boldsymbol{x}} \\
&= \int_{\partial D} ((\nabla^x \boldsymbol{v})^T \boldsymbol{v}) \cdot \boldsymbol{n} \, dA_{\boldsymbol{x}} - \int_D (\nabla^x \boldsymbol{v}) : (\nabla^x \boldsymbol{v}) \, dV_{\boldsymbol{x}} \\
&= -\int_D (\nabla^x \boldsymbol{v}) : (\nabla^x \boldsymbol{v}) \, dV_{\boldsymbol{x}},
\end{aligned}
$$

where the last line follows from the fact that $\boldsymbol{v} = \boldsymbol{0}$ on ∂D. □

The next result establishes an integral inequality between a smooth vector field and its gradient (see Exercise 13 for an indication of the proof).

Result 6.10 ***Poincaré Inequality.*** *Let D be a regular, bounded region in $\boldsymbol{E}^3$. Then there is a scalar constant $\lambda > 0$, depending only on D, with the property*

$$
\int_D |\boldsymbol{w}|^2 \, dV_{\boldsymbol{x}} \leq \lambda \int_D \nabla \boldsymbol{w} : \nabla \boldsymbol{w} \, dV_{\boldsymbol{x}},
$$

for all smooth vector fields $\boldsymbol{w}$ satisfying $\boldsymbol{w} = \boldsymbol{0}$ on ∂D. □

In the above result notice that the physical dimensions of λ are length squared. In fact, the proof of the above result shows that, for a domain of fixed shape, dilation (rescaling) by a factor of α will cause λ to change to $\alpha^2 \lambda$. Our main result concerning the dissipation of kinetic energy can now be stated.

Result 6.11 ***Kinetic Energy of Newtonian and Ideal Fluids.*** *Consider a body of either Newtonian or ideal fluid occupying a regular, bounded region D with fixed boundary so that $B_t = D$ for all $t \geq 0$. Let $\rho_0 > 0$ be the constant mass density, let $K(t)$ denote the kinetic energy of the body at time t defined by*

$$
K(t) = \int_{B_t} \tfrac{1}{2} \rho |\boldsymbol{v}|^2 \, dV_{\boldsymbol{x}} = \int_D \tfrac{1}{2} \rho_0 |\boldsymbol{v}|^2 \, dV_{\boldsymbol{x}},
$$

and let $K_0 = K(0)$. Moreover, suppose the body is subject to a conservative body force per unit mass $\boldsymbol{b} = -\nabla^x \beta$.

(1) For a body of Newtonian fluid described by the Navier–Stokes equations, the kinetic energy satisfies

$$K(t) \leq e^{-2\mu t/\lambda\rho_0} K_0, \quad \forall t \geq 0.$$

Thus the kinetic energy of a Newtonian fluid body dissipates to zero exponentially fast, for any initial condition.

(2) For a body of ideal fluid described by the Euler equations, the kinetic energy satisfies

$$K(t) = K_0, \qquad \forall t \geq 0.$$

Thus the kinetic energy of an ideal fluid body does not vary with time.

□

Proof By definition of the kinetic energy we have

$$\frac{d}{dt}K(t) = \frac{d}{dt}\int_{B_t} \tfrac{1}{2}\rho|\boldsymbol{v}|^2 \, dV_{\boldsymbol{x}} = \int_{B_t} \tfrac{1}{2}\rho\left(|\boldsymbol{v}|^2\right)^{\bullet} \, dV_{\boldsymbol{x}} = \int_{B_t} \rho\dot{\boldsymbol{v}} \cdot \boldsymbol{v} \, dV_{\boldsymbol{x}},$$

where we have used Result 5.6 for differentiation of the integral. Since $\rho = \rho_0$ (constant) and $B_t = D$, the above expression reduces to

$$\frac{d}{dt}K(t) = \int_D \rho_0\dot{\boldsymbol{v}} \cdot \boldsymbol{v} \, dV_{\boldsymbol{x}}. \tag{6.42}$$

For the Navier–Stokes equations with a conservative body force we have

$$\rho_0\dot{\boldsymbol{v}} = \mu\Delta^x \boldsymbol{v} - \nabla^x \psi, \quad \forall \boldsymbol{x} \in D, \ t \geq 0,$$

where $\psi = p + \rho_0\beta$. For the Euler equations we have the same expression, but with $\mu = 0$. Substituting for $\rho_0\dot{\boldsymbol{v}}$ in (6.42) yields

$$\frac{d}{dt}K(t) = \int_D \left[\mu(\Delta^x \boldsymbol{v}) \cdot \boldsymbol{v} - \nabla^x \psi \cdot \boldsymbol{v}\right] dV_{\boldsymbol{x}}. \tag{6.43}$$

We next show that the second term on the right-hand side vanishes. In particular, since $\nabla^x \cdot \boldsymbol{v} = 0$ for both the Navier–Stokes and Euler equations, we have

$$\nabla^x \cdot (\psi\boldsymbol{v}) = \nabla^x \psi \cdot \boldsymbol{v} + (\nabla^x \cdot \boldsymbol{v})\psi = \nabla^x \psi \cdot \boldsymbol{v}.$$

Moreover, since $\boldsymbol{v} \cdot \boldsymbol{n} = 0$ on ∂D for both the Navier–Stokes and Euler equations, we obtain

$$\int_D \nabla^x \psi \cdot \boldsymbol{v} \, dV_{\boldsymbol{x}} = \int_D \nabla^x \cdot (\psi\boldsymbol{v}) \, dV_{\boldsymbol{x}} = \int_{\partial D} \psi\boldsymbol{v} \cdot \boldsymbol{n} \, dA_{\boldsymbol{x}} = 0.$$

For the remaining term on the right-hand side of (6.43) we can apply Result 6.9 in the Navier–Stokes case, or set $\mu = 0$ in the Euler case, and either way we obtain

$$\frac{d}{dt}K(t) = -\mu \int_D \nabla^x \boldsymbol{v} : \nabla^x \boldsymbol{v} \, dV_{\boldsymbol{x}}, \quad \forall t \geq 0. \tag{6.44}$$

For the Navier–Stokes case we apply the Poincaré inequality to the right-hand side of (6.44) and obtain

$$\frac{d}{dt}K(t) \leq -\frac{\mu}{\lambda} \int_D |\boldsymbol{v}|^2 \, dV_{\boldsymbol{x}} = -\frac{2\mu}{\lambda\rho_0} K(t), \tag{6.45}$$

where λ depends on the domain D. Multiplying (6.45) by the factor $e^{2\mu t/\lambda\rho_0}$ we get

$$\frac{d}{ds}\left(e^{2\mu s/\lambda\rho_0} K(s)\right) \leq 0,$$

and integrating from $s = 0$ to $s = t$ yields the first result, namely

$$K(t) \leq e^{-2\mu t/\lambda\rho_0} K_0, \quad \forall t \geq 0.$$

For the Euler case we set $\mu = 0$ in (6.44) and obtain

$$\frac{d}{dt}K(t) = 0, \quad \forall t \geq 0.$$

This yields the second result, namely $K(t) = K_0$ for all $t \geq 0$. □

Remarks:

(1) The ideal fluid model has no mechanism for the dissipation of kinetic (mechanical) energy, whereas a Newtonian model does. The lack or presence of dissipation can be attributed to the lack or presence of viscous shear stresses as controlled by the fluid viscosity μ.

(2) The above result shows that the kinetic energy in a body of Newtonian fluid tends to zero in the absence of fluid motion on the boundary, provided the body force is conservative. The rate of decay depends upon μ, ρ_0 and λ. For fluids with a high viscosity to density ratio the rate of decay is very fast.

(3) The parameter λ appearing in the decay estimate is entirely determined by the geometry of the domain D. For fixed fluid properties, the rate of decay of kinetic energy decreases as a domain of fixed shape is increased in size. □

Bibliographic Notes

General introductions to the theory of fluid mechanics can be found in Batchelor (2000), Chorin and Marsden (1993) and the article by Serrin (1959). Various engineering applications can be found in White (2006) and Happel and Brenner (1983). The study of waves on liquid surfaces is a classic subject which has a long and rich history. General treatments of this subject are given in Johnson (1997), Whitham (1974) and Stoker (1957), and a more engineering approach is given in Dean and Dalrymple (1984).

The subject of existence and uniqueness of solutions for the Euler equations is discussed in detail in the text of Majda and Bertozzi (2002). For the Navier–Stokes equations analogous questions are overviewed in the book of Constantin and Foias (1988); see also Robinson (2001), Doering and Gibbon (1995), Temam (1984) and Ladyzhenskaya (1969). The subject of existence and uniqueness of solutions to systems of hyperbolic conservations laws, such as those arising for the elastic fluid equations, are overviewed in the texts of Dafermos (2005) and Majda (1984).

The numerical solution of the equations of fluid mechanics plays a central role in both the development of mathematical and physical understanding, and in engineering practice. For an overview of numerical methods for fluid mechanics, see the contribution by Marion and Temam in Volume VI of the book by Ciarlet and Lions (1998). See also Girault and Raviart (1986) for the analysis of methods for the Navier–Stokes equations, and LeVeque (1992) for the analysis of methods for hyperbolic conservation laws.

A general proof of Result 6.2 can be found in Fleming (1977) in the language of differential forms. For an elementary proof which holds in all of space see Marsden and Tromba (1988). A complete proof of Result 6.8 can be found in Gurtin (1981), and a complete proof of Result 6.10 can be found in Oden and Demkowicz (1996).

Exercises

6.1 Consider an ideal fluid with mass density ρ_0 subject to a uniform, constant body force field per unit mass $\boldsymbol{b}$. Find the pressure field $p(\boldsymbol{x})$ assuming the fluid is at rest and that $p(\boldsymbol{x}_*) = p_*$ at some reference point $\boldsymbol{x}_*$.

6.2 Consider an ideal fluid in a region D with mass density ρ_0 subject to zero body force. Suppose D is the unbounded region exterior to a fixed, solid obstacle Ω. Show that, if the motion of the fluid is steady and irrotational, then the resultant force exerted by the fluid on the obstacle is

$$\boldsymbol{r} = \frac{\rho_0}{2} \int_{\partial\Omega} |\boldsymbol{v}|^2 \boldsymbol{n} \, dA_{\boldsymbol{x}},$$

where $\boldsymbol{n}$ is the outward unit normal field on $\partial\Omega$.

6.3 Let $D = \{\boldsymbol{x} \in \mathbb{E}^3 \mid 0 < x_i < 1\}$ and consider the scalar Laplace equation

$$\Delta^x \phi = 0, \quad \forall \boldsymbol{x} \in D,$$

together with the boundary condition

$$\nabla^x \phi \cdot \boldsymbol{n} = g, \quad \forall \boldsymbol{x} \in \partial D,$$

where g is a given function and $\boldsymbol{n}$ is the outward unit normal field on ∂D.

(a) Find ϕ assuming $g = \cos(k\pi x_1)\cos(l\pi x_2)$ on the face with $\boldsymbol{n} = \boldsymbol{e}_3$, and $g = 0$ elsewhere. Here k and l are arbitrary integers.

(b) Find ϕ assuming $g = \sum_{k,l=1}^{\infty} a_{kl} \cos(k\pi x_1)\cos(l\pi x_2)$ on the face with $\boldsymbol{n} = \boldsymbol{e}_3$, and $g = 0$ elsewhere. Here a_{kl} are constants.

(c) Find ϕ assuming $g = 0$ on all faces. What does this case say about irrotational motions of an ideal fluid in D?

6.4 Repeat Exercise 1, but now for an elastic fluid with mass density ρ and constitutive relation $p = \lambda\rho^\gamma$ where $\lambda > 0$ and $\gamma > 1$ are constants. (A relation of this form describes various real gases at moderate conditions, for example atmospheric air.)

6.5 Consider a body of elastic fluid, subject to zero body force, undergoing a steady, irrotational motion with spatial velocity field $\boldsymbol{v} = \nabla^x \phi$ in a fixed region D. Assume the constitutive relation $p = \frac{1}{2}\rho^2$.

(a) Find the associated functions $\gamma(s)$ and $\zeta(s)$ for this fluid.

(b) Show that the velocity potential ϕ satisfies the nonlinear equation

$$c\Delta^x \phi = \tfrac{1}{2}|\nabla^x \phi|^2 \Delta^x \phi + \nabla^x \phi \cdot (\nabla^x \nabla^x \phi)\nabla^x \phi, \quad \forall \boldsymbol{x} \in D,$$

where c is a constant. Here $\nabla^x \nabla^x \phi$ is the second-order tensor with components $[\nabla^x \nabla^x \phi]_{ij} = \phi_{,ij}$.

6.6 Let $c > 0$ be constant and consider the wave equation for the density disturbance field in an elastic fluid

$$\frac{\partial^2}{\partial t^2}\rho^{(1)} = c^2 \Delta^x \rho^{(1)}.$$

Show that

$$\rho^{(1)}(\boldsymbol{x}, t) = f(\boldsymbol{k}\cdot\boldsymbol{x} - ct) + g(\boldsymbol{k}\cdot\boldsymbol{x} + ct)$$

is a solution of the wave equation for any functions $f, g : \mathbb{R} \to \mathbb{R}$ and any unit vector $\boldsymbol{k}$. Solutions of this form are called **plane waves.** $f(\boldsymbol{k}\cdot\boldsymbol{x} - ct)$ is a wave profile which moves with speed c in the direction $\boldsymbol{k}$, and $g(\boldsymbol{k}\cdot\boldsymbol{x} + ct)$ is a profile which moves with speed c in the direction $-\boldsymbol{k}$.

6.7 Consider a body of elastic fluid with reference configuration B and equation of state $p = \pi(\rho)$ between the spatial pressure and spatial mass density.

(a) Show that the first Piola–Kirchhoff stress tensor for the fluid can be written in the form

$$\boldsymbol{P} = \boldsymbol{F}\overline{\boldsymbol{\Sigma}}(\boldsymbol{C}, \rho_0),$$

where $\overline{\boldsymbol{\Sigma}}(\boldsymbol{C}, \rho_0)$ is a certain function depending on the state function π. Here $\boldsymbol{C}$ is the Cauchy–Green strain tensor and ρ_0 is the reference mass density.

(b) Based on the result in (a) show that an elastic fluid may also be viewed as an isotropic elastic material as considered in Chapter 7, with stress response function depending on the reference mass density.

6.8 Let $\boldsymbol{v}$ be a smooth vector field satisfying the condition $\nabla^x \cdot \boldsymbol{v} = 0$ in a regular, bounded region D. Assuming $\boldsymbol{v}\cdot\boldsymbol{n} = 0$ on ∂D show

$$\int_D (\nabla^x \boldsymbol{v})\boldsymbol{v}\, dV_{\boldsymbol{x}} = \boldsymbol{0}.$$

6.9 Show that the Cauchy reactive stress $\boldsymbol{S}^{(r)}$ in either an ideal or Newtonian fluid body does no work in any motion compatible with the incompressibility constraint. In particular, show that its contribution to the stress power vanishes, that is

$$\boldsymbol{S}^{(r)} : \boldsymbol{L} = 0, \quad \forall \boldsymbol{x} \in B_t,\ t \geq 0.$$

Here $\boldsymbol{L} = \mathrm{sym}(\nabla^x \boldsymbol{v})$ is the rate of strain tensor and $\boldsymbol{v}$ is the spatial velocity field.

6.10 Consider the Navier–Stokes equations with zero body force, and consider solutions for which the velocity field $\boldsymbol{v}$ is of the form

$$\boldsymbol{v}(\boldsymbol{x}, t) = v_1(x_1, x_2, t)\boldsymbol{e}_1 + v_2(x_1, x_2, t)\boldsymbol{e}_2.$$

Such solutions are called **planar**. Moreover, consider velocity fields which can be represented in the form

$$\boldsymbol{v} = \frac{\partial \psi}{\partial x_2}\boldsymbol{e}_1 - \frac{\partial \psi}{\partial x_1}\boldsymbol{e}_2 =: \nabla^{x\perp} \psi,$$

where $\psi(x_1, x_2, t)$ is an arbitrary scalar-valued function called a **streamfunction**. (The notation $\nabla^{x\perp}$ is motivated by the fact that $\nabla^{x\perp}\psi \cdot \nabla^x \psi = 0$ for any ψ. The operator $\nabla^{x\perp}$ is sometimes referred to as a **skew gradient**.) Here we show that the streamfunction ψ for a planar solution of the Navier–Stokes equations satisfies a relatively simple equation.

(a) Show that the incompressibility constraint (conservation of mass equation) is automatically satisfied for any ψ.

(b) Let $\boldsymbol{w} = \nabla^x \times \boldsymbol{v}$ denote the vorticity field associated with $\boldsymbol{v}$. Show that

$$\boldsymbol{w} = w\boldsymbol{e}_3 \quad \text{where} \quad w = -\Delta^x \psi.$$

(c) Use the balance of linear momentum equation to show that w must satisfy

$$\frac{\partial w}{\partial t} + \boldsymbol{v} \cdot \nabla^x w = \nu \Delta^x w,$$

where $\nu = \mu/\rho_0$. Hence deduce that ψ must satisfy the equation

$$(\Delta^x)^2 \psi = \frac{1}{\nu}\Big[\frac{\partial}{\partial t}(\Delta^x \psi) + \nabla^{x\perp}\psi \cdot \nabla^x(\Delta^x \psi)\Big].$$

Remark: A good approximation to the above equation when $\nu \gg 1$ is

$$(\Delta^x)^2 \psi = 0.$$

This equation is usually called the **biharmonic equation**. It is also a good approximation when the flow is steady and gradients of ψ are small, so that a linear approximation is valid.

6.11 Consider a Newtonian fluid subject to zero body force occupying the unbounded region of space

$$D = \{\boldsymbol{x} \in \mathbb{E}^3 \mid \quad 0 < x_2 < h, \quad -\infty < x_1, x_3 < \infty\}.$$

Here we find steady solutions of the Navier–Stokes equations under various different boundary conditions assuming the velocity field has the simple, planar form

$$\boldsymbol{v}(\boldsymbol{x}) = v_1(x_1, x_2)\boldsymbol{e}_1.$$

(a) Show that, when written in components, the balance of linear momentum and conservation of mass equations reduce to

$$\mu\frac{\partial^2 v_1}{\partial x_2^2} = \frac{\partial p}{\partial x_1}, \quad \frac{\partial p}{\partial x_2} = 0, \quad \frac{\partial p}{\partial x_3} = 0, \quad \frac{\partial v_1}{\partial x_1} = 0.$$

Moreover, show that the general solution of these equations is

$$v_1 = \frac{\alpha}{2\mu}x_2^2 + \beta x_2 + \gamma, \quad p = \alpha x_1 + \delta,$$

where α, β, γ, δ are arbitrary constants.

(b) Find the velocity and pressure fields assuming the no-slip boundary condition $\boldsymbol{v} = \boldsymbol{0}$ at $x_2 = 0$ and $\boldsymbol{v} = \vartheta\boldsymbol{e}_1$ at $x_2 = h$. This solution describes a fluid between two infinite parallel plates. One plate is at $x_2 = 0$ and is stationary, the other is at $x_2 = h$ and is moving with speed ϑ in the x_1-direction. Further assumptions on the pressure are needed to fix a unique solution.

(c) Find the velocity and pressure fields assuming the no-slip boundary condition $\boldsymbol{v} = \boldsymbol{0}$ at $x_2 = 0$ and assuming a **traction boundary condition** of the form $\boldsymbol{S}\boldsymbol{n} = a\boldsymbol{e}_1 + b\boldsymbol{e}_2$ at $x_2 = h$. Here $\boldsymbol{S}$ is the Cauchy stress, $\boldsymbol{n}$ is the outward unit normal on ∂D and a, b are given constants. This solution describes a film of fluid of constant thickness on a flat plate. The plate is at $x_2 = 0$ and the free surface of the film is at $x_2 = h$.

6.12 Consider a steady velocity field of the form $\boldsymbol{u}(\boldsymbol{x}) = \boldsymbol{A}\nabla^x\psi(\boldsymbol{x})$, where ψ is a scalar function and $\boldsymbol{A}$ is a second-order tensor defined in a given frame by

$$\psi(\boldsymbol{x}) = \sin(x_1)\sin(x_2), \quad [\boldsymbol{A}] = \begin{pmatrix} 0 & 1 & 0 \\ -1 & 0 & 0 \\ 0 & 0 & 1 \end{pmatrix}.$$

(a) Show that $(\nabla^x \boldsymbol{u})\boldsymbol{u} = -\nabla^x \phi$, where

$$\phi(\boldsymbol{x}) = \tfrac{1}{4}\cos(2x_1) + \tfrac{1}{4}\cos(2x_2).$$

(b) Show that, for an appropriate choice of pressure field $q(\boldsymbol{x})$, the pair $(\boldsymbol{u}, q)$ is a steady solution of the Euler equations (6.1) with zero body force.

(c) Let $\boldsymbol{v}(\boldsymbol{x}, t) = e^{-\lambda t}\boldsymbol{u}(\boldsymbol{x})$. Show that, for an appropriate choice of pressure field $p(\boldsymbol{x}, t)$ and constant λ, the pair $(\boldsymbol{v}, p)$ is a non-steady solution of the Navier–Stokes equations (6.39) with zero body force.

Remark: The results in (b) and (c) are known as the **Taylor–Green solutions**. They are simple examples of exact solutions to the Euler and Navier–Stokes equations for bodies occupying all of space.

6.13 Let D be a regular, bounded region. Assume the eigenvalue problem

$$\begin{cases} -\Delta^x \boldsymbol{\eta} = \lambda \boldsymbol{\eta}, & \forall \boldsymbol{x} \in D, \\ \boldsymbol{\eta} = \boldsymbol{0}, & \forall \boldsymbol{x} \in \partial D, \end{cases}$$

has an infinite set of smooth eigenfunctions $\{\boldsymbol{\eta}_i\}_{i=1}^{\infty}$ and corresponding positive eigenvalues $\{\lambda_i\}_{i=1}^{\infty}$, ordered so that

$$0 < \lambda_1 \le \lambda_2 \le \cdots.$$

Since the eigenvalue problem is self-adjoint, the eigenfunctions may be chosen so that

$$\int_D \boldsymbol{\eta}_i \cdot \boldsymbol{\eta}_j \, dV_{\boldsymbol{x}} = \delta_{ij}, \quad i, j = 1, 2, \ldots.$$

Prove the Poincaré Inequality (Result 6.10) assuming that every smooth vector field $\boldsymbol{w}$ on D can be represented by an eigenfunction expansion

$$\boldsymbol{w} = \sum_{i=1}^{\infty} \alpha_i \boldsymbol{\eta}_i,$$

where $\{\alpha_i\}_{i=1}^{\infty}$ are constants depending on $\boldsymbol{w}$.

6.14 Let D be a given region. For any two smooth vector fields $\boldsymbol{v}, \boldsymbol{b}$ and any real number $s > 0$ show that

$$\int_D |\boldsymbol{b} \cdot \boldsymbol{v}| \, dV_{\boldsymbol{x}} \leq \frac{s^2}{2} \int_D |\boldsymbol{b}|^2 dV_{\boldsymbol{x}} + \frac{1}{2s^2} \int_D |\boldsymbol{v}|^2 dV_{\boldsymbol{x}}.$$

6.15 Let $K(t)$ denote the kinetic energy of a Newtonian fluid body with mass density ρ_0 and viscosity μ occupying a fixed region D. Here we generalize the estimate in Result 6.11 to the case of a steady, *non-conservative* body force field per unit mass $\boldsymbol{b}$.

(a) Use the result of Exercise 14 to show

$$\frac{d}{dt} K(t) + \mu \int_D \nabla^x \boldsymbol{v} : \nabla^x \boldsymbol{v} \, dV_{\boldsymbol{x}} \leq \frac{\alpha^2}{2} \int_D \rho_0 |\boldsymbol{b}|^2 \, dV_{\boldsymbol{x}} + \frac{1}{\alpha^2} K(t),$$

where $\boldsymbol{v}$ is the spatial velocity field and $\alpha > 0$ is an arbitrary constant.

(b) Use part (a) and the Poincaré inequality to show

$$K(t) \leq K_0 e^{-\mu t/\lambda \rho_0} + \frac{\lambda \rho_0 g}{\mu} \left[1 - e^{-\mu t/\lambda \rho_0}\right], \quad \forall t \geq 0.$$

Here $\lambda > 0$ is the constant from the Poincaré inequality, $g \geq 0$ is a constant defined by $g = \frac{\lambda \rho_0}{2\mu} \int_D \rho_0 |\boldsymbol{b}|^2 \, dV_{\boldsymbol{x}}$ and K_0 is the initial kinetic energy.

(c) Show that

$$K(t) \leq \max\left(K_0, \frac{\lambda \rho_0 g}{\mu}\right), \quad \forall t \geq 0.$$

Remark: The above result shows that the kinetic energy of a Newtonian fluid subject to a steady, non-conservative body force cannot grow without bound. In particular, power delivered by the body force cannot all go towards the production of kinetic energy; it must eventually be balanced by viscous dissipation.

Answers to Selected Exercises

6.1 From the balance of linear momentum equation with $\boldsymbol{v}$ identically zero (fluid is at rest) we get $\nabla^x p = \rho_0 \boldsymbol{b}$. Since $\rho_0 \boldsymbol{b}$ is independent of $\boldsymbol{x}$, the general solution of this equation is $p(\boldsymbol{x}) = \rho_0 \boldsymbol{b} \cdot \boldsymbol{x} + c$, where c is an arbitrary constant. Since $p(\boldsymbol{x}_*) = p_*$ we get

$$p(\boldsymbol{x}) = p_* + \rho_0 \boldsymbol{b} \cdot (\boldsymbol{x} - \boldsymbol{x}_*).$$

6.3 Since $\Delta^x \phi = \phi_{,ii}$, the Laplace equation is

$$\frac{\partial^2 \phi}{\partial {x_1}^2} + \frac{\partial^2 \phi}{\partial {x_2}^2} + \frac{\partial^2 \phi}{\partial {x_3}^2} = 0, \quad 0 < x_1, x_2, x_3 < 1,$$

and since the outward unit normal field on ∂D is piecewise constant with values $\boldsymbol{n} = \pm \boldsymbol{e}_1, \pm \boldsymbol{e}_2, \pm \boldsymbol{e}_3$ and $\nabla^x \phi = \phi_{,i} \boldsymbol{e}_i$, the boundary conditions are

$$\begin{aligned} \phi_{,x_1} &= 0 \quad \text{for } x_1 = 0, 1\ , \\ \phi_{,x_2} &= 0 \quad \text{for } x_2 = 0, 1\ , \\ \phi_{,x_3} &= 0 \quad \text{for } x_3 = 0\ , \\ \phi_{,x_3} &= g \quad \text{for } x_3 = 1. \end{aligned}$$

(a) We are given $g = \cos(k\pi x_1)\cos(l\pi x_2)$. Using (x, y, z) in place of (x_1, x_2, x_3), we follow the method of separation of variables and assume $\phi(x, y, z) = X(x)Y(y)Z(z)$. Then from the Laplace equation and the first four boundary conditions we get, after separating variables

$$\begin{aligned} &\frac{X''}{X} + \frac{Y''}{Y} + \frac{Z''}{Z} = 0, \quad 0 < x, y, z < 1, \\ &X'(0) = X'(1) = 0, \quad Y'(0) = Y'(1) = 0. \end{aligned}$$

Since x, y and z are independent the first equation implies that each of its terms must be constant. Thus

$$\frac{X''}{X} = -w_1, \quad \frac{Y''}{Y} = -w_2, \quad \text{and} \quad \frac{Z''}{Z} = -w_3,$$

where $w_1 + w_2 + w_3 = 0$. The equations for X and Y are then

$$\begin{aligned} X'' + w_1 X = 0, &\quad X'(0) = X'(1) = 0, \\ Y'' + w_2 Y = 0, &\quad Y'(0) = Y'(1) = 0. \end{aligned}$$

The non-trivial solutions of these (eigenvalue) problems are

$$w_1 = m^2\pi^2,\ X = \cos(m\pi x), \qquad w_2 = n^2\pi^2,\ Y = \cos(n\pi y),$$

where $m, n \geq 0$ are integers. The boundary condition $XYZ' = g$

at $z = 1$ requires $m = k$, $n = l$ and $Z'(1) = 1$. Thus we deduce that Z must satisfy

$$Z'' + w_3 Z = 0, \quad Z'(0) = 0, \quad Z'(1) = 1,$$

where $w_3 = -(w_1 + w_2) = -(k^2 + l^2)\pi^2$. Solving this system we get

$$Z(z) = \frac{\cosh(\pi z\sqrt{k^2 + l^2})}{\pi\sqrt{k^2 + l^2}\sinh(\pi\sqrt{k^2 + l^2})}, \quad k^2 + l^2 \neq 0.$$

Thus the solution is

$$\phi = \frac{\cos(k\pi x)\cos(l\pi y)\cosh(\pi z\sqrt{k^2 + l^2})}{\pi\sqrt{k^2 + l^2}\sinh(\pi\sqrt{k^2 + l^2})}, \quad k^2 + l^2 \neq 0.$$

(b) Let ϕ^{kl} be the solution found in part (a) with boundary condition $g^{kl} = \cos(k\pi x_1)\cos(l\pi x_2)$. (For brevity we suppose g^{kl} is defined on all faces and only indicate the nonzero value.) Then $\Delta^x \phi^{kl} = 0$ in D and $\nabla^x \phi^{kl} \cdot \boldsymbol{n} = g^{kl}$ on ∂D. If the boundary condition is changed to

$$g = \sum_{k,l=1}^{\infty} a_{kl} g^{kl},$$

then, by linearity of the Laplacian and gradient, a corresponding solution will be

$$\phi = \sum_{k,l=1}^{\infty} a_{kl} \phi^{kl}.$$

In particular, assuming the series can be differentiated term-by-term, we have

$$\Delta^x \phi = \Delta^x \left(\sum_{k,l=1}^{\infty} a_{kl}\phi^{kl} \right) = \sum_{k,l=1}^{\infty} a_{kl}(\Delta^x \phi^{kl}) = 0, \quad \forall \boldsymbol{x} \in D.$$

Moreover

$$\begin{aligned}\nabla^x \phi \cdot \boldsymbol{n} &= \nabla^x \left(\sum_{k,l=1}^{\infty} a_{kl}\phi^{kl} \right) \cdot \boldsymbol{n} \\ &= \sum_{k,l=1}^{\infty} a_{kl}(\nabla^x \phi^{kl} \cdot \boldsymbol{n}) = \sum_{k,l=1}^{\infty} a_{kl} g^{kl} = g, \quad \forall \boldsymbol{x} \in \partial D.\end{aligned}$$

(c) Using the result from (b) we notice that $g = 0$ on all faces corresponds to $a_{kl} = 0$ for all k, l, which yields the trivial solution $\phi = 0$. Alternatively, a direct approach with separation of variables shows that the general solution is $\phi = c$, where c is an arbitrary constant. This general solution is not contained in the result in (b) because the series given for g (hence ϕ) does not have a constant term. In either case, the resulting velocity field associated with ϕ vanishes. Thus, assuming a conservative body force, the only irrotational motion of an ideal fluid in D is the trivial one with zero velocity.

6.5 (a) We have $p = \pi(\rho)$ where $\pi(s) = \frac{1}{2}s^2$. From the definition of $\gamma(s)$ we get

$$\gamma(s) = \int_a^s \frac{\pi'(\xi)}{\xi}\, d\xi = s - a, \tag{6.46}$$

where $a > 0$ is an arbitrary constant. Since ζ is the inverse function of γ we find that $\zeta(s) = s + a$.

(b) For steady, irrotational motion with a conservative body force per unit mass, the velocity potential ϕ must satisfy

$$\nabla^x \cdot (\rho \nabla^x \phi) = 0, \quad \forall \boldsymbol{x} \in D,$$

where ρ is given by

$$\rho = \zeta\Big(f - \tfrac{1}{2}|\nabla^x \phi|^2 - \beta\Big).$$

Here f is a constant, and because the body force field is zero, β must also be a constant. Since $\zeta(s) = s + a$ we get

$$\rho = f - \tfrac{1}{2}|\nabla^x \phi|^2 - \beta + a = c - \tfrac{1}{2}|\nabla^x \phi|^2,$$

where $c = f - \beta + a$. Expanding the equation for ϕ we have

$$\rho \Delta^x \phi + \nabla^x \rho \cdot \nabla^x \phi = 0.$$

Using the fact that the gradient of c is zero we get

$$\begin{aligned} \nabla^x \rho = -\nabla^x\big(\tfrac{1}{2}|\nabla^x \phi|^2\big) &= -\big(\tfrac{1}{2}|\nabla^x \phi|^2\big)_{,i} \boldsymbol{e}_i \\ &= -\big(\tfrac{1}{2}\phi_{,j}\phi_{,j}\big)_{,i} \boldsymbol{e}_i \\ &= -\phi_{,ij}\phi_{,j} \boldsymbol{e}_i \\ &= -(\nabla^x \nabla^x \phi)\nabla^x \phi. \end{aligned}$$

Substituting the expressions for ρ and $\nabla^x \rho$ into the equation for ϕ gives the desired result

$$c\Delta^x \phi = \tfrac{1}{2}|\nabla^x \phi|^2 \Delta^x \phi + \nabla^x \phi \cdot (\nabla^x \nabla^x \phi)\nabla^x \phi.$$

6.7 (a) By definition, the Cauchy stress in an elastic fluid is $\boldsymbol{S}(\boldsymbol{x},t) = -\pi(\rho(\boldsymbol{x},t))\boldsymbol{I}$. In the material description this reads

$$\boldsymbol{S}_m(\boldsymbol{X},t) = -\pi(\rho_m(\boldsymbol{X},t))\boldsymbol{I}.$$

By the conservation of mass equation in Lagrangian form we have $\rho_m \det \boldsymbol{F} = \rho_0$, where ρ_0 is the mass density in the reference configuration B. Using this expression we may eliminate ρ_m and obtain

$$\boldsymbol{S}_m(\boldsymbol{X},t) = -\pi\left(\frac{\rho_0(\boldsymbol{X})}{\det \boldsymbol{F}(\boldsymbol{X},t)}\right)\boldsymbol{I}.$$

Using the definition $\boldsymbol{P} = (\det \boldsymbol{F})\boldsymbol{S}_m\boldsymbol{F}^{-T}$ we get

$$\begin{aligned}\boldsymbol{P} &= -(\det \boldsymbol{F})\pi\Big(\frac{\rho_0}{\det \boldsymbol{F}}\Big)\boldsymbol{F}^{-T}\\ &= -\boldsymbol{F}(\det \boldsymbol{F})\pi\Big(\frac{\rho_0}{\det \boldsymbol{F}}\Big)\boldsymbol{F}^{-1}\boldsymbol{F}^{-T}\\ &= -\boldsymbol{F}\sqrt{\det \boldsymbol{C}}\,\pi\Big(\frac{\rho_0}{\sqrt{\det \boldsymbol{C}}}\Big)\boldsymbol{C}^{-1} = \boldsymbol{F}h(\det \boldsymbol{C},\rho_0)\boldsymbol{C}^{-1},\end{aligned}$$

where $\boldsymbol{C} = \boldsymbol{F}^T\boldsymbol{F}$ is the Cauchy–Green strain tensor and h is the function defined by

$$h(z,r) = -\sqrt{z}\,\pi\Big(\frac{r}{\sqrt{z}}\Big).$$

Thus $\boldsymbol{P} = \boldsymbol{F}\overline{\boldsymbol{\Sigma}}(\boldsymbol{C},\rho_0)$, where $\overline{\boldsymbol{\Sigma}}(\boldsymbol{C},\rho_0) = h(\det \boldsymbol{C},\rho_0)\boldsymbol{C}^{-1}$.

(b) As discussed in Chapter 7, isotropic elastic materials are characterized by constitutive relations of the form

$$\boldsymbol{P} = \boldsymbol{F}[\gamma_0\boldsymbol{I} + \gamma_1\boldsymbol{C} + \gamma_2\boldsymbol{C}^{-1}],$$

where $\gamma_i : \mathbb{R}^3 \to \mathbb{R}$ $(i = 0,1,2)$ are functions of the principal invariants of $\boldsymbol{C}$. The result in (a) is precisely of this form with $\gamma_0 = \gamma_1 = 0$ and $\gamma_2 = h(\det \boldsymbol{C},\rho_0)$. Thus an elastic fluid is an isotropic elastic material.

6.9 The reactive stress in either an ideal or Newtonian fluid is $\boldsymbol{S}^{(r)} = -p\boldsymbol{I}$, where p is the pressure field associated with the incompressibility constraint $\nabla^x \cdot \boldsymbol{v} = 0$. Using the fact that $\boldsymbol{I} : \boldsymbol{A} = \operatorname{tr} \boldsymbol{A}$ and $\operatorname{tr}(\operatorname{sym}(\boldsymbol{A})) = \operatorname{tr} \boldsymbol{A}$ for any second-order tensor $\boldsymbol{A}$ we get

$$\begin{aligned}\boldsymbol{S}^{(r)} : \boldsymbol{L} &= -p\boldsymbol{I} : \operatorname{sym}(\nabla^x \boldsymbol{v})\\ &= -p\operatorname{tr}(\operatorname{sym}(\nabla^x \boldsymbol{v}))\\ &= -p\operatorname{tr}(\nabla^x \boldsymbol{v}) = -p\nabla^x \cdot \boldsymbol{v} = 0, \quad \forall \boldsymbol{x} \in B_t,\ t \geq 0.\end{aligned}$$

6.11 (a) The steady Navier–Stokes equations with zero body force are

$$\rho_0(\nabla^x \boldsymbol{v})\boldsymbol{v} = \mu\Delta^x \boldsymbol{v} - \nabla^x p, \quad \nabla^x \cdot \boldsymbol{v} = 0.$$

Working in components, and using the fact that $\boldsymbol{v} = v_1(x_1, x_2)\boldsymbol{e}_1$, we have

$$[\boldsymbol{v}] = \begin{Bmatrix} v_1 \\ 0 \\ 0 \end{Bmatrix}, \quad [\nabla^x \boldsymbol{v}] = \begin{pmatrix} v_{1,1} & v_{1,2} & 0 \\ 0 & 0 & 0 \\ 0 & 0 & 0 \end{pmatrix},$$

$$[\Delta^x \boldsymbol{v}] = \begin{Bmatrix} v_{1,11} + v_{1,22} \\ 0 \\ 0 \end{Bmatrix}, \quad [\nabla^x p] = \begin{Bmatrix} p_{,1} \\ p_{,2} \\ p_{,3} \end{Bmatrix}.$$

Thus, in components, the steady Navier–Stokes equations are

$$\rho_0 \begin{Bmatrix} v_{1,1}v_1 \\ 0 \\ 0 \end{Bmatrix} = \mu \begin{Bmatrix} v_{1,11} + v_{1,22} \\ 0 \\ 0 \end{Bmatrix} - \begin{Bmatrix} p_{,1} \\ p_{,2} \\ p_{,3} \end{Bmatrix}, \quad v_{1,1} = 0.$$

Exploiting the fact that $v_{1,1} = 0$ we obtain the desired result, namely

$$\mu v_{1,22} = p_{,1}, \quad p_{,2} = 0, \quad p_{,3} = 0, \quad v_{1,1} = 0.$$

The equation $v_{1,1} = 0$ implies v_1 is independent of x_1, so we have $v_1 = v_1(x_2)$. The equations $p_{,2} = 0$ and $p_{,3} = 0$ imply p is independent of x_2 and x_3, so we have $p = p(x_1)$. Since x_1 and x_2 are independent, the equation $\mu v_{1,22} = p_{,1}$ implies that each side must be constant. Upon setting $\mu v_{1,22} = \alpha$ and $p_{,1} = \alpha$ and integrating we obtain

$$v_1 = \frac{\alpha}{2\mu}x_2^2 + \beta x_2 + \gamma, \quad p = \alpha x_1 + \delta,$$

where α, β, γ, δ are arbitrary constants.

(b) $v_1 = 0$ at $x_2 = 0$ requires $\gamma = 0$, and $v_1 = \vartheta$ at $x_2 = h$ requires $\frac{\alpha}{2\mu}h^2 + \beta h = \vartheta$. No conditions are given on the pressure. Thus the resulting solution is

$$v_1 = \frac{\alpha}{2\mu}x_2^2 + \frac{1}{h}\Big(\vartheta - \frac{\alpha}{2\mu}h^2\Big)x_2, \quad p = \alpha x_1 + \delta,$$

where α and δ are arbitrary constants.

(c) The no-slip boundary condition $v_1 = 0$ at $x_2 = 0$ requires $\gamma =$

0. By definition, the Cauchy stress is $\boldsymbol{S} = -p\boldsymbol{I} + 2\mu\,\mathrm{sym}(\nabla^x \boldsymbol{v})$. Using results from part (a) we have

$$[\boldsymbol{S}] = \begin{pmatrix} -p & \mu v_{1,2} & 0 \\ \mu v_{1,2} & -p & 0 \\ 0 & 0 & -p \end{pmatrix}.$$

At $x_2 = h$ we have $\boldsymbol{n} = \boldsymbol{e}_2$ and the traction boundary condition $\boldsymbol{S}\boldsymbol{n} = a\boldsymbol{e}_1 + b\boldsymbol{e}_2$ implies

$$\begin{Bmatrix} \mu v_{1,2} \\ -p \\ 0 \end{Bmatrix} = \begin{Bmatrix} a \\ b \\ 0 \end{Bmatrix}, \quad x_2 = h, \quad -\infty < x_1 < \infty.$$

The condition $p = -b$ for all x_1 implies $\alpha = 0$ and $\delta = -b$. Moreover, the condition $\mu v_{1,2} = a$ implies $\beta = a/\mu$. Thus we obtain the unique solution

$$v_1 = \frac{a}{\mu} x_2, \quad p = -b.$$

6.13 Let $\boldsymbol{w}$ be an arbitrary, smooth vector field with $\boldsymbol{w} = \boldsymbol{0}$ on ∂D. Using the identity $\nabla^x \cdot (\boldsymbol{S}^T \boldsymbol{w}) = (\nabla^x \cdot \boldsymbol{S}) \cdot \boldsymbol{w} + \boldsymbol{S} : \nabla^x \boldsymbol{w}$, where $\boldsymbol{S}$ is any second-order tensor field, we have

$$\nabla^x \cdot (\nabla^x \boldsymbol{w}^T \boldsymbol{w}) = (\Delta^x \boldsymbol{w}) \cdot \boldsymbol{w} + \nabla^x \boldsymbol{w} : \nabla^x \boldsymbol{w},$$

and by the Divergence Theorem we find

$$\int_D [\nabla^x \boldsymbol{w} : \nabla^x \boldsymbol{w} + \Delta^x \boldsymbol{w} \cdot \boldsymbol{w}] \, dV_{\boldsymbol{x}} = \int_{\partial D} (\nabla^x \boldsymbol{w}^T \boldsymbol{w}) \cdot \boldsymbol{n} \, dA_{\boldsymbol{x}} = \boldsymbol{0},$$

or equivalently

$$\int_D \nabla^x \boldsymbol{w} : \nabla^x \boldsymbol{w} \, dV_{\boldsymbol{x}} = -\int_D \Delta^x \boldsymbol{w} \cdot \boldsymbol{w} \, dV_{\boldsymbol{x}}.$$

By assumption, $\boldsymbol{w}$ can be written in the form

$$\boldsymbol{w} = \sum_{i=1}^{\infty} \alpha_i \boldsymbol{\eta}_i,$$

where $\{\alpha_i\}_{i=1}^{\infty}$ are constants and $\{\boldsymbol{\eta}_i\}_{i=1}^{\infty}$ are the given eigenfunctions with corresponding positive eigenvalues $\{\lambda_i\}_{i=1}^{\infty}$, ordered so that $0 < \lambda_1 \leq \lambda_2 \leq \cdots$. Since the eigenfunctions satisfy $-\Delta^x \boldsymbol{\eta}_i = \lambda_i \boldsymbol{\eta}_i$ (no summation convention) we have

$$-\Delta^x \boldsymbol{w} = \sum_{i=1}^{\infty} \alpha_i \lambda_i \boldsymbol{\eta}_i.$$

Taking the dot product of this expression with $\boldsymbol{w}$, integrating the result over D and using the condition $\int_D \boldsymbol{\eta}_i \cdot \boldsymbol{\eta}_j \, dV_{\boldsymbol{x}} = \delta_{ij}$ we get

$$-\int_D \Delta^x \boldsymbol{w} \cdot \boldsymbol{w} \, dV_{\boldsymbol{x}} = \sum_{i=1}^{\infty} \alpha_i^2 \lambda_i \geq \lambda_1 \sum_{i=1}^{\infty} \alpha_i^2,$$

where the inequality follows from the ordering of the eigenvalues. From the condition $\int_D \boldsymbol{\eta}_i \cdot \boldsymbol{\eta}_j \, dV_{\boldsymbol{x}} = \delta_{ij}$ we also get

$$\int_D \boldsymbol{w} \cdot \boldsymbol{w} \, dV_{\boldsymbol{x}} = \sum_{i=1}^{\infty} \alpha_i^2,$$

and substituting this result into the above inequality gives

$$-\int_D \Delta^x \boldsymbol{w} \cdot \boldsymbol{w} \, dV_{\boldsymbol{x}} \geq \lambda_1 \int_D \boldsymbol{w} \cdot \boldsymbol{w} \, dV_{\boldsymbol{x}}.$$

Combining this with the integral identity for $\boldsymbol{w}$ gives

$$\int_D \nabla^x \boldsymbol{w} : \nabla^x \boldsymbol{w} \, dV_{\boldsymbol{x}} \geq \lambda_1 \int_D \boldsymbol{w} \cdot \boldsymbol{w} \, dV_{\boldsymbol{x}}.$$

The desired result follows with $\lambda = 1/\lambda_1$.

6.15 (a) As in the proof of the decay estimate in the text we have

$$\frac{d}{dt} K(t) = \int_D \rho_0 \dot{\boldsymbol{v}} \cdot \boldsymbol{v} \, dV_{\boldsymbol{x}},$$

and from the balance of linear momentum equation we have

$$\rho_0 \dot{\boldsymbol{v}} = \mu \Delta^x \boldsymbol{v} - \nabla^x p + \rho_0 \boldsymbol{b}.$$

Substituting for $\rho_0 \dot{\boldsymbol{v}}$ in the integral we get

$$\frac{d}{dt} K(t) = \int_D \left[\mu(\Delta^x \boldsymbol{v}) \cdot \boldsymbol{v} - \nabla^x p \cdot \boldsymbol{v} + \rho_0 \boldsymbol{b} \cdot \boldsymbol{v} \right] \, dV_{\boldsymbol{x}}.$$

Integrating the first two terms by parts using the fact that $\boldsymbol{v} = \boldsymbol{0}$ on ∂D we get

$$\frac{d}{dt} K(t) + \mu \int_D \nabla^x \boldsymbol{v} : \nabla^x \boldsymbol{v} \, dV_{\boldsymbol{x}} = \int_D \rho_0 \boldsymbol{b} \cdot \boldsymbol{v} \, dV_{\boldsymbol{x}}.$$

Writing $\rho_0 \boldsymbol{b} \cdot \boldsymbol{v}$ as $\sqrt{\rho_0}\boldsymbol{b} \cdot \sqrt{\rho_0}\boldsymbol{v}$ and using the result of Exercise 14 we obtain, for any $\alpha > 0$,

$$\frac{d}{dt}K(t) + \mu \int_D \nabla^x \boldsymbol{v} : \nabla^x \boldsymbol{v} \, dV_{\boldsymbol{x}} \le \frac{\alpha^2}{2} \int_D \rho_0 |\boldsymbol{b}|^2 \, dV_{\boldsymbol{x}} + \frac{1}{2\alpha^2} \int_D \rho_0 |\boldsymbol{v}|^2 \, dV_{\boldsymbol{x}},$$

which is the desired result.

(b) Choosing $\alpha^2 = \lambda\rho_0/\mu$ in the result from part (a) yields

$$\frac{d}{dt}K(t) + \mu \int_D \nabla^x \boldsymbol{v} : \nabla^x \boldsymbol{v} \, dV_{\boldsymbol{x}} \le g + \frac{\mu}{\lambda\rho_0} K(t).$$

From this expression and the Poincaré inequality we deduce

$$\frac{d}{dt}K(t) + \frac{\mu}{\lambda\rho_0}K(t) \le g.$$

Multiplying by the factor $e^{\mu t/\lambda\rho_0}$ we get

$$\frac{d}{ds}\left(e^{\mu s/\lambda\rho_0} K(s)\right) \le e^{\mu s/\lambda\rho_0} g,$$

and integrating from $s = 0$ to $s = t$ we obtain the desired result

$$K(t) \le K_0 e^{-\mu t/\lambda\rho_0} + \frac{\lambda\rho_0 g}{\mu}\left[1 - e^{-\mu t/\lambda\rho_0}\right], \quad \forall t \ge 0.$$

(c) Suppose $K_0 > \frac{\lambda\rho_0 g}{\mu}$. Then from (b) and the fact $e^{-\mu t/\lambda\rho_0} \le 1$ for all $t \ge 0$ we get

$$K(t) \le \left(K_0 - \frac{\lambda\rho_0 g}{\mu}\right) e^{-\mu t/\lambda\rho_0} + \frac{\lambda\rho_0 g}{\mu} \le K_0.$$

Suppose $K_0 \le \frac{\lambda\rho_0 g}{\mu}$. Then from (b) and the fact that $e^{-\mu t/\lambda\rho_0} > 0$ for all $t \ge 0$ we get

$$K(t) \le \left(K_0 - \frac{\lambda\rho_0 g}{\mu}\right) e^{-\mu t/\lambda\rho_0} + \frac{\lambda\rho_0 g}{\mu} \le \frac{\lambda\rho_0 g}{\mu}.$$

Thus we always have

$$K(t) \le \max\left(K_0, \frac{\lambda\rho_0 g}{\mu}\right), \quad \forall t \ge 0.$$

7
Isothermal Solid Mechanics

In this chapter we consider several applications of the Lagrangian balance laws introduced in Chapter 5. As in the previous chapter, we neglect thermal effects both because it is physically reasonable to do so in many situations, and because it is mathematically simpler. Thermal effects will be considered in Chapter 9. Considering only the isothermal case there are 15 basic unknown fields in the Lagrangian description of a continuum body:

$\varphi_i(\boldsymbol{X}, t)$	3 components of motion
$V_i(\boldsymbol{X}, t)$	3 components of velocity
$P_{ij}(\boldsymbol{X}, t)$	9 components of stress.

To determine these unknown fields we have the following 9 equations:

$V_i = \dot{\varphi}_i$	3 kinematical
$\rho_0 \dot{V}_i = P_{ij,j} + \rho_0 \left[b_i\right]_m$	3 linear momentum
$P_{ik} F_{jk} = F_{ik} P_{jk}$	3 independent angular momentum.

Thus 6 additional equations are required to balance the number of equations and unknowns. These are provided by constitutive equations that relate the independent components of the first Piola–Kirchhoff stress field $\boldsymbol{P}$ to the variables $\boldsymbol{\varphi}$ and $\boldsymbol{V}$. Any such constitutive equation must be consistent with the axiom of material frame-indifference. Moreover, assuming a material is energetically passive, the constitutive equation must also satisfy the mechanical energy inequality as discussed in Chapter 5.

In this chapter we study constitutive models that relate the stress $\boldsymbol{P}$ to the deformation gradient $\boldsymbol{F} = \nabla^X \boldsymbol{\varphi}$. Such models are typically used to describe the behavior of various types of solids. Because these models are independent of $\boldsymbol{V}$, this variable may be eliminated by substitution of the kinematical equation into the balance of linear momentum equation. Thus a closed system for $\boldsymbol{\varphi}$ and $\boldsymbol{P}$ is provided by the balance equations for linear and angular momentum, together with the constitutive relation. In contrast to the formulations of the previous chapter, the balance of mass equation is not considered because the spatial mass density does not appear. Throughout our developments we express results in terms of either the first Piola–Kirchhoff stress $\boldsymbol{P}$, the Cauchy stress $\boldsymbol{S}_m$ (material description) or the second Piola–Kirchhoff stress $\boldsymbol{\Sigma}$, whichever seems most convenient. Any result stated in terms of one of these stress tensors can always be expressed in terms of the others (see Definition 5.14).

The constitutive models considered in this chapter are those for: (i) general elastic solids, (ii) hyperelastic solids and (iii) linear elastic solids. We outline a result which shows that the general models in (i) satisfy the mechanical energy inequality if and only if they are hyperelastic as in (ii). Moreover, we use the technique of linearization on the general models in (i) to motivate the linear models in (iii). For each model we analyze the consequences of material frame-indifference, outline a typical initial-boundary value problem, and study various qualitative properties. We also introduce the notion of isotropy for elastic models and study its implications for constitutive equations.

7.1 Elastic Solids

In this section we study the constitutive model for an elastic solid. We study consequences of the axiom of material frame-indifference, outline a standard initial-boundary value problem and discuss the notion of isotropy. The mechanical energy inequality for an elastic solid will be analyzed in Section 7.2.

7.1.1 Definition

A continuum body with reference configuration B is said to be an **elastic solid** if:

(1) The Cauchy stress field is of the form

$$\boldsymbol{S}_m(\boldsymbol{X},t) = \widehat{\boldsymbol{S}}(\boldsymbol{F}(\boldsymbol{X},t),\boldsymbol{X}), \quad \forall \boldsymbol{X} \in B,\ t \geq 0,$$

where $\widehat{\boldsymbol{S}} : \mathcal{V}^2 \times B \to \mathcal{V}^2$ is a given function called a **stress response function** for $\boldsymbol{S}_m$.

(2) The function $\widehat{\boldsymbol{S}}$ has the property

$$\widehat{\boldsymbol{S}}(\boldsymbol{F},\boldsymbol{X})^T = \widehat{\boldsymbol{S}}(\boldsymbol{F},\boldsymbol{X}), \quad \forall \boldsymbol{X} \in B,\ \boldsymbol{F} \in \mathcal{V}^2,\ \det \boldsymbol{F} > 0.$$

Property (1) implies that the stress at a point in an elastic solid depends only on a measure of present strain at that point. In particular, it is independent of the past history and rate of strain. This type of relation can be interpreted as a generalization of **Hooke's Law**. It is often called a **stress-strain relation**. Property (2) implies that the Cauchy stress field is necessarily symmetric. Thus the balance equation for angular momentum (Result 5.16) is automatically satisfied and will not be considered further. Because $\det \boldsymbol{F}(\boldsymbol{X},t) > 0$ for any admissible motion, we only consider $\widehat{\boldsymbol{S}}(\boldsymbol{F},\boldsymbol{X})$ for arguments $\boldsymbol{F}$ satisfying $\det \boldsymbol{F} > 0$.

Remarks:

(1) The response function $\widehat{\boldsymbol{S}}(\boldsymbol{F},\boldsymbol{X})$ for an elastic body is an intrinsic property of the body. It depends only on material properties and the reference configuration B, and determines the Cauchy stress in all admissible motions.

(2) An elastic body is said to have a **homogeneous elastic response** if $\widehat{\boldsymbol{S}}(\boldsymbol{F},\boldsymbol{X})$ is independent of $\boldsymbol{X}$. Throughout the remainder of our developments we consider only homogeneous elastic bodies in this sense. This is done for notational simplicity alone; all subsequent results can be generalized in a straightforward way to the inhomogeneous case.

(3) In view of Definition 5.14, response functions for $\boldsymbol{S}_m$, $\boldsymbol{P}$ and $\boldsymbol{\Sigma}$ must satisfy the relations

$$\widehat{\boldsymbol{P}}(\boldsymbol{F}) = (\det \boldsymbol{F})\widehat{\boldsymbol{S}}(\boldsymbol{F})\boldsymbol{F}^{-T}, \quad \widehat{\boldsymbol{\Sigma}}(\boldsymbol{F}) = \boldsymbol{F}^{-1}\widehat{\boldsymbol{P}}(\boldsymbol{F}). \tag{7.1}$$

Thus response functions for all three stress tensors can be determined from the response function of any one.

(4) A classic example of an elastic stress response function is the **St. Venant–Kirchhoff model** given by

$$\widehat{\boldsymbol{\Sigma}}(\boldsymbol{F}) = \lambda(\operatorname{tr}\boldsymbol{G})\boldsymbol{I} + 2\mu\boldsymbol{G}, \tag{7.2}$$

where $\boldsymbol{G} = \frac{1}{2}(\boldsymbol{C} - \boldsymbol{I})$ and $\boldsymbol{C} = \boldsymbol{F}^T\boldsymbol{F}$ is the right Cauchy–Green strain tensor. Here λ and μ are material constants.

(5) Intuition suggests that the stress response function for an elastic body should deliver extreme stresses in the presence of extreme strains. In particular, appropriate components of stress should tend to infinity in absolute value as any principal stretch tends to infinity (infinite extension) or tends to zero (infinite compression). The St. Venant–Kirchhoff model has the first property, but not the second (see Exercise 1). □

7.1.2 Elasticity Equations

Let $\rho_0(\boldsymbol{X})$ denote the mass density of an elastic body in its reference configuration B, and let $\boldsymbol{b}_m(\boldsymbol{X},t)$ denote the material description of a prescribed spatial body force field per unit mass $\boldsymbol{b}(\boldsymbol{x},t)$, that is

$$\boldsymbol{b}_m(\boldsymbol{X},t) = \boldsymbol{b}(\boldsymbol{\varphi}(\boldsymbol{X},t),t).$$

Then, setting $\boldsymbol{P}(\boldsymbol{X},t) = \widehat{\boldsymbol{P}}(\boldsymbol{F}(\boldsymbol{X},t))$ in the balance of linear momentum equation (Result 5.15), we obtain a closed system for the motion $\boldsymbol{\varphi}$ of an elastic body. In particular, the motion must satisfy the following equation for all $\boldsymbol{X} \in B$ and $t \geq 0$

$$\boxed{\rho_0 \ddot{\boldsymbol{\varphi}} = \nabla^X \cdot (\widehat{\boldsymbol{P}}(\boldsymbol{F})) + \rho_0 \boldsymbol{b}_m.} \tag{7.3}$$

This is known as the **Elastodynamics Equation.**

Remarks:

(1) By definition of the divergence of a second-order tensor we have $[\nabla^X \cdot \widehat{\boldsymbol{P}}]_i = \widehat{P}_{ij,j}$, and by the chain rule and the assumption of homogeneity we get

$$[\nabla^X \cdot \widehat{\boldsymbol{P}}]_i = \frac{\partial \widehat{P}_{ij}}{\partial F_{kl}} \frac{\partial F_{kl}}{\partial X_j}.$$

Since $F_{kl} = \varphi_{k,l}$, the equation in (7.3) can be written in components as

$$\rho_0 \frac{\partial^2 \varphi_i}{\partial t^2} = \mathsf{A}_{ijkl} \frac{\partial^2 \varphi_k}{\partial X_j \partial X_l} + \rho_0 b_i,$$

where $\mathsf{A}_{ijkl} = \partial \widehat{P}_{ij} / \partial F_{kl}$ and b_i denote the components of $\boldsymbol{b}_m$. Thus (7.3) is a system of second-order partial differential equations for the components of $\boldsymbol{\varphi}$.

(2) Equation (7.3) is typically nonlinear in φ. Nonlinearities arise through the functional form of the stress response function $\widehat{\boldsymbol{P}}(\boldsymbol{F})$ and the body force $\boldsymbol{b}_m$. Under appropriate conditions on the response function $\widehat{\boldsymbol{P}}(\boldsymbol{F})$, we expect (7.3) to admit a full range of wave-like solutions.

(3) Setting $\ddot{\varphi}$ equal to zero in (7.3) yields

$$\nabla^X \cdot (\widehat{\boldsymbol{P}}(\boldsymbol{F})) + \rho_0 \boldsymbol{b}_m = \boldsymbol{0}. \tag{7.4}$$

This is known as the **Elastostatics Equation**. It is the Lagrangian form of the local equilibrium equation introduced in Section 3.4, specialized to an elastic solid. □

7.1.3 Frame-Indifference Considerations

Here we show that the stress response functions for an elastic solid must necessarily satisfy certain restrictions in order to comply with the axiom of material frame-indifference.

Result 7.1 Frame-Indifferent Stress Response. *The Cauchy stress field in an elastic body is frame-indifferent if and only if the stress response function $\widehat{\boldsymbol{S}}$ can be written in the form*

$$\widehat{\boldsymbol{S}}(\boldsymbol{F}) = \boldsymbol{F}\overline{\boldsymbol{S}}(\boldsymbol{C})\boldsymbol{F}^T, \tag{7.5}$$

for some function $\overline{\boldsymbol{S}} : \mathcal{V}^2 \to \mathcal{V}^2$, where $\boldsymbol{C} = \boldsymbol{F}^T\boldsymbol{F}$. Equivalently, the response functions $\widehat{\boldsymbol{P}}$ and $\widehat{\boldsymbol{\Sigma}}$ must satisfy

$$\widehat{\boldsymbol{P}}(\boldsymbol{F}) = \boldsymbol{F}\overline{\boldsymbol{\Sigma}}(\boldsymbol{C}) \quad \textit{and} \quad \widehat{\boldsymbol{\Sigma}}(\boldsymbol{F}) = \overline{\boldsymbol{\Sigma}}(\boldsymbol{C}),$$

for some function $\overline{\boldsymbol{\Sigma}} : \mathcal{V}^2 \to \mathcal{V}^2$. In particular, $\overline{\boldsymbol{\Sigma}}(\boldsymbol{C}) = \sqrt{\det \boldsymbol{C}}\, \overline{\boldsymbol{S}}(\boldsymbol{C})$.

□

Proof Let $\boldsymbol{x} = \varphi(\boldsymbol{X}, t)$ be an arbitrary motion and let $\boldsymbol{x}^* = \varphi^*(\boldsymbol{X}, t)$ be a second motion defined by

$$\boldsymbol{x}^* = \boldsymbol{g}(\boldsymbol{x}, t) = \boldsymbol{Q}(t)\boldsymbol{x} + \boldsymbol{c}(t),$$

where $\boldsymbol{Q}(t)$ is an arbitrary rotation tensor and $\boldsymbol{c}(t)$ is an arbitrary vector. Moreover, let $\boldsymbol{S}(\boldsymbol{x}, t)$ and $\boldsymbol{S}^*(\boldsymbol{x}^*, t)$ denote the Cauchy stress fields in B_t

and B_t^*. Then the axiom of material frame-indifference (Axiom 5.24) requires

$$\boldsymbol{Q}^T(t)\boldsymbol{S}^*(\boldsymbol{x}^*,t)\boldsymbol{Q}(t) = \boldsymbol{S}(\boldsymbol{x},t), \quad \forall \boldsymbol{x} \in B_t,\ t \geq 0,$$

where $\boldsymbol{x}^* = \boldsymbol{g}(\boldsymbol{x},t)$. In terms of the material descriptions $\boldsymbol{S}_m(\boldsymbol{X},t) = \boldsymbol{S}(\boldsymbol{\varphi}(\boldsymbol{X},t),t)$ and $\boldsymbol{S}_m^*(\boldsymbol{X},t) = \boldsymbol{S}^*(\boldsymbol{\varphi}^*(\boldsymbol{X},t),t)$, the above condition can be written as

$$\boldsymbol{Q}^T(t)\boldsymbol{S}_m^*(\boldsymbol{X},t)\boldsymbol{Q}(t) = \boldsymbol{S}_m(\boldsymbol{X},t), \quad \forall \boldsymbol{X} \in B,\ t \geq 0. \tag{7.6}$$

Since the Cauchy stress in any admissible motion is delivered by the response function $\widehat{\boldsymbol{S}}$ we have

$$\boldsymbol{S}_m(\boldsymbol{X},t) = \widehat{\boldsymbol{S}}(\boldsymbol{F}(\boldsymbol{X},t)) \quad \text{and} \quad \boldsymbol{S}_m^*(\boldsymbol{X},t) = \widehat{\boldsymbol{S}}(\boldsymbol{F}^*(\boldsymbol{X},t)).$$

Moreover, by Result 5.22 we have $\boldsymbol{F}^* = \boldsymbol{Q}\boldsymbol{F}$. These results together with (7.6) imply that the response function $\widehat{\boldsymbol{S}}$ must satisfy

$$\boldsymbol{Q}^T\widehat{\boldsymbol{S}}(\boldsymbol{Q}\boldsymbol{F})\boldsymbol{Q} = \widehat{\boldsymbol{S}}(\boldsymbol{F}). \tag{7.7}$$

Thus the Cauchy stress field in an elastic body is frame-indifferent if and only if the response function $\widehat{\boldsymbol{S}}$ satisfies (7.7). In particular, this relation must hold for all $\boldsymbol{F}$ with $\det \boldsymbol{F} > 0$ and all rotations $\boldsymbol{Q}$ by the arbitrariness of $\boldsymbol{\varphi}$ and $\boldsymbol{g}$.

We next show that (7.7) holds for all stated $\boldsymbol{F}$ and $\boldsymbol{Q}$ if and only if (7.5) holds. To begin, assume (7.7) is true and let $\boldsymbol{F} = \boldsymbol{R}\boldsymbol{U}$ be the right polar decomposition of $\boldsymbol{F}$. Then choosing $\boldsymbol{Q} = \boldsymbol{R}^T$ in (7.7) yields

$$\widehat{\boldsymbol{S}}(\boldsymbol{F}) = \boldsymbol{R}\widehat{\boldsymbol{S}}(\boldsymbol{U})\boldsymbol{R}^T. \tag{7.8}$$

Define $\boldsymbol{C}^{1/2} = \sqrt{\boldsymbol{C}}$ and $\boldsymbol{C}^{-1/2} = (\sqrt{\boldsymbol{C}})^{-1}$. Then, since $\boldsymbol{U} = \boldsymbol{C}^{1/2}$, we have $\boldsymbol{R} = \boldsymbol{F}\boldsymbol{C}^{-1/2}$. Substitution of these relations into (7.8) leads to the result in (7.5), namely

$$\widehat{\boldsymbol{S}}(\boldsymbol{F}) = \boldsymbol{F}\overline{\boldsymbol{S}}(\boldsymbol{C})\boldsymbol{F}^T \quad \text{where} \quad \overline{\boldsymbol{S}}(\boldsymbol{C}) = \boldsymbol{C}^{-1/2}\widehat{\boldsymbol{S}}(\boldsymbol{C}^{1/2})\boldsymbol{C}^{-1/2}.$$

Conversely, if we assume (7.5) is true, then it is straightforward to verify that (7.7) holds for all stated $\boldsymbol{F}$ and $\boldsymbol{Q}$. Thus we conclude that the Cauchy stress field in an elastic body is frame-indifferent if and only if $\widehat{\boldsymbol{S}}$ can be put into the form (7.5). The results for $\widehat{\boldsymbol{P}}$ and $\widehat{\boldsymbol{\Sigma}}$ are immediate consequences of (7.1) and the relation $\det \boldsymbol{C} = (\det \boldsymbol{F})^2$. □

Remarks:

(1) Roughly speaking, the above result says that the stress response functions for an elastic body are frame-indifferent if and only if they depend on the motion through the right Cauchy–Green strain tensor $\boldsymbol{C}$. Since $\boldsymbol{C} = \boldsymbol{F}^T\boldsymbol{F}$, where $\boldsymbol{F} = \nabla^X \boldsymbol{\varphi}$, we see that frame-indifferent response functions will generally be nonlinear in the motion $\boldsymbol{\varphi}$.

(2) The above result can be used to show the St. Venant–Kirchhoff model introduced in (7.2) is frame-indifferent. In particular, for this model we have $\widehat{\boldsymbol{\Sigma}}(\boldsymbol{F}) = \lambda(\operatorname{tr}\boldsymbol{G})\boldsymbol{I} + 2\mu\boldsymbol{G}$, where $\boldsymbol{G} = \frac{1}{2}(\boldsymbol{C}-\boldsymbol{I})$ and $\boldsymbol{C} = \boldsymbol{F}^T\boldsymbol{F}$. Substituting for $\boldsymbol{G}$ we obtain $\widehat{\boldsymbol{\Sigma}}(\boldsymbol{F}) = \overline{\boldsymbol{\Sigma}}(\boldsymbol{C})$ where

$$\overline{\boldsymbol{\Sigma}}(\boldsymbol{C}) = \frac{\lambda}{2}[\operatorname{tr}(\boldsymbol{C}-\boldsymbol{I})]\boldsymbol{I} + \mu(\boldsymbol{C}-\boldsymbol{I}). \tag{7.9}$$

□

7.1.4 Initial-Boundary Value Problems

An initial-boundary value problem for an elastic body is a set of equations for determining the motion of a given body subject to specified initial conditions in B at time $t = 0$, and boundary conditions on ∂B at times $t \geq 0$. We typically assume B is a bounded open set as shown in Figure 7.1. However, it is also useful in applications to consider unbounded open sets such as the exterior of the region shown in the figure, or the whole of Euclidean space.

A standard initial-boundary value problem for an elastic body with reference configuration B is the following: Find $\boldsymbol{\varphi} : B \times [0,T] \to \mathbb{E}^3$ such that

$$\boxed{\begin{array}{lll} \rho_0\ddot{\boldsymbol{\varphi}} = \nabla^X \cdot [\widehat{\boldsymbol{P}}(\boldsymbol{F})] + \rho_0\boldsymbol{b}_m & \text{in} & B \times [0,T] \\ \boldsymbol{\varphi} = \boldsymbol{g} & \text{in} & \Gamma_d \times [0,T] \\ \widehat{\boldsymbol{P}}(\boldsymbol{F})\boldsymbol{N} = \boldsymbol{h} & \text{in} & \Gamma_\sigma \times [0,T] \\ \boldsymbol{\varphi}(\cdot,0) = \boldsymbol{X} & \text{in} & B \\ \dot{\boldsymbol{\varphi}}(\cdot,0) = \boldsymbol{V}_0 & \text{in} & B. \end{array}} \tag{7.10}$$

In the above system, Γ_d and Γ_σ are subsets of ∂B with the properties $\Gamma_d \cup \Gamma_\sigma = \partial B$ and $\Gamma_d \cap \Gamma_\sigma = \emptyset$, ρ_0 is the reference mass density, $\boldsymbol{b}_m$

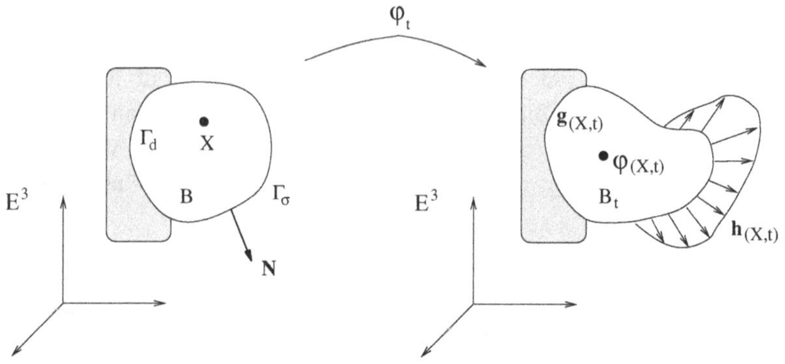

Fig. 7.1 Typical initial-boundary value problem for an elastic body B.

is the material description of a specified spatial body force field per unit mass, $\widehat{\boldsymbol{P}}$ is a frame-indifferent stress response function (see Result 7.1), $\boldsymbol{N}$ is the unit outward normal field on ∂B, and $\boldsymbol{g}$, $\boldsymbol{h}$ and $\boldsymbol{V}_0$ are prescribed fields. Equation $(7.10)_1$ is the balance of linear momentum equation, $(7.10)_2$ is a motion boundary condition on Γ_d, $(7.10)_3$ is a traction boundary condition on Γ_σ, $(7.10)_4$ is the initial condition for the motion $\boldsymbol{\varphi}$ and $(7.10)_5$ is an initial condition for the material velocity field $\dot{\boldsymbol{\varphi}}$. In general, the initial conditions should be compatible with the boundary conditions at time $t = 0$.

Remarks:

(1) The system in (7.10) is a nonlinear initial-boundary value problem for the motion map $\boldsymbol{\varphi}$. Existence and uniqueness of solutions on finite time intervals $[0, T]$ can be proved under suitable assumptions on the stress response function, the prescribed initial and boundary data, the domain B and the subsets Γ_d and Γ_σ. Numerical approximation is generally required to obtain quantitative information about solutions.

(2) The quantity $\boldsymbol{h}$ appearing in $(7.10)_3$ is a prescribed traction field for the first Piola–Kirchhoff stress. It corresponds to the external force on the *current* boundary expressed per unit area of the *reference* boundary. Because physically meaningful external forces are typically expressed per unit area of the current boundary, expressions for $\boldsymbol{h}$ typically depend on the motion $\boldsymbol{\varphi}$. In particular,

surface area elements in the current and reference configurations are related through φ (see Section 4.6).

(3) More general boundary conditions may be considered with (7.10). For example, $\varphi_i(\boldsymbol{X},t)$ may be specified for all $\boldsymbol{X} \in \Gamma_d^i$ $(i=1,2,3)$, and $P_{ij}(\boldsymbol{X},t)N_j(\boldsymbol{X})$ may be specified for all $\boldsymbol{X} \in \Gamma_\sigma^i$ $(i=1,2,3)$, where Γ_d^i and Γ_σ^i are subsets of ∂B with the properties $\Gamma_d^i \cup \Gamma_\sigma^i = \partial B$ and $\Gamma_d^i \cap \Gamma_\sigma^i = \emptyset$ for each $i = 1,2,3$. Thus, various components of the motion or traction may be specified at each point of the boundary.

(4) Any time-independent solution of (7.10) is called an **equilibrium** or **rest configuration** of the body. Such configurations must satisfy the nonlinear boundary-value problem

$$\begin{aligned} &\nabla^X \cdot [\widehat{\boldsymbol{P}}(\boldsymbol{F})] + \rho_0 \boldsymbol{b}_m = \boldsymbol{0} && \text{in} \quad B \\ &\boldsymbol{\varphi} = \boldsymbol{g} && \text{in} \quad \Gamma_d \\ &\widehat{\boldsymbol{P}}(\boldsymbol{F})\boldsymbol{N} = \boldsymbol{h} && \text{in} \quad \Gamma_\sigma. \end{aligned} \tag{7.11}$$

Whereas (7.10) typically has a unique solution, we expect (7.11) to have genuinely non-unique solutions in various circumstances. Indeed, phenomena such as *buckling* can be understood in terms of such non-uniqueness. □

7.1.5 Isotropy, Simplified Response Functions

Here we introduce the notion of isotropy for an elastic body and study the implications for the stress response functions. In particular, we show that the response functions for an isotropic body take particularly simple forms.

Definition 7.2 *An elastic body with reference configuration B and Cauchy stress response function $\widehat{\boldsymbol{S}}$ is said to be* **isotropic** *if*

$$\widehat{\boldsymbol{S}}(\boldsymbol{F}) = \widehat{\boldsymbol{S}}(\boldsymbol{F}\boldsymbol{Q}),$$

for all $\boldsymbol{F} \in \mathcal{V}^2$ with $\det \boldsymbol{F} > 0$ *and all rotation tensors $\boldsymbol{Q}$.* □

The above definition implies that an elastic body is isotropic if the Cauchy stress field is invariant under rotations of the *reference configuration B*. Roughly speaking, an elastic body is isotropic if it has the same "stiffness" in all directions. Not all materials will enjoy this property, but many do. We now relate this property to the property of isotropic tensor functions.

Result 7.3 Isotropic Stress Response. *If the stress response function $\widehat{\boldsymbol{S}}$ for an isotropic elastic body is frame-indifferent, then the stress response functions $\overline{\boldsymbol{S}}$ and $\overline{\boldsymbol{\Sigma}}$ appearing in Result 7.1 satisfy*

$$\overline{\boldsymbol{S}}(\boldsymbol{Q}\boldsymbol{C}\boldsymbol{Q}^T) = \boldsymbol{Q}\overline{\boldsymbol{S}}(\boldsymbol{C})\boldsymbol{Q}^T \quad \text{and} \quad \overline{\boldsymbol{\Sigma}}(\boldsymbol{Q}\boldsymbol{C}\boldsymbol{Q}^T) = \boldsymbol{Q}\overline{\boldsymbol{\Sigma}}(\boldsymbol{C})\boldsymbol{Q}^T,$$

for all rotation tensors $\boldsymbol{Q}$. Thus $\overline{\boldsymbol{S}}$ and $\overline{\boldsymbol{\Sigma}}$ are isotropic tensor functions in the sense of Section 1.5. Moreover, the function $\widehat{\boldsymbol{S}}$ must itself be an isotropic tensor function, that is

$$\widehat{\boldsymbol{S}}(\boldsymbol{Q}\boldsymbol{F}\boldsymbol{Q}^T) = \boldsymbol{Q}\widehat{\boldsymbol{S}}(\boldsymbol{F})\boldsymbol{Q}^T,$$

for all rotation tensors $\boldsymbol{Q}$. □

Proof Since $\boldsymbol{Q}^T$ is a rotation if and only if $\boldsymbol{Q}$ is, we deduce from Definition 7.2 that $\widehat{\boldsymbol{S}}$ must satisfy

$$\widehat{\boldsymbol{S}}(\boldsymbol{F}) = \widehat{\boldsymbol{S}}(\boldsymbol{F}\boldsymbol{Q}^T), \tag{7.12}$$

for all $\boldsymbol{F} \in \mathcal{V}^2$ with $\det \boldsymbol{F} > 0$ and all rotation tensors $\boldsymbol{Q}$. Moreover, since $\widehat{\boldsymbol{S}}$ is frame-indifferent we have

$$\widehat{\boldsymbol{S}}(\boldsymbol{F}) = \boldsymbol{F}\overline{\boldsymbol{S}}(\boldsymbol{C})\boldsymbol{F}^T, \tag{7.13}$$

for some function $\overline{\boldsymbol{S}}(\boldsymbol{C})$, where $\boldsymbol{C} = \boldsymbol{F}^T\boldsymbol{F}$. Combining (7.12) and (7.13) gives

$$\boldsymbol{F}\overline{\boldsymbol{S}}(\boldsymbol{C})\boldsymbol{F}^T = \boldsymbol{F}\boldsymbol{Q}^T\overline{\boldsymbol{S}}(\boldsymbol{Q}\boldsymbol{F}^T\boldsymbol{F}\boldsymbol{Q}^T)\boldsymbol{Q}\boldsymbol{F}^T,$$

from which we deduce, since $\det \boldsymbol{F} > 0$

$$\overline{\boldsymbol{S}}(\boldsymbol{C}) = \boldsymbol{Q}^T\overline{\boldsymbol{S}}(\boldsymbol{Q}\boldsymbol{C}\boldsymbol{Q}^T)\boldsymbol{Q} \quad \text{or} \quad \boldsymbol{Q}\overline{\boldsymbol{S}}(\boldsymbol{C})\boldsymbol{Q}^T = \overline{\boldsymbol{S}}(\boldsymbol{Q}\boldsymbol{C}\boldsymbol{Q}^T). \tag{7.14}$$

Thus $\overline{\boldsymbol{S}}(\boldsymbol{C})$ is an isotropic tensor function as defined in Section 1.5. The result for $\overline{\boldsymbol{\Sigma}}(\boldsymbol{C})$ follows from the above result for $\overline{\boldsymbol{S}}(\boldsymbol{C})$ and the relation $\overline{\boldsymbol{\Sigma}}(\boldsymbol{C}) = \sqrt{\det \boldsymbol{C}}\,\overline{\boldsymbol{S}}(\boldsymbol{C})$. In particular, we have

$$\begin{aligned}\boldsymbol{Q}\overline{\boldsymbol{\Sigma}}(\boldsymbol{C})\boldsymbol{Q}^T &= \sqrt{\det \boldsymbol{C}}\,\boldsymbol{Q}\overline{\boldsymbol{S}}(\boldsymbol{C})\boldsymbol{Q}^T \\ &= \sqrt{\det(\boldsymbol{Q}\boldsymbol{C}\boldsymbol{Q}^T)}\,\overline{\boldsymbol{S}}(\boldsymbol{Q}\boldsymbol{C}\boldsymbol{Q}^T) = \overline{\boldsymbol{\Sigma}}(\boldsymbol{Q}\boldsymbol{C}\boldsymbol{Q}^T).\end{aligned}$$

Thus $\overline{\boldsymbol{\Sigma}}(\boldsymbol{C})$ is also an isotropic tensor function. The result that $\widehat{\boldsymbol{S}}(\boldsymbol{F})$ must itself be an isotropic tensor function is a straightforward consequence of (7.13) and (7.14). □

The above result can now be used to show that the stress response functions for an isotropic elastic body take particularly simple forms.

Result 7.4 ***Simplified Stress Response Functions.*** *The stress response function* $\widehat{\boldsymbol{S}}$ *for an isotropic elastic body is frame-indifferent if and only if it can be written in the form*

$$\widehat{\boldsymbol{S}}(\boldsymbol{F}) = \boldsymbol{F}\,[\,\beta_0(\mathcal{I}_C)\boldsymbol{I} + \beta_1(\mathcal{I}_C)\boldsymbol{C} + \beta_2(\mathcal{I}_C)\boldsymbol{C}^{-1}\,]\boldsymbol{F}^T,$$

for some functions $\beta_0, \beta_1, \beta_2 : \mathbb{R}^3 \to \mathbb{R}$. *Here* $\mathcal{I}_C$ *denotes the three principal invariants of the right Cauchy–Green strain tensor* $\boldsymbol{C}$. *Equivalently, the response functions* $\widehat{\boldsymbol{P}}$ *and* $\widehat{\boldsymbol{\Sigma}}$ *can be written in the form*

$$\widehat{\boldsymbol{P}}(\boldsymbol{F}) = \boldsymbol{F}\,[\,\gamma_0(\mathcal{I}_C)\boldsymbol{I} + \gamma_1(\mathcal{I}_C)\boldsymbol{C} + \gamma_2(\mathcal{I}_C)\boldsymbol{C}^{-1}\,],$$
$$\widehat{\boldsymbol{\Sigma}}(\boldsymbol{F}) = \gamma_0(\mathcal{I}_C)\boldsymbol{I} + \gamma_1(\mathcal{I}_C)\boldsymbol{C} + \gamma_2(\mathcal{I}_C)\boldsymbol{C}^{-1},$$

for some functions $\gamma_0, \gamma_1, \gamma_2 : \mathbb{R}^3 \to \mathbb{R}$. □

Proof If $\widehat{\boldsymbol{S}}$ is a frame-indifferent response function for an isotropic elastic body, then Result 7.1 implies

$$\widehat{\boldsymbol{S}}(\boldsymbol{F}) = \boldsymbol{F}\overline{\boldsymbol{S}}(\boldsymbol{C})\boldsymbol{F}^T,$$

for some function $\overline{\boldsymbol{S}}(\boldsymbol{C})$, where $\boldsymbol{C} = \boldsymbol{F}^T\boldsymbol{F}$, and Result 7.3 implies $\overline{\boldsymbol{S}}(\boldsymbol{C})$ is an isotropic tensor function. Moreover, $\overline{\boldsymbol{S}}(\boldsymbol{C})$ must be symmetric because $\widehat{\boldsymbol{S}}(\boldsymbol{F})$ is symmetric by definition of an elastic solid. By Result 1.15 we find that $\overline{\boldsymbol{S}}(\boldsymbol{C})$ can be written in the form

$$\overline{\boldsymbol{S}}(\boldsymbol{C}) = \beta_0(\mathcal{I}_C)\boldsymbol{I} + \beta_1(\mathcal{I}_C)\boldsymbol{C} + \beta_2(\mathcal{I}_C)\boldsymbol{C}^{-1},$$

for some functions $\beta_0, \beta_1, \beta_2 : \mathbb{R}^3 \to \mathbb{R}$. Thus $\widehat{\boldsymbol{S}}(\boldsymbol{F})$ must necessarily be of the stated form. Conversely, if $\widehat{\boldsymbol{S}}(\boldsymbol{F})$ has the stated form, then it is straightforward to verify that this form satisfies Definition 7.2 and is frame-indifferent. The results for $\widehat{\boldsymbol{P}}$ and $\widehat{\boldsymbol{\Sigma}}$ are immediate consequences of (7.1) and the relation $\det \boldsymbol{C} = (\det \boldsymbol{F})^2$. In particular, if $\widehat{\boldsymbol{S}}(\boldsymbol{F})$ has the stated form, then $\widehat{\boldsymbol{P}}$ and $\widehat{\boldsymbol{\Sigma}}$ have the stated forms with $\gamma_i = \sqrt{\det \boldsymbol{C}}\,\beta_i$ $(i = 0, 1, 2)$. □

Remarks:

(1) The above result implies that the stress response functions for an isotropic elastic body are completely defined by three scalar-valued functions of the principal invariants of $\boldsymbol{C}$.

(2) The St. Venant–Kirchhoff model given in (7.2) is actually a model for an isotropic elastic solid. In particular, for this model we have,

using (7.9)

$$\widehat{\boldsymbol{\Sigma}}(\boldsymbol{F}) = \frac{\lambda}{2}[\mathrm{tr}(\boldsymbol{C}-\boldsymbol{I})]\boldsymbol{I} + \mu(\boldsymbol{C}-\boldsymbol{I})$$
$$= \gamma_0(\mathcal{I}_C)\boldsymbol{I} + \gamma_1(\mathcal{I}_C)\boldsymbol{C} + \gamma_2(\mathcal{I}_C)\boldsymbol{C}^{-1},$$

where $\gamma_0 = \frac{\lambda}{2}\,\mathrm{tr}\,\boldsymbol{C} - \frac{3\lambda}{2} - \mu$, $\gamma_1 = \mu$ and $\gamma_2 = 0$. Here we have used the fact that $\mathrm{tr}\,\boldsymbol{I} = 3$. □

7.2 Hyperelastic Solids

In this section we study the constitutive model for a hyperelastic solid, which is a special case of an elastic solid introduced in the previous section. We outline a result which shows that an elastic solid satisfies the mechanical energy inequality if and only if it is hyperelastic. We also study the implications of material frame-indifference and establish a result on the balance of total mechanical energy. We end with a brief list of some common hyperelastic models used in practice.

7.2.1 Definition

A continuum body with reference configuration B is said to be a **hyperelastic solid** if:

(1) The body is elastic with stress response functions $\widehat{\boldsymbol{S}}(\boldsymbol{F})$, $\widehat{\boldsymbol{P}}(\boldsymbol{F})$ and $\widehat{\boldsymbol{\Sigma}}(\boldsymbol{F})$ for the stress fields $\boldsymbol{S}_m(\boldsymbol{X},t)$, $\boldsymbol{P}(\boldsymbol{X},t)$ and $\boldsymbol{\Sigma}(\boldsymbol{X},t)$.

(2) The response function $\widehat{\boldsymbol{P}}$ is of the form

$$\widehat{\boldsymbol{P}}(\boldsymbol{F}) = DW(\boldsymbol{F}), \tag{7.15}$$

where $W : \mathcal{V}^2 \to \mathbb{R}$ is a given function called a **strain energy density** for the body. Here $DW(\boldsymbol{F})$ denotes the derivative of W at $\boldsymbol{F}$ (see Definition 2.20).

(3) The function W has the property

$$DW(\boldsymbol{F})\boldsymbol{F}^T = \boldsymbol{F}DW(\boldsymbol{F})^T, \quad \forall \boldsymbol{F} \in \mathcal{V}^2,\ \det \boldsymbol{F} > 0.$$

Property (1) implies that, in any motion, the first Piola–Kirchhoff stress field is given by $\boldsymbol{P}(\boldsymbol{X},t) = \widehat{\boldsymbol{P}}(\boldsymbol{F}(\boldsymbol{X},t))$. Property (2) states that the response function $\widehat{\boldsymbol{P}}$ is the derivative of a scalar strain energy function W. This property has many important consequences. For example, it guarantees that the mechanical energy inequality is satisfied, and also implies that the systems in (7.10) and (7.11) have a variational structure. Property (3) guarantees that the balance of angular momentum equation (Result 5.16) is automatically satisfied, as required by the definition of the response functions of an elastic body.

7.2.2 Frame-Indifference Considerations

Here we outline a result which shows that the strain energy function must be of a particular form in order to satisfy the axiom of material frame-indifference.

Result 7.5 Frame-Indifferent Hyperelastic Stress Response. *The Cauchy stress field in a hyperelastic body is frame-indifferent if and only if the strain energy function W can be expressed as*

$$W(\boldsymbol{F}) = \overline{W}(\boldsymbol{C}), \tag{7.16}$$

for some function $\overline{W} : \mathcal{V}^2 \to \mathbb{R}$, where $\boldsymbol{C} = \boldsymbol{F}^T\boldsymbol{F}$. In this case, the stress response functions $\widehat{\boldsymbol{P}}$, $\widehat{\boldsymbol{\Sigma}}$ and $\widehat{\boldsymbol{S}}$ are given by

$$\begin{gathered}\widehat{\boldsymbol{P}}(\boldsymbol{F}) = 2\boldsymbol{F}D\overline{W}(\boldsymbol{C}), \qquad \widehat{\boldsymbol{\Sigma}}(\boldsymbol{F}) = 2D\overline{W}(\boldsymbol{C}), \\ \widehat{\boldsymbol{S}}(\boldsymbol{F}) = 2(\det \boldsymbol{C})^{-1/2}\boldsymbol{F}D\overline{W}(\boldsymbol{C})\boldsymbol{F}^T,\end{gathered} \tag{7.17}$$

where $D\overline{W}(\boldsymbol{C})$ denotes the derivative of $\overline{W}$ at $\boldsymbol{C}$. □

Proof For brevity we prove that (7.16) is sufficient for frame-indifference and omit the proof of necessity. We begin by showing that (7.16) implies the functional form for $\widehat{\boldsymbol{P}}$ stated in (7.17). From the component version of (7.15) we have

$$\widehat{P}_{ij}(\boldsymbol{F}) = \frac{\partial W}{\partial F_{ij}}(\boldsymbol{F}).$$

Moreover, since $W(\boldsymbol{F}) = \overline{W}(\boldsymbol{C})$ and $\boldsymbol{C} = \boldsymbol{F}^T\boldsymbol{F}$, the chain rule implies

$$\widehat{P}_{ij} = \frac{\partial \overline{W}}{\partial C_{ml}}\frac{\partial C_{ml}}{\partial F_{ij}},$$

where we omit the arguments for clarity. Since $C_{ml} = F_{km}F_{kl}$ we obtain

$$\frac{\partial C_{ml}}{\partial F_{ij}} = \delta_{ki}\delta_{mj}F_{kl} + F_{km}\delta_{ki}\delta_{lj} = \delta_{mj}F_{il} + \delta_{lj}F_{im},$$

which yields

$$\widehat{P}_{ij} = \frac{\partial \overline{W}}{\partial C_{jl}} F_{il} + \frac{\partial \overline{W}}{\partial C_{mj}} F_{im}.$$

Noting that $D\overline{W}(\boldsymbol{C})$ is symmetric because $\boldsymbol{C}$ is, the above expression may be written as

$$\widehat{P}_{ij} = F_{il} \frac{\partial \overline{W}}{\partial C_{lj}} + F_{im} \frac{\partial \overline{W}}{\partial C_{mj}},$$

which in tensor notation becomes

$$\widehat{\boldsymbol{P}}(\boldsymbol{F}) = 2\boldsymbol{F} D\overline{W}(\boldsymbol{C}). \tag{7.18}$$

Thus (7.16) implies the functional form for $\widehat{\boldsymbol{P}}$ stated in (7.17). We next show that this form is sufficient for frame-indifference. From (7.1) and the relation $\det \boldsymbol{C} = (\det \boldsymbol{F})^2$ we have

$$\widehat{\boldsymbol{P}}(\boldsymbol{F}) = (\det \boldsymbol{F})\widehat{\boldsymbol{S}}(\boldsymbol{F})\boldsymbol{F}^{-T} = (\det \boldsymbol{C})^{1/2}\widehat{\boldsymbol{S}}(\boldsymbol{F})\boldsymbol{F}^{-T}.$$

Combining this expression with (7.18) we get

$$2\boldsymbol{F} D\overline{W}(\boldsymbol{C}) = (\det \boldsymbol{C})^{1/2}\widehat{\boldsymbol{S}}(\boldsymbol{F})\boldsymbol{F}^{-T},$$

which implies

$$\widehat{\boldsymbol{S}}(\boldsymbol{F}) = \boldsymbol{F}\left[2(\det \boldsymbol{C})^{-1/2} D\overline{W}(\boldsymbol{C})\right]\boldsymbol{F}^{T}.$$

Thus $\widehat{\boldsymbol{S}}$ has the stated form in (7.17) and frame-indifference follows from Result 7.1. The stated form of $\widehat{\boldsymbol{\Sigma}}$ is an immediate consequence of (7.1). □

Remarks:

(1) An intuitive explanation of the sufficiency of (7.16) for frame-indifference follows from Result 5.22, which shows that $\boldsymbol{C}$ is unaffected by superposed rigid motions. In particular, $W(\boldsymbol{F})$ is a frame-indifferent scalar field when it has the form (7.16).

(2) Property (3) in the definition of a hyperelastic solid is satisfied for any stored energy function of the form (7.16). This is a straightforward consequence of the relation $DW(\boldsymbol{F}) = 2\boldsymbol{F} D\overline{W}(\boldsymbol{C})$ and the symmetry of $D\overline{W}(\boldsymbol{C})$.

(3) The notion of isotropy can also be applied to hyperelastic bodies. In particular, a hyperelastic body is isotropic if and only if the function $\overline{W}(\boldsymbol{C})$ in (7.16) can be expressed in the form

$$\overline{W}(\boldsymbol{C}) = \widehat{W}(\mathcal{I}_{\boldsymbol{C}}),$$

for some function $\widehat{W} : \mathbb{R}^3 \to \mathbb{R}$, where $\mathcal{I}_{\boldsymbol{C}}$ are the principal invariants of $\boldsymbol{C}$ (see Exercises 5 and 7).

(4) The St. Venant–Kirchhoff model given in (7.2) is actually hyperelastic. In particular, from (7.9) we have

$$\widehat{\boldsymbol{\Sigma}}(\boldsymbol{F}) = \frac{\lambda}{2}[\mathrm{tr}(\boldsymbol{C} - \boldsymbol{I})]\boldsymbol{I} + \mu(\boldsymbol{C} - \boldsymbol{I}),$$

and we can verify that $\widehat{\boldsymbol{\Sigma}}(\boldsymbol{F}) = 2D\overline{W}(\boldsymbol{C})$, where

$$\overline{W}(\boldsymbol{C}) = \frac{\lambda}{8}[\mathrm{tr}(\boldsymbol{C} - \boldsymbol{I})]^2 + \frac{\mu}{4}\,\mathrm{tr}[(\boldsymbol{C} - \boldsymbol{I})^2].$$

Moreover, because $\overline{W}(\boldsymbol{C})$ can be shown to depend only on the principal invariants of $\boldsymbol{C}$, the model is isotropic, as verified directly in Section 7.1. □

7.2.3 Mechanical Energy Considerations

Here we assume the isothermal model of an elastic solid is energetically passive as discussed in Section 5.7 and outline a result which shows that it satisfies the mechanical energy inequality (Result 5.29) if and only if it is hyperelastic. We also establish a result on the balance of total mechanical energy.

Result 7.6 ***Mechanical Energy Inequality for Elastic Solids.*** *An elastic solid satisfies the mechanical energy inequality (Result 5.29) if and only if it is hyperelastic, that is*

$$\widehat{\boldsymbol{P}}(\boldsymbol{F}) = DW(\boldsymbol{F}), \tag{7.19}$$

for some strain energy function $W(\boldsymbol{F})$. □

Proof For brevity we prove that (7.19) is sufficient and omit the proof of necessity. Let $\boldsymbol{P}$ denote the first Piola–Kirchhoff stress field and assume that (7.19) holds. Then

$$\boldsymbol{P}(\boldsymbol{X}, t) = \widehat{\boldsymbol{P}}(\boldsymbol{F}(\boldsymbol{X}, t)) = DW(\boldsymbol{F}(\boldsymbol{X}, t)),$$

which implies

$$\begin{aligned} \boldsymbol{P}(\boldsymbol{X},t) : \dot{\boldsymbol{F}}(\boldsymbol{X},t) &= \widehat{\boldsymbol{P}}(\boldsymbol{F}(\boldsymbol{X},t)) : \dot{\boldsymbol{F}}(\boldsymbol{X},t) \\ &= DW(\boldsymbol{F}(\boldsymbol{X},t)) : \dot{\boldsymbol{F}}(\boldsymbol{X},t) \\ &= \frac{\partial}{\partial t} W(\boldsymbol{F}(\boldsymbol{X},t)), \quad \forall \boldsymbol{X} \in B,\ t \geq 0. \end{aligned} \tag{7.20}$$

Integrating the above expression over any arbitrary time interval $[t_0, t_1]$ gives

$$\begin{aligned} \int_{t_0}^{t_1} & \boldsymbol{P}(\boldsymbol{X},t) : \dot{\boldsymbol{F}}(\boldsymbol{X},t)\, dt \\ &= W(\boldsymbol{F}(\boldsymbol{X},t_1)) - W(\boldsymbol{F}(\boldsymbol{X},t_0)), \quad \forall \boldsymbol{X} \in B. \end{aligned} \tag{7.21}$$

For any closed, isothermal process in $[t_0, t_1]$ we have $\boldsymbol{\varphi}(\boldsymbol{X}, t_1) = \boldsymbol{\varphi}(\boldsymbol{X}, t_0)$ for all $\boldsymbol{X} \in B$, which implies $\boldsymbol{F}(\boldsymbol{X}, t_1) = \boldsymbol{F}(\boldsymbol{X}, t_0)$ for all $\boldsymbol{X} \in B$. Thus the right-hand side of (7.21) vanishes and the mechanical energy inequality is satisfied for any strain energy function $W(\boldsymbol{F})$. □

The next result establishes a balance law for the total mechanical energy of a hyperelastic body.

Result 7.7 Hyperelastic Energy Balance. *Let $\boldsymbol{\varphi} : B \times [0, \infty) \to \mathbb{E}^3$ denote a motion of a hyperelastic body with strain energy function $W(\boldsymbol{F})$, and at any time $t \geq 0$ let B_t denote the current configuration of B. Then*

$$\frac{d}{dt}(U[B_t] + K[B_t]) = \mathcal{P}[B_t], \quad \forall t \geq 0,$$

where $U[B_t]$ is the total **strain energy** *defined by*

$$U[B_t] = \int_B W(\boldsymbol{F}(\boldsymbol{X},t))\, dV_{\boldsymbol{X}},$$

$K[B_t]$ is the total kinetic energy and $\mathcal{P}[B_t]$ is the power of external forces on B_t (see Chapter 5). □

Proof From Result 5.17 we have

$$\frac{d}{dt} K[B_t] + \int_B \boldsymbol{P}(\boldsymbol{X},t) : \dot{\boldsymbol{F}}(\boldsymbol{X},t)\, dV_{\boldsymbol{X}} = \mathcal{P}[B_t],$$

and from (7.20) we have

$$\int_B \boldsymbol{P}(\boldsymbol{X},t) : \dot{\boldsymbol{F}}(\boldsymbol{X},t)\, dV_{\boldsymbol{X}} = \frac{d}{dt} \int_B W(\boldsymbol{F}(\boldsymbol{X},t))\, dV_{\boldsymbol{X}}.$$

Combining these two relations yields the desired result. □

In the absence of external forces, Result 7.7 implies that the total energy of a hyperelastic body, defined as the sum of the kinetic and strain energies, is conserved along motions (see Exercise 8). In fact, the equations of motion may be viewed as a separable Hamiltonian system with the strain energy being the potential energy.

7.2.4 Common Models

In view of Result 7.6, many isothermal models of elastic solids employed in practice are hyperelastic. Indeed, the mechanical energy inequality, which is derived from the Clausius–Duhem inequality (Second Law of Thermodynamics), demands that elastic models be hyperelastic. We end this section with a brief list of some common (frame-indifferent, isotropic) hyperelastic models, all defined through their strain energy function $\overline{W}(\boldsymbol{C})$.

(1) **St. Venant–Kirchhoff model.**

$$\overline{W}(\boldsymbol{C}) = \frac{\lambda}{2}(\operatorname{tr}\boldsymbol{G})^2 + \mu\operatorname{tr}(\boldsymbol{G}^2).$$
$$\boldsymbol{G} = \tfrac{1}{2}(\boldsymbol{C} - \boldsymbol{I}). \quad \lambda, \mu > 0.$$

(2) **Neo-Hookean model.**

$$\overline{W}(\boldsymbol{C}) = a\operatorname{tr}\boldsymbol{C} + \Gamma(\sqrt{\det\boldsymbol{C}}).$$
$$\Gamma(s) = cs^2 - d\ln s. \quad a, c, d > 0.$$

(3) **Mooney–Rivlin model.**

$$\overline{W}(\boldsymbol{C}) = a\operatorname{tr}\boldsymbol{C} + b\operatorname{tr}\boldsymbol{C}_* + \Gamma(\sqrt{\det\boldsymbol{C}}).$$
$$\boldsymbol{C}_* = (\det\boldsymbol{C})\boldsymbol{C}^{-1}. \quad a, b > 0.$$

(4) **Ogden model.**

$$\overline{W}(\boldsymbol{C}) = \sum_{i=1}^{M} a_i \operatorname{tr}(\boldsymbol{C}^{\gamma_i/2}) + \sum_{j=1}^{N} b_j \operatorname{tr}(\boldsymbol{C}_*^{\delta_j/2}) + \Gamma(\sqrt{\det\boldsymbol{C}}).$$
$$M, N \geq 1. \quad a_i, b_j > 0. \quad \gamma_i, \delta_j \geq 1.$$

All the above models, most notably (2), (3) and (4), have been used in the modeling of rubber-like materials at large strains. In particular, (2), (3) and (4) have typically been used in the modeling of incompressible materials, in which case the term $\Gamma(\sqrt{\det \boldsymbol{C}})$ is omitted and the constraint $\det \boldsymbol{F} = 1$, or equivalently $\det \boldsymbol{C} = 1$, is considered along with a multiplier (see Section 5.6). In the general compressible case considered here, the term $\Gamma(\sqrt{\det \boldsymbol{C}})$ gives rise to appropriate extreme stress in the case of extreme compression when any principal stretch tends to zero, equivalently $\det \boldsymbol{C} \to 0$. In this sense models (2), (3) and (4) are better suited for extreme strains than model (1), which does not have this property, as previously noted in Section 7.1. All the above models give rise to appropriate extreme stress in the case of extreme extension when any principal stretch tends to infinity.

7.3 Linearization of Elasticity Equations

The equations governing the motion of an elastic solid are, for most constitutive models of interest, too complex to solve exactly. Hence approximations are sometimes made in order to simplify them. In this section we derive a set of simplified equations for an elastic solid under the assumption that deformations are small.

For motivation, consider an elastic body at rest in a stress-free reference configuration B. We say that a reference configuration is **stress-free** if any one, hence all, of the stress response functions vanish when $\boldsymbol{F} = \boldsymbol{I}$, that is

$$\widehat{\boldsymbol{P}}(\boldsymbol{I}) = \boldsymbol{O}, \quad \widehat{\boldsymbol{\Sigma}}(\boldsymbol{I}) = \boldsymbol{O}, \quad \widehat{\boldsymbol{S}}(\boldsymbol{I}) = \boldsymbol{O}. \tag{7.22}$$

Suppose the body is subject to a body force, and boundary and initial conditions as given in (7.10), and assume $\boldsymbol{b}_m$, $\boldsymbol{g}$, $\boldsymbol{h}$ and $\boldsymbol{V}_0$ are all small in the sense that

$$|\boldsymbol{b}_m(\boldsymbol{X},t)|, \quad |\boldsymbol{g}(\boldsymbol{X},t) - \boldsymbol{X}|, \quad |\boldsymbol{h}(\boldsymbol{X},t)|, \quad |\boldsymbol{V}_0(\boldsymbol{X})| = \mathcal{O}(\epsilon), \tag{7.23}$$

where $0 \le \epsilon \ll 1$ is a small parameter. In this case it is reasonable to expect that the body will deviate only slightly from its reference configuration in the sense that

$$|\boldsymbol{\varphi}(\boldsymbol{X},t) - \boldsymbol{X}| = \mathcal{O}(\epsilon). \tag{7.24}$$

Indeed, if $\boldsymbol{b}_m = \boldsymbol{h} = \boldsymbol{V}_0 = \boldsymbol{0}$ and $\boldsymbol{g} = \boldsymbol{X}$, then equations (7.10) are satisfied by $\boldsymbol{\varphi}(\boldsymbol{X},t) = \boldsymbol{X}$ for all $\boldsymbol{X} \in B$ and $t \ge 0$.

Our goal here is to derive a set of simplified equations to describe motions satisfying (7.24) under the assumptions (7.22) and (7.23). We discuss initial-boundary value problems for these equations and study implications of the axiom of material frame-indifference and the assumption of isotropy.

7.3.1 Linearized Equations, Elasticity Tensors

In view of (7.23) we assume that $\boldsymbol{b}_m$, $\boldsymbol{g}$, $\boldsymbol{h}$ and $\boldsymbol{V}_0$ can all be expressed in the form

$$\boldsymbol{b}_m^\epsilon = \epsilon \boldsymbol{b}_m^{(1)}, \quad \boldsymbol{g}^\epsilon = \boldsymbol{X} + \epsilon \boldsymbol{g}^{(1)}, \quad \boldsymbol{h}^\epsilon = \epsilon \boldsymbol{h}^{(1)}, \quad \boldsymbol{V}_0^\epsilon = \epsilon \boldsymbol{V}_0^{(1)}, \tag{7.25}$$

for some functions $\boldsymbol{b}_m^{(1)}$, $\boldsymbol{g}^{(1)}$, $\boldsymbol{h}^{(1)}$ and $\boldsymbol{V}_0^{(1)}$. Then, in view of (7.24), we seek a power series expansion for $\boldsymbol{\varphi}$ of the form

$$\boldsymbol{\varphi}^\epsilon = \boldsymbol{X} + \epsilon \boldsymbol{u}^{(1)} + \mathcal{O}(\epsilon^2), \tag{7.26}$$

where $\boldsymbol{u}^{(1)}$ is an unknown field. If we let $\boldsymbol{u}^\epsilon = \boldsymbol{\varphi}^\epsilon - \boldsymbol{X}$ be the displacement field associated with $\boldsymbol{\varphi}^\epsilon$, then from (7.26) we deduce that $\boldsymbol{u}^\epsilon$ has the expansion

$$\boldsymbol{u}^\epsilon = \boldsymbol{0} + \epsilon \boldsymbol{u}^{(1)} + \mathcal{O}(\epsilon^2). \tag{7.27}$$

We next derive a partial differential equation for the first-order displacement disturbance $\boldsymbol{u}^{(1)}$. From $(7.10)_1$ we deduce that $\boldsymbol{u}^\epsilon = \boldsymbol{\varphi}^\epsilon - \boldsymbol{X}$ satisfies the balance of linear momentum equation

$$\rho_0 \ddot{\boldsymbol{u}}^\epsilon = \nabla^X \cdot [\,\widehat{\boldsymbol{P}}(\boldsymbol{F}^\epsilon)\,] + \rho_0 \boldsymbol{b}_m^\epsilon, \tag{7.28}$$

where $\boldsymbol{F}^\epsilon$ is the deformation gradient associated with $\boldsymbol{\varphi}^\epsilon$, namely

$$\boldsymbol{F}^\epsilon = \nabla^X \boldsymbol{\varphi}^\epsilon = \boldsymbol{I} + \epsilon \nabla^X \boldsymbol{u}^{(1)} + \mathcal{O}(\epsilon^2).$$

The following definition will be helpful in simplifying (7.28).

Definition 7.8 *Let $\widehat{\boldsymbol{P}}(\boldsymbol{F})$, $\widehat{\boldsymbol{\Sigma}}(\boldsymbol{F})$ and $\widehat{\boldsymbol{S}}(\boldsymbol{F})$ be the Piola–Kirchhoff and Cauchy stress response functions for an elastic solid. Then the fourth-order* **elasticity tensors A, B** *and* **C** *associated with $\widehat{\boldsymbol{P}}$, $\widehat{\boldsymbol{\Sigma}}$ and $\widehat{\boldsymbol{S}}$ are*

$$\mathsf{A}_{ijkl} = \frac{\partial \widehat{P}_{ij}}{\partial F_{kl}}(\boldsymbol{I}), \quad \mathsf{B}_{ijkl} = \frac{\partial \widehat{\Sigma}_{ij}}{\partial F_{kl}}(\boldsymbol{I}) \quad \text{and} \quad \mathsf{C}_{ijkl} = \frac{\partial \widehat{S}_{ij}}{\partial F_{kl}}(\boldsymbol{I}).$$

In tensor notation we have

$$\mathsf{A}(\boldsymbol{H}) = \frac{d}{d\alpha} \widehat{\boldsymbol{P}}(\boldsymbol{I} + \alpha \boldsymbol{H})\Big|_{\alpha=0} = D\widehat{\boldsymbol{P}}(\boldsymbol{I})\boldsymbol{H}, \quad \forall \boldsymbol{H} \in \mathcal{V}^2.$$

Similar expressions hold for **B** *and* **C** *(see Definition 2.24).* □

Using the above definition, we expand the stress response function $\widehat{\boldsymbol{P}}(\boldsymbol{F}^\epsilon)$ in a Taylor series about $\epsilon = 0$ to get

$$\widehat{\boldsymbol{P}}(\boldsymbol{F}^\epsilon) = \widehat{\boldsymbol{P}}(\boldsymbol{F}^\epsilon)\Big|_{\epsilon=0} + \epsilon\mathbf{A}(\nabla^X \boldsymbol{u}^{(1)}) + \mathcal{O}(\epsilon^2).$$

The fact that $\widehat{\boldsymbol{P}}(\boldsymbol{I}) = \boldsymbol{O}$ then gives

$$\widehat{\boldsymbol{P}}(\boldsymbol{F}^\epsilon) = \epsilon\mathbf{A}(\nabla^X \boldsymbol{u}^{(1)}) + \mathcal{O}(\epsilon^2). \tag{7.29}$$

When we substitute (7.29), (7.27) and (7.25) into (7.28) and retain only those terms involving the first power of ϵ we obtain

$$\rho_0 \ddot{\boldsymbol{u}}^{(1)} = \nabla^X \cdot [\,\mathbf{A}(\nabla^X \boldsymbol{u}^{(1)})\,] + \rho_0 \boldsymbol{b}_m^{(1)}. \tag{7.30}$$

This is the linearized version of the balance of linear momentum equation $(7.10)_1$. Similar considerations can be used to obtain linearized versions of the boundary and initial conditions in $(7.10)_{2,3,4,5}$.

7.3.2 Initial-Boundary Value Problems

Linearization of (7.10) leads to the following initial-boundary value problem for the first-order displacement disturbance in an elastic solid: Find $\boldsymbol{u}^{(1)} : B \times [0,T] \to \mathcal{V}$ such that

$$\boxed{\begin{array}{lll} \rho_0 \ddot{\boldsymbol{u}}^{(1)} = \nabla^X \cdot [\,\mathbf{A}(\nabla^X \boldsymbol{u}^{(1)})\,] + \rho_0 \boldsymbol{b}_m^{(1)} & \text{in} & B \times [0,T] \\ \boldsymbol{u}^{(1)} = \boldsymbol{g}^{(1)} & \text{in} & \Gamma_d \times [0,T] \\ \mathbf{A}(\nabla^X \boldsymbol{u}^{(1)})\boldsymbol{N} = \boldsymbol{h}^{(1)} & \text{in} & \Gamma_\sigma \times [0,T] \\ \boldsymbol{u}^{(1)}(\cdot,0) = \mathbf{0} & \text{in} & B \\ \dot{\boldsymbol{u}}^{(1)}(\cdot,0) = \boldsymbol{V}_0^{(1)} & \text{in} & B. \end{array}} \tag{7.31}$$

The above equations are typically called the **Linearized Elastodynamics Equations**. They are an approximation to the system in (7.10) that are appropriate for describing small deviations from a stress-free reference configuration in an elastic solid. In the above system, Γ_d and Γ_σ are subsets of ∂B with the properties $\Gamma_d \cup \Gamma_\sigma = \partial B$ and $\Gamma_d \cap \Gamma_\sigma = \emptyset$, ρ_0 is the reference mass density, $\mathbf{A}$ is the fourth-order elasticity tensor associated with the stress response function $\widehat{\boldsymbol{P}}$, $\boldsymbol{N}$ is the unit outward normal field on ∂B, and $\boldsymbol{b}_m^{(1)}$, $\boldsymbol{g}^{(1)}$, $\boldsymbol{h}^{(1)}$ and $\boldsymbol{V}_0^{(1)}$ are prescribed disturbance fields. Any of the more general boundary conditions discussed in connection with (7.10) can also be considered for (7.31).

Remarks:

(1) The system in (7.31) is a linear initial-boundary value problem for the disturbance field $\boldsymbol{u}^{(1)}$. In the case when A satisfies major symmetry and strong ellipticity conditions (see Section 7.4 below), this system has a unique solution in any given time interval $[0,T]$ under mild assumptions on the prescribed initial and boundary data, the domain B and the subsets Γ_d and Γ_σ.

(2) In view of (7.24) the difference between the fixed reference configuration B and the current configuration $B_t = \varphi_t(B)$ is first-order in ϵ for all $t \geq 0$. For this reason, the distinction between material and spatial coordinates is often ignored in developments of elasticity where only the linear response is considered.

(3) Any time-independent solution of (7.31) must satisfy the linear boundary-value problem

$$\begin{aligned} &\nabla^X \cdot \left[\mathsf{A}(\nabla^X \boldsymbol{u}^{(1)})\right] + \rho_0 \boldsymbol{b}_m^{(1)} = \mathbf{0} && \text{in} \quad B \\ &\boldsymbol{u}^{(1)} = \boldsymbol{g}^{(1)} && \text{in} \quad \Gamma_d \\ &\mathsf{A}(\nabla^X \boldsymbol{u}^{(1)})\boldsymbol{N} = \boldsymbol{h}^{(1)} && \text{in} \quad \Gamma_\sigma. \end{aligned} \tag{7.32}$$

These equations are typically called the **Linearized Elastostatics Equations**. In the case when A satisfies major symmetry and positivity conditions (see Section 7.4 below), this system has a unique solution under mild assumptions on the domain and prescribed data provided Γ_d is non-empty. When Γ_d is empty, the prescribed field $\boldsymbol{h}^{(1)}$ must satisfy solvability conditions (resultant force and torque must be zero), and solutions are unique only to within an arbitrary infinitesimally rigid deformation. □

7.3.3 Properties of Elasticity Tensors

Here we study the fourth-order elasticity tensors introduced in the previous section. We show that these tensors are all equal and study the implications of the balance law of angular momentum and the axiom of material frame-indifference. We show that the elasticity tensor for an isotropic elastic body takes a particularly simple form.

Result 7.9 ***Equivalence of Elasticity Tensors.*** *Let A, B and C be the elasticity tensors associated with the stress response functions $\widehat{\boldsymbol{P}}$,*

$\widehat{\boldsymbol{\Sigma}}$ *and* $\widehat{\boldsymbol{S}}$ *of an elastic body. If the reference configuration of the body is stress-free, then* $\mathsf{A} = \mathsf{B} = \mathsf{C}$. □

Proof For brevity we only prove $\mathsf{A} = \mathsf{C}$. The proof that $\mathsf{B} = \mathsf{C}$ is similar (see Exercise 10). To begin, from (7.1) we have

$$\widehat{\boldsymbol{P}}(\boldsymbol{F}) = (\det \boldsymbol{F})\widehat{\boldsymbol{S}}(\boldsymbol{F})\boldsymbol{F}^{-T}.$$

Differentiating both sides about $\boldsymbol{F} = \boldsymbol{I}$, using Definition 7.8 for the left-hand side, we obtain

$$\begin{aligned}\mathsf{A}(\boldsymbol{H}) &= \frac{d}{d\alpha}\Big\{\det(\boldsymbol{I}+\alpha\boldsymbol{H})\widehat{\boldsymbol{S}}(\boldsymbol{I}+\alpha\boldsymbol{H})\,(\boldsymbol{I}+\alpha\boldsymbol{H})^{-T}\Big\}\Big|_{\alpha=0} \\ &= \frac{d}{d\alpha}\{\det(\boldsymbol{I}+\alpha\boldsymbol{H})\}\Big|_{\alpha=0}\widehat{\boldsymbol{S}}(\boldsymbol{I})\boldsymbol{I}^{-T} \\ &\qquad + \det(\boldsymbol{I})\frac{d}{d\alpha}\Big\{\widehat{\boldsymbol{S}}(\boldsymbol{I}+\alpha\boldsymbol{H})\Big\}\Big|_{\alpha=0}\boldsymbol{I}^{-T} \\ &\qquad\qquad + \det(\boldsymbol{I})\widehat{\boldsymbol{S}}(\boldsymbol{I})\frac{d}{d\alpha}\{(\boldsymbol{I}+\alpha\boldsymbol{H})^{-T}\}\Big|_{\alpha=0},\end{aligned}$$

for all $\boldsymbol{H} \in \mathcal{V}^2$. Since $\widehat{\boldsymbol{S}}(\boldsymbol{I}) = \boldsymbol{O}$ for a stress-free reference configuration and $\det \boldsymbol{I} = 1$ we obtain

$$\mathsf{A}(\boldsymbol{H}) = \frac{d}{d\alpha}\Big\{\widehat{\boldsymbol{S}}(\boldsymbol{I}+\alpha\boldsymbol{H})\Big\}\Big|_{\alpha=0} = \mathsf{C}(\boldsymbol{H}), \qquad \forall \boldsymbol{H} \in \mathcal{V}^2,$$

which establishes the result. □

Remarks:

(1) The above result implies that an elastic solid with a stress-free reference configuration has a unique elasticity tensor, which we usually denote by C. This tensor can be determined from knowledge of any one of the stress response functions $\widehat{\boldsymbol{P}}$, $\widehat{\boldsymbol{\Sigma}}$ or $\widehat{\boldsymbol{S}}$.

(2) The above result can also be interpreted as saying that the response functions $\widehat{\boldsymbol{P}}$, $\widehat{\boldsymbol{\Sigma}}$ and $\widehat{\boldsymbol{S}}$ have the same linearization at $\boldsymbol{F} = \boldsymbol{I}$, namely

$$\begin{aligned}\widehat{\boldsymbol{P}}(\boldsymbol{F}^{\epsilon}) &= \epsilon\mathsf{C}(\nabla^X \boldsymbol{u}^{(1)}) + \mathcal{O}(\epsilon^2), \\ \widehat{\boldsymbol{\Sigma}}(\boldsymbol{F}^{\epsilon}) &= \epsilon\mathsf{C}(\nabla^X \boldsymbol{u}^{(1)}) + \mathcal{O}(\epsilon^2), \\ \widehat{\boldsymbol{S}}(\boldsymbol{F}^{\epsilon}) &= \epsilon\mathsf{C}(\nabla^X \boldsymbol{u}^{(1)}) + \mathcal{O}(\epsilon^2).\end{aligned}$$

Thus, in the linear approximation, the two Piola–Kirchhoff stress fields and the Cauchy stress field are indistinguishable. □

The next result establishes the implications of the balance law of angular momentum on C.

Result 7.10 ***Left Minor Symmetry of Elasticity Tensor.*** *Consider an elastic body with a stress-free reference configuration and an associated elasticity tensor* C. *Then the balance law of angular momentum implies that* $\mathsf{C}(\boldsymbol{H})$ *is symmetric for all* $\boldsymbol{H} \in \mathcal{V}^2$. *Thus* C *has a left minor symmetry in the sense that* $\mathsf{C}_{ijkl} = \mathsf{C}_{jikl}$, *equivalently (see Section 1.4.5)*

$$\boldsymbol{A} : \mathsf{C}(\boldsymbol{B}) = \mathrm{sym}(\boldsymbol{A}) : \mathsf{C}(\boldsymbol{B}), \quad \forall \boldsymbol{A}, \boldsymbol{B} \in \mathcal{V}^2.$$

□

Proof The balance law of angular momentum implies $\widehat{\boldsymbol{S}}(\boldsymbol{F})^T = \widehat{\boldsymbol{S}}(\boldsymbol{F})$ for all $\boldsymbol{F} \in \mathcal{V}^2$ with $\det \boldsymbol{F} > 0$. By definition of C we have, for arbitrary $\boldsymbol{H} \in \mathcal{V}^2$

$$\begin{aligned}
\mathsf{C}(\boldsymbol{H})^T &= \left(\frac{d}{d\alpha} \widehat{\boldsymbol{S}}(\boldsymbol{I} + \alpha \boldsymbol{H}) \Big|_{\alpha=0} \right)^T \\
&= \frac{d}{d\alpha} \widehat{\boldsymbol{S}}(\boldsymbol{I} + \alpha \boldsymbol{H})^T \Big|_{\alpha=0} \\
&= \frac{d}{d\alpha} \widehat{\boldsymbol{S}}(\boldsymbol{I} + \alpha \boldsymbol{H}) \Big|_{\alpha=0} \\
&= \mathsf{C}(\boldsymbol{H}),
\end{aligned}$$

where the equality between the second and third lines is justified by the fact that $\det(\boldsymbol{I} + \alpha \boldsymbol{H}) > 0$ for α sufficiently small. Thus $\mathsf{C}(\boldsymbol{H})$ is symmetric for all $\boldsymbol{H} \in \mathcal{V}^2$. From Result 1.13 we deduce

$$\boldsymbol{G} : \mathsf{C}(\boldsymbol{H}) = \mathrm{sym}(\boldsymbol{G}) : \mathsf{C}(\boldsymbol{H}), \quad \forall \boldsymbol{G}, \boldsymbol{H} \in \mathcal{V}^2,$$

which, by definition, implies that C has a left minor symmetry. The condition on the components C_{ijkl} follows from Result 1.14. □

The next result establishes the implications of the axiom of material frame-indifference on the elasticity tensor C.

Result 7.11 ***Right Minor Symmetry of Elasticity Tensor.*** *Let $\widehat{\boldsymbol{S}}$ be the Cauchy stress response function for an elastic body with stress-free reference configuration and elasticity tensor C. If $\widehat{\boldsymbol{S}}$ is frame-indifferent, then $\mathsf{C}(\boldsymbol{W}) = \boldsymbol{O}$ for any skew tensor $\boldsymbol{W} \in \mathcal{V}^2$. Thus C has a right minor symmetry in the sense that $\mathsf{C}_{ijkl} = \mathsf{C}_{ijlk}$, equivalently (see Section 1.4.5)*

$$\boldsymbol{A} : \mathsf{C}(\boldsymbol{B}) = \boldsymbol{A} : \mathsf{C}(\mathrm{sym}(\boldsymbol{B})), \quad \forall \boldsymbol{A}, \boldsymbol{B} \in \mathcal{V}^2.$$

□

Proof By frame-indifference we have, from (7.7)

$$\widehat{\boldsymbol{S}}(\boldsymbol{QF}) = \boldsymbol{Q}\widehat{\boldsymbol{S}}(\boldsymbol{F})\boldsymbol{Q}^T,$$

for all $\boldsymbol{F} \in \mathcal{V}^2$ with $\det \boldsymbol{F} > 0$ and all rotations $\boldsymbol{Q} \in \mathcal{V}^2$. Taking $\boldsymbol{F} = \boldsymbol{I}$ and using the stress-free assumption $\widehat{\boldsymbol{S}}(\boldsymbol{I}) = \boldsymbol{O}$ we obtain

$$\widehat{\boldsymbol{S}}(\boldsymbol{Q}) = \boldsymbol{O}, \tag{7.33}$$

for any rotation $\boldsymbol{Q}$. Next, let $\boldsymbol{W} \in \mathcal{V}^2$ be an arbitrary skew-symmetric tensor. From Result 1.7 we find that $\exp(\alpha \boldsymbol{W})$ is a rotation for any $\alpha \in \mathbb{R}$ and hence (7.33) implies

$$\widehat{\boldsymbol{S}}(\exp(\alpha \boldsymbol{W})) = \boldsymbol{O}, \quad \forall \alpha \in \mathbb{R}. \tag{7.34}$$

By definition of C we have

$$\begin{aligned} \mathsf{C}(\boldsymbol{W}) &= \frac{d}{d\alpha}\widehat{\boldsymbol{S}}(\boldsymbol{I} + \alpha \boldsymbol{W})\Big|_{\alpha=0} \\ &= \frac{d}{d\alpha}\widehat{\boldsymbol{S}}(\boldsymbol{I} + \alpha \boldsymbol{W} + \tfrac{1}{2}\alpha^2 \boldsymbol{W}^2 + \tfrac{1}{6}\alpha^3 \boldsymbol{W}^3 + \cdots)\Big|_{\alpha=0} \\ &= \frac{d}{d\alpha}\widehat{\boldsymbol{S}}(\exp(\alpha \boldsymbol{W}))\Big|_{\alpha=0}. \end{aligned} \tag{7.35}$$

Combining (7.35) and (7.34) we obtain $\mathsf{C}(\boldsymbol{W}) = \boldsymbol{O}$ for any skew tensor $\boldsymbol{W} \in \mathcal{V}^2$. From this we deduce

$$\mathsf{C}(\boldsymbol{H}) = \mathsf{C}(\mathrm{sym}(\boldsymbol{H})), \quad \forall \boldsymbol{H} \in \mathcal{V}^2,$$

which, by definition, implies that C has a right minor symmetry. The condition on the components C_{ijkl} follows from Result 1.14. □

The next result shows that the elasticity tensor C takes a particularly simple form when the body is isotropic.

Result 7.12 ***Elasticity Tensor for Isotropic Solid.*** *Let $\widehat{S}$ be a frame-indifferent Cauchy stress response function for an elastic body with stress-free reference configuration and elasticity tensor $\mathbf{C}$. If the body is isotropic, then the fourth-order tensor $\mathbf{C}$ is isotropic in the sense of Section 1.5 and takes the form*

$$\mathbf{C}(\boldsymbol{H}) = \lambda(\operatorname{tr}\boldsymbol{H})\boldsymbol{I} + 2\mu\,\operatorname{sym}(\boldsymbol{H}), \quad \forall \boldsymbol{H} \in \mathcal{V}^2,$$

where λ and μ are constants. □

Proof For an isotropic elastic body we have, by Result 7.3

$$\boldsymbol{Q}\widehat{\boldsymbol{S}}(\boldsymbol{F})\boldsymbol{Q}^T = \widehat{\boldsymbol{S}}(\boldsymbol{Q}\boldsymbol{F}\boldsymbol{Q}^T),$$

for all $\boldsymbol{F} \in \mathcal{V}^2$ with $\det \boldsymbol{F} > 0$ and all rotations $\boldsymbol{Q} \in \mathcal{V}^2$. Let $\boldsymbol{H} \in \mathcal{V}^2$ be arbitrary. Then by definition of $\mathbf{C}$ we have

$$\begin{aligned}
\boldsymbol{Q}\mathbf{C}(\boldsymbol{H})\boldsymbol{Q}^T &= \boldsymbol{Q}\frac{d}{d\alpha}\widehat{\boldsymbol{S}}(\boldsymbol{I}+\alpha\boldsymbol{H})\Big|_{\alpha=0}\boldsymbol{Q}^T \\
&= \frac{d}{d\alpha}\boldsymbol{Q}\widehat{\boldsymbol{S}}(\boldsymbol{I}+\alpha\boldsymbol{H})\boldsymbol{Q}^T\Big|_{\alpha=0} \\
&= \frac{d}{d\alpha}\widehat{\boldsymbol{S}}(\boldsymbol{Q}(\boldsymbol{I}+\alpha\boldsymbol{H})\boldsymbol{Q}^T)\Big|_{\alpha=0} \\
&= \frac{d}{d\alpha}\widehat{\boldsymbol{S}}(\boldsymbol{I}+\alpha\boldsymbol{Q}\boldsymbol{H}\boldsymbol{Q}^T)\Big|_{\alpha=0} \\
&= \mathbf{C}(\boldsymbol{Q}\boldsymbol{H}\boldsymbol{Q}^T),
\end{aligned}$$

where the equality between the second and third lines is justified by the fact that $\det(\boldsymbol{I}+\alpha\boldsymbol{H}) > 0$ for α sufficiently small. Thus $\mathbf{C}$ is isotropic in the sense of Section 1.5. Moreover, by virtue of Result 7.10, the tensor $\mathbf{C}(\boldsymbol{H}) \in \mathcal{V}^2$ is symmetric for any symmetric $\boldsymbol{H} \in \mathcal{V}^2$, and by Result 7.11 we have $\mathbf{C}(\boldsymbol{W}) = \boldsymbol{O}$ for any skew-symmetric $\boldsymbol{W} \in \mathcal{V}^2$. Thus $\mathbf{C}$ satisfies the hypotheses of Result 1.16, which establishes the claim. □

7.3.4 Equation of Linearized, Isotropic Elasticity

In view of Results 7.9 and 7.12, the linearized balance of linear momentum equation (7.30) takes a simple form when a body is isotropic. In particular, we have

$$\mathbf{C}(\nabla^X\boldsymbol{u}^{(1)}) = \lambda(\operatorname{tr}\nabla^X\boldsymbol{u}^{(1)})\boldsymbol{I} + 2\mu\,\operatorname{sym}(\nabla^X\boldsymbol{u}^{(1)}),$$

and taking the divergence gives

$$\begin{aligned}\nabla^X \cdot \mathsf{C}(\nabla^X \boldsymbol{u}^{(1)}) &= \nabla^X \cdot (\,\lambda \operatorname{tr}(\nabla^X \boldsymbol{u}^{(1)})\boldsymbol{I} + 2\mu \operatorname{sym}(\nabla^X \boldsymbol{u}^{(1)})\,) \\ &= \nabla^X \cdot (\,[\lambda u^{(1)}_{k,k}\delta_{ij} + \mu u^{(1)}_{i,j} + \mu u^{(1)}_{j,i}]\,\boldsymbol{e}_i \otimes \boldsymbol{e}_j\,) \\ &= [\lambda u^{(1)}_{k,k}\delta_{ij} + \mu u^{(1)}_{i,j} + \mu u^{(1)}_{j,i}]_{,j}\,\boldsymbol{e}_i \\ &= \lambda u^{(1)}_{k,ki}\,\boldsymbol{e}_i + \mu u^{(1)}_{i,jj}\,\boldsymbol{e}_i + \mu u^{(1)}_{j,ji}\,\boldsymbol{e}_i \\ &= \lambda \nabla^X (\nabla^X \cdot \boldsymbol{u}^{(1)}) + \mu \Delta^X \boldsymbol{u}^{(1)} + \mu \nabla^X (\nabla^X \cdot \boldsymbol{u}^{(1)}) \\ &= \mu \Delta^X \boldsymbol{u}^{(1)} + (\lambda + \mu)\nabla^X (\nabla^X \cdot \boldsymbol{u}^{(1)}).\end{aligned}$$

Thus (7.30) can be written as

$$\boxed{\rho_0 \ddot{\boldsymbol{u}}^{(1)} = \mu \Delta^X \boldsymbol{u}^{(1)} + (\lambda + \mu)\nabla^X (\nabla^X \cdot \boldsymbol{u}^{(1)}) + \rho_0 \boldsymbol{b}^{(1)}_m.} \tag{7.36}$$

Equation (7.36) is sometimes called the **Navier Equation** for the displacement disturbance $\boldsymbol{u}^{(1)}$. It is an approximation to the balance of linear momentum equation appropriate for describing small deviations from a stress-free reference configuration in an isotropic elastic solid.

7.4 Linear Elastic Solids

In the previous section we derived linearized equations for an elastic solid beginning from nonlinear ones. In this section we outline a class of models which yield similar equations. However, here the linear equations are the result of constitutive assumptions rather than linearization.

7.4.1 Definition

A continuum body with reference configuration B is said to be a **linear elastic solid** if:

(1) The first Piola–Kirchhoff stress field is of the form

$$\boldsymbol{P}(\boldsymbol{X}, t) = \widehat{\boldsymbol{P}}(\boldsymbol{F}(\boldsymbol{X}, t)), \quad \forall \boldsymbol{X} \in B,\ t \geq 0,$$

where $\widehat{\boldsymbol{P}} : \mathcal{V}^2 \to \mathcal{V}^2$ is a given response function.

(2) The response function $\widehat{\boldsymbol{P}}$ is of the form

$$\widehat{\boldsymbol{P}}(\boldsymbol{F}) = \mathsf{C}(\nabla^X \boldsymbol{u}), \tag{7.37}$$

where $\nabla^X \boldsymbol{u} = \boldsymbol{F} - \boldsymbol{I}$ is the displacement gradient and C is a given fourth-order tensor called the **elasticity tensor** for the body.

(3) The tensor $\mathbf{C}$ satisfies left and right minor symmetry conditions.

Property (1) implies that the stress at a point in a linear elastic solid depends only on a measure of present strain at that point. Property (2) and the left minor symmetry in Property (3) imply that the first Piola–Kirchhoff stress response is symmetric. Property (2) and the right minor symmetry condition in Property (3) imply that this stress response is a linear function of the infinitesimal strain tensor $\boldsymbol{E} = \text{sym}(\nabla^X \boldsymbol{u})$, namely

$$\widehat{\boldsymbol{P}}(\boldsymbol{F}) = \mathbf{C}(\boldsymbol{E}).$$

When (7.37) is substituted into the elasticity equations in (7.10), the resulting system has the same form as the linearized equations in (7.31), with $\boldsymbol{u}^{(1)}$ replaced by $\boldsymbol{u}$ and so on. In particular, the resulting equations are linear in the displacement field $\boldsymbol{u}$. (Strictly speaking, we must also assume that the prescribed body force and boundary data are linear in $\boldsymbol{u}$.) Thus the exact equations for a linear elastic solid have the same form as the linearized equations of a nonlinear elastic solid with stress-free reference configuration.

Remarks:

(1) The balance law of angular momentum will in general not be satisfied in a linear elastic solid. In particular, by (7.1), symmetry of the first Piola–Kirchhoff stress field generally does not imply symmetry for the Cauchy stress field. However, in view of Results 7.9 and 7.10, balance of angular momentum is satisfied approximately and is manifest in the left minor symmetry of $\mathbf{C}$.

(2) The constitutive model for a linear elastic solid in general does not satisfy the axiom of material frame-indifference. In particular, by Result 7.1, frame-indifference requires $\widehat{\boldsymbol{P}}(\boldsymbol{F}) = \boldsymbol{F}\overline{\boldsymbol{\Sigma}}(\boldsymbol{C})$ for some function $\overline{\boldsymbol{\Sigma}}(\boldsymbol{C})$, which in general is incompatible with the assumption of linearity. However, in view of Results 7.9 and 7.11, frame-indifference is satisfied approximately and is manifest in the right minor symmetry of $\mathbf{C}$.

(3) A classic example of a linear elastic solid is the **isotropic model** defined by

$$\mathbf{C}(\boldsymbol{E}) = \lambda(\text{tr}\,\boldsymbol{E})\boldsymbol{I} + 2\mu\,\text{sym}(\boldsymbol{E}). \tag{7.38}$$

In this case λ and μ are called the **Lamé constants** for the body (see Exercise 11). In view of Results 7.9 and 7.12, this model is consistent with the linearization of any nonlinear, frame-indifferent, isotropic model with stress-free reference configuration. In particular, the balance of linear momentum equation takes the form in (7.36). □

7.4.2 General Properties

Here we study some general properties of linear elastic solids. We show that a linear elastic solid is hyperelastic provided the elasticity tensor has major symmetry. We also introduce conditions on the elasticity tensor under which appropriate existence and uniqueness results can be proved for the Linear Elastodynamics and Elastostatics equations in (7.31) and (7.32).

Definition 7.13 *A linear elastic solid is called* **hyperelastic** *if the response function* $\widehat{\boldsymbol{P}}$ *satisfies*

$$\widehat{\boldsymbol{P}}(\boldsymbol{F}) = DW(\boldsymbol{F}),$$

for some function $W : \mathcal{V}^2 \to \mathbb{R}$ *called a* **strain energy density** *for the body.* □

The above definition is consistent with the notion of a general hyperelastic solid. However, in contrast to the general case, a symmetry condition on $DW(\boldsymbol{F})\boldsymbol{F}^T$ is not required. The next result gives a sufficient condition for a linear elastic solid to be hyperelastic. In fact, this condition is also necessary (see Exercise 12).

Result 7.14 ***Condition for Linear Hyperelastic Solids.*** *If the elasticity tensor* C *has major symmetry, then a linear elastic solid is hyperelastic. In this case the function*

$$W(\boldsymbol{F}) = \tfrac{1}{2}\nabla^X \boldsymbol{u} : \mathsf{C}(\nabla^X \boldsymbol{u}), \tag{7.39}$$

where $\nabla^X \boldsymbol{u} = \boldsymbol{F} - \boldsymbol{I}$, *is a strain energy density for the body. Equivalently, by the minor symmetries of* C

$$W(\boldsymbol{F}) = \tfrac{1}{2}\boldsymbol{E} : \mathsf{C}(\boldsymbol{E}),$$

where $\boldsymbol{E} = \mathrm{sym}(\nabla^X \boldsymbol{u})$ *is the infinitesimal strain tensor.* □

Proof For a linear elastic solid we have $\widehat{\boldsymbol{P}}(\boldsymbol{F}) = \mathbf{C}(\boldsymbol{F} - \boldsymbol{I})$. Thus, assuming $\mathbf{C}$ has major symmetry, we seek to show that $\mathbf{C}(\boldsymbol{F} - \boldsymbol{I}) = DW(\boldsymbol{F})$, where $W(\boldsymbol{F})$ is given by (7.39). By Definition 2.20 we have

$$DW(\boldsymbol{F}) : \boldsymbol{A} = \frac{d}{d\alpha} W(\boldsymbol{F} + \alpha \boldsymbol{A})\Big|_{\alpha=0}, \quad \forall \boldsymbol{A} \in \mathcal{V}^2,$$

and from (7.39), using $\nabla^X \boldsymbol{u} = \boldsymbol{F} - \boldsymbol{I}$, we get

$$\begin{aligned} \frac{d}{d\alpha} W(\boldsymbol{F} + \alpha \boldsymbol{A})\Big|_{\alpha=0} & \\ &= \frac{d}{d\alpha}\Big[\tfrac{1}{2}(\boldsymbol{F} + \alpha\boldsymbol{A} - \boldsymbol{I}) : \mathbf{C}(\boldsymbol{F} + \alpha\boldsymbol{A} - \boldsymbol{I}) \Big]\Big|_{\alpha=0} \\ &= \Big[\tfrac{1}{2}\boldsymbol{A} : \mathbf{C}(\boldsymbol{F} + \alpha\boldsymbol{A} - \boldsymbol{I}) + \tfrac{1}{2}(\boldsymbol{F} + \alpha\boldsymbol{A} - \boldsymbol{I}) : \mathbf{C}(\boldsymbol{A}) \Big]\Big|_{\alpha=0} \\ &= \tfrac{1}{2}\boldsymbol{A} : \mathbf{C}(\boldsymbol{F} - \boldsymbol{I}) + \tfrac{1}{2}(\boldsymbol{F} - \boldsymbol{I}) : \mathbf{C}(\boldsymbol{A}). \end{aligned}$$

Using the major symmetry of $\mathbf{C}$, and the fact that $\boldsymbol{A} : \boldsymbol{B} = \boldsymbol{B} : \boldsymbol{A}$ for any $\boldsymbol{A}, \boldsymbol{B} \in \mathcal{V}^2$, we obtain

$$DW(\boldsymbol{F}) : \boldsymbol{A} = \frac{d}{d\alpha} W(\boldsymbol{F} + \alpha\boldsymbol{A})\Big|_{\alpha=0} = \mathbf{C}(\boldsymbol{F} - \boldsymbol{I}) : \boldsymbol{A}, \quad \forall \boldsymbol{A} \in \mathcal{V}^2,$$

which implies $DW(\boldsymbol{F}) = \mathbf{C}(\boldsymbol{F} - \boldsymbol{I})$ as required. □

Remarks:

(1) It is straightforward to verify that the elasticity tensor for the isotropic model has major symmetry. Thus the isotropic model is hyperelastic.

(2) Hyperelasticity has many important consequences. In particular, just as for general elastic solids, it guarantees that the mechanical energy inequality is satisfied. It also implies that an energy balance is satisfied as in Result 7.7. Furthermore, it implies that the Linear Elastodynamics and Elastostatics equations in (7.31) and (7.32) have a variational structure.

(3) A general fourth-order tensor $\mathbf{C}$ contains 81 independent parameters. If $\mathbf{C}$ satisfies the left and right minor symmetry conditions required for a linear elastic solid, then the number of independent parameters is reduced to 36. Moreover, if $\mathbf{C}$ has major symmetry, then the number of independent parameters is reduced to 21 (see Exercise 14). Furthermore, if $\mathbf{C}$ is isotropic, then the number of independent parameters is reduced to 2. □

We next introduce various conditions on the elasticity tensor which guarantee that the equations for linear elastic solids are well-posed. We then give a result which characterizes these conditions for the isotropic model.

Definition 7.15 *A fourth-order elasticity tensor* C *is called* **strongly elliptic** *if*

$$\boldsymbol{A} : \mathsf{C}(\boldsymbol{A}) > 0, \quad \forall \boldsymbol{A} \in \mathcal{V}^2, \quad \boldsymbol{A} = \boldsymbol{a} \otimes \boldsymbol{b}, \quad \boldsymbol{a} \neq \boldsymbol{0}, \quad \boldsymbol{b} \neq \boldsymbol{0}.$$

Here $\boldsymbol{a} \otimes \boldsymbol{b}$ *denotes the usual dyadic product (see Section 1.3.4). It is called* **positive** *if*

$$\boldsymbol{A} : \mathsf{C}(\boldsymbol{A}) > 0, \quad \forall \boldsymbol{A} \in \mathcal{V}^2, \quad \boldsymbol{A}^T = \boldsymbol{A} \neq \boldsymbol{O}.$$

□

Result 7.16 ***Strong Ellipticity, Positivity of Isotropic Model.*** *Consider an isotropic linear elastic solid with Lamé constants* λ *and* μ. *Then:*

(i) C *is strongly elliptic if and only if* $\mu > 0$ *and* $\lambda + 2\mu > 0$,

(ii) C *is positive if and only if* $\mu > 0$ *and* $3\lambda + 2\mu > 0$. □

Proof To establish (i) consider an arbitrary tensor $\boldsymbol{A}$ of the form $\boldsymbol{A} = \boldsymbol{a} \otimes \boldsymbol{b}$ where $\boldsymbol{a}, \boldsymbol{b} \neq \boldsymbol{0}$. Then, by (7.38), using the facts that $\boldsymbol{I} : \boldsymbol{A} = \operatorname{tr} \boldsymbol{A}$, $\operatorname{tr}(\boldsymbol{a} \otimes \boldsymbol{b}) = \boldsymbol{a} \cdot \boldsymbol{b}$ and $(\boldsymbol{a} \otimes \boldsymbol{b}) : (\boldsymbol{c} \otimes \boldsymbol{d}) = (\boldsymbol{a} \cdot \boldsymbol{c})(\boldsymbol{b} \cdot \boldsymbol{d})$, we have

$$\begin{aligned} \boldsymbol{A} : \mathsf{C}(\boldsymbol{A}) &= \lambda(\operatorname{tr} \boldsymbol{A})^2 + 2\mu \operatorname{sym}(\boldsymbol{A}) : \boldsymbol{A} \\ &= \lambda(\boldsymbol{a} \cdot \boldsymbol{b})^2 + \mu(\boldsymbol{a} \otimes \boldsymbol{b} + \boldsymbol{b} \otimes \boldsymbol{a}) : (\boldsymbol{a} \otimes \boldsymbol{b}) \\ &= (\lambda + \mu)(\boldsymbol{a} \cdot \boldsymbol{b})^2 + \mu(\boldsymbol{a} \cdot \boldsymbol{a})(\boldsymbol{b} \cdot \boldsymbol{b}). \end{aligned}$$

Introducing the perpendicular projection $\boldsymbol{b}^\perp = \boldsymbol{b} - \frac{(\boldsymbol{b} \cdot \boldsymbol{a})}{(\boldsymbol{a} \cdot \boldsymbol{a})} \boldsymbol{a}$ and substituting for $\boldsymbol{b}$ in the second term above gives

$$\boldsymbol{A} : \mathsf{C}(\boldsymbol{A}) = (\lambda + 2\mu)(\boldsymbol{a} \cdot \boldsymbol{b})^2 + \mu(\boldsymbol{a} \cdot \boldsymbol{a})(\boldsymbol{b}^\perp \cdot \boldsymbol{b}^\perp).$$

From this result we see that $\lambda + 2\mu > 0$ and $\mu > 0$ imply that C is strongly elliptic. Conversely, if C is strongly elliptic, then by choosing $\boldsymbol{b}$ such that $\boldsymbol{b} \cdot \boldsymbol{a} = 0$ we get $\mu > 0$, and by choosing $\boldsymbol{b}$ such that $\boldsymbol{b}^\perp = \boldsymbol{0}$ we get $\lambda + 2\mu > 0$.

To establish (ii) consider an arbitrary symmetric tensor $\boldsymbol{A}$ and introduce a symmetric tensor $\boldsymbol{H}$ by the relation $\boldsymbol{H} = \boldsymbol{A} - \alpha \boldsymbol{I}$, where

$\alpha = \frac{1}{3}\operatorname{tr}(\boldsymbol{A})$. Then, by (7.38), using the facts that $\operatorname{sym}(\boldsymbol{A}) = \boldsymbol{A}$, $\boldsymbol{A} = \boldsymbol{H} + \alpha\boldsymbol{I}$ and $\boldsymbol{I} : \boldsymbol{H} = \operatorname{tr}(\boldsymbol{H}) = 0$, we have

$$\begin{aligned}\boldsymbol{A} : \mathbf{C}(\boldsymbol{A}) &= \lambda(\operatorname{tr}\boldsymbol{A})^2 + 2\mu\boldsymbol{A} : \boldsymbol{A}\\ &= \lambda(3\alpha)^2 + 2\mu(\boldsymbol{H} + \alpha\boldsymbol{I}) : (\boldsymbol{H} + \alpha\boldsymbol{I})\\ &= 9\lambda\alpha^2 + 2\mu(\boldsymbol{H} : \boldsymbol{H} + 2\alpha\boldsymbol{I} : \boldsymbol{H} + \alpha^2\boldsymbol{I} : \boldsymbol{I})\\ &= 9\lambda\alpha^2 + 2\mu(\boldsymbol{H} : \boldsymbol{H} + 3\alpha^2)\\ &= 3\alpha^2(3\lambda + 2\mu) + 2\mu(\boldsymbol{H} : \boldsymbol{H}).\end{aligned}$$

From this result we see that $3\lambda + 2\mu > 0$ and $\mu > 0$ imply that $\mathbf{C}$ is positive. Conversely, if $\mathbf{C}$ is positive, then by choosing $\boldsymbol{A}$ such that $\alpha = 0$ we get $\mu > 0$, and by choosing $\boldsymbol{A}$ such that $\boldsymbol{H} = \boldsymbol{O}$ we get $3\lambda + 2\mu > 0$.

□

Remarks:

(1) The condition of strong ellipticity, together with major symmetry, guarantee appropriate existence and uniqueness results for the Linear Elastodynamics equations in (7.31). In this case the equations are of hyperbolic type and admit a full-range of wave-like solutions. These conditions hold for the isotropic model when $\mu > 0$ and $\lambda + 2\mu > 0$ (see Exercise 15).

(2) The condition of positivity, together with major symmetry, guarantee appropriate existence and uniqueness results for the Linear Elastostatics equations in (7.32). These conditions hold for the isotropic model when $\mu > 0$ and $3\lambda + 2\mu > 0$.

(3) It is straightforward to verify that, if an elasticity tensor $\mathbf{C}$ is positive, then it is strongly elliptic. This follows from the minor symmetries of $\mathbf{C}$ by considering symmetric second-order tensors of the form $\boldsymbol{A} = \operatorname{sym}(\boldsymbol{a} \otimes \boldsymbol{b})$.

□

Bibliographic Notes

General references for the theory of elasticity can be found in Antman (1995), Marsden and Hughes (1983), Gurtin (1981) and Malvern (1969). More comprehensive treatments can be found in the treatises by Gurtin (1972), Truesdell and Noll (1965), Truesdell and Toupin (1960) and Love

(1944). The text by Ogden (1984) contains an excellent modern treatment of the theory, with special emphasis on constitutive models for describing large-scale deformations in rubber-like materials.

The theory presented here pertains to three-dimensional bodies in three-dimensional space. However, theories for lower-dimensional bodies such as rods, plates and shells in three-dimensional space can also be considered. Such theories can be derived either by postulating balance laws analogous to those in Chapter 5, or by applying various asymptotic approximation techniques to the theory presented here. Theories for such lower-dimensional bodies can be found, for example, in the texts by Antman (1972, 1995), Ciarlet (1988, Volumes 2,3) and Naghdi (1972).

A large part of the mathematical theory of elasticity is concerned with general existence issues, and the analysis of uniqueness, stability and bifurcation in various contexts. Various existence and uniqueness issues are treated in Ciarlet (1988, Volume 1), Marsden and Hughes (1983), Fichera (1972), and Knops and Payne (1971). For the analysis of bifurcations in elasticity the reader is referred to Antman (1995) and Ogden (1984).

As for fluid mechanics, the numerical solution of the equations of solid mechanics plays a central role in both the development of mathematical and physical understanding, and in engineering practice. For an overview of numerical methods in elasticity see the contribution by Le Tallec in Volume III of the book by Ciarlet and Lions (1994).

An elastic material can be described as one for which the current stress at a point is completely determined by a measure of strain at that point. This assumption can be generalized in a number of different ways leading, for example, to the notion of plastic, viscoplastic and viscoelastic materials. For a general account of such inelastic constitutive models the reader is referred to the texts by Holzapfel (2000), Lubliner (1990) and Fung (1965). The subject of computational plasticity is described in Simo and Hughes (1998). Non-Newtonian fluids, including a variety of viscoelastic models, is discussed from a computational perspective in Owens and Phillips (2002).

Complete proofs for Results 7.5 and 7.6 can be found in Gurtin (1981) and the exercises therein.

Exercises

7.1 Let $B = \{\boldsymbol{X} \in \mathbb{E}^3 \mid 0 < X_i < 1\}$ be the reference configuration of an elastic solid with stress response given by the St. Venant–

Kirchhoff model with material constants $\lambda, \mu > 0$. Consider a time-independent uniaxial deformation $\boldsymbol{x} = \boldsymbol{\varphi}(\boldsymbol{X})$ of the form $x_1 = X_1$, $x_2 = qX_2$, $x_3 = X_3$ where $q > 0$ is a constant.

(a) Find the components of the deformation gradient $\boldsymbol{F}$ and the **Green–St. Venant** strain tensor $\boldsymbol{G} = \frac{1}{2}(\boldsymbol{C} - \boldsymbol{I})$.

(b) Find the components of the two Piola–Kirchhoff stress fields $\boldsymbol{P}$ and $\boldsymbol{\Sigma}$, and the Cauchy stress field $\boldsymbol{S}_m$.

(c) Find the resultant force on each face of the deformed configuration $B' = \boldsymbol{\varphi}(B)$. These are the forces required to maintain the deformation. Is there a non-zero force on each face for all $q \neq 1$?

(d) What happens to the forces from part (c) in the limits $q \to \infty$ and $q \downarrow 0$? Can the body be compressed to zero volume with finite forces?

7.2 Consider an elastic body with reference configuration B and elastic response function $\widehat{\boldsymbol{P}}(\boldsymbol{F})$.

(a) Show that any time-independent, homogeneous deformation $\boldsymbol{x} = \boldsymbol{\varphi}(\boldsymbol{X})$ satisfies the equilibrium equation (7.4) with zero body force.

(b) Show that the external surface force, per unit deformed area, required to maintain a homogeneous equilibrium configuration is of the form

$$\boldsymbol{t}(\boldsymbol{x}) = \boldsymbol{A}\boldsymbol{n}(\boldsymbol{x}),$$

where $\boldsymbol{A}$ is a constant tensor depending on $\widehat{\boldsymbol{P}}$ and $\boldsymbol{\varphi}$, and $\boldsymbol{n}(\boldsymbol{x})$ is the unit outward normal field on the deformed body.

7.3 Show that the three principal invariants $I_i(\boldsymbol{S})$ of a second-order tensor $\boldsymbol{S}$ are isotropic in the sense of Section 1.5, namely

$$I_i(\boldsymbol{S}) = I_i(\boldsymbol{Q}\boldsymbol{S}\boldsymbol{Q}^T),$$

for all $\boldsymbol{S} \in \mathcal{V}^2$ and all rotations $\boldsymbol{Q} \in \mathcal{V}^2$.

7.4 Show that, if the three principal invariants of two second-order tensors $\boldsymbol{A}$ and $\boldsymbol{B}$ are equal, then $\boldsymbol{A}$ and $\boldsymbol{B}$ have the same set of eigenvalues.

7.5 Show that a function $\overline{W} : \mathcal{V}^2 \to I\!R$ has the property

$$\overline{W}(\boldsymbol{C}) = \overline{W}(\boldsymbol{Q}\boldsymbol{C}\boldsymbol{Q}^T),$$

for all symmetric $\boldsymbol{C}$ and all rotations $\boldsymbol{Q}$, if and only if there is a function $\widehat{W} : I\!R^3 \to I\!R$ such that

$$\overline{W}(\boldsymbol{C}) = \widehat{W}(\mathcal{I}_{\boldsymbol{C}}),$$

for all symmetric $\boldsymbol{C}$. Here $\mathcal{I}_{\boldsymbol{C}}$ denotes the three principal invariants of $\boldsymbol{C}$.

7.6 Let $I_i(\boldsymbol{S})$ denote the three principal invariants of a second-order tensor $\boldsymbol{S}$. Show that

$$DI_1(\boldsymbol{S}) = \boldsymbol{I} \quad \text{and} \quad DI_2(\boldsymbol{S}) = \mathrm{tr}(\boldsymbol{S})\boldsymbol{I} - \boldsymbol{S}^T.$$

Remark: By Result 2.22 we have

$$DI_3(\boldsymbol{S}) = \det(\boldsymbol{S})\boldsymbol{S}^{-T}.$$

Thus the derivatives of all three principal invariants of a second-order tensor can be computed explicitly.

7.7 Show that, if the strain energy density for a hyperelastic solid takes the isotropic form

$$W(\boldsymbol{F}) = \widehat{W}(\mathcal{I}_{\boldsymbol{C}}),$$

then the stress response function $\widehat{\boldsymbol{P}}(\boldsymbol{F})$ takes the isotropic form

$$\widehat{\boldsymbol{P}}(\boldsymbol{F}) = \boldsymbol{F}\,[\,\gamma_0(\mathcal{I}_{\boldsymbol{C}})\boldsymbol{I} + \gamma_1(\mathcal{I}_{\boldsymbol{C}})\boldsymbol{C} + \gamma_2(\mathcal{I}_{\boldsymbol{C}})\boldsymbol{C}^{-1}\,],$$

for appropriate functions γ_0, γ_1 and γ_2 which you should determine.

7.8 Consider a hyperelastic body with reference configuration B undergoing a motion φ in the absence of body forces, and suppose the motion satisfies the boundary condition $\varphi(\boldsymbol{X},t) = \boldsymbol{X}$ for all $\boldsymbol{X} \in \partial B$ and $t \geq 0$. Show that the sum of kinetic and strain energy is constant in the sense that

$$\frac{d}{dt}(K[B_t] + U[B_t]) = 0, \quad \forall t \geq 0.$$

7.9 Consider a hyperelastic body under the same conditions as in Exercise 8 and let

$$U^*[B_t] = \int_B W^*(F(\boldsymbol{X},t))\, dV_{\boldsymbol{X}},$$

for some $W^* : \mathcal{V}^2 \to \mathbb{R}$ with the property that

$$\nabla^X \cdot [DW^*(\boldsymbol{F}(\boldsymbol{X},t))] = \boldsymbol{0}, \quad \forall \boldsymbol{X} \in B,\ t \geq 0.$$

Show that

$$\frac{d}{dt} U^*[B_t] = 0, \quad \forall t \geq 0.$$

Remark: The strain energy density W^* can be referred to as a **null energy density** since it does not upset the energy balance established in Exercise 8. In particular, the modified energy $K[B_t]+U[B_t]+\alpha U^*[B_t]$ is also conserved for any scalar constant α.

7.10 Complete the proof of Result 7.9. In particular, show $\mathsf{B} = \mathsf{C}$, where B and C are the elasticity tensors associated with the second Piola–Kirchhoff and Cauchy stress response functions $\widehat{\boldsymbol{\Sigma}}(\boldsymbol{F})$ and $\widehat{\boldsymbol{S}}(\boldsymbol{F})$.

7.11 Show that the components of the fourth-order elasticity tensor C for an isotropic, linear elastic material are

$$\mathsf{C}_{ijkl} = \lambda \delta_{ij}\delta_{kl} + \mu(\delta_{ik}\delta_{jl} + \delta_{jk}\delta_{il}).$$

7.12 Show that a linear elastic solid is hyperelastic if and only if the fourth-order elasticity tensor C has major symmetry.

7.13 Consider an isotropic, linear elastic body under the same conditions as in Exercise 8.

(a) Show that the result in Exercise 8 implies

$$\frac{1}{2}\frac{d}{dt}\int_B \left[\rho_0 \left|\dot{\boldsymbol{u}}\right|^2 + \nabla^X \boldsymbol{u} : \mathsf{C}(\nabla^X \boldsymbol{u})\right] dV_{\boldsymbol{X}} = 0, \quad \forall t \geq 0,$$

where $\boldsymbol{u}(\boldsymbol{X},t) = \boldsymbol{\varphi}(\boldsymbol{X},t) - \boldsymbol{X}$ is the displacement, ρ_0 is the reference mass density and $\mathsf{C}(\boldsymbol{H}) = \lambda(\operatorname{tr}\boldsymbol{H})\boldsymbol{I} + \mu(\boldsymbol{H} + \boldsymbol{H}^T)$ is the elasticity tensor.

(b) Supposing $\lambda = -\mu$, use the balance of linear momentum equation (with zero body force) $\rho_0 \ddot{\boldsymbol{u}} = \nabla^X \cdot [\mathbf{C}(\nabla^X \boldsymbol{u})]$ to show

$$\frac{1}{2}\frac{d}{dt}\int_B \left[\rho_0 \left|\dot{\boldsymbol{u}}\right|^2 + \mu \nabla^X \boldsymbol{u} : \nabla^X \boldsymbol{u}\right] dV_{\boldsymbol{X}} = 0, \quad \forall t \geq 0.$$

(c) Use the results from (a) and (b) to conclude that B possesses a null energy density $W^*(\boldsymbol{F})$, which up to multiplicative and additive constants, can be written as

$$W^*(\boldsymbol{F}) = -\mu(\mathrm{tr}(\nabla^X \boldsymbol{u}))^2 + \mu \nabla^X \boldsymbol{u} : \nabla^X \boldsymbol{u}^T,$$

where $\nabla^X \boldsymbol{u} = \boldsymbol{F} - \boldsymbol{I}$.

7.14 Consider a general linear elastic body with fourth-order elasticity tensor $\mathbf{C}$.

(a) Show that $\mathbf{C}$ can be represented by a matrix $[[\mathbf{C}]] \in \mathbb{R}^{6\times 6}$. (Hint: Use the left and right minor symmetries to show that the transformation $\mathbf{C} : \mathcal{V}^2 \to \mathcal{V}^2$ reduces to $\mathbf{C} : \mathrm{sym}(\mathcal{V}^2) \to \mathrm{sym}(\mathcal{V}^2)$, where $\mathrm{sym}(\mathcal{V}^2) \subset \mathcal{V}^2$ is the subspace of symmetric second-order tensors.)

(b) Each $\boldsymbol{A} \in \mathrm{sym}(\mathcal{V}^2)$ can be uniquely represented by a column vector $[[\boldsymbol{A}]] \in \mathbb{R}^6$ in a number of different ways. Use the representation

$$[[\boldsymbol{A}]] = (A_{11}, A_{22}, A_{33}, \sqrt{2}A_{12}, \sqrt{2}A_{13}, \sqrt{2}A_{23})^T$$

to show that major symmetry of $\mathbf{C}$ is equivalent to symmetry of its matrix representation $[[\mathbf{C}]]$.

(c) Use the results from (a) and (b) to show that a general linear elastic material model is defined by 36 independent elastic constants (components of $\mathbf{C}$). Moreover, if this model has major symmetry (thus hyperelastic), then the number of independent elastic constants is reduced to 21.

7.15 Consider the linear momentum equation for an isotropic, linear elastic body with constant mass density $\rho_0 > 0$, namely

$$\rho_0 \ddot{\boldsymbol{u}} = \mu \Delta^X \boldsymbol{u} + (\mu + \lambda)\nabla^X (\nabla^X \cdot \boldsymbol{u}), \tag{7.40}$$

where $\boldsymbol{u}(\boldsymbol{X}, t)$ is the displacement field and λ, μ are the Lamé constants. Here we study **plane** or **progressive wave** solutions of the form

$$\boldsymbol{u}(\boldsymbol{X}, t) = \sigma \boldsymbol{a} \phi(\gamma \boldsymbol{k} \cdot \boldsymbol{X} - \omega t), \tag{7.41}$$

where $\phi : \mathbb{R} \to \mathbb{R}$ is an arbitrary function which defines the profile of the wave (we assume ϕ'' is not identically zero), $\boldsymbol{a}$ and $\boldsymbol{k}$ are unit vectors called the direction of displacement and propagation, and σ, γ and ω are constants that determine the amplitude, spatial scaling and speed of the wave. In particular, the wave speed is given by ω/γ.

(a) Show that (7.41) satisfies (7.40) if and only if $\boldsymbol{a}$, $\boldsymbol{k}$, σ, γ, ω satisfy

$$\sigma\gamma^2[\mu\boldsymbol{a} + (\mu + \lambda)(\boldsymbol{a}\cdot\boldsymbol{k})\boldsymbol{k}] = \sigma\rho_0\omega^2\boldsymbol{a}. \tag{7.42}$$

Notice that these equations are satisfied by the trivial wave defined by $\sigma = 0$.

(b) Suppose $\lambda + \mu \neq 0$. For any given $\boldsymbol{k}$, $\sigma \neq 0$ and $\gamma \neq 0$ show that the only independent solutions $(\boldsymbol{a}, \omega)$ to (7.42) are:

$$\begin{aligned} &(1) \quad \boldsymbol{a} = \pm\boldsymbol{k}, \quad \omega^2 = \gamma^2(\lambda + 2\mu)/\rho_0, \\ &(2) \quad \boldsymbol{a}\cdot\boldsymbol{k} = 0, \quad \omega^2 = \gamma^2\mu/\rho_0. \end{aligned}$$

What happens in the case when $\lambda + \mu = 0$?

Remark: The above results show that in an isotropic, linear elastic solid with $\lambda + \mu \neq 0$ only two types of plane waves are possible for any given direction $\boldsymbol{k}$: **longitudinal waves** with $\boldsymbol{a}$ parallel to $\boldsymbol{k}$ as in (1), and **transverse waves** with $\boldsymbol{a}$ perpendicular to $\boldsymbol{k}$ as in (2). The speeds of these two types of waves are different. Notice that these speeds are real in view of the strong ellipticity conditions in Result 7.16. For non-isotropic models the situation is more complicated.

7.16 A **viscoelastic** material is one which exhibits both viscous and elastic characteristics. A simple viscoelastic constitutive assumption is that there exists a stress response function $\widehat{\boldsymbol{P}} : \mathcal{V}^2 \times \mathcal{V}^2 \to \mathcal{V}^2$ such that

$$\boldsymbol{P}(\boldsymbol{X}, t) = \widehat{\boldsymbol{P}}(\boldsymbol{F}(\boldsymbol{X}, t), \dot{\boldsymbol{F}}(\boldsymbol{X}, t)).$$

Suppose such a body undergoes a motion φ in which the body deviates only slightly from its reference configuration in the sense that

$$\begin{aligned} \varphi(\boldsymbol{X}, t) &= \boldsymbol{X} + \epsilon\boldsymbol{u}(\boldsymbol{X}, t) + \mathcal{O}(\epsilon^2), \\ \boldsymbol{F}(\boldsymbol{X}, t) &= \boldsymbol{I} + \epsilon\nabla^X\boldsymbol{u}(\boldsymbol{X}, t) + \mathcal{O}(\epsilon^2), \end{aligned}$$

for some scalar parameter $0 \leq \epsilon \ll 1$. Moreover, suppose the reference configuration is stress-free and that there are no body forces.

(a) Show that, to leading order in ϵ, the field $\boldsymbol{u}$ satisfies

$$\rho_0 \ddot{\boldsymbol{u}} = \nabla^X \cdot \mathsf{C}_1(\nabla^X \boldsymbol{u}) + \nabla^X \cdot \mathsf{C}_2(\nabla^X \dot{\boldsymbol{u}}),$$

where C_1 and C_2 are fourth-order tensors which you should specify.

(b) Show that, if C_1 and C_2 are isotropic in the sense that

$$\left.\begin{aligned} \mathsf{C}_1(\boldsymbol{H}) &= \lambda_1 \operatorname{tr}(\boldsymbol{H})\boldsymbol{I} + 2\mu_1 \operatorname{sym}(\boldsymbol{H}) \\ \mathsf{C}_2(\boldsymbol{H}) &= \lambda_2 \operatorname{tr}(\boldsymbol{H})\boldsymbol{I} + 2\mu_2 \operatorname{sym}(\boldsymbol{H}) \end{aligned}\right\} \qquad \forall \boldsymbol{H} \in \mathcal{V}^2,$$

then $\boldsymbol{u}$ satisfies

$$\begin{aligned} \rho_0 \ddot{\boldsymbol{u}} = {} & \mu_1 \Delta^X \boldsymbol{u} + (\lambda_1 + \mu_1)\nabla^X(\nabla^X \cdot \boldsymbol{u}) \\ & + \mu_2 \Delta^X \dot{\boldsymbol{u}} + (\lambda_2 + \mu_2)\nabla^X(\nabla^X \cdot \dot{\boldsymbol{u}}). \end{aligned}$$

7.17 In one spatial dimension, the simplest model of linear viscoelasticity as introduced in Exercise 16 yields an equation of the form

$$\rho_0 \frac{\partial^2}{\partial t^2} u = \eta_1 \frac{\partial^2}{\partial x^2} u + \eta_2 \frac{\partial}{\partial t} \frac{\partial^2}{\partial x^2} u, \tag{7.43}$$

where ρ_0, η_1 and η_2 are scalar constants, $u(x,t)$ is the scalar displacement field and (x,t) are the space and time coordinates. Here we study solutions of (7.43) of the form

$$u(x,t) = \exp[i(x - ct)] \quad \text{where} \quad i^2 = -1.$$

(a) Consider the **elastic limit** defined by $\eta_1 > 0$ and $\eta_2 = 0$. Show that (7.43) admits wave-like solutions which propagate, but do not damp out in time.

(b) Consider the **viscous limit** defined by $\eta_1 = 0$ and $\eta_2 > 0$. Show that (7.43) admits stationary solutions which damp exponentially in time.

(c) Describe mathematically the form of solutions for the case $\eta_1 > 0$ and $\eta_2 > 0$. Distinguish between the cases $4\rho_0\eta_1 > \eta_2^2$ and $4\rho_0\eta_1 < \eta_2^2$.

7.18 In small deformation elasticity the infinitesimal strain field $\boldsymbol{E}$ is defined by

$$\boldsymbol{E} = \tfrac{1}{2}(\nabla^x \boldsymbol{u} + \nabla^x \boldsymbol{u}^T) \quad \text{or} \quad E_{ij} = \tfrac{1}{2}(u_{i,j} + u_{j,i}), \qquad (7.44)$$

where $\boldsymbol{u}$ is the displacement field. In this question we address when (7.44) can be solved for $\boldsymbol{u}$ in terms of $\boldsymbol{E}$. That is, given a strain field $\boldsymbol{E}$, when is it possible to solve (7.44) as a differential equation for $\boldsymbol{u}$? Notice that, by symmetry, (7.44) yields six independent equations for the three unknown components of $\boldsymbol{u}$. Since there are more equations than unknowns, $\boldsymbol{E}$ must satisfy certain **compatibility conditions** for (7.44) to be solvable.

(a) Show that, if (7.44) possesses a smooth solution $\boldsymbol{u}$ for a given $\boldsymbol{E}$, then $\boldsymbol{E}$ must necessarily satisfy the following six compatibility equations

$$\begin{aligned}
E_{11,22} + E_{22,11} &= 2E_{12,12}, \\
E_{22,33} + E_{33,22} &= 2E_{23,23}, \\
E_{33,11} + E_{11,33} &= 2E_{31,31}, \\
E_{11,23} &= (-E_{23,1} + E_{31,2} + E_{12,3})_{,1}, \\
E_{22,31} &= (-E_{31,2} + E_{12,3} + E_{23,1})_{,2}, \\
E_{33,12} &= (-E_{12,3} + E_{23,1} + E_{31,2})_{,3}.
\end{aligned}$$

Hint: Use the fact that the mixed partial derivatives of a smooth solution $\boldsymbol{u}$ must commute.

(b) Show that the six compatibility equations can be written succinctly in index notation as

$$E_{nj,km} + E_{km,jn} - E_{kn,jm} - E_{mj,kn} = 0.$$

7.19 Define two curl operations for a second order tensor $\boldsymbol{S}$ as

$$\nabla \times \boldsymbol{S} = \epsilon_{pkq} S_{km,p}\, \boldsymbol{e}_q \otimes \boldsymbol{e}_m \quad \text{and} \quad \boldsymbol{S} \times \nabla = \epsilon_{kpq} S_{rk,p}\, \boldsymbol{e}_r \otimes \boldsymbol{e}_q.$$

(a) Using the definitions above, show that

$$-[\nabla \times (\boldsymbol{S}^T)]^T = \boldsymbol{S} \times \nabla \quad \text{and} \quad \nabla \times (\boldsymbol{S} \times \nabla) = (\nabla \times \boldsymbol{S}) \times \nabla.$$

(b) Show that the compatibility conditions from Exercise 18 may be written as

$$\nabla \times (\boldsymbol{E} \times \nabla) = \boldsymbol{O}.$$

7.20 Consider a linear elastic body with mass density ρ_0, first Piola–Kirchhoff stress field $\boldsymbol{P}$ and infinitesimal strain field $\boldsymbol{E}$. Suppose the body is isotropic with Lamé constants λ and μ, and suppose the body is subject to a body force per unit mass $\boldsymbol{b}_m$. (Here and throughout the remaining exercises we shall omit the subscript m on $\boldsymbol{b}_m$ for clarity.)

(a) Show that the isotropic, linear elastic stress-strain relation $\boldsymbol{P} = \lambda(\mathrm{tr}\,\boldsymbol{E})\boldsymbol{I} + 2\mu\boldsymbol{E}$ can be inverted as

$$\boldsymbol{E} = -\frac{\nu}{\gamma}(\mathrm{tr}\,\boldsymbol{P})\boldsymbol{I} + \Big(\frac{1+\nu}{\gamma}\Big)\boldsymbol{P},$$

where the **Young's modulus** γ and **Poisson's ratio** ν are related to the Lamé constants λ and μ by

$$\mu = \frac{\gamma}{2(1+\nu)}, \qquad \lambda = \frac{\nu\gamma}{(1+\nu)(1-2\nu)}.$$

(b) Show that, if the body is in equilibrium, then

$$P_{ik,jk} + (\rho_0 b_i)_{,j} = 0.$$

7.21 Consider a linear elastic, isotropic body as in Exercise 20. As before, we suppose the body is in equilibrium.

(a) Show that, when expressed in terms of the stress tensor P_{ij}, the compatibility equations of Exercise 18 can be reduced to

$$P_{ij,kk} + P_{kk,ij} - P_{ik,jk} - P_{jk,ik} = \Big(\frac{\nu}{1+\nu}\Big)[\delta_{ij}P_{pp,kk} + P_{pp,ij}].$$

(b) Use the results from part (a) and Exercise 20(b) to show that P_{ij} satisfies the equation

$$\Delta^X P_{ij} + \Big(\frac{1}{1+\nu}\Big)\psi_{,ij} - \Big(\frac{\nu}{1+\nu}\Big)\delta_{ij}\Delta^X\psi = -(\rho_0 b_i)_{,j} - (\rho_0 b_j)_{,i},$$

where $\psi = \mathrm{tr}(\boldsymbol{P})$.

(c) Again use the results from part (a) and Exercise 20(b) to show that P_{ij} satisfies the equation

$$P_{ij,ij} = \Big(\frac{1-\nu}{1+\nu}\Big)\Delta^X\psi \quad \text{and hence} \quad \Delta^X\psi = -\Big(\frac{1+\nu}{1-\nu}\Big)\beta,$$

where $\beta = \nabla^X\cdot(\rho_0\boldsymbol{b})$.

(d) Combine the results from parts (b) and (c) to show that

$$\Delta^X P_{ij} + \left(\frac{1}{1+\nu}\right)\psi_{,ij} = -\left(\frac{\nu}{1-\nu}\right)\delta_{ij}\beta - (\rho_0 b_i)_{,j} - (\rho_0 b_j)_{,i}. \tag{7.45}$$

Remark: The equations in (7.45) are known as the **Beltrami–Michell compatibility equations**. While six distinct equations (due to symmetry) are represented in (7.45), only three of these are functionally independent. When (7.45) is combined with the equilibrium equations $P_{ij,j} + \rho_0 b_i = 0$, we obtain six independent equations for the six independent components of the stress tensor P_{ij} (given its symmetry). Thus the equilibrium stresses in an isotropic, linear elastic body can in principle be determined without explicit knowledge of the displacement field.

7.22 Consider a linear elastic, isotropic body in equilibrium as in Exercise 20, and suppose the body experiences a **planar deformation** in the sense that

$$\boldsymbol{u}(x, y, z) = u_x(x, y)\boldsymbol{e}_1 + u_y(x, y)\boldsymbol{e}_2,$$

where for convenience we use (x, y, z) to denote the material coordinates (X_1, X_2, X_3), and we use subscripts x, y and z to denote components, not partial derivatives.

(a) Show that the infinitesimal strain tensor $\boldsymbol{E}$ for a planar deformation has a matrix representation of the form

$$[\boldsymbol{E}] = \begin{pmatrix} \epsilon_x & \epsilon_{xy} & 0 \\ \epsilon_{xy} & \epsilon_y & 0 \\ 0 & 0 & 0 \end{pmatrix},$$

for some functions ϵ_x, ϵ_y and ϵ_{xy}.

(b) Under the assumption of isotropy, show that the first Piola–Kirchhoff stress tensor $\boldsymbol{P}$ for a planar deformation has a matrix representation of the form

$$[\boldsymbol{P}] = \begin{pmatrix} \sigma_x & \tau_{xy} & 0 \\ \tau_{xy} & \sigma_y & 0 \\ 0 & 0 & \sigma_z \end{pmatrix},$$

where

$$\begin{aligned} \sigma_x &= \lambda e + 2\mu\epsilon_x, & \tau_{xy} &= 2\mu\epsilon_{xy}, \\ \sigma_y &= \lambda e + 2\mu\epsilon_y, & \sigma_z &= \lambda e, \end{aligned}$$

and $e = \epsilon_x + \epsilon_y$. Moreover, show that $\sigma_z = \nu(\sigma_x + \sigma_y)$, where Poisson's ratio ν is defined as

$$\nu = \frac{\lambda}{2(\lambda + \mu)}.$$

(c) Assuming the body is subject to a planar body force $\boldsymbol{b} = b_x(x,y)\boldsymbol{e}_1 + b_y(x,y)\boldsymbol{e}_2$, the equilibrium equations and the compatibility equation in Exercise 21(c) yield three independent equations for the three independent stress components σ_x, σ_y and τ_{xy}, namely

$$\begin{aligned}
&\frac{\partial \sigma_x}{\partial x} + \frac{\partial \tau_{xy}}{\partial y} + \rho_0 b_x = 0,\\
&\frac{\partial \tau_{xy}}{\partial x} + \frac{\partial \sigma_y}{\partial y} + \rho_0 b_y = 0,\\
&\Delta(\sigma_x + \sigma_y) = -\left(\frac{1}{1-\nu}\right)\left(\frac{\partial(\rho_0 b_x)}{\partial x} + \frac{\partial(\rho_0 b_y)}{\partial y}\right).
\end{aligned}$$

Assuming zero body force show that the above equations are satisfied by

$$\sigma_x = \frac{\partial^2 \phi}{\partial y^2}, \quad \sigma_y = \frac{\partial^2 \phi}{\partial x^2}, \quad \tau_{xy} = -\frac{\partial^2 \phi}{\partial x \partial y},$$

provided that ϕ satisfies the biharmonic equation

$$\Delta^2 \phi = \frac{\partial^4 \phi}{\partial x^4} + 2\frac{\partial^4 \phi}{\partial x^2 \partial y^2} + \frac{\partial^4 \phi}{\partial y^4} = 0.$$

The field ϕ is typically called an **Airy stress function**.

Answers to Selected Exercises

7.1 (a) We have $[\boldsymbol{F}] = \text{diag}(1, q, 1)$ and $[\boldsymbol{C}] = [\boldsymbol{F}]^T[\boldsymbol{F}] = \text{diag}(1, q^2, 1)$ which gives

$$[\boldsymbol{G}] = \tfrac{1}{2}([\boldsymbol{C}] - [\boldsymbol{I}]) = \begin{pmatrix} 0 & 0 & 0 \\ 0 & \frac{1}{2}(q^2-1) & 0 \\ 0 & 0 & 0 \end{pmatrix}.$$

(b) From the definition $[\boldsymbol{\Sigma}] = \lambda(\text{tr}[\boldsymbol{G}])[\boldsymbol{I}] + 2\mu[\boldsymbol{G}]$ we get

$$[\boldsymbol{\Sigma}] = \begin{pmatrix} \frac{\lambda}{2}(q^2-1) & 0 & 0 \\ 0 & (\frac{\lambda}{2}+\mu)(q^2-1) & 0 \\ 0 & 0 & \frac{\lambda}{2}(q^2-1) \end{pmatrix}.$$

From the relation $\boldsymbol{\Sigma} = \boldsymbol{F}^{-1}\boldsymbol{P}$ we deduce $[\boldsymbol{P}] = [\boldsymbol{F}][\boldsymbol{\Sigma}]$, which gives

$$[\boldsymbol{P}] = \begin{pmatrix} \frac{\lambda}{2}(q^2-1) & 0 & 0 \\ 0 & (\frac{\lambda}{2}+\mu)(q^3-q) & 0 \\ 0 & 0 & \frac{\lambda}{2}(q^2-1) \end{pmatrix}.$$

From $\boldsymbol{P} = (\det \boldsymbol{F})\boldsymbol{S}_m \boldsymbol{F}^{-T}$ we deduce $[\boldsymbol{S}_m] = (\det[\boldsymbol{F}])^{-1}[\boldsymbol{P}][\boldsymbol{F}]^T$, which gives

$$[\boldsymbol{S}_m] = \begin{pmatrix} \frac{\lambda}{2}(q-q^{-1}) & 0 & 0 \\ 0 & (\frac{\lambda}{2}+\mu)(q^3-q) & 0 \\ 0 & 0 & \frac{\lambda}{2}(q-q^{-1}) \end{pmatrix}.$$

(c) Just like the reference configuration B, the deformed configuration $B' = \varphi(B)$ is a rectangular solid with faces perpendicular to the unit vectors $\pm\boldsymbol{e}_i$. Denote the faces of B and B' by $\pm\Gamma_i$ and $\pm\Gamma_i'$, and denote the resultant forces on $\pm\Gamma_i'$ by $\boldsymbol{f}^{(i,\pm)}$. Then, by definition of the Cauchy and first Piola–Kirchhoff stress fields, and the fact that the faces $\pm\Gamma_i$ have unit area

$$\begin{aligned} \boldsymbol{f}^{(1,\pm)} &= \int_{\pm\Gamma_1'} \boldsymbol{Sn}\, dA_{\boldsymbol{x}} \\ &= \int_{\pm\Gamma_1} \boldsymbol{PN}\, dA_{\boldsymbol{X}} = \pm\boldsymbol{P}\boldsymbol{e}_1 \,\text{area}(\Gamma_1) = \pm\frac{\lambda}{2}(q^2-1)\boldsymbol{e}_1, \\ \boldsymbol{f}^{(2,\pm)} &= \int_{\pm\Gamma_2'} \boldsymbol{Sn}\, dA_{\boldsymbol{x}} \\ &= \int_{\pm\Gamma_2} \boldsymbol{PN}\, dA_{\boldsymbol{X}} = \pm\Big(\frac{\lambda}{2}+\mu\Big)(q^3-q)\boldsymbol{e}_2, \\ \boldsymbol{f}^{(3,\pm)} &= \int_{\pm\Gamma_3'} \boldsymbol{Sn}\, dA_{\boldsymbol{x}} \\ &= \int_{\pm\Gamma_3} \boldsymbol{PN}\, dA_{\boldsymbol{X}} = \pm\frac{\lambda}{2}(q^2-1)\boldsymbol{e}_3. \end{aligned}$$

Notice that all the resultants are non-zero when $q \neq 1$. We interpret $\boldsymbol{f}^{(2,\pm)}$ as the force required to deform the body along the $\boldsymbol{e}_2$ direction. The forces $\boldsymbol{f}^{(1,\pm)}$ and $\boldsymbol{f}^{(3,\pm)}$ are required to maintain the body in a specified rectangular shape.

(d) In the limit $q \to \infty$ we get $|\boldsymbol{f}^{(i,\pm)}| \to \infty$ for each face $\pm\Gamma_i'$. Thus extreme forces are required to maintain extreme extensions. In the limit $q \downarrow 0$ we find that $|\boldsymbol{f}^{(i,\pm)}|$ is bounded for each face $\pm\Gamma_i'$. Remarkably, the resultant forces on $\pm\Gamma_2'$ tend to zero. Thus extreme forces are not required to deform the body to zero volume, which implies the model is unrealistic for this type of extreme deformation.

7.3 The first principal invariant is $I_1(\boldsymbol{S}) = \mathrm{tr}(\boldsymbol{S})$. Using the fact that $\mathrm{tr}(\boldsymbol{A}\boldsymbol{B}^T) = A_{ij}B_{ij} = \mathrm{tr}(\boldsymbol{B}^T\boldsymbol{A})$, we get

$$I_1(\boldsymbol{Q}\boldsymbol{S}\boldsymbol{Q}^T) = \mathrm{tr}(\boldsymbol{Q}\boldsymbol{S}\boldsymbol{Q}^T) = \mathrm{tr}(\boldsymbol{Q}^T\boldsymbol{Q}\boldsymbol{S}) = \mathrm{tr}(\boldsymbol{S}) = I_1(\boldsymbol{S}).$$

The second principal invariant is $I_2(\boldsymbol{S}) = \frac{1}{2}((\mathrm{tr}(\boldsymbol{S}))^2 - \mathrm{tr}(\boldsymbol{S}^2))$. Using the same property of the trace function as above gives

$$\begin{aligned} I_2(\boldsymbol{Q}\boldsymbol{S}\boldsymbol{Q}^T) &= \tfrac{1}{2}((\mathrm{tr}(\boldsymbol{Q}\boldsymbol{S}\boldsymbol{Q}^T))^2 - \mathrm{tr}(\boldsymbol{Q}\boldsymbol{S}\boldsymbol{Q}^T\boldsymbol{Q}\boldsymbol{S}\boldsymbol{Q}^T)) \\ &= \tfrac{1}{2}((\mathrm{tr}(\boldsymbol{S}))^2 - \mathrm{tr}(\boldsymbol{Q}\boldsymbol{S}^2\boldsymbol{Q}^T)) \\ &= \tfrac{1}{2}((\mathrm{tr}(\boldsymbol{S}))^2 - \mathrm{tr}(\boldsymbol{S}^2)) \\ &= I_2(\boldsymbol{S}). \end{aligned}$$

The third principal invariant is $I_3(\boldsymbol{S}) = \det \boldsymbol{S}$. Since $\det(\boldsymbol{A}\boldsymbol{B}) = (\det \boldsymbol{A})(\det \boldsymbol{B})$ and $\det(\boldsymbol{A}) = \det(\boldsymbol{A}^T)$ by properties of the determinant, and since $\det(\boldsymbol{Q}) = 1$ because $\boldsymbol{Q}$ is a rotation, we have

$$I_3(\boldsymbol{Q}\boldsymbol{S}\boldsymbol{Q}^T) = \det(\boldsymbol{Q}\boldsymbol{S}\boldsymbol{Q}^T) = \det(\boldsymbol{Q})\det(\boldsymbol{S})\det(\boldsymbol{Q}^T) = I_3(\boldsymbol{S}).$$

7.5 Given $\overline{W}(\boldsymbol{C})$ suppose there exists $\widehat{W}(\mathcal{I}_{\boldsymbol{C}})$ such that $\overline{W}(\boldsymbol{C}) = \widehat{W}(\mathcal{I}_{\boldsymbol{C}})$ for all symmetric $\boldsymbol{C}$. Then, because the principal invariants are isotropic functions, we have

$$\overline{W}(\boldsymbol{Q}\boldsymbol{C}\boldsymbol{Q}^T) = \widehat{W}(\mathcal{I}_{\boldsymbol{Q}\boldsymbol{C}\boldsymbol{Q}^T}) = \widehat{W}(\mathcal{I}_{\boldsymbol{C}}) = \overline{W}(\boldsymbol{C}),$$

for all rotations $\boldsymbol{Q}$. Thus $\overline{W}(\boldsymbol{C})$ is isotropic for all symmetric $\boldsymbol{C}$.

Conversely, suppose the function $\overline{W}(\boldsymbol{C})$ is isotropic in the sense that $\overline{W}(\boldsymbol{C}) = \overline{W}(\boldsymbol{Q}\boldsymbol{C}\boldsymbol{Q}^T)$ for all symmetric $\boldsymbol{C}$ and all rotations $\boldsymbol{Q}$. Let $\boldsymbol{Q}_c$ be the rotation matrix that diagonalizes $\boldsymbol{C}$. Then

$$\overline{W}(\boldsymbol{C}) = \overline{W}(\boldsymbol{Q}_c\boldsymbol{C}\boldsymbol{Q}_c^T) = \overline{W}\left(\sum_{i=1}^{3} \lambda_i \boldsymbol{e}_i \otimes \boldsymbol{e}_i\right),$$

where λ_i are the eigenvalues of $\boldsymbol{C}$, and $\boldsymbol{e}_i$ are arbitrary basis vectors that are independent of $\boldsymbol{C}$. For any choice of a frame $\{\boldsymbol{e}_i\}$ define a function $\widetilde{W} : \mathbb{R}^3 \to \mathbb{R}$ by

$$\widetilde{W}(\lambda_1, \lambda_2, \lambda_3) = \overline{W}\left(\sum_{i=1}^{3} \lambda_i \boldsymbol{e}_i \otimes \boldsymbol{e}_i\right).$$

Because the eigenvalues λ_i are roots of the characteristic polynomial

$$p_{\boldsymbol{C}}(\lambda) = \det(\boldsymbol{C} - \lambda\boldsymbol{I}) = -\lambda^3 + I_1(\boldsymbol{C})\lambda^2 - I_2(\boldsymbol{C})\lambda + I_3(\boldsymbol{C}),$$

and the roots of a polynomial are continuous functions of its coefficients, there is a function $\Phi : \mathbb{R}^3 \to \mathbb{R}^3$ (real-valued because $\boldsymbol{C}$ is symmetric) such that

$$(\lambda_1, \lambda_2, \lambda_3) = \Phi(I_1(\boldsymbol{C}), I_2(\boldsymbol{C}), I_3(\boldsymbol{C})).$$

Thus there is a function $\widehat{W}(\mathcal{I}_{\boldsymbol{C}})$ such that

$$\overline{W}(\boldsymbol{C}) = \widehat{W}(\mathcal{I}_{\boldsymbol{C}}).$$

In particular, $\widehat{W} = \widetilde{W} \circ \Phi$.

7.7 Let $\overline{W}(\boldsymbol{C}) = \widehat{W}(\mathcal{I}_{\boldsymbol{C}})$. Then from the result on frame-indifferent strain energy functions we have

$$\widehat{\boldsymbol{P}}(\boldsymbol{F}) = 2\boldsymbol{F} D\overline{W}(\boldsymbol{C}),$$

and by definition of $\overline{W}(\boldsymbol{C})$ and the chain rule

$$\frac{\partial \overline{W}}{\partial C_{ij}} = \frac{\partial \widehat{W}}{\partial I_1}\frac{\partial I_1}{\partial C_{ij}} + \frac{\partial \widehat{W}}{\partial I_2}\frac{\partial I_2}{\partial C_{ij}} + \frac{\partial \widehat{W}}{\partial I_3}\frac{\partial I_3}{\partial C_{ij}}.$$

Using the results from Exercise 6 we obtain

$$\begin{aligned} D\overline{W}(\boldsymbol{C}) &= \frac{\partial \widehat{W}}{\partial I_1}[\boldsymbol{I}] + \frac{\partial \widehat{W}}{\partial I_2}[\mathrm{tr}(\boldsymbol{C})\boldsymbol{I} - \boldsymbol{C}^T] + \frac{\partial \widehat{W}}{\partial I_3}[\det(\boldsymbol{C})\boldsymbol{C}^{-T}] \\ &= \left[\frac{\partial \widehat{W}}{\partial I_1} + \frac{\partial \widehat{W}}{\partial I_2}\,\mathrm{tr}(\boldsymbol{C})\right]\boldsymbol{I} - \left[\frac{\partial \widehat{W}}{\partial I_2}\right]\boldsymbol{C} + \left[\frac{\partial \widehat{W}}{\partial I_3}\det(\boldsymbol{C})\right]\boldsymbol{C}^{-1}, \end{aligned}$$

where the symmetry of $\boldsymbol{C}$ has been used. Using the fact that $I_1 = \mathrm{tr}(\boldsymbol{C})$ and $I_3 = \det(\boldsymbol{C})$ the result follows with

$$\gamma_0(\mathcal{I}_C) = 2\frac{\partial \widehat{W}}{\partial I_1} + 2\frac{\partial \widehat{W}}{\partial I_2} I_1, \qquad \gamma_1(\mathcal{I}_C) = -2\frac{\partial \widehat{W}}{\partial I_2},$$

and

$$\gamma_2(\mathcal{I}_C) = 2\frac{\partial \widehat{W}}{\partial I_3} I_3.$$

7.9 From the definition of $U^*[B_t]$ and the relation $F_{kl} = \varphi_{k,l}$ we have

$$\begin{aligned}\frac{d}{dt}U^*[B_t] &= \frac{d}{dt}\int_B W^*(\boldsymbol{F})\, dV_{\boldsymbol{X}} \\ &= \int_B \frac{\partial W^*}{\partial F_{kl}}\frac{\partial F_{kl}}{\partial t}\, dV_{\boldsymbol{X}} = \int_B \frac{\partial W^*}{\partial F_{kl}}\frac{\partial \dot{\varphi}_k}{\partial X_l}\, dV_{\boldsymbol{X}}.\end{aligned}$$

By the divergence theorem we obtain

$$\begin{aligned}\frac{d}{dt}U^*[B_t] &= -\int_B \frac{\partial}{\partial X_l}\left(\frac{\partial W^*}{\partial F_{kl}}\right)\dot{\varphi}_k\, dV_{\boldsymbol{X}} + \int_{\partial B}\frac{\partial W^*}{\partial F_{kl}}\dot{\varphi}_k N_l\, dA_{\boldsymbol{X}} \\ &= -\int_B [\nabla^X \cdot DW^*]\cdot\dot{\boldsymbol{\varphi}}\, dV_{\boldsymbol{X}} + \int_{\partial B}[DW^*\boldsymbol{N}]\cdot\dot{\boldsymbol{\varphi}}\, dA_{\boldsymbol{X}}.\end{aligned}$$

Since by assumption $\nabla^X \cdot [DW^*] = \boldsymbol{0}$ in B and $\dot{\boldsymbol{\varphi}} = \boldsymbol{0}$ on ∂B for all $t \geq 0$ we get

$$\frac{d}{dt}U^*[B_t] = 0, \quad \forall t \geq 0.$$

7.11 From the definition $\mathsf{C}(\boldsymbol{H}) = \lambda(\operatorname{tr}\boldsymbol{H})\boldsymbol{I} + 2\mu\,\operatorname{sym}(\boldsymbol{H})$ we deduce

$$\begin{aligned}[\mathsf{C}(\boldsymbol{H})]_{ij} &= \lambda H_{kk}\delta_{ij} + \mu(H_{ij} + H_{ji}) \\ &= (\lambda\delta_{ij}\delta_{kl} + \mu\delta_{ik}\delta_{jl} + \mu\delta_{jk}\delta_{il})H_{kl}.\end{aligned}$$

The desired result follows upon noting that $[\mathsf{C}(\boldsymbol{H})]_{ij} = \mathsf{C}_{ijkl}H_{kl}$.

7.13 (a) An isotropic, linear elastic body is hyperelastic with strain energy density function

$$W(\boldsymbol{F}) = \tfrac{1}{2}\nabla^X\boldsymbol{u} : \mathsf{C}(\nabla^X\boldsymbol{u}).$$

Since

$$U[B_t] = \int_B W(\boldsymbol{F})\, dV_{\boldsymbol{X}} = \int_B \frac{1}{2}\nabla^X\boldsymbol{u} : \mathsf{C}(\nabla^X\boldsymbol{u})\, dV_{\boldsymbol{X}}$$

and

$$K[B_t] = \int_B \frac{1}{2}\rho_0\,|\dot{\boldsymbol{\varphi}}|^2\, dV_{\boldsymbol{X}} = \int_B \frac{1}{2}\rho_0\,|\dot{\boldsymbol{u}}|^2\, dV_{\boldsymbol{X}},$$

the result of Exercise 8 implies

$$\frac{1}{2}\frac{d}{dt}\int_B \left[\rho_0\,|\dot{\boldsymbol{u}}|^2 + \nabla^X\boldsymbol{u} : \mathsf{C}(\nabla^X\boldsymbol{u})\right] dV_{\boldsymbol{X}} = 0. \tag{7.46}$$

(b) By definition of C, the balance of linear momentum equation in the absence of a body force is

$$\rho_0\ddot{\boldsymbol{u}} = \mu\Delta^X\boldsymbol{u} + (\mu + \lambda)\nabla^X(\nabla^X\cdot\boldsymbol{u}),$$

and the assumption $\lambda = -\mu$ gives

$$\rho_0 \ddot{\boldsymbol{u}} = \mu \Delta^X \boldsymbol{u}. \tag{7.47}$$

Consider taking the dot product of (7.47) with $\dot{\boldsymbol{u}}$ and integrating the result over B. For the left-hand side we obtain

$$\int_B \rho_0 \ddot{\boldsymbol{u}} \cdot \dot{\boldsymbol{u}}\, dV_{\boldsymbol{X}} = \frac{1}{2}\frac{d}{dt} \int_B \rho_0 \left|\dot{\boldsymbol{u}}\right|^2 \, dV_{\boldsymbol{X}}, \tag{7.48}$$

and for the right-hand side we obtain

$$\begin{aligned} \int_B \mu\, \Delta^X \boldsymbol{u} \cdot \dot{\boldsymbol{u}}\, dV_{\boldsymbol{X}} &= \int_B \mu\, u_{i,jj} \dot{u}_i \, dV_{\boldsymbol{X}} \\ &= -\int_B \mu\, u_{i,j} \dot{u}_{i,j}\, dV_X + \int_{\partial B} \mu\, u_{i,j} \dot{u}_i N_j \, dA_{\boldsymbol{X}} \\ &= -\int_B \mu\, \nabla^X \boldsymbol{u} : \nabla^X \dot{\boldsymbol{u}}\, dV_{\boldsymbol{X}}, \end{aligned} \tag{7.49}$$

where we have used the fact that $\boldsymbol{u} = \boldsymbol{0}$ on ∂B. Moreover, since

$$\frac{\partial}{\partial t} (\nabla^X \boldsymbol{u} : \nabla^X \boldsymbol{u}) = 2 \nabla^X \boldsymbol{u} : \nabla^X \dot{\boldsymbol{u}},$$

we find that

$$\int_B \mu\, \nabla^X \boldsymbol{u} : \nabla^X \dot{\boldsymbol{u}}\, dV_{\boldsymbol{X}} = \frac{1}{2}\frac{d}{dt} \int_B \mu\, \nabla^X \boldsymbol{u} : \nabla^X \boldsymbol{u}\, dV_{\boldsymbol{X}}. \tag{7.50}$$

Combining (7.50), (7.49) and (7.48) gives

$$\frac{1}{2}\frac{d}{dt} \int_B \left[\rho_0 \left|\dot{\boldsymbol{u}}\right|^2 + \mu\, \nabla^X \boldsymbol{u} : \nabla^X \boldsymbol{u}\right] dV_{\boldsymbol{X}} = 0, \tag{7.51}$$

which is the desired result.

(c) Comparing (7.51) and (7.46) we find that the function

$$W^*(\boldsymbol{F}) = \nabla^X \boldsymbol{u} : \mathbf{C}(\nabla^X \boldsymbol{u}) - \mu\, \nabla^X \boldsymbol{u} : \nabla^X \boldsymbol{u}$$

must be a null energy density for B. From the definition of $\mathbf{C}$ we get

$$\begin{aligned} W^*(\boldsymbol{F}) &= \nabla^X \boldsymbol{u} : (\mathbf{C}(\nabla^X \boldsymbol{u}) - \mu \nabla^X \boldsymbol{u}) \\ &= \nabla^X \boldsymbol{u} : (-\mu \operatorname{tr}(\nabla^X \boldsymbol{u}) \boldsymbol{I} + \mu \nabla^X \boldsymbol{u}^T) \\ &= -\mu (\operatorname{tr}(\nabla^X \boldsymbol{u}))^2 + \mu \nabla^X \boldsymbol{u} : \nabla^X \boldsymbol{u}^T. \end{aligned}$$

7.15 (a) Since $u_i = \sigma a_i \phi(\gamma k_l X_l - \omega t)$ we have, using a prime to denote the derivative of ϕ

$$u_{i,j} = \sigma a_i \phi'(\gamma k_l X_l - \omega t)(\gamma k_l X_l)_{,j} = \sigma \gamma a_i k_j \phi'(\gamma k_l X_l - \omega t),$$

and differentiating again we get

$$u_{i,jm} = \sigma \gamma^2 a_i k_j k_m \phi''(\gamma k_l X_l - \omega t).$$

From these results we obtain, using the fact that $\boldsymbol{k}$ is a unit vector

$$\begin{aligned}\Delta^X \boldsymbol{u} = u_{i,jj} \boldsymbol{e}_i &= \sigma \gamma^2 a_i k_j k_j \phi''(\gamma k_l X_l - \omega t) \boldsymbol{e}_i \\ &= \sigma \gamma^2 \boldsymbol{a}(\boldsymbol{k} \cdot \boldsymbol{k}) \phi''(\gamma \boldsymbol{k} \cdot \boldsymbol{X} - \omega t) \\ &= \sigma \gamma^2 \boldsymbol{a} \phi''(\gamma \boldsymbol{k} \cdot \boldsymbol{X} - \omega t),\end{aligned}$$

and

$$\begin{aligned}\nabla^X (\nabla^X \cdot \boldsymbol{u}) = u_{i,ij} \boldsymbol{e}_j &= \sigma \gamma^2 a_i k_i k_j \phi''(\gamma k_l X_l - \omega t) \boldsymbol{e}_j \\ &= \sigma \gamma^2 (\boldsymbol{a} \cdot \boldsymbol{k}) \boldsymbol{k} \phi''(\gamma \boldsymbol{k} \cdot \boldsymbol{X} - \omega t).\end{aligned}$$

In a similar manner we obtain

$$\ddot{\boldsymbol{u}} = \sigma \omega^2 \boldsymbol{a} \phi''(\gamma \boldsymbol{k} \cdot \boldsymbol{X} - \omega t).$$

Substituting the above results into the linear momentum equation and collecting terms gives

$$\left[\sigma \gamma^2 \mu \boldsymbol{a} + \sigma \gamma^2 (\mu + \lambda)(\boldsymbol{a} \cdot \boldsymbol{k}) \boldsymbol{k} - \sigma \rho_0 \omega^2 \boldsymbol{a}\right] \phi''(\gamma \boldsymbol{k} \cdot \boldsymbol{X} - \omega t) = \boldsymbol{0}.$$

From this result, and the fact that ϕ'' is not identically zero, we deduce that the momentum equation is satisfied for all $\boldsymbol{X}$ and t if and only if $\boldsymbol{a}$, $\boldsymbol{k}$, σ, γ, ω satisfy

$$\sigma \gamma^2 [\mu \boldsymbol{a} + (\mu + \lambda)(\boldsymbol{a} \cdot \boldsymbol{k}) \boldsymbol{k}] = \sigma \rho_0 \omega^2 \boldsymbol{a}, \tag{7.52}$$

which is the desired result.

(b) For any given $\boldsymbol{k}$, $\sigma \neq 0$ and $\gamma \neq 0$ the equation in (7.52) can be written, after dividing through by $\sigma \gamma^2$, as an eigenvalue problem

$$\boldsymbol{A} \boldsymbol{a} = \alpha \boldsymbol{a} \quad \text{where} \quad \boldsymbol{A} = \mu \boldsymbol{I} + (\mu + \lambda) \boldsymbol{k} \otimes \boldsymbol{k}, \quad \alpha = \frac{\rho_0 \omega^2}{\gamma^2}.$$

Let $\{\boldsymbol{e}_i\}$ be any frame with $\boldsymbol{e}_1 = \boldsymbol{k}$. Then the matrix representation of $\boldsymbol{A}$ is

$$[\boldsymbol{A}] = \begin{pmatrix} 2\mu + \lambda & 0 & 0 \\ 0 & \mu & 0 \\ 0 & 0 & \mu \end{pmatrix}.$$

Under the condition $\lambda + \mu \neq 0$ we find that this matrix has two distinct eigenvalues: $\alpha = 2\mu + \lambda$ and $\alpha = \mu$. Any vector parallel to $\boldsymbol{e}_1$ is an eigenvector for the first, and any vector perpendicular to $\boldsymbol{e}_1$ is an eigenvector for the second. From the definition of α and the fact that $\boldsymbol{e}_1 = \boldsymbol{k}$ we find that the only independent solutions $(\boldsymbol{a}, \omega)$ of (7.52) are

$$\boldsymbol{a} = \pm\boldsymbol{k},\ \omega^2 = \gamma^2(\lambda + 2\mu)/\rho_0 \quad \text{and} \quad \boldsymbol{a}\cdot\boldsymbol{k} = 0,\ \omega^2 = \gamma^2\mu/\rho_0,$$

which is the desired result. In the degenerate case when $\lambda+\mu = 0$ we get $\boldsymbol{A} = \mu\boldsymbol{I}$. In this case the solutions of (7.52) are

$$\boldsymbol{a} \text{ arbitrary},\ \omega^2 = \gamma^2\mu/\rho_0.$$

7.17 (a) For $\eta_1 > 0$ and $\eta_2 = 0$ the given equation reduces to

$$\frac{\partial^2 u}{\partial t^2} = \frac{\eta_1}{\rho_0}\frac{\partial^2 u}{\partial x^2}, \tag{7.53}$$

which is the classic wave equation. This equation admits wave-like solutions which do not damp out in time. To see this, consider a solution to (7.53) of the form

$$u(x,t) = e^{i(x-ct)}. \tag{7.54}$$

Then (7.54) satisfies (7.53) if and only if

$$c = \pm\sqrt{\frac{\eta_1}{\rho_0}}.$$

Recall that, by Euler's formula $e^{i\theta} = \cos(\theta) + i\sin(\theta)$, each independent complex solution yields a pair of independent real solutions. Thus (7.54) yields the real solution

$$\begin{aligned} u(x,t) = A_1\cos(x + t\sqrt{\eta_1/\rho_0}) + A_2\sin(x + t\sqrt{\eta_1/\rho_0}) \\ + B_1\cos(x - t\sqrt{\eta_1/\rho_0}) + B_2\sin(x - t\sqrt{\eta_1/\rho_0}), \end{aligned}$$

where A_1, A_2, B_1 and B_2 are arbitrary constants. This solution is wave like since it is the superposition of simple traveling waves. Moreover, it does not damp out in time, that is, $u(x,t) \not\to 0$ as $t \to \infty$.

(b) For $\eta_1 = 0$ and $\eta_2 > 0$ we have

$$\rho_0\frac{\partial^2}{\partial t^2}u = \eta_2\frac{\partial}{\partial t}\frac{\partial^2}{\partial x^2}u. \tag{7.55}$$

If we define $v = \partial u/\partial t$, then v satisfies

$$\rho_0 \frac{\partial v}{\partial t} = \eta_2 \frac{\partial^2 v}{\partial x^2}, \tag{7.56}$$

which is the classic heat equation. Because this equation admits stationary (non-wave-like) solutions v, we expect the same for u. To see this, consider

$$u(x,t) = e^{i(x-ct)}. \tag{7.57}$$

Then (7.57) satisfies (7.55) if and only if

$$c = 0 \quad \text{or} \quad c = -i\frac{\eta_2}{\rho_0},$$

and Euler's formula leads to the real solution

$$\begin{aligned} u(x,t) = &A_1 \cos(x) + A_2 \sin(x) \\ &+ B_1 e^{-\eta_2 t/\rho_0} \cos(x) + B_2 e^{-\eta_2 t/\rho_0} \sin(x). \end{aligned}$$

In contrast to the solution found in part (a), this solution does not translate as time progresses and hence is stationary. Moreover, the solution damps exponentially fast to a steady-state (time-independent) one.

(c) For η_1, $\eta_2 > 0$ we have

$$\rho_0 \frac{\partial^2}{\partial t^2} u = \eta_1 \frac{\partial^2}{\partial x^2} u + \eta_2 \frac{\partial}{\partial t} \frac{\partial^2}{\partial x^2} u. \tag{7.58}$$

A solution of the form $u(x,t) = e^{i(x-ct)}$ satisfies (7.58) if and only if $\rho_0 c^2 + i\eta_2 c - \eta_1 = 0$, which yields

$$c = \frac{-i\eta_2 \pm [4\rho_0\eta_1 - \eta_2^2]^{\frac{1}{2}}}{2\rho_0}.$$

Case (i) If $4\rho_0\eta_1 > \eta_2^2$, then $c = -i\alpha \pm \beta$, where

$$\alpha = \frac{\eta_2}{2\rho_0} > 0, \quad \beta = \frac{(4\rho_0\eta_1 - \eta_2^2)^{\frac{1}{2}}}{2\rho_0} > 0.$$

This together with Euler's formula leads to the real solution

$$\begin{aligned} u(x,t) = &A_1 e^{-\alpha t} \cos(x + \beta t) + A_2 e^{-\alpha t} \sin(x + \beta t) \\ &+ B_1 e^{-\alpha t} \cos(x - \beta t) + B_2 e^{-\alpha t} \sin(x - \beta t). \end{aligned}$$

This solution simultaneously propagates and damps out in time.

Case (ii) If $4\rho_0\eta_1 < \eta_2^2$, then $c = -i(\alpha \pm \beta)$, where

$$\alpha = \frac{\eta_2}{2\rho_0} > 0, \quad \beta = \frac{(\eta_2^2 - 4\rho_0\eta_1)^{\frac{1}{2}}}{2\rho_0} > 0.$$

Substituting this result into the expression for u and applying Euler's formula leads to the real solution

$$\begin{aligned} u(x,t) &= A_1 e^{-(\alpha+\beta)t}\cos(x) + A_2 e^{-(\alpha+\beta)t}\sin(x) \\ &\quad + B_1 e^{-(\alpha-\beta)t}\cos(x) + B_2 e^{-(\alpha-\beta)t}\sin(x). \end{aligned}$$

Since $\eta_2^2 > \eta_2^2 - 4\rho_0\eta_1 > 0$ we have $\eta_2 > (\eta_2^2 - 4\rho_0\eta_1)^{\frac{1}{2}}$, which implies $\alpha - \beta > 0$. Moreover, since $\alpha > 0$ and $\beta > 0$ we have $\alpha + \beta > 0$. Thus in this case the solution damps in time without propagating.

7.19 (a) By definition we have $\nabla \times (\boldsymbol{S}^T) = \epsilon_{pkq} S_{mk,p}\, \boldsymbol{e}_q \otimes \boldsymbol{e}_m$. Thus

$$\begin{aligned} -[\nabla \times (\boldsymbol{S}^T)]^T &= -[\epsilon_{pkq} S_{mk,p}\, \boldsymbol{e}_q \otimes \boldsymbol{e}_m]^T \\ &= -\epsilon_{pkq} S_{mk,p}\, \boldsymbol{e}_m \otimes \boldsymbol{e}_q \\ &= \epsilon_{kpq} S_{mk,p}\, \boldsymbol{e}_m \otimes \boldsymbol{e}_q = \boldsymbol{S} \times \nabla, \end{aligned}$$

which establishes the first result. Next, let $\boldsymbol{L} = \nabla \times \boldsymbol{S}$ and $\boldsymbol{R} = \boldsymbol{S} \times \nabla$ so that $L_{qm} = \epsilon_{pkq} S_{km,p}$ and $R_{rq} = \epsilon_{spq} S_{rs,p}$. Then

$$\begin{aligned} \boldsymbol{L} \times \nabla = \epsilon_{mrs} L_{qm,r} \boldsymbol{e}_q \otimes \boldsymbol{e}_s &= \epsilon_{mrs}\epsilon_{pkq} S_{km,pr} \boldsymbol{e}_q \otimes \boldsymbol{e}_s \\ &= \epsilon_{abj}\epsilon_{cli} S_{la,cb} \boldsymbol{e}_i \otimes \boldsymbol{e}_j, \end{aligned}$$

and

$$\begin{aligned} \nabla \times \boldsymbol{R} = \epsilon_{krm} R_{rq,k} \boldsymbol{e}_m \otimes \boldsymbol{e}_q &= \epsilon_{krm}\epsilon_{spq} S_{rs,pk} \boldsymbol{e}_m \otimes \boldsymbol{e}_q \\ &= \epsilon_{cli}\epsilon_{abj} S_{la,bc} \boldsymbol{e}_i \otimes \boldsymbol{e}_j. \end{aligned}$$

Since mixed partial derivatives commute we have $S_{la,cb} = S_{la,bc}$, which leads to the result

$$(\nabla \times \boldsymbol{S}) \times \nabla = \boldsymbol{L} \times \nabla = \nabla \times \boldsymbol{R} = \nabla \times (\boldsymbol{S} \times \nabla).$$

(b) Note that, if $\nabla \times (\boldsymbol{E} \times \nabla) = \boldsymbol{O}$, then

$$\epsilon_{mrs}\epsilon_{pkq} E_{km,pr} = 0.$$

Choosing (m, r, p, k) from the permutations of $(1, 2, 1, 2)$ gives

$$E_{21,12} + E_{12,21} - E_{11,22} - E_{22,11} = 0,$$

so that, by symmetry of $\boldsymbol{E}$

$$E_{11,22} + E_{22,11} = 2E_{12,12}.$$

Similar expressions result if we choose from (3, 1, 3, 1) and (2, 3, 2, 3). If we choose (m, r, p, k) from the permutations of (1, 2, 1, 3), we obtain

$$E_{11,23} = -E_{23,11} + E_{31,21} + E_{12,31}.$$

Similar permutations yield the remaining conditions.

7.21 (a) Using the symmetry of $\boldsymbol{E}$ the compatibility equations can be written as

$$E_{ij,km} + E_{km,ij} - E_{ik,jm} - E_{jm,ik} = 0.$$

Making the substitution $E_{ij} = \frac{-\nu}{\gamma} P_{pp}\delta_{ij} + (\frac{1+\nu}{\gamma})P_{ij}$ in this expression gives

$$\left(\frac{1+\nu}{\gamma}\right)[P_{ij,km} + P_{km,ij} - P_{ik,jm} - P_{jm,ik}]$$
$$= \frac{\nu}{\gamma}[\delta_{ij}P_{pp,km} + \delta_{km}P_{pp,ij} - \delta_{ik}P_{pp,jm} - \delta_{jm}P_{pp,ik}]. \tag{7.59}$$

Setting $k = m$ and summing we obtain

$$\begin{aligned} P_{ij,kk} &+ P_{kk,ij} - P_{ik,jk} - P_{jk,ik} \\ &= \left(\frac{\nu}{1+\nu}\right)[\delta_{ij}P_{pp,kk} + \delta_{kk}P_{pp,ij} - \delta_{ik}P_{pp,jk} - \delta_{jk}P_{pp,ik}] \\ &= \left(\frac{\nu}{1+\nu}\right)[\delta_{ij}P_{pp,kk} + 3P_{pp,ij} - P_{pp,ji} - P_{pp,ij}] \\ &= \left(\frac{\nu}{1+\nu}\right)[\delta_{ij}P_{pp,kk} + P_{pp,ij}], \end{aligned}$$

which is the desired result.

(b) Since $P_{kk,ij} = P_{pp,ij}$ we can rewrite the result from part (a) in the form

$$P_{ij,kk} + \left(\frac{1}{1+\nu}\right)P_{pp,ij} - \left(\frac{\nu}{1+\nu}\right)\delta_{ij}P_{pp,kk} = P_{ik,jk} + P_{jk,ik}.$$

In view of the result in Exercise 20(b) we obtain

$$\Delta^X P_{ij} + \left(\frac{1}{1+\nu}\right)\psi_{,ij} - \left(\frac{\nu}{1+\nu}\right)\delta_{ij}\Delta^X \psi = -(\rho_0 b_i)_{,j} - (\rho_0 b_j)_{,i}, \tag{7.60}$$

where $\psi = P_{pp} = \mathrm{tr}(\boldsymbol{P})$.

(c) Setting $k = i$ and $m = j$ in (7.59) and summing gives

$$\begin{aligned}(1+\nu)[P_{ij,ij} + P_{ij,ij} - P_{ii,jj} - P_{jj,ii}]\\ = \nu[P_{pp,ii} + P_{pp,ii} - \delta_{ii}P_{pp,jj} - \delta_{jj}P_{pp,ii}].\end{aligned}$$

By collecting terms and dividing by $1+\nu$ we obtain

$$P_{ij,ij} = \Big(\frac{1-\nu}{1+\nu}\Big)P_{pp,ii} = \Big(\frac{1-\nu}{1+\nu}\Big)\Delta^{X}\psi. \tag{7.61}$$

From Exercise 20(b) we have $P_{ij,jk} + (\rho_0 b_i)_{,k} = 0$, which implies $P_{ij,ij} = -(\rho_0 b_i)_{,i}$. Using this result in (7.61) then gives

$$\Delta^{X}\psi = -\Big(\frac{1+\nu}{1-\nu}\Big)\beta, \tag{7.62}$$

where $\beta = \nabla^{X}\cdot(\rho_0 \boldsymbol{b})$.

(d) Substitution of (7.62) into (7.60) gives the desired result.

8
Thermal Fluid Mechanics

In this chapter we consider applications of the Eulerian balance laws as in Chapter 6. However, in contrast to the isothermal models introduced there, here we study models that include thermal effects. Considering the full thermo-mechanical case there are 22 basic unknown fields in the Eulerian description of a continuum body:

$\varphi_i(\boldsymbol{X}, t)$	3 components of motion
$v_i(\boldsymbol{x}, t)$	3 components of velocity
$\rho(\boldsymbol{x}, t)$	1 mass density
$S_{ij}(\boldsymbol{x}, t)$	9 components of stress
$\theta(\boldsymbol{x}, t)$	1 temperature
$q_i(\boldsymbol{x}, t)$	3 components of heat flux
$\phi(\boldsymbol{x}, t)$	1 internal energy per unit mass
$\eta(\boldsymbol{x}, t)$	1 entropy per unit mass.

To determine these unknown fields we have the following 11 equations:

$[v_i]_m = \frac{\partial}{\partial t}\varphi_i$	3 kinematical
$\frac{\partial}{\partial t}\rho + (\rho v_i)_{,i} = 0$	1 mass
$\rho \dot{v}_i = S_{ij,j} + \rho b_i$	3 linear momentum
$S_{ij} = S_{ji}$	3 independent angular momentum
$\rho \dot{\phi} = S_{ij} v_{i,j} - q_{i,i} + \rho r$	1 energy.

Thus 11 additional equations are required to balance the number of equations and unknowns. These are provided by constitutive equations that relate $(\boldsymbol{S}, \boldsymbol{q}, \phi, \eta)$ to $(\rho, \boldsymbol{v}, \theta, \boldsymbol{\varphi})$. Any such constitutive equation must be consistent with the axiom of material frame-indifference and the Clausius–Duhem inequality (Second Law of Thermodynamics) as discussed in Chapter 5.

In this chapter we study constitutive models for different types of fluids. Because the models we consider are independent of $\boldsymbol{\varphi}$, a closed system of equations for the spatial fields $\boldsymbol{S}$, $\boldsymbol{q}$, ϕ, η, ρ, $\boldsymbol{v}$ and θ is provided by the constitutive equations, together with the balance equations for mass, linear momentum, angular momentum and energy. In particular, these equations provide a complete description of bodies occupying regions of space with prescribed boundaries. When boundaries are not known a-priori, then $\boldsymbol{\varphi}$ and the kinematical equation must in general also be considered as discussed in Chapter 6.

The specific constitutive models considered in this chapter are those for a (compressible) perfect gas and a compressible Newtonian fluid. We summarize the governing equations for each model, show that they are consistent with the axiom of material frame-indifference and the Clausius–Duhem inequality, and discuss various qualitative properties.

8.1 Perfect Gases

In this section we study the constitutive model of a perfect gas. We show that the model satisfies the axiom of material frame-indifference and establish conditions under which it satisfies the Clausius–Duhem inequality. Moreover, we establish conditions under which the model of a perfect gas reduces to that of an elastic fluid as studied in Chapter 6.

8.1.1 Definition

A continuum body with reference configuration B is said to be a **perfect gas** if:

(1) The Cauchy stress field $\boldsymbol{S}(\boldsymbol{x},t)$ is spherical or Eulerian (see Section 3.5), which means that there is a scalar field $p(\boldsymbol{x},t)$ called the pressure such that

$$\boldsymbol{S} = -p\boldsymbol{I}.$$

(2) The pressure $p(\boldsymbol{x},t)$ is related to the mass density $\rho(\boldsymbol{x},t)$ and temperature $\theta(\boldsymbol{x},t)$ by

$$p = \widehat{p}(\rho,\theta),$$

where $\widehat{p} : \mathbb{R}^2 \to \mathbb{R}$ is a given function called the **pressure response function**.

(3) The internal energy $\phi(\boldsymbol{x}, t)$ and entropy $\eta(\boldsymbol{x}, t)$ are related to the mass density $\rho(\boldsymbol{x}, t)$ and temperature $\theta(\boldsymbol{x}, t)$ by

$$\phi = \widehat{\phi}(\rho, \theta), \quad \eta = \widehat{\eta}(\rho, \theta),$$

where $\widehat{\phi}, \widehat{\eta} : \mathbb{R}^2 \to \mathbb{R}$ are given functions called the **internal energy** and **entropy response functions**.

(4) The functions $\widehat{\phi}$ and $\widehat{\eta}$ have the property

$$\frac{\partial \widehat{\phi}}{\partial \theta}(\rho, \theta) > 0 \quad \text{and} \quad \frac{\partial \widehat{\eta}}{\partial \theta}(\rho, \theta) > 0, \quad \forall \rho > 0,\ \theta > 0.$$

(5) The heat flux vector field $\boldsymbol{q}(\boldsymbol{x}, t)$ is identically zero, that is

$$\boldsymbol{q} = \boldsymbol{0}.$$

Properties (1) and (2) imply that the Cauchy stress at a point in a perfect gas is entirely determined by the current mass density and temperature at that point. In particular, it is independent of the past histories and rates of change of these quantities, and does not contain any viscous contribution arising from the spatial velocity gradient at that point. This type of relation is similar to the constitutive relation for an elastic fluid (see Chapter 6), but with thermal effects included. Property (1) also implies that the stress field is necessarily symmetric. Thus the balance equation for angular momentum (Result 5.8) is automatically satisfied and will not be considered further. Because $\rho(\boldsymbol{x}, t) > 0$ and $\theta(\boldsymbol{x}, t) > 0$ for any admissible thermo-mechanical process, we only consider $\widehat{p}(\rho, \theta)$ for arguments $\rho > 0$ and $\theta > 0$.

Property (3) implies that, just as for pressure, the internal energy and entropy per unit mass at a point are entirely determined by the current mass density and temperature at that point. The three relations

$$p = \widehat{p}(\rho, \theta), \quad \phi = \widehat{\phi}(\rho, \theta), \quad \eta = \widehat{\eta}(\rho, \theta), \tag{8.1}$$

are typically referred to as the **thermal equations of state**. Just as for $\widehat{p}(\rho, \theta)$, we only consider $\widehat{\phi}(\rho, \theta)$ and $\widehat{\eta}(\rho, \theta)$ for arguments $\rho > 0$ and $\theta > 0$.

Property (4) implies that, for each fixed value of ρ, the functions $\widehat{\phi}(\rho, \theta)$ and $\widehat{\eta}(\rho, \theta)$ are strictly increasing in θ. Physically, this means that when density is held fixed (or volume in the case of a homogeneous body), internal energy and entropy increase with temperature. Mathematically, these two conditions mean that either ϕ or η can replace θ as an independent variable. For example, the condition $\frac{\partial\hat{\eta}}{\partial\theta}(\rho,\theta) > 0$ guarantees that the relation $\eta = \widehat{\eta}(\rho,\theta)$ in $(8.1)_3$ can be inverted as $\theta = \overline{\theta}(\rho,\eta)$. When this relation is substituted into $(8.1)_{1,2}$ we obtain

$$p = \overline{p}(\rho,\eta), \quad \phi = \overline{\phi}(\rho,\eta), \quad \theta = \overline{\theta}(\rho,\eta). \tag{8.2}$$

These relations are typically referred to as the **caloric equations of state**. Below we use this form of the constitutive equations to establish conditions under which the model of a perfect gas satisfies the Clausius–Duhem inequality. The function $\frac{\partial\hat{\phi}}{\partial\theta}(\rho,\theta) > 0$ is typically called the **specific heat** at constant density or volume.

Property (5) states that the Fourier–Stokes heat flux vector field vanishes at every point in a perfect gas. From the definition of net heating (see Chapter 5), this implies that a body of perfect gas can experience body heating, but no surface heating. In other words, heat can be produced or consumed throughout the volume of a perfect gas, but it cannot be transferred from one part to another. In particular, there is no heat conduction.

Remarks:

(1) In general, the assumption on the heat flux is not essential. It is made for simplicity and is consistent with the assumption that the Cauchy stress is spherical with no viscous contribution. Indeed, at a microscopic level, heat conduction and viscosity in a gas both arise from similar mechanisms (molecular collisions), and are of the same order of magnitude. Thus if one is to be neglected, then it is often natural to neglect the other.

(2) In addition to the conditions in Property (4), we typically assume that the function $\overline{p}(\rho,\eta)$ satisfies $\partial\overline{p}/\partial\rho > 0$ for all $\rho > 0$ and η. The quantity $c = \sqrt{\partial\overline{p}/\partial\rho}$ is typically called the **local speed of sound** in a perfect gas. The technique of linearization can be used to motive this terminology. In particular, small disturbances in pressure from a uniform state can be shown to propagate with speed c.

(3) A classic example of a perfect gas is an **ideal gas** defined by the thermal equations of state

$$\begin{aligned} p &= \widehat{p}(\rho,\theta) = \rho R\theta, \\ \phi &= \widehat{\phi}(\rho,\theta) = c_v\theta, \\ \eta &= \widehat{\eta}(\rho,\theta) = c_v \ln\theta - R\ln\rho + \eta_*, \end{aligned} \tag{8.3}$$

where $R > 0$, $c_v > 0$ and η_* are constants. Here c_v corresponds to the specific heat at constant volume, and η_* determines the entropy in a given reference state.

(4) The conditions in Property (4) are satisfied by the ideal gas model since

$$\frac{\partial\widehat{\phi}}{\partial\theta} = c_v > 0, \qquad \frac{\partial\widehat{\eta}}{\partial\theta} = \frac{c_v}{\theta} > 0.$$

In particular, we can solve for θ in $(8.3)_3$ and obtain the caloric equations of state

$$\begin{aligned} p &= \overline{p}(\rho,\eta) = Re^{(\eta-\eta_*)/c_v}\,\rho^\gamma, \\ \phi &= \overline{\phi}(\rho,\eta) = c_v e^{(\eta-\eta_*)/c_v}\,\rho^{\gamma-1}, \\ \theta &= \overline{\theta}(\rho,\eta) = e^{(\eta-\eta_*)/c_v}\,\rho^{\gamma-1}, \end{aligned} \tag{8.4}$$

where we introduce the constant $\gamma = 1 + \frac{R}{c_v}$ for convenience.

(5) Equations of state such as those for an ideal gas in (8.3) and (8.4) are typically based on theories or experiments for a gas in equilibrium. For the models introduced here, it is assumed that such equations hold not only for a gas in equilibrium, but also for a gas undergoing flow and deformation.

□

8.1.2 Gas Dynamics Equations

A closed system of equations for the variables $(\rho, \boldsymbol{v}, \eta)$ in a body of perfect gas are provided by the caloric equations of state (8.2), the conservation of mass equation (Result 5.5), the balance of linear momentum equation (Result 5.7) and the balance of energy equation (Result 5.10).

Setting $\boldsymbol{S} = -p\boldsymbol{I}$ in the momentum and energy equations, and using the fact that $\boldsymbol{q} = \boldsymbol{0}$, we obtain

$$\rho\dot{\boldsymbol{v}} = \nabla^x \cdot (-p\boldsymbol{I}) + \rho\boldsymbol{b} \quad \text{and} \quad \rho\dot{\phi} = (-p\boldsymbol{I}) : \boldsymbol{L} + \rho r,$$

where $\boldsymbol{L} = \text{sym}(\nabla^x \boldsymbol{v})$ is the rate of strain tensor, $\boldsymbol{b}$ is a prescribed body force per unit mass and r is a prescribed heat supply per unit mass. In view of Result 4.7 we have

$$\dot{\boldsymbol{v}} = \frac{\partial}{\partial t}\boldsymbol{v} + (\nabla^x \boldsymbol{v})\boldsymbol{v} \quad \text{and} \quad \dot{\phi} = \frac{\partial}{\partial t}\phi + \nabla^x \phi \cdot \boldsymbol{v}.$$

Also, it is straightforward to verify that

$$\nabla^x \cdot (-p\boldsymbol{I}) = -\nabla^x p \quad \text{and} \quad (-p\boldsymbol{I}) : \boldsymbol{L} = -p\nabla^x \cdot \boldsymbol{v}.$$

Thus, setting $p = \overline{p}(\rho, \eta)$ and $\phi = \overline{\phi}(\rho, \eta)$, we find that the spatial density, velocity and entropy fields in a body of perfect gas with reference configuration B must satisfy the following equations for all $\boldsymbol{x} \in B_t$ and $t \geq 0$

$$\boxed{\begin{aligned} &\frac{\partial}{\partial t}\rho + \nabla^x \cdot (\rho \boldsymbol{v}) = 0, \\ &\rho \left[\frac{\partial}{\partial t}\boldsymbol{v} + (\nabla^x \boldsymbol{v})\boldsymbol{v}\right] + \nabla^x \overline{p}(\rho, \eta) = \rho \boldsymbol{b}, \\ &\rho \left[\frac{\partial}{\partial t}\overline{\phi}(\rho, \eta) + \nabla^x \overline{\phi}(\rho, \eta) \cdot \boldsymbol{v}\right] + \overline{p}(\rho, \eta)\nabla^x \cdot \boldsymbol{v} = \rho r. \end{aligned}} \tag{8.5}$$

These equations are known as the **Gas Dynamics Equations**. By an admissible process for a perfect gas we mean fields $(\rho, \boldsymbol{v}, \eta)$ satisfying these equations.

Remarks:

(1) The equations in (8.5) are a system of coupled, nonlinear partial differential equations for $(\rho, \boldsymbol{v}, \eta)$. Alternatively, use of the thermal equations of state (8.1) in place of (8.2) leads to a system of the same form, but with $(\rho, \boldsymbol{v}, \theta)$ as the variables. The two formulations are equivalent provided all fields involved are sufficiently smooth. For any choice of variables, these equations can also be written in so-called **conservation** or **divergence form** as in Exercise 5.9 of Chapter 5.

(2) Just as for the elastic fluid equations considered in Chapter 6, solutions of (8.5) may not be smooth due to the appearance of shock waves. These are surfaces across which one or more components of a solution has a jump discontinuity. A proper analysis of such solutions requires the introduction of an appropriate weak form of the balance laws. Here we assume that solutions are

continuously differentiable as many times as are required for the calculations we perform.

(3) The **local Mach number** $M = |\boldsymbol{v}|/c$, where c is the local speed of sound, plays an important role in the study of gas flows described by equations (8.5). Regions of flow in which $M < 1$ or $M > 1$ are called **subsonic** or **supersonic**. Dramatic physical and mathematical differences between subsonic and supersonic flows makes the study of initial-boundary value problems for (8.5) difficult. In particular, there are various issues surrounding the specification of appropriate boundary conditions for different types of gas flow problems.

□

8.1.3 Frame-Indifference Considerations

Consider a body of perfect gas with reference configuration B. Let $\boldsymbol{x} = \boldsymbol{\varphi}(\boldsymbol{X}, t)$ be an arbitrary motion and let $\boldsymbol{x}^* = \boldsymbol{\varphi}^*(\boldsymbol{X}, t)$ be a second motion defined by

$$\boldsymbol{x}^* = \boldsymbol{g}(\boldsymbol{x}, t) = \boldsymbol{Q}(t)\boldsymbol{x} + \boldsymbol{c}(t),$$

where $\boldsymbol{Q}(t)$ is an arbitrary rotation tensor and $\boldsymbol{c}(t)$ is an arbitrary vector. Let ρ, η, ϕ, θ and $\boldsymbol{S}$ denote the spatial density, entropy, internal energy, temperature and Cauchy stress fields in B_t, and let ρ^*, η^*, ϕ^*, θ^* and $\boldsymbol{S}^*$ denote the corresponding spatial fields in B_t^*. (We do not consider the spatial heat flux field since it is identically zero in B_t and B_t^* by definition of a perfect gas.) Then the axiom of material frame-indifference (Axiom 5.24) postulates the following for all $\boldsymbol{x} \in B_t$ and $t \geq 0$

$$\begin{aligned}
\rho^*(\boldsymbol{x}^*, t) &= \rho(\boldsymbol{x}, t), \\
\eta^*(\boldsymbol{x}^*, t) &= \eta(\boldsymbol{x}, t), \\
\phi^*(\boldsymbol{x}^*, t) &= \phi(\boldsymbol{x}, t), \\
\theta^*(\boldsymbol{x}^*, t) &= \theta(\boldsymbol{x}, t), \\
\boldsymbol{S}^*(\boldsymbol{x}^*, t) &= \boldsymbol{Q}(t)\boldsymbol{S}(\boldsymbol{x}, t)\boldsymbol{Q}(t)^T.
\end{aligned} \tag{8.6}$$

Substituting the constitutive equation for the Cauchy stress into (8.6), along with the caloric equations of state (8.2) for the pressure, internal energy and temperature fields, we obtain (omitting the arguments $\boldsymbol{x}$, $\boldsymbol{x}^*$

and t for brevity)

$$\begin{aligned}
\rho^* &= \rho, \\
\eta^* &= \eta, \\
\overline{\phi}(\rho^*, \eta^*) &= \overline{\phi}(\rho, \eta), \\
\overline{\theta}(\rho^*, \eta^*) &= \overline{\theta}(\rho, \eta), \\
-\overline{p}(\rho^*, \eta^*)\boldsymbol{I} &= \boldsymbol{Q}[-\overline{p}(\rho, \eta)\boldsymbol{I}]\boldsymbol{Q}^T.
\end{aligned} \tag{8.7}$$

Eliminating ρ^* and η^* using $(8.7)_{1,2}$, and noting that $\boldsymbol{Q}\boldsymbol{Q}^T = \boldsymbol{I}$, we observe that the remaining relations in $(8.7)_{3-5}$ are satisfied identically. Thus the constitutive model of a perfect gas complies with the axiom of material frame-indifference with no restrictions on the response functions $\overline{p}(\rho, \eta)$, $\overline{\phi}(\rho, \eta)$ and $\overline{\theta}(\rho, \eta)$.

Remarks:

(1) A similar conclusion is obtained if we use the thermal equations of state (8.1) in place of (8.2). In particular, the axiom of material frame-indifference is satisfied with no restrictions on the response functions $\widehat{p}(\rho, \theta)$, $\widehat{\phi}(\rho, \theta)$ and $\widehat{\eta}(\rho, \theta)$.

(2) Different conclusions can be obtained if the form of the response functions are changed. For example, if we consider functions of the more general form $\overline{p}(\rho, \eta, \boldsymbol{x})$, $\overline{\phi}(\rho, \eta, \boldsymbol{x})$ and $\overline{\theta}(\rho, \eta, \boldsymbol{x})$, then the axiom of material frame-indifference is satisfied if and only if these functions are independent of $\boldsymbol{x}$ (see Exercise 2). □

8.1.4 Thermodynamical Considerations

Here we outline a result which shows that the caloric response functions for a perfect gas must satisfy certain conditions in order to comply with the Clausius–Duhem inequality.

Result 8.1 ***Implications of Clausius–Duhem Inequality.*** *The Clausius–Duhem inequality (Result 5.11) is satisfied by every admissible process in a perfect gas if and only if the caloric response functions $\overline{p}$, $\overline{\phi}$ and $\overline{\theta}$ satisfy*

$$\overline{p}(\rho, \eta) = \rho^2 \frac{\partial \overline{\phi}}{\partial \rho}(\rho, \eta) \quad \textit{and} \quad \overline{\theta}(\rho, \eta) = \frac{\partial \overline{\phi}}{\partial \eta}(\rho, \eta). \tag{8.8}$$

□

Proof The Clausius–Duhem inequality in Eulerian form is

$$\rho\dot{\eta} \geq \theta^{-1}\rho r - \nabla^x \cdot (\theta^{-1}\boldsymbol{q}).$$

After multiplying through by $\rho^{-1}\theta > 0$, using $\dot{\eta} = \frac{\partial}{\partial t}\eta + \nabla^x\eta \cdot \boldsymbol{v}$ by Result 4.7, and using $\boldsymbol{q} = \boldsymbol{0}$ and $\theta = \overline{\theta}(\rho, \eta)$ by definition of a perfect gas, the above inequality becomes

$$\overline{\theta}(\rho, \eta)\left[\frac{\partial \eta}{\partial t} + \nabla^x\eta \cdot \boldsymbol{v}\right] \geq r. \tag{8.9}$$

From $(8.5)_3$, after dividing through by $\rho > 0$, we find

$$r = \left[\frac{\partial}{\partial t}\overline{\phi}(\rho, \eta) + \nabla^x\overline{\phi}(\rho, \eta) \cdot \boldsymbol{v}\right] + \frac{1}{\rho}\overline{p}(\rho, \eta)\nabla^x \cdot \boldsymbol{v}. \tag{8.10}$$

Using the chain rule we have (omitting arguments for brevity)

$$\frac{\partial}{\partial t}\overline{\phi}(\rho, \eta) = \frac{\partial\overline{\phi}}{\partial\rho}\frac{\partial\rho}{\partial t} + \frac{\partial\overline{\phi}}{\partial\eta}\frac{\partial\eta}{\partial t},$$

and similarly we find

$$\nabla^x\overline{\phi}(\rho, \eta) = \frac{\partial\overline{\phi}}{\partial\rho}\nabla^x\rho + \frac{\partial\overline{\phi}}{\partial\eta}\nabla^x\eta.$$

Substitution of these expressions into (8.10) gives

$$r = \frac{\partial\overline{\phi}}{\partial\rho}\left[\frac{\partial\rho}{\partial t} + \nabla^x\rho \cdot \boldsymbol{v}\right] + \frac{\partial\overline{\phi}}{\partial\eta}\left[\frac{\partial\eta}{\partial t} + \nabla^x\eta \cdot \boldsymbol{v}\right] + \frac{1}{\rho}\overline{p}\nabla^x \cdot \boldsymbol{v}. \tag{8.11}$$

From $(8.5)_1$, using the fact that $\nabla^x \cdot (\rho\boldsymbol{v}) = \nabla^x\rho \cdot \boldsymbol{v} + \rho\nabla^x \cdot \boldsymbol{v}$, we get

$$\frac{\partial\rho}{\partial t} + \nabla^x\rho \cdot \boldsymbol{v} = -\rho\nabla^x \cdot \boldsymbol{v},$$

and substitution of this result into (8.11) gives

$$r = \frac{\partial\overline{\phi}}{\partial\eta}\left[\frac{\partial\eta}{\partial t} + \nabla^x\eta \cdot \boldsymbol{v}\right] + \left[\frac{1}{\rho}\overline{p} - \rho\frac{\partial\overline{\phi}}{\partial\rho}\right]\nabla^x \cdot \boldsymbol{v}. \tag{8.12}$$

Using (8.12) in (8.9) we find that the Clausius–Duhem inequality is equivalent to

$$\left[\overline{\theta} - \frac{\partial\overline{\phi}}{\partial\eta}\right]\frac{\partial\eta}{\partial t} + \left[\overline{\theta} - \frac{\partial\overline{\phi}}{\partial\eta}\right]\nabla^x\eta \cdot \boldsymbol{v} + \left[\rho\frac{\partial\overline{\phi}}{\partial\rho} - \frac{1}{\rho}\overline{p}\right]\nabla^x \cdot \boldsymbol{v} \geq 0. \tag{8.13}$$

It can now be shown that (8.8) are necessary and sufficient conditions for (8.13) to hold for all admissible processes. Sufficiency of (8.8) is clear, for then equality holds in (8.13). To establish necessity, suppose (8.13) holds for all processes. Notice that the terms in brackets depend only on ρ and η, and are independent of $\frac{\partial}{\partial t}\eta$ and $\boldsymbol{v}$. Because $\frac{\partial}{\partial t}\eta$ can be varied independently of ρ and η, we deduce that the coefficient of $\frac{\partial}{\partial t}\eta$ in (8.13) must vanish, that is, $[\overline{\theta} - \frac{\partial \overline{\phi}}{\partial \eta}] = 0$. Otherwise, we could take $\frac{\partial}{\partial t}\eta$ to be a sufficiently large negative multiple of $[\overline{\theta} - \frac{\partial \overline{\phi}}{\partial \eta}]$ and thereby cause (8.13) to be violated. Thus (8.13) necessarily reduces to the form

$$\left[\rho \frac{\partial \overline{\phi}}{\partial \rho} - \frac{1}{\rho}\overline{p}\right] \nabla^x \cdot \boldsymbol{v} \geq 0.$$

Because $\nabla^x \cdot \boldsymbol{v}$ can be varied independently of ρ and η, a similar argument shows that the coefficient of $\nabla^x \cdot \boldsymbol{v}$ in the above inequality must also vanish, that is, $[\rho\frac{\partial \overline{\phi}}{\partial \rho} - \frac{1}{\rho}\overline{p}] = 0$. Thus the conditions in (8.8) are necessary. □

Remarks:

(1) In the above arguments we implicitly assume that, at any given instant of time, the spatial fields $\frac{\partial}{\partial t}\eta$ and $\boldsymbol{v}$ can be varied independently of ρ and η. In view of the balance of energy and momentum equations in (8.5), arbitrary values for the fields $\frac{\partial}{\partial t}\eta$ and $\boldsymbol{v}$ can be achieved, for given ρ and η, by appropriate choice of heat supply r and body force $\boldsymbol{b}$.

(2) The conditions in Result 8.1 apply to the caloric equations of state in (8.2). A different, but equivalent version of these conditions can also be derived for the thermal equations of state in (8.1). In particular, the Clausius–Duhem inequality holds if and only if $\widehat{p}$, $\widehat{\phi}$ and $\widehat{\eta}$ satisfy (see Exercise 3)

$$\begin{aligned} \frac{\partial \widehat{\eta}}{\partial \theta}(\rho, \theta) &= \frac{1}{\theta}\frac{\partial \widehat{\phi}}{\partial \theta}(\rho, \theta), \\ \widehat{p}(\rho, \theta) &= \rho^2 \frac{\partial \widehat{\phi}}{\partial \rho}(\rho, \theta) - \rho^2 \theta \frac{\partial \widehat{\eta}}{\partial \rho}(\rho, \theta). \end{aligned} \tag{8.14}$$

(3) For the ideal gas model it is straightforward to verify that the caloric equations of state in (8.4) satisfy the conditions in (8.8). Equivalently, it is straightforward to verify that the thermal equations of state in (8.3) satisfy the conditions in (8.14). Thus, either way, we conclude that the constitutive model of an ideal gas satisfies the Clausius–Duhem inequality.

(4) A process in a material body is called **reversible** if equality is achieved in the Clausius–Duhem inequality throughout the process; otherwise, **irreversible**. The proof of Result 8.1 shows that, under the constitutive restrictions in (8.8), a perfect gas can experience only reversible processes. This can be attributed to the fact that viscous and heat conduction effects are neglected in the model.

□

8.1.5 Entropy Formulation, Isentropic Equations

Here use Result 8.1 to derive a different form of the equations in (8.5). We then show that, in the absence of a heat supply, one of the resulting equations can be solved exactly to yield a reduced system for the spatial density and velocity fields.

Result 8.2 ***Balance Equations in Entropy Form.*** *If the caloric response functions satisfy the conditions in Result 8.1, then the balance equations in* (8.5) *can be written in the equivalent form*

$$\begin{aligned}
&\frac{\partial}{\partial t}\rho + \nabla^x \cdot (\rho \boldsymbol{v}) = 0, \\
&\rho\left[\frac{\partial}{\partial t}\boldsymbol{v} + (\nabla^x \boldsymbol{v})\boldsymbol{v}\right] + \nabla^x \overline{p}(\rho, \eta) = \rho \boldsymbol{b}, \\
&\frac{\partial \overline{\phi}}{\partial \eta}(\rho, \eta)\left[\frac{\partial}{\partial t}\eta + \nabla^x \eta \cdot \boldsymbol{v}\right] = r.
\end{aligned} \tag{8.15}$$

□

Proof From the proof of Result 8.1 we observe that $(8.5)_3$ can be written in the form (8.12), namely

$$\frac{\partial \overline{\phi}}{\partial \eta}\left[\frac{\partial \eta}{\partial t} + \nabla^x \eta \cdot \boldsymbol{v}\right] + \left[\frac{1}{\rho}\overline{p} - \rho \frac{\partial \overline{\phi}}{\partial \rho}\right] \nabla^x \cdot \boldsymbol{v} = r.$$

Moreover, by $(8.8)_1$ the coefficient of $\nabla^x \cdot \boldsymbol{v}$ in this expression vanishes. Thus, under the condition in $(8.8)_1$, we find that $(8.15)_3$ is equivalent to $(8.5)_3$. The remaining equations in $(8.15)_{1,2}$ are identical to those in $(8.5)_{1,2}$. □

The next result shows that equation $(8.15)_3$ can be solved explicitly when there is no heat supply and the entropy field in the reference configuration is uniform (constant).

Result 8.3 ***Isentropic Balance Equations.*** *Consider a body of perfect gas with reference configuration* B. *Assume there is no heat supply, so* $r = 0$, *and assume the entropy field at time* $t = 0$ *is uniform, so* $\eta_0(\boldsymbol{X}) = \eta_0$ *(constant) for all* $\boldsymbol{X} \in B$. *Then the unique solution of* $(8.15)_3$ *is* $\eta(\boldsymbol{x}, t) = \eta_0$ *for all* $\boldsymbol{x} \in B_t$ *and* $t \geq 0$. *Moreover, the equations in* (8.15) *reduce to*

$$
\begin{aligned}
&\frac{\partial}{\partial t}\rho + \nabla^x \cdot (\rho \boldsymbol{v}) = 0, \\
&\rho \left[\frac{\partial}{\partial t}\boldsymbol{v} + (\nabla^x \boldsymbol{v})\boldsymbol{v} \right] + \nabla^x \overline{p}(\rho, \eta_0) = \rho \boldsymbol{b}.
\end{aligned}
\tag{8.16}
$$

□

Proof When $r = 0$ equation $(8.15)_3$ becomes

$$
\frac{\partial \overline{\phi}}{\partial \eta}(\rho, \eta) \left[\frac{\partial}{\partial t}\eta + \nabla^x \eta \cdot \boldsymbol{v} \right] = 0, \quad \forall \boldsymbol{x} \in B_t,\ t \geq 0.
$$

By definition of $\overline{\theta}(\rho, \eta)$ and $\overline{\phi}(\rho, \eta)$, together with the chain rule and Property (4) of a perfect gas, we deduce $\frac{\partial \overline{\phi}}{\partial \eta}(\rho, \eta) > 0$. This implies

$$
\frac{\partial}{\partial t}\eta + \nabla^x \eta \cdot \boldsymbol{v} = 0, \quad \forall \boldsymbol{x} \in B_t,\ t \geq 0,
$$

and by definition of the total or material time derivative we obtain

$$
\frac{\partial}{\partial t}\eta(\boldsymbol{\varphi}(\boldsymbol{X}, t), t) = 0, \quad \forall \boldsymbol{X} \in B,\ t \geq 0,
$$

where $\boldsymbol{x} = \boldsymbol{\varphi}(\boldsymbol{X}, t)$ is the motion map associated with the spatial velocity field $\boldsymbol{v}(\boldsymbol{x}, t)$. From this result, together with the convention $\boldsymbol{\varphi}(\boldsymbol{X}, 0) = \boldsymbol{X}$ and the definition of $\eta_0(\boldsymbol{X})$, we get

$$
\eta(\boldsymbol{\varphi}(\boldsymbol{X}, t), t) = \eta(\boldsymbol{X}, 0) = \eta_0(\boldsymbol{X}), \quad \forall \boldsymbol{X} \in B,\ t \geq 0.
$$

Since $\eta_0(\boldsymbol{X}) = \eta_0$ (constant) and $\boldsymbol{x} = \boldsymbol{\varphi}(\boldsymbol{X}, t)$, we obtain a unique solution of $(8.15)_3$, namely $\eta(\boldsymbol{x}, t) = \eta_0$ for all $\boldsymbol{x} \in B_t$ and $t \geq 0$. Substituting this result into $(8.15)_{1,2}$ we obtain the system in (8.16).

□

Remarks:

(1) The equations in (8.16) are a system of coupled, nonlinear partial differential equations for the variables $(\rho, \boldsymbol{v})$. Considering only smooth solutions, this system is equivalent to (8.15) under the

assumptions that the heat supply is zero and the initial entropy field is uniform.

(2) A process in a material body is called **isentropic** if $\eta(\boldsymbol{x}, t) = \eta_0$ (constant) throughout the process. For this reason the equations in (8.16) are typically called the **Isentropic Gas Dynamics Equations**. Under the conditions of Result 8.1, they must be satisfied by any isentropic process in a perfect gas.

(3) A process in a material body is called **adiabatic** if $\boldsymbol{q}(\boldsymbol{x}, t) = \boldsymbol{0}$ and $r(\boldsymbol{x}, t) = 0$ throughout the process. Result 8.3 implies that, under the conditions of Result 8.1, an adiabatic process in a perfect gas is isentropic provided the initial entropy field is uniform.

(4) The equations in (8.16) are of precisely the same form as the equations for an elastic fluid (see Chapter 6) with pressure response function $\pi(\rho) = \overline{p}(\rho, \eta_0)$. Thus an elastic fluid can be interpreted as an isentropic model rather than an isothermal model of fluid (gas) behavior. All results on irrotational motion and Bernoulli's Theorem derived for the elastic fluid equations carry over to the equations in (8.16).

(5) Linearization of the elastic fluid equations (see Chapter 6) shows that small disturbances in density, and hence pressure, from a uniform quiescent state with density ρ_0 propagate at the speed $\sqrt{\pi'(\rho_0)}$. This provides some justification for calling $\sqrt{\partial\overline{p}/\partial\rho}$ the local speed of sound. A rigorous justification is given by the theory of singular surfaces and sound waves, which applies to arbitrary motions of perfect gases, not just isentropic motions as described by (8.16).

□

8.2 Compressible Newtonian Fluids

Here we study the constitutive model for a compressible Newtonian fluid. We show that the model satisfies the axiom of material frame-indifference, establish conditions under which it satisfies the Clausius–Duhem inequality and outline a standard initial-boundary value problem.

8.2.1 Definition

A continuum body with reference configuration B is said to be a **compressible Newtonian fluid** if:

(1) The Cauchy stress field $\boldsymbol{S}(\boldsymbol{x},t)$ is of the form

$$\boldsymbol{S} = -p\boldsymbol{I} + \lambda(\nabla^x \cdot \boldsymbol{v})\boldsymbol{I} + 2\mu\boldsymbol{L},$$

where $p(\boldsymbol{x},t)$ is the **thermodynamic pressure** field, $\boldsymbol{L}(\boldsymbol{x},t)$ is the rate of strain tensor field (see Section 4.5.1), and λ, μ are scalar constants.

(2) The thermodynamic pressure $p(\boldsymbol{x},t)$ is related to the mass density $\rho(\boldsymbol{x},t)$ and temperature $\theta(\boldsymbol{x},t)$ by

$$p = \widehat{p}(\rho, \theta),$$

where $\widehat{p} : \mathbb{R}^2 \to \mathbb{R}$ is a given response function.

(3) The internal energy $\phi(\boldsymbol{x},t)$ and entropy $\eta(\boldsymbol{x},t)$ are related to the mass density $\rho(\boldsymbol{x},t)$ and temperature $\theta(\boldsymbol{x},t)$ by

$$\phi = \widehat{\phi}(\rho, \theta), \quad \eta = \widehat{\eta}(\rho, \theta),$$

where $\widehat{\phi}, \widehat{\eta} : \mathbb{R}^2 \to \mathbb{R}$ are given response functions.

(4) The functions $\widehat{\phi}$ and $\widehat{\eta}$ have the property

$$\frac{\partial \widehat{\phi}}{\partial \theta}(\rho, \theta) > 0 \quad \text{and} \quad \frac{\partial \widehat{\eta}}{\partial \theta}(\rho, \theta) > 0, \quad \forall \rho > 0,\ \theta > 0.$$

(5) The heat flux vector field $\boldsymbol{q}(\boldsymbol{x},t)$ is related to the temperature field $\theta(\boldsymbol{x},t)$ by

$$\boldsymbol{q} = -\kappa \nabla^x \theta,$$

where κ is a scalar constant.

Properties (1) and (2) imply that the Cauchy stress at a point in a compressible Newtonian fluid is entirely determined by the current mass density, temperature and spatial velocity gradient at that point. In particular, it is independent of the past histories and rates of change of these quantities. This type of relation is similar to the constitutive relation for an incompressible Newtonian fluid (see Chapter 6), but with thermal and compressibility effects included. Property (1) also implies that the stress field is necessarily symmetric. Thus the balance equation for angular momentum (Result 5.8) is automatically satisfied and will not be considered further. Because $\rho(\boldsymbol{x},t) > 0$ and $\theta(\boldsymbol{x},t) > 0$ for any admissible thermo-mechanical process, we only consider $\widehat{p}(\rho,\theta)$ for arguments $\rho > 0$ and $\theta > 0$.

Property (3) implies that, just as for pressure, the internal energy and entropy per unit mass at a point are entirely determined by the current mass density and temperature at that point. The three relations

$$p = \widehat{p}(\rho,\theta), \quad \phi = \widehat{\phi}(\rho,\theta), \quad \eta = \widehat{\eta}(\rho,\theta), \tag{8.17}$$

are typically referred to as the thermal equations of state as in the case of a perfect gas. Just as for $\widehat{p}(\rho,\theta)$, we only consider $\widehat{\phi}(\rho,\theta)$ and $\widehat{\eta}(\rho,\theta)$ for arguments $\rho > 0$ and $\theta > 0$.

Property (4) implies that, for each fixed value of ρ, the functions $\widehat{\phi}(\rho,\theta)$ and $\widehat{\eta}(\rho,\theta)$ are strictly increasing in θ. As in the case of a perfect gas, these two conditions imply that either ϕ or η can replace θ as an independent variable. In particular, the relation $\eta = \widehat{\eta}(\rho,\theta)$ in $(8.17)_3$ can be inverted as $\theta = \overline{\theta}(\rho,\eta)$ which leads, as before, to the caloric equations of state

$$p = \overline{p}(\rho,\eta), \quad \phi = \overline{\phi}(\rho,\eta), \quad \theta = \overline{\theta}(\rho,\eta). \tag{8.18}$$

Whereas this form of the constitutive equations was preferred in the case of a perfect gas, here we prefer the form in (8.17) since the heat flux vector depends explicitly on the temperature field. In particular, we will view temperature, rather than entropy, as a primary variable.

Property (5) states that the Fourier–Stokes heat flux vector at any point is proportional to the spatial temperature gradient at that point. This relation may be viewed as an extension of **Fourier's Law** of heat conduction. In particular, whereas Fourier's Law postulates such a relation in a static body, here we assume it also holds in a flowing and deforming fluid.

Remarks:

(1) The constants μ and λ are typically called the **shear** and **bulk viscosities**, and the constant κ the **thermal conductivity** of the fluid. Arguments based on the Clausius–Duhem inequality (see Result 8.4) imply that $\mu \geq 0$, $\lambda \geq -\frac{2}{3}\mu$ and $\kappa \geq 0$.

(2) The quantities μ, λ and κ are assumed constant for simplicity. More generally, they could be viewed as variables, depending on temperature and density for example. Indeed, such dependence is important in many applications.

(3) The compressible Newtonian fluid model can be viewed as a generalization of the perfect gas model to include viscous and heat conduction effects. For this reason, the equations of state for both models are typically assumed to be identical. Notice that a compressible Newtonian fluid model reduces to a perfect gas model when $\mu = 0$, $\lambda = 0$ and $\kappa = 0$.

(4) Just as with a perfect gas (see Exercise 2), arguments based on the axiom of material frame-indifference show that the response functions $(\widehat{p}, \widehat{\phi}, \widehat{\eta})$, equivalently $(\overline{p}, \overline{\phi}, \overline{\theta})$, cannot depend explicitly on the spatial point $\boldsymbol{x}$. Thus the assumption of homogeneity is necessary.

(5) The thermodynamic pressure p is generally distinct from the mechanical pressure or mean normal stress $p_{\mathrm{m}} = -\frac{1}{3}\operatorname{tr}(\boldsymbol{S})$ (see Section 3.5.4). In particular, using the fact that $\boldsymbol{L} = \operatorname{sym}(\nabla^x \boldsymbol{v})$, for a compressible Newtonian fluid we find

$$\begin{aligned} p_{\mathrm{m}} = -\frac{1}{3}\operatorname{tr}[\boldsymbol{S}] &= -\frac{1}{3}\operatorname{tr}[-p\boldsymbol{I} + \lambda(\nabla^x \cdot \boldsymbol{v})\boldsymbol{I} + 2\mu\boldsymbol{L}] \\ &= -\frac{1}{3}[-3p + 3\lambda(\nabla^x \cdot \boldsymbol{v}) + 2\mu\nabla^x \cdot \boldsymbol{v}] \\ &= p - \left(\lambda + \frac{2}{3}\mu\right)\nabla^x \cdot \boldsymbol{v}. \end{aligned}$$

This shows that $p_{\mathrm{m}} = p$ when $\nabla^x \cdot \boldsymbol{v} = 0$ or $\lambda + \frac{2}{3}\mu = 0$. Notice that equality between p_{m} and p is achieved, for any values of λ and μ, when the fluid is at rest or undergoing isochoric motion. Alternatively, equality is achieved, for any possible motion, when $\lambda = -\frac{2}{3}\mu$. In the particular case of a perfect gas we find that p_{m} is always equal to p. □

8.2.2 Compressible Navier–Stokes Equations

A closed system of equations for the variables $(\rho, \boldsymbol{v}, \theta)$ in a body of compressible Newtonian fluid are provided by the thermal equations of state (8.17), the conservation of mass equation (Result 5.5), the balance of linear momentum equation (Result 5.7) and the balance of energy equation (Result 5.10).

The balance of linear momentum and energy equations are

$$\rho\dot{\boldsymbol{v}} = \nabla^x \cdot \boldsymbol{S} + \rho\boldsymbol{b} \quad \text{and} \quad \rho\dot{\phi} = \boldsymbol{S} : \boldsymbol{L} - \nabla^x \cdot \boldsymbol{q} + \rho r,$$

where $\boldsymbol{b}$ is a prescribed body force per unit mass and r is a prescribed heat supply per unit mass. From the definition of $\boldsymbol{S}$ and the fact that $\boldsymbol{L} = \mathrm{sym}(\nabla^x \boldsymbol{v})$, we obtain, assuming λ and μ are constant

$$\nabla^x \cdot \boldsymbol{S} = -\nabla^x p + \lambda \nabla^x (\nabla^x \cdot \boldsymbol{v}) + \mu \nabla^x \cdot (\nabla^x \boldsymbol{v}) + \mu \nabla^x \cdot (\nabla^x \boldsymbol{v})^T.$$

Using the definition of the divergence of a second-order tensor we find $\nabla^x \cdot (\nabla^x \boldsymbol{v}) = \Delta^x \boldsymbol{v}$ and $\nabla^x \cdot (\nabla^x \boldsymbol{v})^T = \nabla^x (\nabla^x \cdot \boldsymbol{v})$. These results imply

$$\nabla^x \cdot \boldsymbol{S} = -\nabla^x p + (\lambda + \mu) \nabla^x (\nabla^x \cdot \boldsymbol{v}) + \mu \Delta^x \boldsymbol{v}.$$

Using the definition of $\boldsymbol{S}$ and the fact that $\boldsymbol{I} : \boldsymbol{L} = \nabla^x \cdot \boldsymbol{v}$ we find

$$\boldsymbol{S} : \boldsymbol{L} = -p \nabla^x \cdot \boldsymbol{v} + \lambda (\nabla^x \cdot \boldsymbol{v})^2 + 2\mu \boldsymbol{L} : \boldsymbol{L} = -p \nabla^x \cdot \boldsymbol{v} + \Phi,$$

where for convenience we introduce the quantity

$$\Phi = \lambda (\nabla^x \cdot \boldsymbol{v})^2 + 2\mu \boldsymbol{L} : \boldsymbol{L}. \tag{8.19}$$

Moreover, using the definition of $\boldsymbol{q}$ and the fact that κ is constant, we have

$$\nabla^x \cdot \boldsymbol{q} = -\kappa \Delta^x \theta.$$

Thus, setting $p = \widehat{p}(\rho, \theta)$ and $\phi = \widehat{\phi}(\rho, \theta)$, and using Result 4.7 for the total or material time derivatives, we find that the spatial density, velocity and temperature fields in a body of compressible Newtonian fluid with reference configuration B must satisfy the following equations for all $\boldsymbol{x} \in B_t$ and $t \geq 0$

$$\boxed{\begin{aligned} &\frac{\partial}{\partial t} \rho + \nabla^x \cdot (\rho \boldsymbol{v}) = 0, \\ &\rho \left[\frac{\partial}{\partial t} \boldsymbol{v} + (\nabla^x \boldsymbol{v}) \boldsymbol{v} \right] = \mu \Delta^x \boldsymbol{v} + (\lambda + \mu) \nabla^x (\nabla^x \cdot \boldsymbol{v}) - \nabla^x \widehat{p}(\rho, \theta) + \rho \boldsymbol{b}, \\ &\rho \left[\frac{\partial}{\partial t} \widehat{\phi}(\rho, \theta) + \nabla^x \widehat{\phi}(\rho, \theta) \cdot \boldsymbol{v} \right] = -\widehat{p}(\rho, \theta) \nabla^x \cdot \boldsymbol{v} + \kappa \Delta^x \theta + \Phi + \rho r. \end{aligned}} \tag{8.20}$$

These equations are typically called the **Compressible Navier–Stokes Equations**. By an admissible process for a compressible Newtonian fluid we mean fields $(\rho, \boldsymbol{v}, \theta)$ satisfying these equations.

Remarks:

(1) The equations in (8.20) are a system of coupled, nonlinear partial differential equations for $(\rho, \boldsymbol{v}, \theta)$. Alternatively, use of the caloric equations of state (8.18) in place of (8.17) leads to a system of the

same form, but with $(\rho, \boldsymbol{v}, \eta)$ as the variables. The two formulations are equivalent provided all fields involved are sufficiently smooth.

(2) In contrast to solutions of the gas dynamics equations considered in the previous section, solutions of (8.20) do not generally develop shock waves (jump discontinuities). However, for appropriately small values of μ, λ and κ, solutions may exhibit rapid transitions across narrow regions called **shock layers**. Shock waves in the gas dynamics equations can be viewed as idealizations of these layers.

(3) An important quantity arising in the study of flows described by equations (8.20) is the **Reynolds number** $Re = \rho_* \vartheta_* \ell_* / \mu$ (a dimensionless constant), where ρ_* is a characteristic density, ϑ_* is a characteristic speed and ℓ_* is a characteristic length associated with the flow. Flows with a low Reynolds number are typically smooth or **laminar**, while flows with a high Reynolds number are typically fluctuating or **turbulent**. □

8.2.3 Frame-Indifference Considerations

Consider a body of compressible Newtonian fluid with reference configuration B. Let $\boldsymbol{x} = \boldsymbol{\varphi}(\boldsymbol{X}, t)$ be an arbitrary motion and let $\boldsymbol{x}^* = \boldsymbol{\varphi}^*(\boldsymbol{X}, t)$ be a second motion defined by

$$\boldsymbol{x}^* = \boldsymbol{g}(\boldsymbol{x}, t) = \boldsymbol{Q}(t)\boldsymbol{x} + \mathbf{c}(t), \tag{8.21}$$

where $\boldsymbol{Q}(t)$ is an arbitrary rotation tensor and $\mathbf{c}(t)$ is an arbitrary vector. Let ρ, θ, ϕ, η, $\boldsymbol{q}$ and $\boldsymbol{S}$ denote the spatial density, temperature, internal energy, entropy, heat flux and Cauchy stress fields in B_t, and let ρ^*, θ^*, ϕ^*, η^*, $\boldsymbol{q}^*$ and $\boldsymbol{S}^*$ denote the corresponding spatial fields in B_t^*. Then the axiom of material frame-indifference (Axiom 5.24) postulates the following for all $\boldsymbol{x} \in B_t$ and $t \geq 0$

$$\begin{aligned}
\rho^*(\boldsymbol{x}^*, t) &= \rho(\boldsymbol{x}, t), \\
\theta^*(\boldsymbol{x}^*, t) &= \theta(\boldsymbol{x}, t), \\
\phi^*(\boldsymbol{x}^*, t) &= \phi(\boldsymbol{x}, t), \\
\eta^*(\boldsymbol{x}^*, t) &= \eta(\boldsymbol{x}, t), \\
\boldsymbol{q}^*(\boldsymbol{x}^*, t) &= \boldsymbol{Q}(t)\boldsymbol{q}(\boldsymbol{x}, t), \\
\boldsymbol{S}^*(\boldsymbol{x}^*, t) &= \boldsymbol{Q}(t)\boldsymbol{S}(\boldsymbol{x}, t)\boldsymbol{Q}(t)^T.
\end{aligned} \tag{8.22}$$

Substituting the constitutive equations for the Cauchy stress and heat flux into (8.22), along with the thermal equations of state (8.17) for the pressure, internal energy and entropy, we obtain (omitting the arguments $\boldsymbol{x}$, $\boldsymbol{x}^*$ and t for brevity)

$$
\begin{aligned}
\rho^* &= \rho, \\
\theta^* &= \theta, \\
\widehat{\phi}(\rho^*, \theta^*) &= \widehat{\phi}(\rho, \theta), \\
\widehat{\eta}(\rho^*, \theta^*) &= \widehat{\eta}(\rho, \theta), \\
-\kappa \nabla^{x^*} \theta^* &= \boldsymbol{Q}[-\kappa \nabla^{x} \theta], \\
-\widehat{p}(\rho^*, \theta^*)\boldsymbol{I} + \lambda(\nabla^{x^*} \cdot \boldsymbol{v}^*)\boldsymbol{I} + 2\mu \boldsymbol{L}^* & \\
&= \boldsymbol{Q}[-\widehat{p}(\rho, \theta)\boldsymbol{I} + \lambda(\nabla^{x} \cdot \boldsymbol{v})\boldsymbol{I} + 2\mu \boldsymbol{L}]\boldsymbol{Q}^T.
\end{aligned}
\tag{8.23}
$$

Eliminating ρ^* and θ^* using $(8.23)_{1,2}$, we observe that the relations in $(8.23)_{3,4}$ are satisfied identically for any functions $\widehat{\phi}$ and $\widehat{\eta}$. Moreover, using the relation $\theta^*(\boldsymbol{x}^*, t) = \theta(\boldsymbol{x}, t)$, together with (8.21) and the chain rule, we find

$$\nabla^{x^*} \theta^* = \boldsymbol{Q} \nabla^{x} \theta.$$

This shows that $(8.23)_5$ is satisfied for any rotation $\boldsymbol{Q}$. Finally, from Result 5.22 we have

$$\nabla^{x^*} \boldsymbol{v}^* = \boldsymbol{Q} \nabla^{x} \boldsymbol{v} \boldsymbol{Q}^T + \dot{\boldsymbol{Q}} \boldsymbol{Q}^T \quad \text{and} \quad \boldsymbol{L}^* = \boldsymbol{Q} \boldsymbol{L} \boldsymbol{Q}^T. \tag{8.24}$$

Differentiating the relation $\boldsymbol{Q}\boldsymbol{Q}^T = \boldsymbol{I}$ with respect to time we find

$$\dot{\boldsymbol{Q}} \boldsymbol{Q}^T = -(\dot{\boldsymbol{Q}} \boldsymbol{Q}^T)^T,$$

which implies $\mathrm{tr}(\dot{\boldsymbol{Q}} \boldsymbol{Q}^T) = 0$. From $(8.24)_1$ and the invariance property of the trace function, namely $\mathrm{tr}(\boldsymbol{Q} \boldsymbol{A} \boldsymbol{Q}^T) = \mathrm{tr}(\boldsymbol{A})$ for all second-order tensors $\boldsymbol{A}$ and rotations $\boldsymbol{Q}$, we get

$$\nabla^{x^*} \cdot \boldsymbol{v}^* = \mathrm{tr}(\nabla^{x^*} \boldsymbol{v}^*) = \mathrm{tr}(\boldsymbol{Q} \nabla^{x} \boldsymbol{v} \boldsymbol{Q}^T) = \mathrm{tr}(\nabla^{x} \boldsymbol{v}) = \nabla^{x} \cdot \boldsymbol{v}. \tag{8.25}$$

Using (8.25) and $(8.24)_2$ we find that $(8.23)_6$ is satisfied for any rotation $\boldsymbol{Q}$. Thus the constitutive model of a compressible Newtonian fluid complies with the axiom of material frame-indifference with no restrictions on the response functions $\widehat{p}(\rho, \theta)$, $\widehat{\phi}(\rho, \theta)$ and $\widehat{\eta}(\rho, \theta)$.

Remarks:

(1) A similar conclusion is obtained if we use the caloric equations of state (8.18) in place of (8.17). In particular, the axiom of material

frame-indifference is satisfied with no restrictions on the response functions $\overline{p}(\rho,\eta)$, $\overline{\phi}(\rho,\eta)$ and $\overline{\theta}(\rho,\eta)$.

(2) Different conclusions can be obtained if the form of the model is changed. For example, if we consider a more general heat flux model of the form $\boldsymbol{q} = -\boldsymbol{K}(\boldsymbol{x},\theta)\nabla^x\theta$, where $\boldsymbol{K}(\boldsymbol{x},\theta)$ is a symmetric second-order tensor, then the axiom of material frame-indifference is satisfied if and only if $\boldsymbol{K}(\boldsymbol{x},\theta) = \kappa(\theta)\boldsymbol{I}$. That is, $\boldsymbol{K}(\boldsymbol{x},\theta)$ must be isotropic and independent of $\boldsymbol{x}$ (see Exercise 5).

□

8.2.4 Thermodynamical Considerations

Here we outline a result which establishes conditions under which a compressible Newtonian fluid model satisfies the Clausius–Duhem inequality.

Result 8.4 ***Implications of Clausius–Duhem Inequality.*** *Suppose the response functions $\widehat{p}$, $\widehat{\phi}$ and $\widehat{\eta}$ satisfy the conditions in* (8.14). *Then the Clausius–Duhem inequality (Result 5.11) is satisfied by every admissible process in a compressible Newtonian fluid if and only if the material constants μ, λ and κ satisfy*

$$\mu \geq 0, \quad \lambda + \frac{2}{3}\mu \geq 0, \quad \kappa \geq 0. \tag{8.26}$$

□

Proof The Clausius–Duhem inequality in Eulerian form is

$$\rho\dot{\eta} \geq \theta^{-1}\rho r - \nabla^x \cdot (\theta^{-1}\boldsymbol{q}).$$

After expanding the divergence term, multiplying through by $\theta > 0$, and using the balance of energy equation, we obtain

$$\rho\theta\dot{\eta} - \rho\dot{\phi} + \boldsymbol{S} : \boldsymbol{L} - \theta^{-1}\boldsymbol{q}\cdot\nabla^x\theta \geq 0. \tag{8.27}$$

Using the equations of state $\eta = \widehat{\eta}(\rho,\theta)$ and $\phi = \widehat{\phi}(\rho,\theta)$, together with Result 4.7 and the chain rule, we get

$$\dot{\eta} = \frac{\partial\widehat{\eta}}{\partial\rho}\dot{\rho} + \frac{\partial\widehat{\eta}}{\partial\theta}\dot{\theta}, \qquad \dot{\phi} = \frac{\partial\widehat{\phi}}{\partial\rho}\dot{\rho} + \frac{\partial\widehat{\phi}}{\partial\theta}\dot{\theta}.$$

Substitution of these expressions into (8.27) and rearranging terms gives (omitting arguments ρ and θ for brevity)

$$\rho\left[\theta\frac{\partial\widehat{\eta}}{\partial\rho}-\frac{\partial\widehat{\phi}}{\partial\rho}\right]\dot{\rho}+\rho\left[\theta\frac{\partial\widehat{\eta}}{\partial\theta}-\frac{\partial\widehat{\phi}}{\partial\theta}\right]\dot{\theta}+\boldsymbol{S}:\boldsymbol{L}-\theta^{-1}\boldsymbol{q}\cdot\nabla^{x}\theta\geq 0. \quad (8.28)$$

From the conservation of mass equation we have

$$\dot{\rho}=-\rho\nabla^{x}\cdot\boldsymbol{v}, \quad (8.29)$$

and from the definition of a compressible Newtonian fluid we get

$$\begin{aligned}\boldsymbol{S}:\boldsymbol{L}&=-p\nabla^{x}\cdot\boldsymbol{v}+\lambda(\nabla^{x}\cdot\boldsymbol{v})^2+2\mu\boldsymbol{L}:\boldsymbol{L},\\ \boldsymbol{q}\cdot\nabla^{x}\theta&=-\kappa|\nabla^{x}\theta|^2.\end{aligned} \quad (8.30)$$

Substituting (8.30) and (8.29) into (8.28), and setting $p=\widehat{p}(\rho,\theta)$, we obtain (omitting arguments ρ and θ for brevity)

$$\begin{aligned}\rho\left[\theta\frac{\partial\widehat{\eta}}{\partial\theta}-\frac{\partial\widehat{\phi}}{\partial\theta}\right]\dot{\theta}-\rho^2\left[\theta\frac{\partial\widehat{\eta}}{\partial\rho}-\frac{\partial\widehat{\phi}}{\partial\rho}+\frac{1}{\rho^2}\widehat{p}\right]\nabla^{x}\cdot\boldsymbol{v}\\ +\lambda(\nabla^{x}\cdot\boldsymbol{v})^2+2\mu\boldsymbol{L}:\boldsymbol{L}+\theta^{-1}\kappa|\nabla^{x}\theta|^2\geq 0.\end{aligned}$$

Assuming $\widehat{p}$, $\widehat{\phi}$ and $\widehat{\eta}$ satisfy the conditions in (8.14), we find that the coefficients of $\dot{\theta}$ and $\nabla^{x}\cdot\boldsymbol{v}$ vanish. Thus the Clausius–Duhem inequality takes the reduced form

$$\lambda(\nabla^{x}\cdot\boldsymbol{v})^2+2\mu\boldsymbol{L}:\boldsymbol{L}+\theta^{-1}\kappa|\nabla^{x}\theta|^2\geq 0. \quad (8.31)$$

To establish the conditions in (8.26) we introduce the deviatoric rate of strain tensor $\boldsymbol{L}_{\text{dev}}=\boldsymbol{L}-\frac{1}{3}(\nabla^{x}\cdot\boldsymbol{v})\boldsymbol{I}$. Then, using the facts that

$$\boldsymbol{L}=\boldsymbol{L}_{\text{dev}}+\frac{1}{3}(\nabla^{x}\cdot\boldsymbol{v})\boldsymbol{I}\quad\text{and}\quad\boldsymbol{I}:\boldsymbol{L}_{\text{dev}}=\text{tr}(\boldsymbol{L}_{\text{dev}})=0,$$

the Clausius–Duhem inequality (8.31) becomes

$$\left(\lambda+\frac{2}{3}\mu\right)(\nabla^{x}\cdot\boldsymbol{v})^2+2\mu\boldsymbol{L}_{\text{dev}}:\boldsymbol{L}_{\text{dev}}+\theta^{-1}\kappa|\nabla^{x}\theta|^2\geq 0. \quad (8.32)$$

It can now be shown that (8.26) are necessary and sufficient conditions for (8.32) to hold for all admissible processes. Sufficiency of (8.26) is clear, for then all terms in (8.32) are non-negative. To establish necessity, suppose (8.32) holds for all processes. Then by choosing $\boldsymbol{v}$ and θ such that $\nabla^{x}\cdot\boldsymbol{v}\neq 0$, $\boldsymbol{L}_{\text{dev}}=\boldsymbol{O}$ and $\nabla^{x}\theta=\boldsymbol{0}$ we get $\lambda+\frac{2}{3}\mu\geq 0$. By choosing $\boldsymbol{v}$ and θ such that $\nabla^{x}\cdot\boldsymbol{v}=0$, $\boldsymbol{L}_{\text{dev}}\neq\boldsymbol{O}$ and $\nabla^{x}\theta=\boldsymbol{0}$ we get $\mu\geq 0$. Finally, by choosing $\boldsymbol{v}$ and θ such that $\nabla^{x}\cdot\boldsymbol{v}=0$, $\boldsymbol{L}_{\text{dev}}=\boldsymbol{O}$ and $\nabla^{x}\theta\neq\boldsymbol{0}$ we get $\kappa\geq 0$. Thus the conditions in (8.26) are necessary. □

Remarks:

(1) In the above arguments we implicitly assume that the spatial fields $\boldsymbol{v}$ and θ can be assigned arbitrary values. In view of the balance of momentum and energy equations, arbitrary values for $\boldsymbol{v}$ and θ can be achieved by appropriate choice of body force $\boldsymbol{b}$ and heat supply r.

(2) As mentioned earlier, a compressible Newtonian fluid model can be interpreted as an extension of a perfect gas model to include viscous and heat conduction effects. Result 8.4 shows that, if a gas model satisfies the Clausius–Duhem inequality as encapsulated in conditions (8.14), then its extension to a Newtonian model will satisfy the Clausius–Duhem inequality if and only if the viscosity and thermal conductivity constants satisfy (8.26).

(3) The proof of Result 8.4 shows that a compressible Newtonian fluid can experience irreversible processes. Indeed, most processes will be irreversible in the case when $\mu > 0$, $\lambda + \frac{2}{3}\mu > 0$ and $\kappa > 0$ since equality will seldom be achieved in (8.32). In contrast, all processes are reversible when $\mu = 0$, $\lambda = 0$ and $\kappa = 0$, which corresponds to the case of a perfect gas.

□

8.2.5 Initial-Boundary Value Problems

An initial-boundary value problem for a body of compressible Newtonian fluid is a set of equations that describe the motion of the body subject to specified initial conditions in B at time $t = 0$, and boundary conditions on ∂B_t for $t \geq 0$. The Eulerian form of the balance laws are particularly well-suited for those problems in which the body occupies a fixed region D of space. In this case, the spatial density, velocity and temperature fields ρ, $\boldsymbol{v}$ and θ can be determined independently of the motion φ. We typically assume D is a bounded open set as shown in Figure 6.1. However, it is also useful in applications to consider unbounded open sets such as the exterior of the region shown in the figure, or the whole of Euclidean space.

A standard initial-boundary value problem for a body of compressible Newtonian fluid occupying a fixed region D can be stated as follows:

Find $\rho, \theta : D \times [0,T] \to I\!R$ and $\boldsymbol{v} : D \times [0,T] \to \mathcal{V}$ such that

$$
\boxed{
\begin{array}{ll}
\frac{\partial}{\partial t}\rho + \nabla^x \cdot (\rho \boldsymbol{v}) = 0 & \text{in} \quad D \times [0,T] \\
\rho \left[\frac{\partial}{\partial t}\boldsymbol{v} + (\nabla^x \boldsymbol{v})\boldsymbol{v} \right] & \\
\qquad = \mu \Delta^x \boldsymbol{v} + (\lambda + \mu)\nabla^x (\nabla^x \cdot \boldsymbol{v}) - \nabla^x p + \rho \boldsymbol{b} & \text{in} \quad D \times [0,T] \\
\rho \left[\frac{\partial}{\partial t}\phi + \nabla^x \phi \cdot \boldsymbol{v} \right] = -p\nabla^x \cdot \boldsymbol{v} - \nabla^x \cdot \boldsymbol{q} + \Phi + \rho r & \text{in} \quad D \times [0,T] \\
\boldsymbol{v} = \boldsymbol{0} & \text{in} \ \partial D \times [0,T] \\
\theta = \xi & \text{in} \quad \Gamma_\theta \times [0,T] \\
\boldsymbol{q} \cdot \boldsymbol{n} = \zeta & \text{in} \quad \Gamma_q \times [0,T] \\
\rho(\cdot, 0) = \rho_0(\cdot) & \text{in} \quad D \\
\boldsymbol{v}(\cdot, 0) = \boldsymbol{v}_0(\cdot) & \text{in} \quad D \\
\theta(\cdot, 0) = \theta_0(\cdot) & \text{in} \quad D.
\end{array}
}
\tag{8.33}
$$

In the above system, μ and λ are the constant viscosities of the fluid, Γ_θ and Γ_q are subsets of ∂D with the properties $\Gamma_\theta \cup \Gamma_q = \partial D$ and $\Gamma_\theta \cap \Gamma_q = \emptyset$, $\boldsymbol{b}$ is a prescribed body force per unit mass, r is a prescribed heat supply per unit mass, $\boldsymbol{n}$ is the unit outward normal field on ∂D, and ξ, ζ, ρ_0, $\boldsymbol{v}_0$ and θ_0 are prescribed fields. The pressure p, internal energy per unit mass ϕ and heat flux $\boldsymbol{q}$ are related to ρ and θ through constitutive relations of the form

$$
p = \widehat{p}(\rho, \theta), \quad \phi = \widehat{\phi}(\rho, \theta), \quad \boldsymbol{q} = -\kappa \nabla^x \theta,
$$

where $\widehat{p}$ and $\widehat{\phi}$ are given response functions and κ is the constant thermal conductivity of the fluid. The quantity Φ is the thermal dissipation function defined in (8.19).

Equation $(8.33)_1$ is the conservation of mass equation, $(8.33)_2$ is the balance of linear momentum equation, $(8.33)_3$ is the balance of energy equation, $(8.33)_4$ is a no-slip boundary condition on ∂D, $(8.33)_5$ is a temperature boundary condition on Γ_θ, $(8.33)_6$ is a heat flux boundary condition on Γ_q, $(8.33)_7$ is an initial condition for the density ρ, $(8.33)_8$ is an initial condition for the velocity $\boldsymbol{v}$ and $(8.33)_9$ is an initial condition for the temperature θ. In general, the initial conditions should be compatible with the boundary conditions at time $t = 0$.

Remarks:

(1) The system in (8.33) is a nonlinear initial-boundary value problem for the fields $(\rho, \boldsymbol{v}, \theta)$. We expect this system to have a solution on some finite time interval $[0, T]$ under mild assumptions on the domain D, response functions $\widehat{p}$ and $\widehat{\phi}$, and the prescribed data of the problem. However, general questions of existence and uniqueness for all time are difficult. Numerical approximation is generally required to obtain quantitative information about solutions.

(2) The boundary condition $\boldsymbol{v} = \boldsymbol{0}$ on ∂D is appropriate for describing the interface between a fixed, impermeable solid and a compressible Newtonian fluid at moderate density. If the solid were not fixed, but instead had a prescribed velocity field $\boldsymbol{\vartheta}$, then an appropriate boundary condition would be $\boldsymbol{v} = \boldsymbol{\vartheta}$ on ∂D. Other boundary conditions can also be considered when the fluid density is expected to be low (in which case the fluid may slip), the solid is permeable, the motion of the boundary is not known a-priori, or when the interface is not fluid-solid. Different conditions on temperature could also be imposed, for example to allow for convective or radiative heat transfer at the boundary.

□

Bibliographic Notes

A general account of the basic principles of compressible as well as incompressible fluid flow with viscous and thermal effects is the classic article by Serrin (1959). An account of the theory from a microscopic point of view (kinetic theory) is given in Chapman and Cowling (1970). Various engineering applications may be found in Anderson (2003) and Saad (1993) for the case of compressible inviscid flow (perfect gases), and in White (2006) for the case of compressible viscous flow (compressible Navier–Stokes).

The equations for compressible inviscid flow have typically been studied in abstract form as a system of hyperbolic conservation laws. A large part of the mathematical theory of such systems is focused on the analysis of shock waves. Classic references for the theory of hyperbolic conservation laws are Courant and Friedrichs (1948) and Lax (1973), and more modern treatments are given in Majda (1984) and Dafermos

(2005). Various aspects of the numerical analysis of hyperbolic conservation laws are given in LeVeque (1992) and Laney (1998).

A large part of the theory of compressible viscous flow is focused on the analysis of boundary layers, stability and turbulence. An analysis of these topics from an engineering point of view is given in White (2006). In contrast to the inviscid case, the mathematical theory of compressible viscous flow is less developed. Various aspects of the numerical treatment of compressible viscous flow are discussed in Hirsch (1988).

Exercises

8.1 Consider a body of perfect gas occupying a fixed region D so that $B_t = D$ for all $t \geq 0$. Assume there is no body force or heat supply, and define the total energy of the gas by

$$E(t) = \int_{B_t} \tfrac{1}{2}\rho|\boldsymbol{v}|^2 + \rho\phi \, dV_{\boldsymbol{x}}.$$

Show that this total energy is conserved in any smooth motion satisfying $\boldsymbol{v} \cdot \boldsymbol{n} = 0$ on ∂B_t, that is

$$\frac{d}{dt}E(t) = 0.$$

8.2 Consider a perfect gas with caloric response functions $\overline{\phi}(\rho, \eta, \boldsymbol{x})$, $\overline{\theta}(\rho, \eta, \boldsymbol{x})$ and $\overline{p}(\rho, \eta, \boldsymbol{x})$. Show that the axiom of material frame-indifference is satisfied if and only if these functions are independent of $\boldsymbol{x}$.

8.3 Consider a perfect gas with thermal state equations $p = \widehat{p}(\rho, \theta)$, $\phi = \widehat{\phi}(\rho, \theta)$ and $\eta = \widehat{\eta}(\rho, \theta)$. Show that the Clausius–Duhem inequality holds if and only if $\widehat{p}$, $\widehat{\phi}$ and $\widehat{\eta}$ satisfy

$$\frac{\partial \widehat{\eta}}{\partial \theta} = \frac{1}{\theta}\frac{\partial \widehat{\phi}}{\partial \theta}, \qquad \widehat{p} = \rho^2 \frac{\partial \widehat{\phi}}{\partial \rho} - \rho^2 \theta \frac{\partial \widehat{\eta}}{\partial \rho}.$$

8.4 Consider a body of perfect gas in smooth, steady, isentropic motion with body force $\boldsymbol{b} = \boldsymbol{0}$, heat supply $r = 0$, density $\rho(\boldsymbol{x})$, velocity $\boldsymbol{v}(\boldsymbol{x})$, and entropy η_0 (constant). Let $\boldsymbol{y}(s)$, $s \in \mathbb{R}$, denote an arbitrary streamline defined by the equation $\boldsymbol{y}(s)' = \boldsymbol{v}(\boldsymbol{y}(s))$, and let $\vartheta(s) = |\boldsymbol{v}(\boldsymbol{y}(s))|$ denote the local flow speed and $m(s) = \rho(\boldsymbol{y}(s))\vartheta(s)$ the local mass flux along the streamline.

(a) For any streamline show

$$\frac{d\rho}{ds} = \nabla^x \rho \cdot \boldsymbol{v}, \qquad \vartheta \frac{d\vartheta}{ds} = \boldsymbol{v} \cdot (\nabla^x \boldsymbol{v})\boldsymbol{v}.$$

(b) Use balance of linear momentum to show

$$\frac{d\rho}{ds} = -\frac{\rho\vartheta}{c^2}\frac{d\vartheta}{ds}, \qquad \frac{dm}{ds} = \rho[1 - M^2]\frac{d\vartheta}{ds},$$

where $c(s)$ is the local sound speed and $M(s)$ is the local Mach number along the streamline.

Remark: The first result in (b) shows that mass density always decreases with increasing flow speed. The second result shows that mass flux may increase or decrease depending on the local Mach number. In subsonic regions of a flow ($M < 1$), mass flux increases with increasing flow speed. However, in supersonic regions of flow ($M > 1$), mass flux decreases with increasing flow speed. Many phenomena in the dynamics of perfect gases can be attributed to this change in behavior of the mass flux between subsonic and supersonic speeds.

8.5 Consider a compressible Newtonian fluid with the more general heat flux model

$$\boldsymbol{q} = -\boldsymbol{K}(\boldsymbol{x}, \theta)\nabla^x \theta,$$

where $\boldsymbol{K}(\boldsymbol{x}, \theta)$ is a symmetric, second-order tensor function. Show that the axiom of material frame-indifference is satisfied if and only if $\boldsymbol{K}(\boldsymbol{x}, \theta)$ is isotropic and independent of $\boldsymbol{x}$, that is

$$\boldsymbol{K}(\boldsymbol{x}, \theta) = \kappa(\theta)\boldsymbol{I},$$

for some scalar function $\kappa(\theta)$.

8.6 Consider a body of compressible Newtonian fluid with viscosities μ and λ and conductivity κ occupying a fixed region D so that $B_t = D$ for all $t \geq 0$. Assume there is no body force or heat supply, and define the total energy of the fluid by

$$E(t) = \int_{B_t} \tfrac{1}{2}\rho|\boldsymbol{v}|^2 + \rho\phi \, dV_{\boldsymbol{x}}.$$

Show that this total energy is conserved in any smooth motion satisfying $\boldsymbol{v} = \boldsymbol{0}$ and $\boldsymbol{q} \cdot \boldsymbol{n} = 0$ on ∂B_t, that is

$$\frac{d}{dt}E(t) = 0.$$

Does this result also hold in the case of non-constant viscosities and conductivity?

8.7 Let Γ be a surface in $\mathbb{E}^3$ with closed, simple, piecewise smooth boundary curve $C = \partial\Gamma$. Let $\boldsymbol{n}$ be a unit normal field on Γ and $\boldsymbol{\nu}$ a unit tangent field along C, compatibly oriented according to Stokes' Theorem. For any smooth vector field $\boldsymbol{v}$ on $\mathbb{E}^3$ show

$$\int_C \boldsymbol{v} \times \boldsymbol{\nu}\, ds = \int_\Gamma \boldsymbol{n}(\nabla^x \cdot \boldsymbol{v}) + \tfrac{1}{2}\boldsymbol{w} \times \boldsymbol{n} - \boldsymbol{L}\boldsymbol{n}\, dA_{\boldsymbol{x}},$$

where $\boldsymbol{w} = \nabla^x \times \boldsymbol{v}$ is the vorticity field and $\boldsymbol{L} = \mathrm{sym}(\nabla^x \boldsymbol{v})$ is the rate of strain field associated with $\boldsymbol{v}$.

8.8 Consider a body of compressible Newtonian fluid with viscosities μ and λ, pressure field p, and velocity field $\boldsymbol{v}$ occupying a fixed region D so that $B_t = D$ for all $t \geq 0$.

(a) Assuming $\boldsymbol{v} = \boldsymbol{0}$ for all $\boldsymbol{x} \in \partial B_t$ show

$$\left.\begin{aligned} &\boldsymbol{w} \cdot \boldsymbol{n} = 0 \\ &\boldsymbol{n}(\nabla^x \cdot \boldsymbol{v}) + \tfrac{1}{2}\boldsymbol{w} \times \boldsymbol{n} - \boldsymbol{L}\boldsymbol{n} = \boldsymbol{0} \end{aligned}\right\} \quad \forall \boldsymbol{x} \in \partial B_t,$$

where $\boldsymbol{w} = \nabla^x \times \boldsymbol{v}$ is the vorticity field and $\boldsymbol{L} = \mathrm{sym}(\nabla^x \boldsymbol{v})$ is the rate of strain field associated with $\boldsymbol{v}$.

(b) Assuming $\boldsymbol{v} = \boldsymbol{0}$ for all $\boldsymbol{x} \in \partial B_t$ show

$$\boldsymbol{t} = [-p + (\lambda + 2\mu)\nabla^x \cdot \boldsymbol{v}]\boldsymbol{n} + \mu(\boldsymbol{w} \times \boldsymbol{n}), \quad \forall \boldsymbol{x} \in \partial B_t,$$

where $\boldsymbol{t} = \boldsymbol{S}\boldsymbol{n}$ is the traction field on ∂B_t.

Remark: The results in (a) are purely kinematic and imply, among other things, that the local vorticity at a fixed boundary must be tangential to the boundary. The result in (b) is particular to Newtonian fluids. It says that tangential (shearing) stresses at a fixed boundary depend only on the local vorticity, while normal stresses depend only on the local pressure and volume expansion. A similar result holds in the incompressible case.

Answers to Selected Exercises

8.1 From the result on the time derivative of integrals relative to a mass density (see Chapter 5) we have

$$\frac{d}{dt}\int_{B_t} \Phi(\boldsymbol{x},t)\rho(\boldsymbol{x},t)\, dV_{\boldsymbol{x}} = \int_{B_t} \dot{\Phi}(\boldsymbol{x},t)\rho(\boldsymbol{x},t)\, dV_{\boldsymbol{x}},$$

where $\Phi(\boldsymbol{x},t)$ is any spatial field. Applying this to $E(t)$, noting by the chain rule that $\frac{d}{dt}|\boldsymbol{v}|^2 = 2\boldsymbol{v}\cdot\dot{\boldsymbol{v}}$, we have

$$\frac{d}{dt}E(t) = \int_{B_t} \rho\dot{\boldsymbol{v}}\cdot\boldsymbol{v} + \rho\dot{\phi}\, dV_{\boldsymbol{x}}.$$

In terms of material time derivatives, the balance of linear momentum and energy equations for a perfect gas are

$$\rho\dot{\boldsymbol{v}} = -\nabla^x p + \rho\boldsymbol{b}, \qquad \rho\dot{\phi} = -p\nabla^x\cdot\boldsymbol{v} + \rho r.$$

Substituting these into the above integral, using the fact that $\nabla^x\cdot(p\boldsymbol{v}) = \nabla^x p\cdot\boldsymbol{v} + p\nabla^x\cdot\boldsymbol{v}$, and applying the Divergence Theorem, we find

$$\frac{d}{dt}E(t) = \int_{B_t} \rho\boldsymbol{b}\cdot\boldsymbol{v} + \rho r\, dV_{\boldsymbol{x}} - \int_{\partial B_t} p\boldsymbol{v}\cdot\boldsymbol{n}\, dA_{\boldsymbol{x}}.$$

In the case when $\boldsymbol{b} = \boldsymbol{0}$, $r = 0$ and $\boldsymbol{v}\cdot\boldsymbol{n} = 0$ on ∂B_t we get

$$\frac{d}{dt}E(t) = 0.$$

8.3 The Clausius–Duhem inequality in Eulerian form is

$$\rho\dot{\eta} \geq \theta^{-1}\rho r - \nabla^x\cdot(\theta^{-1}\boldsymbol{q}).$$

After multiplying through by $\theta > 0$, setting $\boldsymbol{q} = \boldsymbol{0}$ (by definition of a perfect gas), and using the balance of energy equation we obtain

$$\rho\theta\dot{\eta} \geq \rho\dot{\phi} + p\nabla^x\cdot\boldsymbol{v}. \tag{8.34}$$

Using the equations of state $\eta = \widehat{\eta}(\rho,\theta)$ and $\phi = \widehat{\phi}(\rho,\theta)$, together with the chain rule, we deduce

$$\dot{\eta} = \frac{\partial\widehat{\eta}}{\partial\rho}\dot{\rho} + \frac{\partial\widehat{\eta}}{\partial\theta}\dot{\theta}, \qquad \dot{\phi} = \frac{\partial\widehat{\phi}}{\partial\rho}\dot{\rho} + \frac{\partial\widehat{\phi}}{\partial\theta}\dot{\theta}.$$

Substitution of these results into (8.34) and rearranging terms gives (omitting arguments ρ and θ for brevity)

$$\rho\left[\theta\frac{\partial\widehat{\eta}}{\partial\rho} - \frac{\partial\widehat{\phi}}{\partial\rho}\right]\dot{\rho} + \rho\left[\theta\frac{\partial\widehat{\eta}}{\partial\theta} - \frac{\partial\widehat{\phi}}{\partial\theta}\right]\dot{\theta} - p\nabla^x\cdot\boldsymbol{v} \geq 0. \tag{8.35}$$

From the conservation of mass equation we find

$$\dot{\rho} = -\rho \nabla^x \cdot \boldsymbol{v}. \tag{8.36}$$

Substituting (8.36) into (8.35), and setting $p = \widehat{p}(\rho, \theta)$, we obtain (omitting arguments ρ and θ for brevity)

$$\rho \left[\theta \frac{\partial \widehat{\eta}}{\partial \theta} - \frac{\partial \widehat{\phi}}{\partial \theta} \right] \dot{\theta} - \rho^2 \left[\theta \frac{\partial \widehat{\eta}}{\partial \rho} - \frac{\partial \widehat{\phi}}{\partial \rho} + \frac{1}{\rho^2} \widehat{p} \right] \nabla^x \cdot \boldsymbol{v} \geq 0. \tag{8.37}$$

This inequality holds for all admissible processes defined by ρ, $\boldsymbol{v}$ and θ if and only if the coefficients of $\dot{\theta}$ and $\nabla^x \cdot \boldsymbol{v}$ vanish. Sufficiency is clear. Necessity follows from the fact that the terms in brackets are independent of the quantities $\dot{\theta}$ and $\nabla^x \cdot \boldsymbol{v}$, each of which can take any possible value.

8.5 Sufficiency follows from the result in the text. To establish necessity let $\boldsymbol{x}^* = \boldsymbol{Q}(t)\boldsymbol{x} + \boldsymbol{c}(t)$ where $\boldsymbol{Q}(t)$ is an arbitrary rotation tensor and $\boldsymbol{c}(t)$ is an arbitrary vector. Then, considering only temperature and heat flux, the axiom of material-frame indifference implies

$$\theta^*(\boldsymbol{x}^*, t) = \theta(\boldsymbol{x}, t), \qquad \boldsymbol{q}^*(\boldsymbol{x}^*, t) = \boldsymbol{Q}(t)\boldsymbol{q}(\boldsymbol{x}, t),$$

for all $\boldsymbol{x}$ and t. Substituting the given model for the heat flux this becomes (omitting inessential arguments $\boldsymbol{x}^*$, $\boldsymbol{x}$ and t for brevity)

$$\theta^* = \theta, \qquad \boldsymbol{K}(\boldsymbol{x}^*, \theta^*)\nabla^{x^*}\theta^* = \boldsymbol{Q}\boldsymbol{K}(\boldsymbol{x}, \theta)\nabla^x \theta. \tag{8.38}$$

Using the relation $\theta^*(\boldsymbol{x}^*, t) = \theta(\boldsymbol{x}, t)$ together with the relation $\boldsymbol{x}^* = \boldsymbol{Q}(t)\boldsymbol{x} + \boldsymbol{c}(t)$ and the chain rule we find $\nabla^{x^*}\theta^* = \boldsymbol{Q}\nabla^x \theta$. Eliminating θ^* and $\nabla^{x^*}\theta^*$ from (8.38) we get

$$\boldsymbol{K}(\boldsymbol{x}^*, \theta)\boldsymbol{Q}\nabla^x \theta = \boldsymbol{Q}\boldsymbol{K}(\boldsymbol{x}, \theta)\nabla^x \theta,$$

which, by the arbitrariness of $\nabla^x \theta$, implies

$$\boldsymbol{K}(\boldsymbol{x}, \theta) = \boldsymbol{Q}^T \boldsymbol{K}(\boldsymbol{x}^*, \theta)\boldsymbol{Q}.$$

Since for each fixed $\boldsymbol{x}$ we may choose $\boldsymbol{c}(t)$ to make $\boldsymbol{x}^* = \boldsymbol{0}$ we obtain

$$\boldsymbol{K}(\boldsymbol{x}, \theta) = \boldsymbol{Q}^T \boldsymbol{K}(\boldsymbol{0}, \theta)\boldsymbol{Q},$$

which implies that $\boldsymbol{K}(\boldsymbol{x}, \theta)$ must be independent of $\boldsymbol{x}$. Writing

$\boldsymbol{K}(\theta)$ in place of $\boldsymbol{K}(\boldsymbol{x},\theta)$, and considering matrix representations in any arbitrary frame, we have

$$[\boldsymbol{K}(\theta)] = [\boldsymbol{Q}]^T[\boldsymbol{K}(\theta)][\boldsymbol{Q}], \tag{8.39}$$

which must hold for all rotation tensors $\boldsymbol{Q}$. By the Spectral Decomposition Theorem (see Chapter 1), for each fixed value of θ there is a rotation matrix $\boldsymbol{R}$ which diagonalizes $\boldsymbol{K}(\theta)$. Taking $\boldsymbol{Q} = \boldsymbol{R}$ in (8.39) we find

$$[\boldsymbol{K}(\theta)] = \begin{pmatrix} \kappa_1(\theta) & 0 & 0 \\ 0 & \kappa_2(\theta) & 0 \\ 0 & 0 & \kappa_3(\theta) \end{pmatrix}, \tag{8.40}$$

where $\kappa_1(\theta)$, $\kappa_2(\theta)$ and $\kappa_3(\theta)$ are the eigenvalues of $\boldsymbol{K}(\theta)$. Let $\boldsymbol{P}$ be the rotation tensor defined by the representation

$$[\boldsymbol{P}] = \begin{pmatrix} 0 & 1 & 0 \\ -1 & 0 & 0 \\ 0 & 0 & 1 \end{pmatrix}.$$

Then taking $\boldsymbol{Q} = \boldsymbol{P}$ in (8.39) and using the result in (8.40) we find

$$[\boldsymbol{K}(\theta)] = \begin{pmatrix} \kappa_2(\theta) & 0 & 0 \\ 0 & \kappa_1(\theta) & 0 \\ 0 & 0 & \kappa_3(\theta) \end{pmatrix}. \tag{8.41}$$

Comparing (8.40) and (8.41) we deduce $\kappa_1(\theta) = \kappa_2(\theta)$. Similarly, by considering the rotation tensor

$$[\boldsymbol{P}] = \begin{pmatrix} 1 & 0 & 0 \\ 0 & 0 & 1 \\ 0 & -1 & 0 \end{pmatrix},$$

we deduce $\kappa_2(\theta) = \kappa_3(\theta)$. Denoting the common value by $\kappa(\theta)$ we get

$$[\boldsymbol{K}(\theta)] = \begin{pmatrix} \kappa(\theta) & 0 & 0 \\ 0 & \kappa(\theta) & 0 \\ 0 & 0 & \kappa(\theta) \end{pmatrix} = [\kappa(\theta)\boldsymbol{I}],$$

which implies $\boldsymbol{K}(\theta) = \kappa(\theta)\boldsymbol{I}$.

8.7 Let $\boldsymbol{u} = \boldsymbol{a} \times \boldsymbol{v}$ where $\boldsymbol{a}$ is an arbitrary constant vector. Then Stokes' Theorem states

$$\int_C \boldsymbol{u} \cdot \boldsymbol{\nu}\, ds = \int_\Gamma (\nabla^x \times \boldsymbol{u}) \cdot \boldsymbol{n}\, dA_{\boldsymbol{x}}. \tag{8.42}$$

In components we have $u_i = a_p v_q \epsilon_{ipq}$. Using the definition of the curl operation (see Chapter 2) and the epsilon-delta identities (see Chapter 1) we get

$$\begin{aligned}[\nabla^x \times \boldsymbol{u}]_j &= u_{i,k}\epsilon_{ijk} \\ &= (a_p v_q \epsilon_{ipq})_{,k}\epsilon_{ijk} \\ &= a_p v_{q,k}\epsilon_{ipq}\epsilon_{ijk} \\ &= a_p v_{q,k}(\delta_{pj}\delta_{qk} - \delta_{pk}\delta_{qj}) \\ &= a_j v_{k,k} - a_k v_{j,k},\end{aligned}$$

which in tensor notation becomes

$$\nabla^x \times \boldsymbol{u} = (\nabla^x \cdot \boldsymbol{v})\boldsymbol{a} - (\nabla^x \boldsymbol{v})\boldsymbol{a}.$$

Substituting this into (8.42), and using the definition of $\boldsymbol{u}$, we get

$$\int_C (\boldsymbol{a} \times \boldsymbol{v}) \cdot \boldsymbol{\nu}\, ds = \int_\Gamma [(\nabla^x \cdot \boldsymbol{v})\boldsymbol{a} - (\nabla^x \boldsymbol{v})\boldsymbol{a}] \cdot \boldsymbol{n}\, dA_{\boldsymbol{x}}. \qquad (8.43)$$

By permutation properties of the triple scalar product (see Chapter 1) we have $(\boldsymbol{a} \times \boldsymbol{v}) \cdot \boldsymbol{\nu} = \boldsymbol{a} \cdot (\boldsymbol{v} \times \boldsymbol{\nu})$, and by definition of the transpose we have $(\nabla^x \boldsymbol{v})\boldsymbol{a} \cdot \boldsymbol{n} = \boldsymbol{a} \cdot (\nabla^x \boldsymbol{v}^T)\boldsymbol{n}$. Using these results in (8.43) we find, since $\boldsymbol{a}$ is constant

$$\boldsymbol{a} \cdot \int_C \boldsymbol{v} \times \boldsymbol{\nu}\, ds = \boldsymbol{a} \cdot \int_\Gamma (\nabla^x \cdot \boldsymbol{v})\boldsymbol{n} - (\nabla^x \boldsymbol{v}^T)\boldsymbol{n}\, dA_{\boldsymbol{x}}.$$

By the arbitrariness of $\boldsymbol{a}$ this implies

$$\int_C \boldsymbol{v} \times \boldsymbol{\nu}\, ds = \int_\Gamma (\nabla^x \cdot \boldsymbol{v})\boldsymbol{n} - (\nabla^x \boldsymbol{v}^T)\boldsymbol{n}\, dA_{\boldsymbol{x}}. \qquad (8.44)$$

From the definition of the curl $\boldsymbol{w} = \nabla^x \times \boldsymbol{v}$ (see Chapter 2) we have $\boldsymbol{w} \times \boldsymbol{n} = (\nabla^x \boldsymbol{v} - \nabla^x \boldsymbol{v}^T)\boldsymbol{n}$, and by definition of $\boldsymbol{L} = \mathrm{sym}(\nabla^x \boldsymbol{v})$ we have $\boldsymbol{L}\boldsymbol{n} = \frac{1}{2}(\nabla^x \boldsymbol{v} + \nabla^x \boldsymbol{v}^T)\boldsymbol{n}$. Combining these two relations we find

$$\tfrac{1}{2}\boldsymbol{w} \times \boldsymbol{n} - \boldsymbol{L}\boldsymbol{n} = -(\nabla^x \boldsymbol{v}^T)\boldsymbol{n}.$$

Substitution of this into (8.44) gives the desired result.

9
Thermal Solid Mechanics

In this chapter we consider applications of the Lagrangian balance laws as in Chapter 7. However, in contrast to the isothermal models introduced there, here we study models that include thermal effects. Considering the full thermo-mechanical case there are 21 basic unknown fields in the Lagrangian description of a continuum body:

$\varphi_i(\boldsymbol{X}, t)$	3 components of motion
$V_i(\boldsymbol{X}, t)$	3 components of velocity
$P_{ij}(\boldsymbol{X}, t)$	9 components of stress
$\Theta(\boldsymbol{X}, t)$	1 temperature
$Q_i(\boldsymbol{X}, t)$	3 components of heat flux
$\Phi(\boldsymbol{X}, t)$	1 internal energy per unit mass
$\eta_m(\boldsymbol{X}, t)$	1 entropy per unit mass.

To determine these unknown fields we have the following 10 equations:

$V_i = \frac{\partial}{\partial t}\varphi_i$	3 kinematical
$\rho_0 \dot{V}_i = P_{ij,j} + \rho_0 \left[b_i\right]_m$	3 linear momentum
$P_{ik}F_{jk} = F_{ik}P_{jk}$	3 independent angular momentum
$\rho_0\dot{\Phi} = P_{ij}\dot{F}_{ij} - Q_{i,i} + \rho_0 R$	1 energy.

Thus 11 additional equations are required to balance the number of equations and unknowns. These are provided by constitutive equations that relate $(\boldsymbol{P}, \boldsymbol{Q}, \Phi, \eta_m)$ to $(\boldsymbol{\varphi}, \boldsymbol{V}, \Theta)$. Any such constitutive equation must be consistent with the axiom of material frame-indifference and the Clausius–Duhem inequality (Second Law of Thermodynamics) as discussed in Chapter 5.

In this chapter we study a general constitutive model that relates $\boldsymbol{P}$, $\boldsymbol{Q}$, Φ and η_m to the deformation gradient $\boldsymbol{F} = \nabla^X \boldsymbol{\varphi}$, temperature Θ and temperature gradient $\nabla^X \Theta$. Such models are typically used to describe the behavior of various types of solids. Because such models are independent of $\boldsymbol{V}$, this variable may be eliminated by substitution of the kinematical equation into the balance of linear momentum equation. Thus a closed system for $\boldsymbol{P}$, $\boldsymbol{Q}$, Φ, η_m, $\boldsymbol{\varphi}$ and Θ is provided by the constitutive equations, together with the balance equations for linear momentum, angular momentum and energy. As with the isothermal Lagrangian formulations of Chapter 7, the balance of mass equation is not considered because the spatial mass density does not appear.

The general constitutive model considered in this chapter is that for a thermoelastic solid. We study various qualitative properties, outline a standard initial-boundary value problem, and use the technique of linearization to derive an approximate system of balance equations appropriate for describing small disturbances from a reference state.

9.1 Thermoelastic Solids

In this section we introduce the constitutive model for a thermoelastic solid. We show that the model satisfies the Clausius–Duhem inequality and study consequences of the axiom of material frame-indifference. We also discuss various notions of isotropy and outline a standard initial-boundary value problem.

9.1.1 Definition

A continuum body with reference configuration B is said to be a **thermoelastic solid** if:

(1) The first Piola–Kirchhoff stress $\boldsymbol{P}(\boldsymbol{X}, t)$ is related to the deformation gradient $\boldsymbol{F}(\boldsymbol{X}, t)$ and temperature $\Theta(\boldsymbol{X}, t)$ by

$$\boldsymbol{P} = \widehat{\boldsymbol{P}}(\boldsymbol{F}, \Theta),$$

where $\widehat{\boldsymbol{P}} : \mathcal{V}^2 \times \mathbb{R} \to \mathcal{V}^2$ is a given function called the **stress response function**.

(2) The function $\widehat{\boldsymbol{P}}$ has the property

$$\widehat{\boldsymbol{P}}(\boldsymbol{F}, \Theta)\boldsymbol{F}^T = \boldsymbol{F}\widehat{\boldsymbol{P}}(\boldsymbol{F}, \Theta)^T,$$

for all $(\boldsymbol{F}, \Theta)$ with $\det \boldsymbol{F} > 0$ and $\Theta > 0$.

(3) The internal energy $\Phi(\boldsymbol{X},t)$ and entropy $\eta_m(\boldsymbol{X},t)$ are related to the deformation gradient $\boldsymbol{F}(\boldsymbol{X},t)$ and temperature $\Theta(\boldsymbol{X},t)$ by

$$\Phi = \widehat{\Phi}(\boldsymbol{F},\Theta) \quad \text{and} \quad \eta_m = \widehat{\eta}_m(\boldsymbol{F},\Theta),$$

where $\widehat{\Phi}, \widehat{\eta}_m : \mathcal{V}^2 \times \mathbb{R} \to \mathbb{R}$ are given functions called the **internal energy** and **entropy response functions**.

(4) The functions $\widehat{\Phi}$ and $\widehat{\eta}_m$ have the property

$$\frac{\partial \widehat{\Phi}}{\partial \Theta}(\boldsymbol{F},\Theta) > 0 \quad \text{and} \quad \frac{\partial \widehat{\eta}_m}{\partial \Theta}(\boldsymbol{F},\Theta) > 0,$$

for all $(\boldsymbol{F},\Theta)$ with $\det \boldsymbol{F} > 0$ and $\Theta > 0$.

(5) The material heat flux vector $\boldsymbol{Q}(\boldsymbol{X},t)$ is related to the deformation gradient $\boldsymbol{F}(\boldsymbol{X},t)$ and temperature $\Theta(\boldsymbol{X},t)$ by

$$\boldsymbol{Q} = -\widehat{\boldsymbol{K}}(\boldsymbol{F},\Theta)\, \nabla^{X} \Theta,$$

where $\widehat{\boldsymbol{K}} : \mathcal{V}^2 \times \mathbb{R} \to \mathcal{V}^2$ is a given function called the **thermal conductivity function**.

Property (1) implies that the stress at a point in a thermoelastic solid depends only on a measure of the current strain and temperature at that point. In particular, it is independent of the past histories and rates of change of these quantities. This type of relation is similar to the stress-strain relation for an elastic solid (see Chapter 7), but with thermal effects included. Property (2) implies that the balance equation for angular momentum (Result 5.16) is automatically satisfied. Thus this balance equation will not be considered further. Because $\det \boldsymbol{F}(\boldsymbol{X},t) > 0$ and $\Theta(\boldsymbol{X},t) > 0$ for any admissible thermo-mechanical process, we only consider $\widehat{\boldsymbol{P}}(\boldsymbol{F},\Theta)$ for arguments $(\boldsymbol{F},\Theta)$ satisfying $\det \boldsymbol{F} > 0$ and $\Theta > 0$.

Property (3) implies that, just as for the Piola–Kirchhoff stress, the internal energy and entropy per unit mass at a point are entirely determined by the current deformation gradient and temperature at that point. The three relations

$$\boldsymbol{P} = \widehat{\boldsymbol{P}}(\boldsymbol{F},\Theta), \quad \Phi = \widehat{\Phi}(\boldsymbol{F},\Theta), \quad \eta_m = \widehat{\eta}_m(\boldsymbol{F},\Theta), \tag{9.1}$$

are typically referred to as the **thermal equations of state**. Just as for $\widehat{\boldsymbol{P}}(\boldsymbol{F},\Theta)$, we only consider $\widehat{\Phi}(\boldsymbol{F},\Theta)$ and $\widehat{\eta}_m(\boldsymbol{F},\Theta)$ for arguments $(\boldsymbol{F},\Theta)$ satisfying $\det \boldsymbol{F} > 0$ and $\Theta > 0$.

Property (4) implies that, for each fixed value of $\boldsymbol{F}$, the functions $\widehat{\Phi}(\boldsymbol{F},\Theta)$ and $\widehat{\eta}_m(\boldsymbol{F},\Theta)$ are strictly increasing in Θ. Physically, this means that when the configuration of a body is held fixed, internal energy and entropy increase with temperature. Mathematically, these two conditions mean that either Φ or η_m can replace Θ as an independent variable. In particular, the relation $\eta_m = \widehat{\eta}_m(\boldsymbol{F}, \Theta)$ in $(9.1)_3$ can be inverted as $\Theta = \widetilde{\Theta}(\boldsymbol{F}, \eta_m)$. When this relation is substituted into $(9.1)_{1,2}$ we obtain

$$\boldsymbol{P} = \widetilde{\boldsymbol{P}}(\boldsymbol{F}, \eta_m), \quad \Phi = \widetilde{\Phi}(\boldsymbol{F}, \eta_m), \quad \Theta = \widetilde{\Theta}(\boldsymbol{F}, \eta_m). \tag{9.2}$$

These relations are typically referred to as the **caloric equations of state**. While some authors prefer this form, we prefer to work with the form in (9.1) since the heat flux vector depends explicitly on the temperature field. As in the case of perfect gases and Newtonian fluids, the function

$$\alpha(\boldsymbol{F}, \Theta) = \frac{\partial \widehat{\Phi}}{\partial \Theta}(\boldsymbol{F}, \Theta) > 0 \tag{9.3}$$

is typically called the **specific heat** at constant volume.

Throughout our developments it will be convenient to consider the free energy $\Psi(\boldsymbol{X}, t)$ defined by (see Chapter 5)

$$\Psi = \Phi - \Theta \eta_m. \tag{9.4}$$

Rather than specify an internal energy response function $\widehat{\Phi}(\boldsymbol{F}, \Theta)$, we could alternatively specify a **free energy response function** $\widehat{\Psi}(\boldsymbol{F}, \Theta)$ such that

$$\Psi = \widehat{\Psi}(\boldsymbol{F}, \Theta). \tag{9.5}$$

In particular, from (9.4) we see that knowledge of one implies the other once the entropy response function $\widehat{\eta}_m(\boldsymbol{F}, \Theta)$ is known. Notice that, because the free energy at a point depends only on a measure of current strain and temperature at that point, a thermoelastic solid is energetically passive as defined in Chapter 5. As with the other response functions, we only consider $\widehat{\Psi}(\boldsymbol{F}, \Theta)$ for arguments $(\boldsymbol{F}, \Theta)$ with $\det \boldsymbol{F} > 0$ and $\Theta > 0$.

Property (5) implies that the material heat flux vector at a point is linearly related to the material temperature gradient at that point. This relation may be viewed as an extension of **Fourier's Law**. In particular, whereas Fourier's Law postulates such a relation in a static body, here we assume it also holds in a deforming body. As with all other response functions, we only consider $\widehat{\boldsymbol{K}}(\boldsymbol{F}, \Theta)$ for arguments $(\boldsymbol{F}, \Theta)$ with $\det \boldsymbol{F} > 0$ and $\Theta > 0$.

Remarks:

(1) Arguments based on the Clausius–Duhem inequality (see Result 9.1) imply that $\boldsymbol{a} \cdot \widehat{\boldsymbol{K}}(\boldsymbol{F}, \Theta)\boldsymbol{a} \geq 0$ for all vectors $\boldsymbol{a}$. In particular, $\widehat{\boldsymbol{K}}(\boldsymbol{F}, \Theta)$ must be positive semi-definite. Physically, this means that the heat flux vector must make an obtuse angle with the temperature gradient, so that heat flows from hot to cold. Independent of this condition, the tensor $\widehat{\boldsymbol{K}}(\boldsymbol{F}, \Theta)$ is often assumed to be symmetric. This assumption, however, is not essential.

(2) Given a scalar-valued function such as $\widehat{\Phi}(\boldsymbol{F}, \Theta)$ we employ the notation $D_\Theta \widehat{\Phi}$ or $\partial \widehat{\Phi}/\partial \Theta$ to denote the partial derivative with respect to the scalar Θ, and $D_{\boldsymbol{F}} \widehat{\Phi}$ or $\partial \widehat{\Phi}/\partial \boldsymbol{F}$ to denote the partial derivative with respect to the second-order tensor $\boldsymbol{F}$. Thus $\partial \widehat{\Phi}/\partial \Theta$ is a scalar-valued function, $\partial \widehat{\Phi}/\partial \boldsymbol{F}$ and $\partial^2 \widehat{\Phi}/\partial \Theta \partial \boldsymbol{F}$ are second-order tensor-valued functions and $\partial^2 \widehat{\Phi}/\partial \boldsymbol{F}^2$ is a fourth-order tensor-valued function.

(3) As in Chapter 7, all our response functions are assumed to be homogeneous in the sense that they do not explicitly depend on the point $\boldsymbol{X}$. This is done for notational simplicity alone. All subsequent results can be generalized to the inhomogeneous case.

□

9.1.2 Thermoelasticity Equations

Let $\rho_0(\boldsymbol{X})$ denote the mass density of a thermoelastic body in its reference configuration B. Moreover, let $\boldsymbol{b}_m(\boldsymbol{X}, t)$ and $R(\boldsymbol{X}, t)$ denote the material descriptions of prescribed spatial body force and heat supply fields per unit mass $\boldsymbol{b}(\boldsymbol{x}, t)$ and $r(\boldsymbol{x}, t)$, that is

$$\boldsymbol{b}_m(\boldsymbol{X}, t) = \boldsymbol{b}(\boldsymbol{\varphi}(\boldsymbol{X}, t), t) \quad \text{and} \quad R(\boldsymbol{X}, t) = r(\boldsymbol{\varphi}(\boldsymbol{X}, t), t).$$

Then, setting $\boldsymbol{P} = \widehat{\boldsymbol{P}}(\boldsymbol{F}, \Theta)$, $\Phi = \widehat{\Phi}(\boldsymbol{F}, \Theta)$ and $\boldsymbol{Q} = -\widehat{\boldsymbol{K}}(\boldsymbol{F}, \Theta)\nabla^X \Theta$ in the balance of linear momentum equation (Result 5.15) and balance of energy equation (Result 5.18), we find that the motion and temperature fields in a thermoelastic body must satisfy the following equations for all $\boldsymbol{X} \in B$ and $t \geq 0$

$$\boxed{\begin{aligned} \rho_0 \frac{\partial^2 \boldsymbol{\varphi}}{\partial t^2} &= \nabla^X \cdot (\widehat{\boldsymbol{P}}(\boldsymbol{F}, \Theta)) + \rho_0 \boldsymbol{b}_m, \\ \rho_0 \frac{\partial}{\partial t} \widehat{\Phi}(\boldsymbol{F}, \Theta) &= \widehat{\boldsymbol{P}}(\boldsymbol{F}, \Theta) : \frac{\partial \boldsymbol{F}}{\partial t} + \nabla^X \cdot (\widehat{\boldsymbol{K}}(\boldsymbol{F}, \Theta)\nabla^X \Theta) + \rho_0 R. \end{aligned}} \tag{9.6}$$

These are known as the **Thermoelastodynamics Equations**. By an admissible process for a thermoelastic solid we mean fields $(\boldsymbol{\varphi}, \Theta)$ satisfying these equations.

Remarks:

(1) By definition of the divergence of a second-order tensor we have $[\nabla^X \cdot \widehat{\boldsymbol{P}}]_i = \widehat{P}_{ij,j}$, and by the chain rule and the assumption of homogeneity we get

$$[\nabla^X \cdot \widehat{\boldsymbol{P}}]_i = \frac{\partial \widehat{P}_{ij}}{\partial F_{kl}} \frac{\partial F_{kl}}{\partial X_j} + \frac{\partial \widehat{P}_{ij}}{\partial \Theta} \frac{\partial \Theta}{\partial X_j}.$$

Similarly, by the chain rule we get

$$\frac{\partial}{\partial t} \widehat{\Phi} = \frac{\partial \widehat{\Phi}}{\partial F_{kl}} \frac{\partial F_{kl}}{\partial t} + \frac{\partial \widehat{\Phi}}{\partial \Theta} \frac{\partial \Theta}{\partial t}.$$

Since $F_{kl} = \partial \varphi_k / \partial X_l$, the equations in (9.6) can be written in components as

$$\begin{aligned} \rho_0 \frac{\partial^2 \varphi_i}{\partial t^2} &= \mathsf{A}_{ijkl} \frac{\partial^2 \varphi_k}{\partial X_l \partial X_j} + D_{ij} \frac{\partial \Theta}{\partial X_j} + \rho_0 b_i, \\ \rho_0 \alpha \frac{\partial \Theta}{\partial t} &= H_{kl} \frac{\partial^2 \varphi_k}{\partial X_l \partial t} + \frac{\partial}{\partial X_i} \left(\widehat{K}_{ij} \frac{\partial \Theta}{\partial X_j} \right) + \rho_0 R, \end{aligned}$$

where $\mathsf{A}_{ijkl} = \frac{\partial \widehat{P}_{ij}}{\partial F_{kl}}$, $D_{ij} = \frac{\partial \widehat{P}_{ij}}{\partial \Theta}$, $H_{kl} = \widehat{P}_{kl} - \rho_0 \frac{\partial \widehat{\Phi}}{\partial F_{kl}}$ and $\alpha = \frac{\partial \widehat{\Phi}}{\partial \Theta}$. Here b_i denote the components of $\boldsymbol{b}_m$. Thus (9.6) is a coupled system of second-order partial differential equations for $\boldsymbol{\varphi}$ and Θ. These equations are typically nonlinear.

(2) Setting all time derivatives equal to zero in (9.6) yields

$$\begin{aligned} \nabla^X \cdot (\widehat{\boldsymbol{P}}(\boldsymbol{F}, \Theta)) + \rho_0 \boldsymbol{b}_m &= \boldsymbol{0}, \\ \nabla^X \cdot (\widehat{\boldsymbol{K}}(\boldsymbol{F}, \Theta) \nabla^X \Theta) + \rho_0 R &= 0. \end{aligned}$$

These equations are called the **Thermoelastostatics Equations**. They must be satisfied by every static (time-independent) deformation and temperature field of a thermoelastic body.

□

9.1.3 Thermodynamical Considerations

In this section we outline a result which shows that the stress, entropy and conductivity response functions for a thermoelastic solid must satisfy certain conditions in order to comply with the Clausius–Duhem inequality. Using this result we derive a simplified form for the balance of energy equation.

Result 9.1 ***Implications of Clausius–Duhem Inequality.*** *The Clausius–Duhem inequality (Result 5.20) is satisfied by every admissible process in a thermoelastic solid if and only if:*

$$(i)\ \widehat{\boldsymbol{P}}(\boldsymbol{F},\Theta)=\rho_0 D_{\boldsymbol{F}}\widehat{\Psi}(\boldsymbol{F},\Theta)\ \text{and}\ \widehat{\eta}_m(\boldsymbol{F},\Theta)=-D_\Theta\widehat{\Psi}(\boldsymbol{F},\Theta), \tag{9.7}$$

$$(ii)\ \boldsymbol{a}\cdot\widehat{\boldsymbol{K}}(\boldsymbol{F},\Theta)\boldsymbol{a}\geq 0\ \text{for all vectors}\ \boldsymbol{a}. \tag{9.8}$$

□

Proof The (reduced) Clausius–Duhem inequality in Lagrangian form is

$$\rho_0(\eta_m\dot{\Theta}+\dot{\Psi})-\boldsymbol{P}:\dot{\boldsymbol{F}}+\Theta^{-1}\boldsymbol{Q}\cdot\nabla^X\Theta\leq 0. \tag{9.9}$$

From the relation $\Psi=\widehat{\Psi}(\boldsymbol{F},\Theta)$ and the chain rule we have

$$\dot{\Psi}=D_{\boldsymbol{F}}\widehat{\Psi}(\boldsymbol{F},\Theta):\dot{\boldsymbol{F}}+D_\Theta\widehat{\Psi}(\boldsymbol{F},\Theta)\dot{\Theta}.$$

Substituting this result and the relations $\boldsymbol{P}=\widehat{\boldsymbol{P}}(\boldsymbol{F},\Theta)$, $\eta_m=\widehat{\eta}_m(\boldsymbol{F},\Theta)$, and $\boldsymbol{Q}=-\widehat{\boldsymbol{K}}(\boldsymbol{F},\Theta)\nabla^X\Theta$ into (9.9), we find that the Clausius–Duhem inequality is equivalent to (omitting arguments $\boldsymbol{F}$ and Θ for brevity)

$$\rho_0\left[\widehat{\eta}_m+D_\Theta\widehat{\Psi}\right]\dot{\Theta}+\left[\rho_0 D_{\boldsymbol{F}}\widehat{\Psi}-\widehat{\boldsymbol{P}}\right]:\dot{\boldsymbol{F}}-\Theta^{-1}\nabla^X\Theta\cdot\widehat{\boldsymbol{K}}\nabla^X\Theta\leq 0. \tag{9.10}$$

It can now be shown that (9.7) and (9.8) are necessary and sufficient conditions for (9.10) to hold for all admissible processes. Sufficiency of (9.7) and (9.8) is clear, for then each term in (9.10) is either zero or non-positive. To establish necessity, suppose (9.10) holds for all processes. Notice that the terms in brackets and the heat flux term depend only on the fields $\boldsymbol{F}$ and Θ, and are independent of $\dot{\boldsymbol{F}}$ and $\dot{\Theta}$. Because $\dot{\Theta}$ can be varied independently of $\boldsymbol{F}$ and Θ, we deduce that the coefficient of $\dot{\Theta}$ in (9.10) must vanish, that is, $[\widehat{\eta}_m+D_\Theta\widehat{\Psi}]=0$. Otherwise, we could take $\dot{\Theta}$ to be a large positive multiple of $[\widehat{\eta}_m+D_\Theta\widehat{\Psi}]$ and thereby cause (9.10) to be violated. By similar arguments, we find that the coefficient of $\dot{\boldsymbol{F}}$ in (9.10) must also vanish, that is, $[\rho_0 D_{\boldsymbol{F}}\widehat{\Psi}-\widehat{\boldsymbol{P}}]=\boldsymbol{O}$. Thus (9.10) reduces to the form

$$-\Theta^{-1}\nabla^X\Theta\cdot\widehat{\boldsymbol{K}}\nabla^X\Theta\leq 0, \tag{9.11}$$

which, by the positivity of Θ and the arbitrariness of $\nabla^X \Theta$, implies that $\widehat{\boldsymbol{K}}$ must be positive semi-definite. Thus the conditions in (9.7) and (9.8) are necessary. □

Remarks:

(1) In the above arguments we implicitly assume that, at any given instant of time, the material fields $\dot{\Theta}$ and $\dot{\boldsymbol{F}}$ can be varied independently of Θ and $\boldsymbol{F}$. Using (9.6) it is possible to show that arbitrary values for $\dot{\Theta}$ and $\dot{\boldsymbol{F}}$ can be achieved, for given Θ and $\boldsymbol{F}$, by appropriate choice of heat supply R and body force $\boldsymbol{b}_m$.

(2) The conditions in (9.7) apply to the thermal equations of state in (9.1), where the free energy was introduced for convenience. A different, but equivalent version of these conditions can also be derived for the caloric equations of state in (9.2). In particular, using the relation between $\widehat{\Psi}(\boldsymbol{F},\Theta)$ and $\widehat{\Phi}(\boldsymbol{F},\Theta)$, the definition of $\widetilde{\Phi}(\boldsymbol{F},\eta_m)$, and the chain rule, we deduce that the conditions in (9.7) are equivalent to

$$\begin{aligned}\widetilde{\boldsymbol{P}}(\boldsymbol{F},\eta_m) &= \rho_0 D_{\boldsymbol{F}}\widetilde{\Phi}(\boldsymbol{F},\eta_m),\\ \widetilde{\Theta}(\boldsymbol{F},\eta_m) &= D_{\eta_m}\widetilde{\Phi}(\boldsymbol{F},\eta_m).\end{aligned} \tag{9.12}$$

(3) Result 9.1 can be viewed as an extension of Result 7.6 to the non-isothermal case. In both cases, we see that the first Piola–Kirchhoff stress response function must be the derivative of a scalar function in order to comply with the Clausius–Duhem inequality. For other implications of this inequality see Exercise 1.

(4) By differentiating (9.7), and using the fact that mixed partial derivatives are equal, we deduce that the response functions $\widehat{\boldsymbol{P}}$ and $\widehat{\eta}_m$ satisfy

$$\frac{\partial \widehat{\boldsymbol{P}}}{\partial \Theta} = -\rho_0 \frac{\partial \widehat{\eta}_m}{\partial \boldsymbol{F}}. \tag{9.13}$$

This equation is sometimes called a **Maxwell relation**. It will be used below when we linearize the equations of thermoelasticity.

(5) By substituting response functions into (9.4), differentiating with respect to Θ, and employing $(9.7)_2$, we deduce that the specific heat α defined in (9.3) satisfies

$$\alpha = -\Theta \frac{\partial^2 \widehat{\Psi}}{\partial \Theta^2}. \tag{9.14}$$

From this we deduce that $\alpha > 0$ if and only if $\partial^2\widehat{\Psi}/\partial\Theta^2 < 0$. Thus, the assumed positivity of the specific heat is equivalent to a convexity condition on the free energy.

(6) A process in a material body is called **isothermal** if $\Theta(\boldsymbol{X}, t) = \Theta_0$ (constant) throughout the process. From $(9.7)_1$ we see that, in any isothermal process, a thermoelastic model reduces to a hyperelastic model with strain energy function $W(\boldsymbol{F}) = \rho_0\widehat{\Psi}(\boldsymbol{F}, \Theta_0)$ (see Chapter 7).

(7) A process in a material body is called **reversible** if equality is achieved in the Clausius–Duhem inequality throughout the process; otherwise, **irreversible**. The proof of Result 9.1 shows that a thermoelastic solid can experience irreversible processes. Indeed, most processes will be irreversible when $\widehat{\boldsymbol{K}} \neq \boldsymbol{O}$ since equality will seldom be achieved in (9.11). In contrast, special processes such as isothermal ones are reversible. □

We next use Result 9.1 to simplify or reduce the balance of energy equation.

Result 9.2 ***Reduced Energy Equation.*** *If the stress and entropy response functions satisfy the conditions in Result 9.1, then the balance of energy equation (Result 5.18) takes the reduced form*

$$\rho_0\Theta\dot{\eta}_m + \nabla^X \cdot \boldsymbol{Q} = \rho_0 R.$$

□

Proof The balance of energy equation in Lagrangian form is

$$\rho_0\dot{\Phi} = \boldsymbol{P} : \dot{\boldsymbol{F}} - \nabla^X \cdot \boldsymbol{Q} + \rho_0 R. \tag{9.15}$$

Taking the time derivative of (9.4) we get

$$\dot{\Phi} = \dot{\Psi} + \dot{\eta}_m\Theta + \eta_m\dot{\Theta}, \tag{9.16}$$

and taking the time derivative of (9.5), using the chain rule, and substituting from (9.7), we get

$$\dot{\Psi} = D_{\boldsymbol{F}}\widehat{\Psi} : \dot{\boldsymbol{F}} + D_\Theta\widehat{\Psi}\,\dot{\Theta} = \rho_0^{-1}\boldsymbol{P} : \dot{\boldsymbol{F}} - \eta_m\dot{\Theta}. \tag{9.17}$$

The desired result follows by substituting (9.16) and (9.17) into (9.15) and canceling terms. □

Remarks:

(1) A process in a material body is called **adiabatic** if $\boldsymbol{Q}(\boldsymbol{X},t) = \boldsymbol{0}$ and $R(\boldsymbol{X},t) = 0$ throughout the process, and is called **isentropic** if $\eta_m(\boldsymbol{X},t) = \eta_0$ (constant). Result 9.2 implies that, under the conditions of Result 9.1, an adiabatic process in a thermoelastic solid is isentropic provided the initial entropy field is uniform.

(2) From $(9.12)_1$ we see that, in any isentropic process, a thermoelastic model reduces to a hyperelastic model with strain energy function $W(\boldsymbol{F}) = \rho_0 \widetilde{\Phi}(\boldsymbol{F}, \eta_0)$ (see Chapter 7). □

9.1.4 Frame-Indifference Considerations

Consider a thermoelastic body with reference configuration B. Let $\boldsymbol{x} = \boldsymbol{\varphi}(\boldsymbol{X},t)$ be an arbitrary motion and let $\boldsymbol{x}^* = \boldsymbol{\varphi}^*(\boldsymbol{X},t)$ be a second motion defined by

$$\boldsymbol{x}^* = \boldsymbol{g}(\boldsymbol{x},t) = \boldsymbol{\Lambda}(t)\boldsymbol{x} + \boldsymbol{c}(t),$$

where $\boldsymbol{\Lambda}(t)$ is an arbitrary rotation tensor and $\boldsymbol{c}(t)$ is an arbitrary vector. (We avoid using the notation $\boldsymbol{Q}(t)$ for the rotation to avoid confusion with the heat flux vector.) Let θ, ϕ, η, $\boldsymbol{q}$ and $\boldsymbol{S}$ denote the spatial temperature, internal energy, entropy, heat flux and Cauchy stress fields in B_t, and let θ^*, ϕ^*, η^*, $\boldsymbol{q}^*$ and $\boldsymbol{S}^*$ denote the corresponding spatial fields in B_t^*. Then, in the material description, the axiom of material frame-indifference (Axiom 5.24) postulates the following for all $\boldsymbol{X} \in B$ and $t \geq 0$

$$\begin{aligned}
\theta_m^*(\boldsymbol{X},t) &= \theta_m(\boldsymbol{X},t), \\
\phi_m^*(\boldsymbol{X},t) &= \phi_m(\boldsymbol{X},t), \\
\eta_m^*(\boldsymbol{X},t) &= \eta_m(\boldsymbol{X},t), \\
\boldsymbol{q}_m^*(\boldsymbol{X},t) &= \boldsymbol{\Lambda}(t)\boldsymbol{q}_m(\boldsymbol{X},t), \\
\boldsymbol{S}_m^*(\boldsymbol{X},t) &= \boldsymbol{\Lambda}(t)\boldsymbol{S}_m(\boldsymbol{X},t)\boldsymbol{\Lambda}(t)^T.
\end{aligned} \tag{9.18}$$

From Chapter 5 we recall the relations $\theta_m = \Theta$, $\phi_m = \Phi$, $\boldsymbol{q}_m = (\det \boldsymbol{F})^{-1}\boldsymbol{F}\boldsymbol{Q}$ and $\boldsymbol{S}_m = (\det \boldsymbol{F})^{-1}\boldsymbol{P}\boldsymbol{F}^T$, and by Result 5.22 we recall that $\boldsymbol{F}^* = \boldsymbol{\Lambda}\boldsymbol{F}$ and $\boldsymbol{C}^* = \boldsymbol{C}$. Substituting these relations and the thermoelastic response functions into (9.18) (omitting the arguments $\boldsymbol{X}$

and t for brevity) we obtain

$$\begin{aligned}\Theta^* &= \Theta,\\ \widehat{\Phi}(\boldsymbol{F}^*,\Theta^*) &= \widehat{\Phi}(\boldsymbol{F},\Theta),\\ \widehat{\eta}_m(\boldsymbol{F}^*,\Theta^*) &= \widehat{\eta}_m(\boldsymbol{F},\Theta),\\ (\det\boldsymbol{F}^*)^{-1}\boldsymbol{F}^*\widehat{\boldsymbol{K}}(\boldsymbol{F}^*,\Theta^*)\nabla^X\Theta^* &= \boldsymbol{\Lambda}(\det\boldsymbol{F})^{-1}\boldsymbol{F}\widehat{\boldsymbol{K}}(\boldsymbol{F},\Theta)\nabla^X\Theta,\\ (\det\boldsymbol{F}^*)^{-1}\widehat{\boldsymbol{P}}(\boldsymbol{F}^*,\Theta^*)\boldsymbol{F}^{*T} &= \boldsymbol{\Lambda}(\det\boldsymbol{F})^{-1}\widehat{\boldsymbol{P}}(\boldsymbol{F},\Theta)\boldsymbol{F}^T\boldsymbol{\Lambda}^T.\end{aligned}$$

Eliminating Θ^*, using the facts that $\boldsymbol{F}^* = \boldsymbol{\Lambda}\boldsymbol{F}$ and $\det\boldsymbol{F}^* = \det\boldsymbol{F}$, and using (9.4) to introduce the free energy function $\widehat{\Psi}$ in place of $\widehat{\Phi}$, we find that the above equations reduce to

$$\begin{aligned}\widehat{\Psi}(\boldsymbol{\Lambda}\boldsymbol{F},\Theta) &= \widehat{\Psi}(\boldsymbol{F},\Theta),\\ \widehat{\eta}_m(\boldsymbol{\Lambda}\boldsymbol{F},\Theta) &= \widehat{\eta}_m(\boldsymbol{F},\Theta),\\ \widehat{\boldsymbol{K}}(\boldsymbol{\Lambda}\boldsymbol{F},\Theta)\nabla^X\Theta &= \widehat{\boldsymbol{K}}(\boldsymbol{F},\Theta)\nabla^X\Theta,\\ \widehat{\boldsymbol{P}}(\boldsymbol{\Lambda}\boldsymbol{F},\Theta) &= \boldsymbol{\Lambda}\widehat{\boldsymbol{P}}(\boldsymbol{F},\Theta).\end{aligned} \tag{9.19}$$

Thus, in order to comply with the axiom of material frame-indifference, the response functions $\widehat{\Psi}$, $\widehat{\eta}_m$, $\widehat{\boldsymbol{K}}$ and $\widehat{\boldsymbol{P}}$ must satisfy (9.19) for all rotations $\boldsymbol{\Lambda}$ and all admissible values of $\boldsymbol{F}$, Θ and $\nabla^X\Theta$. The following result establishes conditions under which equations (9.19) are satisfied.

Result 9.3 ***Frame-Indifferent Thermoelastic Response.*** *Suppose the conditions in Result 9.1 hold. Then the constitutive model of a thermoelastic solid is frame-indifferent if and only if the functions* $\widehat{\Psi}$ *and* $\widehat{\boldsymbol{K}}$ *can be expressed as*

$$\widehat{\Psi}(\boldsymbol{F},\Theta) = \overline{\Psi}(\boldsymbol{C},\Theta) \quad \textit{and} \quad \widehat{\boldsymbol{K}}(\boldsymbol{F},\Theta) = \overline{\boldsymbol{K}}(\boldsymbol{C},\Theta), \tag{9.20}$$

for some functions $\overline{\Psi}$ *and* $\overline{\boldsymbol{K}}$*, where* $\boldsymbol{C} = \boldsymbol{F}^T\boldsymbol{F}$ *is the Cauchy–Green strain tensor. In this case, the response functions* $\widehat{\boldsymbol{P}}$ *and* $\widehat{\eta}_m$ *are given by*

$$\widehat{\boldsymbol{P}}(\boldsymbol{F},\Theta) = 2\rho_0\boldsymbol{F}D_{\boldsymbol{C}}\overline{\Psi}(\boldsymbol{C},\Theta) \quad \textit{and} \quad \widehat{\eta}_m(\boldsymbol{F},\Theta) = -D_\Theta\overline{\Psi}(\boldsymbol{C},\Theta). \tag{9.21}$$

□

Proof Assume the frame-indifference relations in (9.19) hold. Then, using the arbitrariness of $\nabla^X\Theta$, we have

$$\widehat{\Psi}(\boldsymbol{\Lambda}\boldsymbol{F},\Theta) = \widehat{\Psi}(\boldsymbol{F},\Theta) \quad \text{and} \quad \widehat{\boldsymbol{K}}(\boldsymbol{\Lambda}\boldsymbol{F},\Theta) = \widehat{\boldsymbol{K}}(\boldsymbol{F},\Theta), \tag{9.22}$$

for all rotations $\boldsymbol{\Lambda}$ and all $(\boldsymbol{F},\Theta)$ with $\det\boldsymbol{F} > 0$ and $\Theta > 0$. Let

$\boldsymbol{F} = \boldsymbol{R}\boldsymbol{U}$ be the right polar decomposition of $\boldsymbol{F}$. Choosing $\boldsymbol{\Lambda} = \boldsymbol{R}^T$ in (9.22) yields

$$\widehat{\Psi}(\boldsymbol{F}, \Theta) = \widehat{\Psi}(\boldsymbol{U}, \Theta) \quad \text{and} \quad \widehat{\boldsymbol{K}}(\boldsymbol{F}, \Theta) = \widehat{\boldsymbol{K}}(\boldsymbol{U}, \Theta).$$

Since $\boldsymbol{U} = \sqrt{\boldsymbol{C}}$ we have $\widehat{\Psi}(\boldsymbol{F}, \Theta) = \overline{\Psi}(\boldsymbol{C}, \Theta)$ and $\widehat{\boldsymbol{K}}(\boldsymbol{F}, \Theta) = \overline{\boldsymbol{K}}(\boldsymbol{C}, \Theta)$, where $\overline{\Psi}(\boldsymbol{C}, \Theta) = \widehat{\Psi}(\sqrt{\boldsymbol{C}}, \Theta)$ and $\overline{\boldsymbol{K}}(\boldsymbol{C}, \Theta) = \widehat{\boldsymbol{K}}(\sqrt{\boldsymbol{C}}, \Theta)$. Thus (9.19) implies (9.20).

Conversely, assume (9.20) holds. Then, since $\boldsymbol{C} = \boldsymbol{F}^T\boldsymbol{F}$ and $\boldsymbol{\Lambda}^T\boldsymbol{\Lambda} = \boldsymbol{I}$, we immediately find that $(9.19)_{1,3}$ hold since

$$\widehat{\Psi}(\boldsymbol{\Lambda}\boldsymbol{F}, \Theta) = \overline{\Psi}(\boldsymbol{F}^T\boldsymbol{\Lambda}^T\boldsymbol{\Lambda}\boldsymbol{F}, \Theta) = \overline{\Psi}(\boldsymbol{F}^T\boldsymbol{F}, \Theta) = \widehat{\Psi}(\boldsymbol{F}, \Theta),$$

and

$$\widehat{\boldsymbol{K}}(\boldsymbol{\Lambda}\boldsymbol{F}, \Theta) = \overline{\boldsymbol{K}}(\boldsymbol{F}^T\boldsymbol{\Lambda}^T\boldsymbol{\Lambda}\boldsymbol{F}, \Theta) = \overline{\boldsymbol{K}}(\boldsymbol{F}^T\boldsymbol{F}, \Theta) = \widehat{\boldsymbol{K}}(\boldsymbol{F}, \Theta).$$

Using the condition in Result 9.1 and $(9.19)_1$ we find that $(9.19)_2$ holds since

$$\widehat{\eta}_m(\boldsymbol{\Lambda}\boldsymbol{F}, \Theta) = -D_\Theta\widehat{\Psi}(\boldsymbol{\Lambda}\boldsymbol{F}, \Theta) = -D_\Theta\widehat{\Psi}(\boldsymbol{F}, \Theta) = \widehat{\eta}_m(\boldsymbol{F}, \Theta).$$

From the proof of Result 7.5 we observe that the relation $\widehat{\Psi}(\boldsymbol{F}, \Theta) = \overline{\Psi}(\boldsymbol{C}, \Theta)$ implies

$$D_{\boldsymbol{F}}\widehat{\Psi}(\boldsymbol{F}, \Theta) = 2\boldsymbol{F}D_{\boldsymbol{C}}\overline{\Psi}(\boldsymbol{C}, \Theta). \tag{9.23}$$

Using this result and the condition in Result 9.1 we find that $(9.19)_4$ holds. In particular, using the notation $\boldsymbol{F}^* = \boldsymbol{\Lambda}\boldsymbol{F}$ and $\boldsymbol{C}^* = \boldsymbol{F}^{*T}\boldsymbol{F}^* = \boldsymbol{C}$ for convenience, we have

$$\begin{aligned} \widehat{\boldsymbol{P}}(\boldsymbol{F}^*, \Theta) &= \rho_0 D_{\boldsymbol{F}^*}\widehat{\Psi}(\boldsymbol{F}^*, \Theta) \\ &= 2\rho_0\boldsymbol{F}^*D_{\boldsymbol{C}^*}\overline{\Psi}(\boldsymbol{C}^*, \Theta) \\ &= \boldsymbol{\Lambda}[2\rho_0\boldsymbol{F}D_{\boldsymbol{C}}\overline{\Psi}(\boldsymbol{C}, \Theta)] = \boldsymbol{\Lambda}\widehat{\boldsymbol{P}}(\boldsymbol{F}, \Theta). \end{aligned}$$

Thus, under the conditions in Result 9.1, we find that (9.20) implies (9.19). The relations in (9.21) follow from $(9.20)_1$ and (9.23). □

Remarks:

(1) Results 9.1 and 9.3 show that a thermoelastic model which complies with the Clausius–Duhem inequality and the axiom of material frame-indifference is completely defined by a free energy function $\overline{\Psi}(\boldsymbol{C}, \Theta)$ and a thermal conductivity function $\overline{\boldsymbol{K}}(\boldsymbol{C}, \Theta)$. The function $\overline{\Psi}$ characterizes the stress, entropy, free energy and internal energy. The function $\overline{\boldsymbol{K}}$ characterizes the heat flux.

(2) Under the conditions of Results 9.1 and 9.3 we find that Property (2) in the definition of a thermoelastic solid is satisfied for any free energy function $\overline{\Psi}$. This is a straightforward consequence of the relation in $(9.21)_1$ and the symmetry of $D_C\overline{\Psi}$. In particular, $\widehat{\boldsymbol{P}}\boldsymbol{F}^T = 2\rho_0 \boldsymbol{F} D_C\overline{\Psi}\boldsymbol{F}^T$ is symmetric.

(3) Just as for elastic solids, a thermoelastic solid is said to be **(mechanically) isotropic** if the Cauchy stress field $\boldsymbol{S}_m$ is invariant under rotations of the reference configuration (see Chapter 7). Under the conditions of Results 9.1 and 9.3, a thermoelastic solid is mechanically isotropic if and only if the free energy function $\overline{\Psi}$ can be expressed in the form

$$\overline{\Psi}(\boldsymbol{C}, \Theta) = \breve{\Psi}(\mathcal{I}_C, \Theta),$$

for some function $\breve{\Psi}$. Here $\mathcal{I}_C$ are the principal invariants of $\boldsymbol{C}$. In this case, $\widehat{\boldsymbol{P}}$ takes the form

$$\widehat{\boldsymbol{P}}(\boldsymbol{F}, \Theta) = \boldsymbol{F}\,[\,\gamma_0(\mathcal{I}_C, \Theta)\boldsymbol{I} + \gamma_1(\mathcal{I}_C, \Theta)\boldsymbol{C} + \gamma_2(\mathcal{I}_C, \Theta)\boldsymbol{C}^{-1}\,],$$

for some scalar-valued functions γ_0, γ_1 and γ_2.

(4) A thermoelastic solid is said to be **thermally isotropic** if the Fourier–Stokes heat flux vector field $\boldsymbol{q}_m$ is invariant under rotations of the reference configuration. Under the conditions of Result 9.3, a thermoelastic solid is thermally isotropic if and only if the thermal conductivity $\overline{\boldsymbol{K}}$ is an isotropic tensor function of $\boldsymbol{C}$ in the sense of Section 1.5. In this case, $\overline{\boldsymbol{K}}$ takes the form

$$\overline{\boldsymbol{K}}(\boldsymbol{C}, \Theta) = \kappa_0(\mathcal{I}_C, \Theta)\boldsymbol{I} + \kappa_1(\mathcal{I}_C, \Theta)\boldsymbol{C} + \kappa_2(\mathcal{I}_C, \Theta)\boldsymbol{C}^{-1},$$

for some scalar-valued functions κ_0, κ_1 and κ_2 (see Exercise 2).

□

9.1.5 Initial-Boundary Value Problems

An initial-boundary value problem for a thermoelastic body is a set of equations for determining the motion φ and temperature Θ of a given body subject to specified initial conditions in B at time $t = 0$, and boundary conditions on ∂B at times $t \geq 0$. We typically assume B is a bounded open set as shown in Figure 7.1. However, it is also useful in applications to consider unbounded open sets such as the exterior of the region shown in the figure, or the whole of Euclidean space.

A standard initial-boundary value problem for a thermoelastic body with reference configuration B is the following: Find $\boldsymbol{\varphi} : B \times [0,T] \to \mathbb{E}^3$ and $\Theta : B \times [0,T] \to \mathbb{R}$ such that

$$\boxed{\begin{array}{ll} \rho_0 \ddot{\boldsymbol{\varphi}} = \nabla^X \cdot \boldsymbol{P} + \rho_0 \boldsymbol{b}_m & \text{in} \quad B \times [0,T] \\ \rho_0 \Theta \dot{\eta}_m + \nabla^X \cdot \boldsymbol{Q} = \rho_0 R & \text{in} \quad B \times [0,T] \\ \boldsymbol{\varphi} = \boldsymbol{g} & \text{in} \quad \Gamma_d \times [0,T] \\ \boldsymbol{P}\boldsymbol{N} = \boldsymbol{h} & \text{in} \quad \Gamma_\sigma \times [0,T] \\ \Theta = \xi & \text{in} \quad \Gamma_\theta \times [0,T] \\ \boldsymbol{Q} \cdot \boldsymbol{N} = \zeta & \text{in} \quad \Gamma_q \times [0,T] \\ \boldsymbol{\varphi}(\cdot, 0) = \boldsymbol{X} & \text{in} \quad B \\ \dot{\boldsymbol{\varphi}}(\cdot, 0) = \boldsymbol{V}_0 & \text{in} \quad B \\ \Theta(\cdot, 0) = \Theta_0 & \text{in} \quad B. \end{array}} \tag{9.24}$$

In the above system, Γ_d and Γ_σ are subsets of ∂B with the properties $\Gamma_d \cup \Gamma_\sigma = \partial B$ and $\Gamma_d \cap \Gamma_\sigma = \emptyset$, Γ_θ and Γ_q are subsets of ∂B with similar properties, ρ_0 is the reference mass density, $\boldsymbol{b}_m$ is a material body force field per unit reference mass, R is a material heat supply field per unit reference mass, $\boldsymbol{N}$ is the unit outward normal field on ∂B, and $\boldsymbol{g}$, $\boldsymbol{h}$, ξ, ζ, $\boldsymbol{V}_0$ and Θ_0 are prescribed fields. The first Piola–Kirchhoff stress $\boldsymbol{P}$, material heat flux $\boldsymbol{Q}$ and material entropy η_m are related to $\boldsymbol{\varphi}$ and Θ through frame-indifferent constitutive relations of the form

$$\boldsymbol{P} = 2\rho_0 \boldsymbol{F} D_{\boldsymbol{C}} \overline{\Psi}(\boldsymbol{C}, \Theta), \quad \eta_m = -D_\Theta \overline{\Psi}(\boldsymbol{C}, \Theta),$$

$$\boldsymbol{Q} = -\overline{\boldsymbol{K}}(\boldsymbol{C}, \Theta) \nabla^X \Theta,$$

where $\overline{\Psi}(\boldsymbol{C}, \Theta)$ is the free energy response function, $\overline{\boldsymbol{K}}(\boldsymbol{C}, \Theta)$ is the thermal conductivity function and $\boldsymbol{C} = \boldsymbol{F}^T \boldsymbol{F}$ is the Cauchy–Green strain tensor.

Equation $(9.24)_1$ is the balance of linear momentum equation, $(9.24)_2$ is the reduced balance of energy equation (Result 9.2), $(9.24)_3$ is a motion boundary condition on Γ_d, $(9.24)_4$ is a traction boundary condition on Γ_σ, $(9.24)_5$ is a temperature boundary condition on Γ_θ, $(9.24)_6$ is a heat flux boundary condition on Γ_q, $(9.24)_7$ is an initial condition for the motion $\boldsymbol{\varphi}$, $(9.24)_8$ is an initial condition for the material velocity $\dot{\boldsymbol{\varphi}}$

and $(9.24)_9$ is an initial condition for the temperature Θ. In general, the initial conditions should be compatible with the boundary conditions at time $t = 0$.

Remarks:

(1) The system in (9.24) is a coupled, nonlinear initial-boundary value problem for $\boldsymbol{\varphi}$ and Θ. Existence and uniqueness of solutions on finite time intervals $[0, T]$, where T depends on the problem, can be proved under suitable assumptions on the constitutive response functions, the prescribed initial and boundary data, the domain B and the subsets Γ_d, Γ_σ, Γ_θ and Γ_q. As in the case of (isothermal) elastic solids, questions of existence for all time are difficult. Numerical approximation is generally required to obtain quantitative information about solutions.

(2) The quantity $\boldsymbol{h}$ appearing in $(9.24)_4$ is a prescribed traction field for the first Piola–Kirchhoff stress. It corresponds to the external force on the current boundary expressed per unit area of the reference boundary. The quantity ζ appearing in $(9.24)_6$ represents a specified flux for the material heat flux vector field. It corresponds to the rate of heat transfer across the current boundary expressed per unit area of the reference boundary. Because physically meaningful forces and fluxes are typically expressed per unit area of the current boundary, expressions for $\boldsymbol{h}$ and ζ typically depend on the motion $\boldsymbol{\varphi}$. In particular, surface area elements in the current and reference configurations are related through $\boldsymbol{\varphi}$ (see Section 4.6).

(3) More general mechanical boundary conditions may be considered with (9.24). For example, $\varphi_i(\boldsymbol{X}, t)$ may be specified for all $\boldsymbol{X} \in \Gamma_d^i$ $(i = 1, 2, 3)$, and $P_{ij}(\boldsymbol{X}, t)N_j(\boldsymbol{X})$ may be specified for all $\boldsymbol{X} \in \Gamma_\sigma^i$ $(i = 1, 2, 3)$, where Γ_d^i and Γ_σ^i are subsets of ∂B with the properties $\Gamma_d^i \cup \Gamma_\sigma^i = \partial B$ and $\Gamma_d^i \cap \Gamma_\sigma^i = \emptyset$ for each $i = 1, 2, 3$. Thus, various components of the motion or traction may be specified at each point of the boundary.

(4) Any time-independent solution of (9.24) is called a **thermoelastic equilibrium** of the body. Such equilibria must satisfy

the nonlinear boundary-value problem

$$\begin{aligned} \nabla^X \cdot \boldsymbol{P} + \rho_0 \boldsymbol{b}_m &= \boldsymbol{0} && \text{in} \quad B \\ \nabla^X \cdot \boldsymbol{Q} - \rho_0 R &= 0 && \text{in} \quad B \\ \boldsymbol{\varphi} &= \boldsymbol{g} && \text{in} \quad \Gamma_d \\ \boldsymbol{P}\boldsymbol{N} &= \boldsymbol{h} && \text{in} \quad \Gamma_\sigma \\ \Theta &= \xi && \text{in} \quad \Gamma_\theta \\ \boldsymbol{Q} \cdot \boldsymbol{N} &= \zeta && \text{in} \quad \Gamma_q . \end{aligned} \tag{9.25}$$

Whereas (9.24) typically has a unique solution, we expect (9.25) to have genuinely non-unique solutions in various circumstances, just as in the purely elastic case. □

9.2 Linearization of Thermoelasticity Equations

As in the purely elastic case, the equations governing the motion of a thermoelastic solid are too complex to solve exactly for most constitutive models of interest. Hence approximations are sometimes made in order to simplify them. Here we use the technique of linearization to derive an approximate system of balance equations appropriate for describing small disturbances from a uniform reference state.

9.2.1 Preliminaries

Consider a thermoelastic body with uniform mass density $\rho_0(\boldsymbol{X}) = \rho_*$ (constant) and uniform temperature $\Theta(\boldsymbol{X}, 0) = \Theta_*$ (constant) at rest in a reference configuration B. Let $\widehat{\boldsymbol{P}}(\boldsymbol{F}, \Theta)$, $\widehat{\eta}_m(\boldsymbol{F}, \Theta)$ and $\widehat{\boldsymbol{K}}(\boldsymbol{F}, \Theta)$ denote the stress, entropy and thermal conductivity response functions for the body. Assume these functions satisfy the conditions in Results 9.1 and 9.3, and assume the reference configuration is uniform and stress-free in the sense that

$$\widehat{\boldsymbol{P}}(\boldsymbol{I}, \Theta_*) = \boldsymbol{O}, \quad \widehat{\eta}_m(\boldsymbol{I}, \Theta_*) = \eta_*, \quad \widehat{\boldsymbol{K}}(\boldsymbol{I}, \Theta_*) = \boldsymbol{K}_*, \tag{9.26}$$

where η_* is a given scalar (constant) and $\boldsymbol{K}_*$ is a given positive semi-definite tensor (constant).

Suppose the body is subject to a body force, heat supply, and boundary and initial conditions as given in (9.24). Assuming $\boldsymbol{b}_m$, $\boldsymbol{g}$, $\boldsymbol{h}$, $\boldsymbol{V}_0$ and R, ξ, ζ, Θ_0 are all small in the sense that

$$|\boldsymbol{b}_m(\boldsymbol{X}, t)|, \quad |\boldsymbol{g}(\boldsymbol{X}, t) - \boldsymbol{X}|, \quad |\boldsymbol{h}(\boldsymbol{X}, t)|, \quad |\boldsymbol{V}_0(\boldsymbol{X})| = \mathcal{O}(\epsilon), \tag{9.27}$$

and

$$|R(\boldsymbol{X},t)|,\ |\xi(\boldsymbol{X},t)-\Theta_*|,\ |\zeta(\boldsymbol{X},t)|,\ |\Theta_0(\boldsymbol{X})-\Theta_*| = \mathcal{O}(\epsilon), \tag{9.28}$$

for some small parameter $0 \le \epsilon \ll 1$, it is reasonable to expect that the body will deviate only slightly from its reference state in the sense that

$$|\boldsymbol{\varphi}(\boldsymbol{X},t)-\boldsymbol{X}|, \quad |\Theta(\boldsymbol{X},t)-\Theta_*| = \mathcal{O}(\epsilon). \tag{9.29}$$

Indeed, if $\epsilon = 0$, then equations (9.24) are satisfied by $\boldsymbol{\varphi}(\boldsymbol{X},t) = \boldsymbol{X}$ and $\Theta(\boldsymbol{X},t) = \Theta_*$ for all $\boldsymbol{X} \in B$ and $t \ge 0$.

Our goal here is to derive a set of simplified equations to describe processes satisfying (9.29) under the assumptions (9.26), (9.27) and (9.28). We discuss initial-boundary value problems for these equations and study some implications of the axiom of material frame-indifference and the assumption of isotropy.

9.2.2 Linearized Equations, Thermoelasticity Tensors

In view of (9.27) we assume that $\boldsymbol{b}_m$, $\boldsymbol{g}$, $\boldsymbol{h}$ and $\boldsymbol{V}_0$ can all be expressed in the form

$$\boldsymbol{b}_m^\epsilon = \epsilon \boldsymbol{b}_m^{(1)}, \quad \boldsymbol{g}^\epsilon = \boldsymbol{X} + \epsilon \boldsymbol{g}^{(1)}, \quad \boldsymbol{h}^\epsilon = \epsilon \boldsymbol{h}^{(1)}, \quad \boldsymbol{V}_0^\epsilon = \epsilon \boldsymbol{V}_0^{(1)},$$

for some functions $\boldsymbol{b}_m^{(1)}$, $\boldsymbol{g}^{(1)}$, $\boldsymbol{h}^{(1)}$ and $\boldsymbol{V}_0^{(1)}$. Similarly, in view of (9.28) we assume that R, ξ, ζ and Θ_0 can all be expressed in the form

$$R^\epsilon = \epsilon R^{(1)}, \quad \xi^\epsilon = \Theta_* + \epsilon \xi^{(1)}, \quad \zeta^\epsilon = \epsilon \zeta^{(1)}, \quad \Theta_0^\epsilon = \Theta_* + \epsilon \Theta_0^{(1)},$$

for some functions $R^{(1)}$, $\xi^{(1)}$, $\zeta^{(1)}$ and $\Theta_0^{(1)}$. Then, in accord with (9.29), we seek power series expansions for $\boldsymbol{\varphi}$ and Θ of the form

$$\boldsymbol{\varphi}^\epsilon = \boldsymbol{X} + \epsilon \boldsymbol{u}^{(1)} + \mathcal{O}(\epsilon^2) \quad \text{and} \quad \Theta^\epsilon = \Theta_* + \epsilon \Theta^{(1)} + \mathcal{O}(\epsilon^2), \tag{9.30}$$

where $\boldsymbol{u}^{(1)}$ and $\Theta^{(1)}$ are unknown fields.

We next derive partial differential equations for the first-order displacement disturbance $\boldsymbol{u}^{(1)}$ and temperature disturbance $\Theta^{(1)}$. From $(9.24)_1$ we deduce that $\boldsymbol{\varphi}^\epsilon$ satisfies the balance of linear momentum equation

$$\rho_* \ddot{\boldsymbol{\varphi}}^\epsilon = \nabla^X \cdot \boldsymbol{P}^\epsilon + \rho_* \boldsymbol{b}_m^\epsilon, \tag{9.31}$$

where $\boldsymbol{P}^\epsilon$ is the stress field given by $\boldsymbol{P}^\epsilon = \widehat{\boldsymbol{P}}(\boldsymbol{F}^\epsilon, \Theta^\epsilon)$. Here $\boldsymbol{F}^\epsilon$ is the deformation gradient field associated with $\boldsymbol{\varphi}^\epsilon$, namely

$$\boldsymbol{F}^\epsilon = \nabla^X \boldsymbol{\varphi}^\epsilon = \boldsymbol{I} + \epsilon \nabla^X \boldsymbol{u}^{(1)} + \mathcal{O}(\epsilon^2).$$

Similarly, from $(9.24)_2$ we deduce that Θ^ϵ satisfies the balance of energy equation

$$\rho_* \Theta^\epsilon \dot{\eta}_m^\epsilon + \nabla^X \cdot \boldsymbol{Q}^\epsilon = \rho_* R^\epsilon, \tag{9.32}$$

where η_m^ϵ is the material entropy field given by $\eta_m^\epsilon = \widehat{\eta}_m(\boldsymbol{F}^\epsilon, \Theta^\epsilon)$ and $\boldsymbol{Q}^\epsilon$ is the material heat flux field given by $\boldsymbol{Q}^\epsilon = -\widehat{\boldsymbol{K}}(\boldsymbol{F}^\epsilon, \Theta^\epsilon)\nabla^X \Theta^\epsilon$. Here $\nabla^X \Theta^\epsilon$ is the gradient field associated with Θ^ϵ, namely

$$\nabla^X \Theta^\epsilon = \epsilon \nabla^X \Theta^{(1)} + \mathcal{O}(\epsilon^2).$$

The following definition will be helpful in simplifying (9.31) and (9.32).

Definition 9.4 *Let $\widehat{\boldsymbol{P}}(\boldsymbol{F}, \Theta)$ and $\widehat{\eta}_m(\boldsymbol{F}, \Theta)$ be the stress and entropy response functions for a thermoelastic body. Then the fourth-order* **elasticity tensor** *for the body is*

$$\mathsf{A}_* = \frac{\partial \widehat{\boldsymbol{P}}}{\partial \boldsymbol{F}}(\boldsymbol{I}, \Theta_*) \quad or \quad \mathsf{A}^*_{ijkl} = \frac{\partial \widehat{P}_{ij}}{\partial F_{kl}}(\boldsymbol{I}, \Theta_*),$$

the second-order **thermal stress tensor** *is*

$$\boldsymbol{D}_* = \frac{\partial \widehat{\boldsymbol{P}}}{\partial \Theta}(\boldsymbol{I}, \Theta_*) \quad or \quad D^*_{ij} = \frac{\partial \widehat{P}_{ij}}{\partial \Theta}(\boldsymbol{I}, \Theta_*),$$

and the scalar **entropy parameter** *is*

$$\beta_* = \frac{\partial \widehat{\eta}_m}{\partial \Theta}(\boldsymbol{I}, \Theta_*).$$

□

Expanding the stress response function $\widehat{\boldsymbol{P}}(\boldsymbol{F}^\epsilon, \Theta^\epsilon)$ in a Taylor series about $\epsilon = 0$ we get

$$\widehat{P}_{ij}(\boldsymbol{F}^\epsilon, \Theta^\epsilon) = \widehat{P}_{ij}\Big|_{\epsilon=0} + \epsilon\left[\frac{\partial \widehat{P}_{ij}}{\partial F_{kl}}\frac{\partial F^\epsilon_{kl}}{\partial \epsilon} + \frac{\partial \widehat{P}_{ij}}{\partial \Theta}\frac{\partial \Theta^\epsilon}{\partial \epsilon}\right]\Bigg|_{\epsilon=0} + \mathcal{O}(\epsilon^2).$$

Using the stress-free assumption $\widehat{\boldsymbol{P}}(\boldsymbol{I}, \Theta_*) = \boldsymbol{O}$ together with the above definition we get

$$\widehat{\boldsymbol{P}}(\boldsymbol{F}^\epsilon, \Theta^\epsilon) = \epsilon[\mathsf{A}_*(\nabla^X \boldsymbol{u}^{(1)}) + \boldsymbol{D}_*\Theta^{(1)}] + \mathcal{O}(\epsilon^2). \tag{9.33}$$

When we substitute (9.33) into (9.31) and retain only those terms involving the first power of ϵ we obtain

$$\rho_* \ddot{\boldsymbol{u}}^{(1)} = \nabla^X \cdot [\mathsf{A}_*(\nabla^X \boldsymbol{u}^{(1)}) + \boldsymbol{D}_*\Theta^{(1)}] + \rho_* \boldsymbol{b}_m^{(1)}.$$

This is the linearized version of the balance of linear momentum equation $(9.24)_1$.

Expanding the entropy response function $\widehat{\eta}_m(\boldsymbol{F}^\epsilon, \Theta^\epsilon)$ in a Taylor series about $\epsilon = 0$ we get

$$\widehat{\eta}_m(\boldsymbol{F}^\epsilon, \Theta^\epsilon) = \widehat{\eta}_m\Big|_{\epsilon=0} + \epsilon\Big[\frac{\partial\widehat{\eta}_m}{\partial F_{kl}}\frac{\partial F^\epsilon_{kl}}{\partial\epsilon} + \frac{\partial\widehat{\eta}_m}{\partial\Theta}\frac{\partial\Theta^\epsilon}{\partial\epsilon}\Big]\Big|_{\epsilon=0} + \mathcal{O}(\epsilon^2).$$

Using the Maxwell relation (9.13), together with $(9.26)_2$ and Definition 9.4, we find

$$\widehat{\eta}_m(\boldsymbol{F}^\epsilon, \Theta^\epsilon) = \eta_* + \epsilon\Big[-\frac{1}{\rho_*}\boldsymbol{D}_* : \nabla^X \boldsymbol{u}^{(1)} + \beta_*\Theta^{(1)}\Big] + \mathcal{O}(\epsilon^2). \qquad (9.34)$$

Considering the material heat flux $\boldsymbol{Q}^\epsilon = -\widehat{\boldsymbol{K}}(\boldsymbol{F}^\epsilon, \Theta^\epsilon)\nabla^X\Theta^\epsilon$ we expand each factor to obtain

$$\begin{aligned}\boldsymbol{Q}^\epsilon &= -\widehat{\boldsymbol{K}}(\boldsymbol{F}^\epsilon, \Theta^\epsilon)\nabla^X\Theta^\epsilon \\ &= -[\boldsymbol{K}_* + \mathcal{O}(\epsilon)][\epsilon\nabla^X\Theta^{(1)} + \mathcal{O}(\epsilon^2)] \\ &= -\epsilon\boldsymbol{K}_*\nabla^X\Theta^{(1)} + \mathcal{O}(\epsilon^2).\end{aligned} \qquad (9.35)$$

When we substitute (9.34) and (9.35) into (9.32) and retain only those terms involving the first power of ϵ we obtain

$$\rho_*\Theta_*\Big[-\frac{1}{\rho_*}\boldsymbol{D}_* : \nabla^X\dot{\boldsymbol{u}}^{(1)} + \beta_*\dot{\Theta}^{(1)}\Big] - \nabla^X\cdot[\boldsymbol{K}_*\nabla^X\Theta^{(1)}] = \rho_* R^{(1)},$$

and rearranging terms gives

$$\rho_*\Theta_*\beta_*\dot{\Theta}^{(1)} = \nabla^X\cdot[\boldsymbol{K}_*\nabla^X\Theta^{(1)}] + \Theta_*\boldsymbol{D}_* : \nabla^X\dot{\boldsymbol{u}}^{(1)} + \rho_* R^{(1)}.$$

This is the linearized version of the balance of energy equation $(9.24)_2$. Similar considerations can be used to obtain linearized versions of the boundary and initial conditions in $(9.24)_{3-9}$.

Remarks:

(1) The fourth-order elasticity tensor $\mathsf{A}_* = D_{\boldsymbol{F}}\widehat{\boldsymbol{P}}(\boldsymbol{I}, \Theta_*)$ is the same elasticity tensor that arises in the linearization of an isothermal elastic model (see Chapter 7). In particular, it is the derivative of the first Piola–Kirchhoff stress response function with respect to the deformation gradient at fixed temperature, evaluated at the reference configuration.

(2) Under the conditions of Results 9.1 and 9.3, the elasticity tensor A_*, thermal stress tensor $\boldsymbol{D}_*$ and entropy parameter β_* are completely defined by the free energy function $\overline{\Psi}(\boldsymbol{C}, \Theta)$ (see Exercise 3). Moreover, in this case, the tensor A_* necessarily has major

and minor symmetry (both right and left) and the tensor $\boldsymbol{D}_*$ is necessarily symmetric (see Exercise 4).

(3) Assuming $\widehat{\eta}_m$ satisfies the condition of Result 9.1, we find that the entropy parameter β_* is related to the specific heat $\alpha_* = \alpha(\boldsymbol{I}, \Theta_*)$. In particular, we have $\widehat{\eta}_m(\boldsymbol{F}, \Theta) = -\partial\widehat{\Psi}/\partial\Theta(\boldsymbol{F}, \Theta)$, and from (9.14) we have $\alpha(\boldsymbol{F}, \Theta) = -\Theta\partial^2\widehat{\Psi}/\partial\Theta^2(\boldsymbol{F}, \Theta)$, which implies

$$\alpha_* = -\Theta_* \frac{\partial^2 \widehat{\Psi}}{\partial \Theta^2}(\boldsymbol{I}, \Theta_*) = \Theta_* \beta_*.$$

□

9.2.3 Initial-Boundary Value Problems

Linearization of (9.24) leads to the following initial-boundary value problem for the first-order displacement and temperature disturbances in a thermoelastic solid: Find $\boldsymbol{u}^{(1)} : B \times [0, T] \to \mathcal{V}$ and $\Theta^{(1)} : B \times [0, T] \to \boldsymbol{R}$ such that

$$\boxed{\begin{array}{lll}
\rho_* \ddot{\boldsymbol{u}}^{(1)} = \nabla^X \cdot [\mathsf{A}_*(\nabla^X \boldsymbol{u}^{(1)}) + \boldsymbol{D}_* \Theta^{(1)}] + \rho_* \boldsymbol{b}_m^{(1)} & \text{in} & B \times [0, T] \\
\rho_* \alpha_* \dot{\Theta}^{(1)} = \nabla^X \cdot [\boldsymbol{K}_* \nabla^X \Theta^{(1)}] & & \\
\qquad\qquad + \Theta_* \boldsymbol{D}_* : \nabla^X \dot{\boldsymbol{u}}^{(1)} + \rho_* R^{(1)} & \text{in} & B \times [0, T] \\
\boldsymbol{u}^{(1)} = \boldsymbol{g}^{(1)} & \text{in} & \Gamma_d \times [0, T] \\
[\mathsf{A}_*(\nabla^X \boldsymbol{u}^{(1)}) + \boldsymbol{D}_* \Theta^{(1)}] \boldsymbol{N} = \boldsymbol{h}^{(1)} & \text{in} & \Gamma_\sigma \times [0, T] \\
\Theta^{(1)} = \xi^{(1)} & \text{in} & \Gamma_\theta \times [0, T] \\
\boldsymbol{K}_* \nabla^X \Theta^{(1)} \cdot \boldsymbol{N} = -\zeta^{(1)} & \text{in} & \Gamma_q \times [0, T] \\
\boldsymbol{u}^{(1)}(\cdot, 0) = \boldsymbol{0} & \text{in} & B \\
\dot{\boldsymbol{u}}^{(1)}(\cdot, 0) = \boldsymbol{V}_0^{(1)} & \text{in} & B \\
\Theta^{(1)}(\cdot, 0) = \Theta_0^{(1)} & \text{in} & B.
\end{array}} \tag{9.36}$$

The above equations are typically called the **Linearized Thermoelastodynamics Equations**. They are an approximation to the system in (9.24) that are appropriate for describing small deviations from a uniform, stress-free reference state in a thermoelastic solid. In the above system, Γ_d and Γ_σ are subsets of ∂B with the properties $\Gamma_d \cup \Gamma_\sigma = \partial B$ and $\Gamma_d \cap \Gamma_\sigma = \emptyset$, Γ_θ and Γ_q are subsets of ∂B with similar properties,

ρ_* is the reference mass density, Θ_* is the reference temperature, α_* is the reference specific heat, A_* is the elasticity tensor, $\boldsymbol{D}_*$ is the thermal stress tensor, $\boldsymbol{K}_*$ is the thermal conductivity tensor, $\boldsymbol{N}$ is the unit outward normal field on ∂B, and $\boldsymbol{b}_m^{(1)}$, $\boldsymbol{g}^{(1)}$, $\boldsymbol{h}^{(1)}$, $\boldsymbol{V}_0^{(1)}$, $R^{(1)}$, $\xi^{(1)}$, $\zeta^{(1)}$ and $\Theta_0^{(1)}$ are prescribed disturbance fields. Any of the more general boundary conditions discussed in connection with (9.24) can also be considered for (9.36).

Remarks:

(1) The system in (9.36) is a coupled, linear initial-boundary value problem for the disturbance fields $\boldsymbol{u}^{(1)}$ and $\Theta^{(1)}$. This system has a unique solution in any given time interval $[0,T]$ under mild assumptions on the tensors A_*, $\boldsymbol{D}_*$ and $\boldsymbol{K}_*$, the prescribed initial and boundary data, the domain B, and the subsets Γ_d, Γ_σ, Γ_θ and Γ_q.

(2) When $\boldsymbol{D}_* = \boldsymbol{O}$ the system in (9.36) decouples. In particular, $\boldsymbol{u}^{(1)}$ satisfies the equation of linearized isothermal elasticity independent of $\Theta^{(1)}$, and $\Theta^{(1)}$ satisfies a standard, time-dependent heat equation independent of $\boldsymbol{u}^{(1)}$ (see Exercise 5). When $\boldsymbol{D}_* \neq \boldsymbol{O}$ the system does not decouple in general.

(3) Any time-independent solution of (9.36) must satisfy the linear boundary-value problem

$$\begin{aligned}
&\nabla^x \cdot [\mathsf{A}_*(\nabla^x \boldsymbol{u}^{(1)}) + \boldsymbol{D}_*\Theta^{(1)}] + \rho_*\boldsymbol{b}_m^{(1)} = \boldsymbol{0} && \text{in} \quad B\\
&\boldsymbol{u}^{(1)} = \boldsymbol{g}^{(1)} && \text{in} \quad \Gamma_d\\
&[\mathsf{A}_*(\nabla^x \boldsymbol{u}^{(1)}) + \boldsymbol{D}_*\Theta^{(1)}]\boldsymbol{N} = \boldsymbol{h}^{(1)} && \text{in} \quad \Gamma_\sigma,
\end{aligned}$$

$$\begin{aligned}
&\nabla^x \cdot [\boldsymbol{K}_*\nabla^x \Theta^{(1)}] + \rho_* R^{(1)} = 0 && \text{in} \quad B\\
&\Theta^{(1)} = \xi^{(1)} && \text{in} \quad \Gamma_\theta\\
&\boldsymbol{K}_*\nabla^x \Theta^{(1)} \cdot \boldsymbol{N} = -\zeta^{(1)} && \text{in} \quad \Gamma_q.
\end{aligned}$$

These equations are typically called the **Linearized Thermoelastostatics Equations**. The system for $\Theta^{(1)}$ is a steady-state heat equation which can be solved independently of $\boldsymbol{u}^{(1)}$. The system for $\boldsymbol{u}^{(1)}$ has the same form as the linearized isothermal elastostatics equations, but with a superimposed thermal stress field $\boldsymbol{D}_*\Theta^{(1)}$.

(4) The tensors A_*, $\boldsymbol{D}_*$ and $\boldsymbol{K}_*$ take simple forms when the underlying model is isotropic. In particular, the linearization of any

mechanically and thermally isotropic model gives

$$\mathsf{A}_*(\boldsymbol{H}) = \lambda_*(\operatorname{tr}\boldsymbol{H})\boldsymbol{I} + 2\mu_* \operatorname{sym}(\boldsymbol{H}), \quad \forall \boldsymbol{H} \in \mathcal{V}^2,$$

$$\boldsymbol{D}_* = m_* \boldsymbol{I}, \quad \boldsymbol{K}_* = \kappa_* \boldsymbol{I},$$

where λ_* and μ_* are the Lamé constants, m_* is the **thermal stress coefficient** and κ_* is the **thermal conductivity** for the material (see Exercises 6 and 7).

(5) Assuming A_* has minor symmetry (both right and left), and that $\boldsymbol{D}_*$ is symmetric, the linearized first Piola–Kirchhoff stress field is itself symmetric and is given by

$$\boldsymbol{P}^{(1)} = \mathsf{A}_*(\boldsymbol{E}^{(1)}) + \boldsymbol{D}_*\Theta^{(1)}, \tag{9.37}$$

where $\boldsymbol{E}^{(1)} = \operatorname{sym}(\nabla^X \boldsymbol{u}^{(1)})$ is the infinitesimal strain field associated with $\boldsymbol{u}^{(1)}$. Assuming the above equation can be solved for $\boldsymbol{E}^{(1)}$ we get

$$\boldsymbol{E}^{(1)} = \mathsf{G}_*(\boldsymbol{P}^{(1)}) + \boldsymbol{H}_*\Theta^{(1)},$$

where G_* is called the **compliance tensor** and $\boldsymbol{H}_*$ the **thermal strain tensor**. Notice that $\boldsymbol{D}_*\Theta^{(1)}$ gives the stress in the absence of strain, whereas $\boldsymbol{H}_*\Theta^{(1)}$ gives the strain in the absence of stress (see Exercise 8). The linear thermoelastic constitutive relation (9.37) is typically called the **Duhamel–Neumann model**.

(6) In view of $(9.30)_1$ we see that the difference between the fixed reference configuration B and the current configuration $B_t = \varphi_t(B)$ is first-order in ϵ for all $t \geq 0$. For this reason, the distinction between spatial and material coordinates is typically ignored in (9.36). Moreover, from $(9.30)_2$ we see that the field $\Theta^{(1)}$ measures deviations from the uniform reference temperature Θ_*. Thus $\Theta^{(1)}$ is not an absolute temperature field and can take positive and negative values.

□

Bibliographic Notes

General accounts of the theory of thermoelasticity can be found in Antman (1995), Marsden and Hughes (1983) and Truesdell and Noll (1965). A concise account of the linearized theory is given in Carlson (1972). Various engineering applications may be found in Holzapfel

(2000), Nowacki (1986), Parkus (1976), Fung (1965) and Boley and Weiner (1960). Detailed mathematical analyses of existence, uniqueness and stability in the linearized theory can be found in Dafermos (1968) and Day (1985). Similar analyses for the nonlinear theory can be found in Racke and Jiang (2000) and Slemrod (1981). An introduction to numerical finite element methods for thermoelasticity used in engineering practice is given in Nicholson (2003).

Exercises

9.1 Consider a more general constitutive model for a thermoelastic solid of the form

$$\boldsymbol{P} = \widehat{\boldsymbol{P}}(\boldsymbol{F}, \Theta, \boldsymbol{g}), \quad \boldsymbol{Q} = \widehat{\boldsymbol{Q}}(\boldsymbol{F}, \Theta, \boldsymbol{g}),$$

$$\eta_m = \widehat{\eta}_m(\boldsymbol{F}, \Theta, \boldsymbol{g}), \quad \Psi = \widehat{\Psi}(\boldsymbol{F}, \Theta, \boldsymbol{g}),$$

where $\boldsymbol{F} = \nabla^X \varphi$ and $\boldsymbol{g} = \nabla^X \Theta$. Show that such a model satisfies the Clausius–Duhem inequality if and only if all the following are true:

(i) $\widehat{\boldsymbol{P}}$, $\widehat{\eta}_m$ and $\widehat{\Psi}$ are independent of $\boldsymbol{g}$,

(ii) $\widehat{\boldsymbol{P}}(\boldsymbol{F}, \Theta) = \rho_0 D_{\boldsymbol{F}} \widehat{\Psi}(\boldsymbol{F}, \Theta)$ and $\widehat{\eta}_m(\boldsymbol{F}, \Theta) = -D_\Theta \widehat{\Psi}(\boldsymbol{F}, \Theta)$,

(iii) $\widehat{\boldsymbol{Q}}(\boldsymbol{F}, \Theta, \boldsymbol{g}) \cdot \boldsymbol{g} \leq 0$ for all admissible $(\boldsymbol{F}, \Theta, \boldsymbol{g})$.

9.2 Consider a thermoelastic body with material and Fourier–Stokes heat flux vector fields given by

$$\boldsymbol{Q} = \widehat{\boldsymbol{Q}}(\boldsymbol{F}, \Theta, \boldsymbol{g}), \quad \boldsymbol{q}_m = \widehat{\boldsymbol{q}}_m(\boldsymbol{F}, \Theta, \boldsymbol{g}),$$

where $\boldsymbol{g} = \nabla^X \Theta$. By definition, such a body is thermally isotropic if

$$\widehat{\boldsymbol{q}}_m(\boldsymbol{F}\boldsymbol{\Lambda}, \Theta, \boldsymbol{\Lambda}^T \boldsymbol{g}) = \widehat{\boldsymbol{q}}_m(\boldsymbol{F}, \Theta, \boldsymbol{g}),$$

for all rotation tensors $\boldsymbol{\Lambda}$ and all $(\boldsymbol{F}, \Theta, \boldsymbol{g})$ with $\det \boldsymbol{F} > 0$ and $\Theta > 0$.

(a) Assuming $\widehat{\boldsymbol{Q}}(\boldsymbol{F}, \Theta, \boldsymbol{g}) = -\boldsymbol{K}(\boldsymbol{C}, \Theta)\boldsymbol{g}$, where $\boldsymbol{C} = \boldsymbol{F}^T \boldsymbol{F}$, find an expression for $\widehat{\boldsymbol{q}}_m(\boldsymbol{F}, \Theta, \boldsymbol{g})$.

(b) Show that a thermoelastic solid is thermally isotropic if and only if the thermal conductivity function $\overline{\boldsymbol{K}}$ takes the form

$$\overline{\boldsymbol{K}}(\boldsymbol{C}, \Theta) = \kappa_0(\mathcal{I}_C, \Theta)\boldsymbol{I} + \kappa_1(\mathcal{I}_C, \Theta)\boldsymbol{C} + \kappa_2(\mathcal{I}_C, \Theta)\boldsymbol{C}^{-1},$$

for some scalar-valued functions κ_0, κ_1 and κ_2.

9.3 Consider a thermoelastic body with response functions satisfying the conditions of Results 9.1 and 9.3. Suppose the body has a reference configuration with uniform mass density ρ_* and uniform temperature Θ_*. Moreover, suppose the free energy response function is of the form

$$\overline{\Psi}(\boldsymbol{C},\Theta) = \frac{1}{2}\boldsymbol{G} : \mathsf{J}(\boldsymbol{G}) + \vartheta \boldsymbol{M} : \boldsymbol{G} - \frac{1}{2}c\vartheta^2,$$

where J is a constant fourth-order tensor with major and minor symmetry (both right and left), $\boldsymbol{M}$ is a constant, symmetric second-order tensor, $c > 0$ is a constant scalar, $\boldsymbol{G} = \frac{1}{2}(\boldsymbol{C} - \boldsymbol{I})$ is the Green–St. Venant strain tensor and $\vartheta = \Theta - \Theta_*$ is the temperature deviation.

(a) Show that the response functions $\widehat{\boldsymbol{P}}$ and $\widehat{\eta}_m$ corresponding to $\overline{\Psi}$ are

$$\widehat{\boldsymbol{P}}(\boldsymbol{F},\Theta) = \rho_* \boldsymbol{F}[\mathsf{J}(\boldsymbol{G}) + \vartheta \boldsymbol{M}],$$
$$\widehat{\eta}_m(\boldsymbol{F},\Theta) = c\vartheta - \boldsymbol{M} : \boldsymbol{G}.$$

Moreover, verify that the reference configuration is stress-free.

(b) Show that the elasticity tensor A_*, thermal stress tensor $\boldsymbol{D}_*$ and entropy parameter β_* corresponding to $\overline{\Psi}$ are given by

$$\mathsf{A}_* = \rho_* \mathsf{J}, \quad \boldsymbol{D}_* = \rho_* \boldsymbol{M}, \quad \beta_* = c.$$

9.4 Consider a stress response function of the form $\widehat{\boldsymbol{P}}(\boldsymbol{F},\Theta) = \rho_* D_{\boldsymbol{F}} \widehat{\Psi}(\boldsymbol{F},\Theta)$, where $\widehat{\Psi}(\boldsymbol{F},\Theta) = \overline{\Psi}(\boldsymbol{C},\Theta)$. Assuming a stress-free reference state, show that the elasticity tensor A_* and the thermal stress tensor $\boldsymbol{D}_*$ satisfy

$$\mathsf{A}^*_{ijkl} = \rho_* \frac{\partial^2 \widehat{\Psi}}{\partial F_{ij} \partial F_{kl}}(\boldsymbol{I},\Theta_*) = 4\rho_* \frac{\partial^2 \overline{\Psi}}{\partial C_{ij} \partial C_{kl}}(\boldsymbol{I},\Theta_*),$$

and

$$D^*_{ij} = \rho_* \frac{\partial^2 \widehat{\Psi}}{\partial \Theta \partial F_{ij}}(\boldsymbol{I},\Theta_*) = 2\rho_* \frac{\partial^2 \overline{\Psi}}{\partial \Theta \partial C_{ij}}(\boldsymbol{I},\Theta_*).$$

Deduce that A_* necessarily has major and minor symmetry (both right and left) and that $\boldsymbol{D}_*$ is necessarily symmetric.

9.5 Assuming no coupling ($\boldsymbol{D}_* = \boldsymbol{O}$), consider the linearized thermoelastodynamics equation for the temperature disturbance in a thermoelastic body with reference configuration $B = \{\boldsymbol{X} \in \mathbb{E}^3 \mid 0 < X_i < 1\}$, namely

$$\begin{array}{ll} \rho_* \alpha_* \dot{\Theta}^{(1)} = \nabla^X \cdot [\boldsymbol{K}_* \nabla^X \Theta^{(1)}] + \rho_* R^{(1)} & \text{in} \quad B \times [0,T] \\ \Theta^{(1)} = \xi^{(1)} & \text{in} \quad \Gamma_\theta \times [0,T] \\ \boldsymbol{K}_* \nabla^X \Theta^{(1)} \cdot \boldsymbol{N} = -\zeta^{(1)} & \text{in} \quad \Gamma_q \times [0,T] \\ \Theta^{(1)}(\cdot, 0) = \Theta_0^{(1)} & \text{in} \quad B. \end{array}$$

Suppose that there is no heat supply, so $R^{(1)} = 0$, that the body is thermally isotropic, so $\boldsymbol{K}_* = \kappa_* \boldsymbol{I}$, and that the temperature disturbance vanishes on the entire boundary, so $\xi^{(1)} = 0$, $\Gamma_\theta = \partial B$, and $\Gamma_q = \emptyset$. Moreover, assume that the initial condition can be expanded in a Fourier sine series as

$$\Theta_0^{(1)}(\boldsymbol{X}) = \sum_{k_1,k_2,k_3=1}^{\infty} A_{k_1 k_2 k_3} \sin(k_1 \pi X_1) \sin(k_2 \pi X_2) \sin(k_3 \pi X_3),$$

where $A_{k_1 k_2 k_3}$ are scalar constants. Use separation of variables to find the temperature disturbance field $\Theta^{(1)}(\boldsymbol{X}, t)$ for all $\boldsymbol{X} \in B$ and $t \geq 0$.

9.6 Show that the linearized balance of momentum and energy equations for a mechanically and thermally isotropic body take the forms

$$\begin{aligned} \rho_* \ddot{\boldsymbol{u}}^{(1)} &= \mu_* \Delta^X \boldsymbol{u}^{(1)} + (\lambda_* + \mu_*) \nabla^X (\nabla^X \cdot \boldsymbol{u}^{(1)}) \\ &\qquad + m_* \nabla^X \Theta^{(1)} + \rho_* \boldsymbol{b}_m^{(1)}, \\ \rho_* \alpha_* \dot{\Theta}^{(1)} &= \kappa_* \Delta^X \Theta^{(1)} + \Theta_* m_* \nabla^X \cdot \dot{\boldsymbol{u}}^{(1)} + \rho_* R^{(1)}, \end{aligned}$$

where ρ_* is the mass density, Θ_* is the reference temperature, α_* is the specific heat, λ_* and μ_* are the Lamé constants, m_* is the thermal stress coefficient and κ_* is the thermal conductivity.

9.7 Assuming no body force and no heat supply, consider the equations for a fully isotropic body derived in Exercise 6, namely

$$\begin{aligned} \ddot{\boldsymbol{u}}^{(1)} &= \mu \Delta^X \boldsymbol{u}^{(1)} + (\lambda + \mu) \nabla^X (\nabla^X \cdot \boldsymbol{u}^{(1)}) + m \nabla^X \Theta^{(1)}, \\ \dot{\Theta}^{(1)} &= \kappa \Delta^X \Theta^{(1)} + \nu \nabla^X \cdot \dot{\boldsymbol{u}}^{(1)}, \end{aligned} \tag{9.38}$$

where $\lambda = \lambda_*/\rho_*$, $\mu = \mu_*/\rho_*$, $m = m_*/\rho_*$, $\kappa = \kappa_*/(\rho_* \alpha_*)$ and $\nu = \Theta_* m_*/(\rho_* \alpha_*)$. We suppose $m \neq 0$ and $\nu \neq 0$ so that the equations are fully coupled. Here we study plane progressive wave solutions of the form

$$\begin{aligned}\boldsymbol{u}^{(1)}(\boldsymbol{X},t) &= \sigma \boldsymbol{a} e^{i(\gamma \boldsymbol{k}\cdot\boldsymbol{X}-\omega t)},\\ \Theta^{(1)}(\boldsymbol{X},t) &= \tau e^{i(\gamma \boldsymbol{k}\cdot\boldsymbol{X}-\omega t)},\end{aligned} \tag{9.39}$$

where $\boldsymbol{a}$ and $\boldsymbol{k}$ are unit vectors called the direction of displacement and propagation, σ and τ are real constants called the displacement and thermal amplitude coefficients, $i = \sqrt{-1}$ is the pure imaginary unit, γ is a real constant called the wave number and ω is a complex constant called the wave frequency. The wavelength is given by $2\pi/\gamma$, wave period is given by $2\pi/\operatorname{Re}(\omega)$ and wave speed is given by $\operatorname{Re}(\omega)/\gamma$. The imaginary part $\operatorname{Im}(\omega)$ is called the amplification rate. The solution amplitude decays, remains constant or grows in time according to whether $\operatorname{Im}(\omega)$ is negative, zero or positive.

(a) Show that the functions in (9.39) satisfy (9.38) if and only if $\boldsymbol{a}$, $\boldsymbol{k}$, γ, ω, σ, τ satisfy

$$\begin{aligned}\omega^2\sigma\boldsymbol{a} &= \gamma^2\sigma[\mu\boldsymbol{a} + (\lambda+\mu)(\boldsymbol{a}\cdot\boldsymbol{k})\boldsymbol{k}] - im\gamma\tau\boldsymbol{k},\\ (\kappa\gamma^2 - i\omega)\tau &= \nu\gamma\omega\sigma(\boldsymbol{a}\cdot\boldsymbol{k}).\end{aligned}$$

Notice that these equations are satisfied by the trivial wave defined by $\sigma = 0$ and $\tau = 0$.

(b) Given $\boldsymbol{k}$, $\gamma \neq 0$ show that non-trivial waves with $\boldsymbol{a}\cdot\boldsymbol{k} = 0$ (transverse waves) exist if and only if

$$\omega^2 = \mu\gamma^2.$$

In this case $\tau = 0$ and there is no restriction on σ.

(c) Given $\boldsymbol{k}$, $\gamma \neq 0$ show that non-trivial waves with $\boldsymbol{a}\cdot\boldsymbol{k} = 1$ (longitudinal waves) exist if and only if ω is a solution of the cubic equation

$$[\omega^2 - \gamma^2(\lambda+2\mu)](\kappa\gamma^2 - i\omega) + im\nu\gamma^2\omega = 0.$$

In this case σ and τ are any non-trivial solution of

$$\begin{pmatrix}\omega^2 - \gamma^2(\lambda+2\mu) & im\gamma\\ -\nu\gamma\omega & \kappa\gamma^2 - i\omega\end{pmatrix}\begin{Bmatrix}\sigma\\ \tau\end{Bmatrix} = \begin{Bmatrix}0\\ 0\end{Bmatrix}.$$

Remark: The above results show that transverse waves are independent of thermal effects, whereas longitudinal waves are not. In particular, the result in (b) for transverse waves is essentially identical to that obtained (by different means) in Chapter 7, Exercise 15 for the isothermal case. However, the result in (c) for longitudinal waves is different. Here, longitudinal waves are **damped** in that their frequency is complex, and **dispersed** in that their speed depends on wavenumber. In the isothermal case, both transverse and longitudinal waves are undamped and undispersed.

9.8 Consider the stress-strain relation for a linearized, isotropic, thermoelastic solid with Lamé constants λ_*, μ_* and thermal stress coefficient m_*

$$\boldsymbol{P}^{(1)} = \lambda_*(\mathrm{tr}\,\boldsymbol{E}^{(1)})\boldsymbol{I} + 2\mu_*\boldsymbol{E}^{(1)} + m_*\Theta^{(1)}\boldsymbol{I}.$$

Show that this relation can be inverted as

$$\boldsymbol{E}^{(1)} = a_*(\mathrm{tr}\,\boldsymbol{P}^{(1)})\boldsymbol{I} + 2b_*\boldsymbol{P}^{(1)} + c_*\Theta^{(1)}\boldsymbol{I},$$

where

$$a_* = -\frac{\lambda_*}{2\mu_*(3\lambda_* + 2\mu_*)}, \qquad b_* = \frac{1}{4\mu_*}, \qquad c_* = -\frac{m_*}{3\lambda_* + 2\mu_*}.$$

Remark: The above result shows that, when the elasticity tensor A_* is isotropic and the thermal stress tensor $\boldsymbol{D}_*$ is spherical, so are the compliance tensor G_* and the thermal strain tensor $\boldsymbol{H}_*$. In this case, the thermal stress in the absence of strain is a pressure, whereas the thermal strain in the absence of stress is a dilatation. The constant c_* is typically called the **coefficient of thermal expansion**.

Answers to Selected Exercises

9.1 The Clausius–Duhem inequality in Lagrangian form is

$$\rho_0(\eta_m\dot{\Theta} + \dot{\Psi}) - \boldsymbol{P} : \dot{\boldsymbol{F}} + \Theta^{-1}\boldsymbol{Q}\cdot\nabla^X\Theta \le 0.$$

Sufficiency of conditions (i)–(iii) follows from the result in the text. To prove necessity we substitute the response functions into the above expression and use the chain rule on $\widehat{\Psi}$ to obtain

$$\rho_0\left[\widehat{\eta}_m + D_\Theta\widehat{\Psi}\right]\dot{\Theta} + \left[\rho_0 D_{\boldsymbol{F}}\widehat{\Psi} - \widehat{\boldsymbol{P}}\right] : \dot{\boldsymbol{F}} + \rho_0 D_{\boldsymbol{g}}\widehat{\Psi}\cdot\dot{\boldsymbol{g}} + \Theta^{-1}\widehat{\boldsymbol{Q}}\cdot\boldsymbol{g} \le 0.$$

By definition, all the response functions are functions of $(\boldsymbol{F}, \Theta, \boldsymbol{g})$ and are independent of $(\dot{\boldsymbol{F}}, \dot{\Theta}, \dot{\boldsymbol{g}})$. Because $\dot{\boldsymbol{g}}$ can be varied independently of $(\boldsymbol{F}, \Theta, \boldsymbol{g})$, we deduce that the coefficient of $\dot{\boldsymbol{g}}$ in the above inequality must vanish, that is

$$D_{\boldsymbol{g}}\widehat{\Psi} = \boldsymbol{0}.$$

Otherwise, we could take $\dot{\boldsymbol{g}}$ to be a large positive multiple of $D_{\boldsymbol{g}}\widehat{\Psi}$ and thereby cause the inequality to be violated. By similar arguments we find that the coefficients of $\dot{\boldsymbol{F}}$ and $\dot{\Theta}$ must also vanish, that is

$$[\rho_0 D_{\boldsymbol{F}}\widehat{\Psi} - \widehat{\boldsymbol{P}}] = \boldsymbol{O} \quad \text{and} \quad [\widehat{\eta}_m + D_\Theta \widehat{\Psi}] = 0.$$

Substituting these results into the above inequality we obtain, after multiplication by Θ

$$\widehat{\boldsymbol{Q}} \cdot \boldsymbol{g} \le 0.$$

From this we deduce that conditions (i)–(iii) are necessary.

9.3 (a) We have $\widehat{\boldsymbol{P}}(\boldsymbol{F}, \Theta) = 2\rho_* \boldsymbol{F} D_{\boldsymbol{C}}\overline{\Psi}(\boldsymbol{C}, \Theta)$. To compute the derivative of $\overline{\Psi}$, let $\boldsymbol{C}$ be given and consider the function $\boldsymbol{C}(\alpha) = \boldsymbol{C} + \alpha\boldsymbol{B}$, where $\alpha \in \mathbb{R}$ and $\boldsymbol{B}$ is an arbitrary second-order tensor. Then $D_{\boldsymbol{C}}\overline{\Psi}(\boldsymbol{C}, \Theta)$ is defined by the relation

$$D_{\boldsymbol{C}}\overline{\Psi}(\boldsymbol{C}, \Theta) : \boldsymbol{B} = \frac{d}{d\alpha}\overline{\Psi}(\boldsymbol{C}(\alpha), \Theta)\bigg|_{\alpha=0}, \quad \forall \boldsymbol{B} \in \mathcal{V}^2. \tag{9.40}$$

From the definition of $\overline{\Psi}(\boldsymbol{C}, \Theta)$ we have

$$\overline{\Psi}(\boldsymbol{C}(\alpha), \Theta) = \frac{1}{2}\boldsymbol{G}(\alpha) : \mathsf{J}(\boldsymbol{G}(\alpha)) + \vartheta\boldsymbol{M} : \boldsymbol{G}(\alpha) - \frac{1}{2}c\vartheta^2,$$

where $\boldsymbol{G}(\alpha) = \frac{1}{2}(\boldsymbol{C}(\alpha) - \boldsymbol{I})$. Differentiation with respect to α (denoted by a prime) gives

$$\frac{d}{d\alpha}\overline{\Psi}(\boldsymbol{C}(\alpha), \Theta) = \mathsf{J}(\boldsymbol{G}(\alpha)) : \boldsymbol{G}'(\alpha) + \vartheta\boldsymbol{M} : \boldsymbol{G}'(\alpha), \tag{9.41}$$

where we have used the linearity and major symmetry of J. Combining (9.40) and (9.41), and using the definition of $\boldsymbol{G}(\alpha)$, we get

$$D_{\boldsymbol{C}}\overline{\Psi}(\boldsymbol{C}, \Theta) : \boldsymbol{B} = \tfrac{1}{2}\mathsf{J}(\boldsymbol{G}) : \boldsymbol{B} + \tfrac{1}{2}\vartheta\boldsymbol{M} : \boldsymbol{B}, \quad \forall \boldsymbol{B} \in \mathcal{V}^2,$$

which implies $D_{\boldsymbol{C}}\overline{\Psi}(\boldsymbol{C}, \Theta) = \frac{1}{2}\mathsf{J}(\boldsymbol{G}) + \frac{1}{2}\vartheta\boldsymbol{M}$. Thus we obtain

$$\widehat{\boldsymbol{P}}(\boldsymbol{F}, \Theta) = 2\rho_* \boldsymbol{F} D_{\boldsymbol{C}}\overline{\Psi}(\boldsymbol{C}, \Theta) = \rho_* \boldsymbol{F}[\mathsf{J}(\boldsymbol{G}) + \vartheta\boldsymbol{M}].$$

The result for the entropy response function is a straightforward consequence of the relation $\widehat{\eta}_m(\boldsymbol{F}, \Theta) = -D_\Theta \overline{\Psi}(\boldsymbol{C}, \Theta)$. In the reference configuration we have $\boldsymbol{F} = \boldsymbol{I}$ and $\Theta = \Theta_*$, which gives $\boldsymbol{G} = \boldsymbol{O}$ and $\vartheta = 0$. Thus $\widehat{\boldsymbol{P}}(\boldsymbol{I}, \Theta_*) = \boldsymbol{O}$ and the reference configuration is stress-free.

(b) We have $\mathbf{A}_* = D_{\boldsymbol{F}} \widehat{\boldsymbol{P}}(\boldsymbol{I}, \Theta_*)$. To compute the derivative at $\boldsymbol{F} = \boldsymbol{I}$, consider the function $\boldsymbol{F}(\alpha) = \boldsymbol{I} + \alpha \boldsymbol{B}$, where $\alpha \in \mathbb{R}$ and $\boldsymbol{B}$ is an arbitrary second-order tensor. Then by definition of the derivative we have

$$\mathbf{A}_*(\boldsymbol{B}) = \frac{d}{d\alpha} \widehat{\boldsymbol{P}}(\boldsymbol{F}(\alpha), \Theta_*)\Big|_{\alpha=0}, \quad \forall \boldsymbol{B} \in \mathcal{V}^2. \tag{9.42}$$

Using the result from (a), and the facts that $\boldsymbol{G}(\alpha) = \frac{1}{2}(\boldsymbol{C}(\alpha) - \boldsymbol{I})$ and $\boldsymbol{C}(\alpha) = \boldsymbol{F}(\alpha)^T \boldsymbol{F}(\alpha)$, we find

$$\begin{aligned}\mathbf{A}_*(\boldsymbol{B}) &= \Big[\rho_* \boldsymbol{F}'(\alpha)\mathbf{J}(\boldsymbol{G}(\alpha)) + \rho_* \boldsymbol{F}(\alpha)\mathbf{J}(\boldsymbol{G}'(\alpha))\Big]\Big|_{\alpha=0} \\ &= \rho_* \mathbf{J}(\mathrm{sym}(\boldsymbol{B})) \\ &= \rho_* \mathbf{J}(\boldsymbol{B}),\end{aligned}$$

where the last line follows from the right minor symmetry of $\mathbf{J}$. From this we deduce $\mathbf{A}_* = \rho_* \mathbf{J}$. The results for $\boldsymbol{D}_*$ and β_* follow by direct differentiation of $\widehat{\boldsymbol{P}}$ and $\widehat{\eta}_m$ with respect to Θ.

9.5 Since $\boldsymbol{K}_* = \kappa_* \boldsymbol{I}$ we have

$$\nabla^X \cdot [\boldsymbol{K}_* \nabla^X \Theta^{(1)}] = \nabla^X \cdot [\kappa_* \nabla^X \Theta^{(1)}] = \kappa_* \Delta^X \Theta^{(1)}.$$

After dividing by $\rho_* \alpha_*$ the equations to be solved are

$$\begin{aligned}\dot{\Theta}^{(1)} &= \kappa \Delta^X \Theta^{(1)}, & 0 < X_i < 1, \quad t \geq 0, \\ \Theta^{(1)} &= 0, & X_i = 0, 1, \quad t \geq 0, \\ \Theta^{(1)} &= \Theta_0^{(1)}, & 0 < X_i < 1, \quad t = 0,\end{aligned}$$

where $\kappa = \kappa_*/(\rho_* \alpha_*)$. We seek a solution of the form

$$\Theta^{(1)}(X_1, X_2, X_3, t) = G_1(X_1) G_2(X_2) G_3(X_3) \Gamma(t)$$

Upon substitution into the differential equation we find

$$\Gamma' G_1 G_2 G_3 = \kappa \Gamma [G_1'' G_2 G_3 + G_1 G_2'' G_3 + G_1 G_2 G_3''],$$

which implies

$$\frac{\Gamma'}{\Gamma} = \kappa \left[\frac{G_1''}{G_1} + \frac{G_2''}{G_2} + \frac{G_3''}{G_3}\right]. \tag{9.43}$$

In order for the above equation to hold for all possible values of X_i and t we deduce that each term must be constant. Considering each term on the right-hand side of (9.43) we have

$$\frac{G_i''}{G_i} = -\omega_i^2,$$

where the form of the constant is motivated by the boundary conditions

$$G_i = 0 \quad \text{at} \quad X_i = 0, 1.$$

The above equations for G_i constitute a regular eigenvalue problem. Using standard techniques we find a family of independent solutions (eigenfunctions) of the form

$$G_i = \sin(k_i \pi X_i), \quad \omega_i = k_i \pi,$$

where $k_i \geq 1$ is an integer. Thus there is a family of solutions for each G_i $(i = 1, 2, 3)$. Substituting these solutions into (9.43) then yields the equation

$$\Gamma' = -\kappa[\omega_1^2 + \omega_2^2 + \omega_3^2]\Gamma,$$

whose general solution is

$$\Gamma = C_{k_1 k_2 k_3} e^{-(k_1^2 + k_2^2 + k_3^2)\pi^2 \kappa t},$$

where $C_{k_1 k_2 k_3}$ is a constant. Each choice of k_1, k_2 and k_3 yields an independent solution for $\Theta^{(1)}$. Thus by superposition we arrive at the following general solution

$$\Theta^{(1)}(\boldsymbol{X}, t) = \sum_{k_1=1}^{\infty} \sum_{k_2=1}^{\infty} \sum_{k_3=1}^{\infty} C_{k_1 k_2 k_3} e^{-(k_1^2 + k_2^2 + k_3^2)\pi^2 \kappa t} \cdot \sin(k_1 \pi X_1) \sin(k_2 \pi X_2) \sin(k_3 \pi X_3).$$

The arbitrary constants $C_{k_1 k_2 k_3}$ are determined by the initial condition. In particular, at time $t = 0$ the above expression gives, using a single summation sign for convenience,

$$\Theta^{(1)}(\boldsymbol{X}, 0) = \sum_{k_1, k_2, k_3=1}^{\infty} C_{k_1 k_2 k_3} \sin(k_1 \pi X_1) \sin(k_2 \pi X_2) \sin(k_3 \pi X_3).$$

Thus the initial condition is satisfied by taking $C_{k_1 k_2 k_3} = A_{k_1 k_2 k_3}$, which completes the solution.

9.7 (a) From the expression for $\boldsymbol{u}^{(1)}(\boldsymbol{X},t)$ and the fact that $\boldsymbol{k}$ is a unit vector we get

$$\Delta^X \boldsymbol{u}^{(1)} = -\gamma^2\sigma\boldsymbol{a}e^{i(\gamma\boldsymbol{k}\cdot\boldsymbol{X}-\omega t)}, \quad \nabla^X\cdot\boldsymbol{u}^{(1)} = i\gamma\sigma(\boldsymbol{a}\cdot\boldsymbol{k})e^{i(\gamma\boldsymbol{k}\cdot\boldsymbol{X}-\omega t)},$$

$$\nabla^X(\nabla^X\cdot\boldsymbol{u}^{(1)}) = -\gamma^2\sigma(\boldsymbol{a}\cdot\boldsymbol{k})\boldsymbol{k}e^{i(\gamma\boldsymbol{k}\cdot\boldsymbol{X}-\omega t)},$$

$$\ddot{\boldsymbol{u}}^{(1)} = -\omega^2\sigma\boldsymbol{a}e^{i(\gamma\boldsymbol{k}\cdot\boldsymbol{X}-\omega t)}, \quad \nabla^X\cdot\dot{\boldsymbol{u}}^{(1)} = \gamma\omega\sigma(\boldsymbol{a}\cdot\boldsymbol{k})e^{i(\gamma\boldsymbol{k}\cdot\boldsymbol{X}-\omega t)}.$$

In a similar manner, from the expression for $\Theta^{(1)}(\boldsymbol{X},t)$ we find

$$\nabla^X\Theta^{(1)} = i\gamma\tau\boldsymbol{k}e^{i(\gamma\boldsymbol{k}\cdot\boldsymbol{X}-\omega t)}, \qquad \Delta^X\Theta^{(1)} = -\gamma^2\tau e^{i(\gamma\boldsymbol{k}\cdot\boldsymbol{X}-\omega t)},$$

$$\dot{\Theta}^{(1)} = -i\omega\tau e^{i(\gamma\boldsymbol{k}\cdot\boldsymbol{X}-\omega t)}.$$

Substituting the above results into the differential equations and canceling the exponential factors gives

$$\begin{aligned} \omega^2\sigma\boldsymbol{a} &= \gamma^2\sigma[\mu\boldsymbol{a}+(\lambda+\mu)(\boldsymbol{a}\cdot\boldsymbol{k})\boldsymbol{k}] - im\gamma\tau\boldsymbol{k}, \\ (\kappa\gamma^2 - i\omega)\tau &= \nu\gamma\omega\sigma(\boldsymbol{a}\cdot\boldsymbol{k}), \end{aligned} \tag{9.44}$$

which establishes the result.

(b) Given $\boldsymbol{k}$, $\gamma \neq 0$ we assume $\boldsymbol{a}\cdot\boldsymbol{k} = 0$. Then from (9.44) we obtain, after taking the dot product of $(9.44)_1$ with $\boldsymbol{a}$ and then $\boldsymbol{k}$

$$(\omega^2 - \mu\gamma^2)\sigma = 0, \quad im\gamma\tau = 0, \quad (\kappa\gamma^2 - i\omega)\tau = 0.$$

Since $m\gamma \neq 0$ the second equation holds if and only if $\tau = 0$, and the third equation is also satisfied. Inspection of the first equation shows that a non-trivial wave exists if and only if $\omega^2 = \mu\gamma^2$, in which case σ can be arbitrary.

(c) Given $\boldsymbol{k}$, $\gamma \neq 0$ we assume $\boldsymbol{a}\cdot\boldsymbol{k} = 1$, or equivalently $\boldsymbol{a} = \boldsymbol{k}$. Then from (9.44) we obtain, after taking the dot product of $(9.44)_1$ with $\boldsymbol{k}$

$$\omega^2\sigma = \gamma^2(\lambda+2\mu)\sigma - im\gamma\tau, \quad (\kappa\gamma^2 - i\omega)\tau = \nu\gamma\omega\sigma.$$

This is a homogeneous, square, linear system for σ and τ. In matrix form we have

$$\begin{pmatrix} \omega^2 - \gamma^2(\lambda+2\mu) & im\gamma \\ -\nu\gamma\omega & \kappa\gamma^2 - i\omega \end{pmatrix} \begin{Bmatrix} \sigma \\ \tau \end{Bmatrix} = \begin{Bmatrix} 0 \\ 0 \end{Bmatrix}.$$

Thus non-trivial waves exist if and only if

$$\det\begin{pmatrix} \omega^2 - \gamma^2(\lambda+2\mu) & im\gamma \\ -\nu\gamma\omega & \kappa\gamma^2 - i\omega \end{pmatrix} = 0.$$

Bibliography

Abraham, R. and Marsden, J.E., *Foundations of Mechanics*, Second Edition, Addison-Wesley (1978).

Anderson, J.D., *Modern Compressible Flow: With Historical Perspective*, Third Edition, McGraw-Hill (2003).

Antman, S.S., *The Theory of Rods*, in Handbuch der Physik **6a**(2), Springer-Verlag (1972).

Antman, S.S., *Nonlinear Problems of Elasticity*, Springer-Verlag (1995).

Batchelor, G.K., *An Introduction to Fluid Dynamics*, Cambridge University Press (2000).

Boley, B.A. and Weiner, J.H., *Theory of Thermal Stresses*, Wiley (1960).

Bourne, D.E. and Kendall, P.C., *Vector Analysis and Cartesian Tensors*, Third Edition, Chapman and Hall (1992).

Bowen, R.M. and Wang, C.-C., *Introduction to Vectors and Tensors*, Volumes 1–2, Plenum Press (1976).

Brand, L., *Vector and Tensor Analysis*, Wiley (1947).

Brillouin, L., *Tensors in Mechanics and Elasticity*, Academic Press (1964).

Carlson, D.E., *Linear Thermoelasticity*, in Handbuch der Physik **6a**(2), Springer-Verlag (1972).

Chadwick, P., *Continuum Mechanics: Concise Theory and Problems*, Wiley (1976).

Chapman, S. and Cowling, T.G., *The Mathematical Theory of Non-Uniform Gases*, Third Edition, Cambridge University Press (1970).

Chorin, A.J. and Marsden, J.E., *A Mathematical Introduction to Fluid Mechanics*, Third Edition, Springer-Verlag (1993).

Ciarlet, P.G., *Mathematical Elasticity*, Volumes 1–3, Elsevier Science (1988).

Ciarlet, P.G. and Lions, J.L. (editors), *Handbook of Numerical Analysis III*, North-Holland (1994).

Ciarlet, P.G. and Lions, J.L. (editors), *Handbook of numerical analysis VI*, North-Holland (1998).

Constantin, P. and Foias, C., *Navier–Stokes Equations*, University of Chicago Press (1988).

Courant, R. and Friedrichs, K.O., *Supersonic Flow and Shock Waves*, Interscience (1948).

Crowe, M.J., *A History of Vector Analysis*, University of Notre Dame Press (1967).

Dafermos, C.M., *Hyperbolic Conservation Laws in Continuum Physics*, Second Edition, Springer-Verlag (2005).

Dafermos, C.M., On the existence and asymptotic stability of solutions to the equations of linear thermoelasticity, *Archive for Rational Mechanics and Analysis* **29** (1968) 241–271.

Day, W.A., *Heat Conduction Within Linear Thermoelasticity*, Springer-Verlag (1985).

Dean, R.G. and Dalrymple, R.A., *Water Wave Mechanics for Engineers and Scientists*, Prentice-Hall (1984).

Dodson, C.T.J. and Poston, T., *Tensor Geometry*, Second Edition, Springer-Verlag (1991).

Doering, C.R. and Gibbon, J.D., *Applied Analysis of the Navier–Stokes Equations*, Cambridge University Press (1995).

Fichera, G., *Existence Theorems in Elasticity*, in Handbuch der Physik **6a**(2), Springer-Verlag (1972).

Fleming, W., *Functions of Several Variables*, Second Edition, Springer-Verlag (1977).

Fung, Y.C., *Foundations of Solid Mechanics*, Prentice-Hall (1965).

Girault, V. and Raviart, P.-A., *Finite Element Methods for Navier–Stokes Equations*, Springer-Verlag (1986).

Gurtin, M.E., *The Linear Theory of Elasticity*, in Handbuch der Physik **6a**(2), Springer-Verlag (1972).

Gurtin, M.E., *An Introduction to Continuum Mechanics*, Academic Press (1981).

Halmos, P.R., *Finite-Dimensional Vector Spaces*, Springer-Verlag (1974).

Happel, J. and Brenner, H., *Low Reynolds Number Hydrodynamics*, Martinus Nijhoff (1983).

Hirsch, C., *Numerical Computation of Internal and External Flows*, Volumes 1–2, Wiley (1988).

Hirsch, M.W. and Smale, S., *Differential Equations, Dynamical Systems, and Linear Algebra*, Academic Press (1974).

Holzapfel, G.A., *Nonlinear Solid Mechanics: A Continuum Approach for Engineering*, Wiley (2000).

Hughes, P.C., *Spacecraft Attitude Dynamics*, Wiley (1986).

Johnson, R.S., *A Modern Introduction to the Mathematical Theory of Water Waves*, Cambridge University Press (1997).

Kellogg, O.D., *Foundations of Potential Theory*, Springer-Verlag (1967).

Knops, R.J. and Payne, L.E., *Uniqueness Theorems in Linear Elasticity*, Springer-Verlag (1971).

Knowles, J.K., *Linear Vector Spaces and Cartesian Tensors*, Oxford University Press (1998).

Ladyzhenskaya, O.A., *The Mathematical Theory of Viscous Incompressible Flow*, Second Edition, Gordon and Breach (1969).

Laney, C.B., *Computational Gasdynamics*, Cambridge University Press (1998).

Lax, P.D., *Hyperbolic Systems of Conservation Laws and the Mathematical Theory of Shock Waves*, SIAM (1973).

LeVeque, R.J., *Numerical Methods for Conservation Laws*, Second Edition, Birkhauser (1992).

Love, A.E.H., *A Treatise on the Mathematical Theory of Elasticity*, Fourth Edition, Dover (1944).

Lubliner, J., *Plasticity Theory*, Macmillan (1990).

Majda, A.J., *Compressible Fluid Flow and Systems of Conservation Laws in Several Space Variables*, Springer-Verlag (1984).

Majda, A.J. and Bertozzi, A.L., *Vorticity and Incompressible Flow*, Cambridge University Press (2002).

Malvern, L.E., *Introduction to the Mechanics of a Continuous Medium*, Prentice Hall (1969).

Marsden, J.E. and Hughes, T.J.R., *Mathematical Foundations of Elasticity*, Dover (1983).

Marsden, J.E. and Tromba, A.J., *Vector Calculus*, Third Edition, Freeman (1988).

Mase, G.E., *Continuum Mechanics*, Schaum's Outline Series, McGraw Hill (1970).

Munkres, J.R., *Analysis on Manifolds*, Addison-Wesley (1991).

Naghdi, P.M., *The Theory of Shells*, in Handbuch der Physik **6a**(2), Springer-Verlag (1972).

Nicholson, D.W., *Finite Element Analysis: Thermomechanics of Solids*, CRC Press (2003).

Nowacki, W., *Thermoelasticity*, Second Edition, Pergamon (1986).

Oden, J.T. and Demkowicz, L.F., *Applied Functional Analysis*, CRC Press (1996).

Ogden, R.W., *Non-Linear Elastic Deformations*, Ellis Horwood (1984).

Owens, R.G. and Phillips, T.N., *Computational Rheology*, Imperial College Press (2002).

Parkus, H., *Thermoelasticity,* Second Edition, Springer-Verlag (1976).

Racke, R. and Jiang, S., *Evolution Equations in Thermoelasticity,* CRC Press (2000).

Robinson, J.C., *Infinite-Dimensional Dynamical Systems*, Cambridge University Press (2001).

Saad, M.A., *Compressible Fluid Flow*, Second Edition, Prentice Hall (1993).

Serrin, J., *Mathematical Principles of Classical Fluid Mechanics*, in Handbuch der Physik **8**(1), Springer-Verlag (1959).

Šilhavý, M., *The Mechanics and Thermodynamics of Continuous Media*, Springer-Verlag (1997).

Simmonds, J.G., *A Brief on Tensor Analysis*, Second Edition, Springer-Verlag (1994).

Simo, J.C. and Hughes, T.J.R., *Computational Inelasticity*, Springer-Verlag (1998).

Slemrod, M., Global existence, uniqueness and asymptotic stability of classical smooth solutions in one-dimensional non-linear thermoelasticity, *Archive for Rational Mechanics and Analysis* **76** (1981) 97–134.

Sokolnikoff, I.S., *Tensor Analysis: Theory and Applications*, Wiley (1951).

Spencer, A.J.M., *Continuum Mechanics*, Dover (1992).

Stoker, J.J., *Water Waves: The Mathematical Theory with Applications*, Interscience (1957).

Temam, R., *Navier–Stokes Equations: Theory and Numerical Analysis*, North-Holland (1984).

Truesdell, C., *A First Course in Rational Continuum Mechanics*, Volume 1, Second Edition, Academic Press (1991).

Truesdell, C., *Rational Thermodynamics*, Second Edition, Springer-Verlag (1984).

Truesdell, C. and Noll, W., *The Non-Linear Field Theories of Mechanics*, in Handbuch der Physik **3**(3), Springer-Verlag (1965).

Truesdell, C. and Toupin, R.A., *The Classical Field Theories*, in Handbuch der Physik **3**(1), Springer-Verlag (1960).

Whitham, G.B., *Linear and Nonlinear Waves*, Wiley (1974).

White, F.M., *Viscous Fluid Flow*, Third Edition, McGraw-Hill (2006).

Index

Printed in the United States
By Bookmasters